BIOCHEMICAL REGULATORY MECHANISMS IN EUKARYOTIC CELLS

Biochemical Regulatory Mechanisms in Eukaryotic Cells

Edited by

ERNEST KUN
University of California School of Medicine
San Francisco, California

SANTIAGO GRISOLIA
University of Kansas Medical Center
Kansas City, Kansas

WILEY-INTERSCIENCE,　A Division of John Wiley & Sons, Inc.
New York · London · Sydney · Toronto

Copyright © 1972, by John Wiley & Sons, Inc.

All rights reserved. Published simultaneously in Canada.

No part of this book may be reproduced by any means, nor
transmitted, nor translated into a machine language without the written permission of the publisher.

Library of Congress Cataloging in Publication Data:

Kun, Ernest, 1919–
  Biochemical regulatory mechanisms in eukaryotic cells.

  Includes bibliographical references.
    1. Cytochemistry—Addresses, essays, lectures.
2. Biological control systems—Addresses, essays,
lectures. 3. Eukaryotic cells—Addresses, essays,
lectures. I. Grisolia, Santiago, 1923– joint
author. II. Title.

QH611.K85   574.8′764   75-39321
ISBN 0-471-51065-3

Printed in the United States of America.

10 9 8 7 6 5 4 3 2 1

Preface

This book is composed of essays written by individuals who have as a common bond their scientific interests and their curiosity about how living cells work. Until relatively recently, this question could hardly be raised in a meaningful manner, because cellular function in terms of analytically oriented chemical reactions had first to be discovered. During the last 20 years great successes in classical enzymology brought this goal within sight, but simultaneously even more complex problems emerged. It became obvious that in differentiated multicellular organisms the apparently monotonous enzymatic composition, as recognized in terms of "ubiquitous" metabolic pathways, could hardly explain the remarkably specific physiological functions of organ systems, much less the responses elicited by drugs or hormones. Enzymology had to be reoriented in terms of environmental control through unsuspected kinetic (e.g., allosteric) or direct chemical modifications of enzyme proteins. In the meantime the experimentally almost-unapproachable field of macromolecular metabolism, that is, the biosynthesis and degradation of cellular macromolecules, has also been subject to significant recent developments, which eventually will have to be integrated into cellular metabolism.

A considerable portion of this book (Chapters 1–7) is concerned with the reactions of macromolecular cellular constituents, understanding of which has reached a significant degree of clarity in terms of physicochemical mechanisms. This knowledge brings a new dimension to the functional aspects of cellular biochemistry. Cellular control is also provided by reactions which seem to have only a regulatory role in cellular economy, although this is probably a hard distinction to make. As an example, Chapter 8 deals with polyamines, which may play just this sort of role.

The level of information in most areas of chemistry, including cellular biochemistry, invariably depends on experimental or technical advances. When biochemists learned to deal with soluble enzymes, understanding of both the physicochemical and the molecular biological aspects of water-soluble enzymes progressed significantly over that of enzymes embedded in hydrophobic structures. Attempts to rationalize mitochondrial metabolic and energy-transducing functions are as yet at a beginning stage, as illustrated in Chapters 9 and 10. The strategy of experimentation is critical if aimed at solving problems of bioengineering, as is obviously the case in the study of control mechanisms. Most biochemists depend on intuitive assessment of experimental designs, yet more precise approaches are available if greater cooperation between experimentalists and theoreticians could be achieved. It is hoped that inclusion of a somewhat mathematically oriented chapter (11) will encourage both nonmathematicians and biophysicists to continue dialogs that may eventually bring the field of cellular biology into the realm of rational science.

In terms of cellular biology, it is obvious that functional and molecular genetically oriented biochemical research has much the same purpose; thus distinctions at any level are artificial. Clearly the temporal organization of events, be it replicative or metabolic, will decide the relevance to a chosen area of cellular functions.

ERNEST KUN
SANTIAGO GRISOLIA

San Francisco, California
Kansas City, Kansas
November 1971

Contents

BIOCHEMICAL REGULATORY MECHANISMS IN EUKARYOTIC CELLS

CHAPTER 1

Molecular Control of Hemoglobin Function

AUSTEN RIGGS

Department of Zoology, University of Texas, Austin, Texas

1. GENERAL CONSIDERATIONS

The primary function of blood hemoglobins, oxygen transport, is modulated by the constituents of the red blood cell. We shall consider here certain features of the nature and molecular mechanisms of this regulation in order to gain some insight into the way in which hemoglobin function is modified to meet the changing metabolic needs of animals and the nature of the adaptations in different organisms. These adaptations cover a large spectrum. Some animals have hemoglobins which seem to be totally unaffected by variation in cellular constituents, whereas the hemoglobins of others are very sensitive to even small changes in the hemoglobin environment.

We can view hemoglobin as a protein which may react reversibly with at least four ligands: O_2, H^+, CO_2, and an organic phosphate. The extent to which the last three ligands can modify oxygen binding varies with different hemoglobins. Although some lower vertebrates have hemoglobins which are completely devoid of any effect of CO_2, pH, or organic phosphate, mammalian hemoglobins are generally influenced by all of these agents. The predominant organic phosphate in mammalian red cells is 2,3-diphosphoglycerate (2,3-DPG), but the nucleated red cells of birds and turtles contain inositol hexaphosphate; ATP is the major phosphorylated organic compound in many lower vertebrates. The discussion to follow will be confined largely to mammalian hemoglobins.

Most vertebrate blood hemoglobins consist of four polypeptide chains of two types, α and β, combined to form tetramers, $\alpha_2\beta_2$. Each subunit contains a ferrous heme sandwiched in a hydrophobic pocket. It is now well established that oxygenation of one subunit is associated with changes in the conformation of that subunit such that the relative positions of the amino acid residues change at some of the points of contact between subunits. These shifts induce small changes in the conformation of the neighboring subunit in such a way that the affinity for oxygen of the unoxygenated heme of this neighbor increases. These subunit interactions are often referred to as "positive cooperativity" or as "heme-heme" interaction, although no direct interaction between the hemes occurs. It is apparent that any agent which alters the conformation or is bound more tightly to one form (oxy or deoxy) than to the other will be expected to shift the equilibrium between conformational states and so alter the overall affinity for oxygen. Furthermore it should be recognized that the two kinds of subunit, α and β, may possess intrinsically different kinetic and equilibrium properties. Therefore, any agent which either destroys or modifies the interaction between subunits can be expected to accentuate these differences, which lie concealed in the tetrameric molecule.

Many studies of the properties of the isolated α and β subunits have been carried out, but no assurance exists that these properties can be equated with those of the sub-units in the intact tetramer. To do so we would have to assume that the formation of the tetramer from the isolated subunits is not accompanied by any conformational change within the subunits. If any substance reacts preferentially with one kind of subunit, we should look carefully to see whether this creates a preferential effect on one subunit such that the affinity of this subunit for oxygen differs greatly from that of the other subunits. Such an intramolecular heterogeneity of function can give rise to aberrant oxygen-binding isotherms. It would be wise, therefore, to consider such possible internal heterogeneity before invoking complex theoretical models to explain the data. We will consider these matters in some detail in Section 4 on organic phosphate effects.

The binding of oxygen by hemoglobins can often be approximately described by the Hill equation:

$$\frac{\bar{Y}}{1 - \bar{Y}} = Kp^n, \tag{1}$$

where $\bar{Y}$ is the average fraction of hemes oxygenated, p is the oxygen pressure, and K and n are constants. This equation fits many of the existing data reasonably well between about 10 and 90% oxygenation. The value of n for normal mammalian hemoglobins is about ~ 2.8; any value of $n > 1$ is taken to indicate cooperative interactions between at least some of the four oxygen-binding hemes. However, observation of $n < 1$ does not necessarily imply negative cooperativity or inhibitory interactions *even in supposedly homogeneous preparations*. Neither does it always follow that a decrease in the observed value of n implies a decrease in the overall free-energy change associated with cooperativity. Either $n < 1$ or a decrease in n can also arise from various kinds of intra- and intermolecular heterogeneity. Values of $n < 1$ can readily result from the presence of mixtures of hemoglobin conformations with different affinitites for oxygen which are not at equilibrium with one another through some conformational transition. Equation 1 in logarithmic form gives a straight line of slope n and intercept K. At very high and at very low degrees of oxygenation, equation 1 fails to approximate the data and the curves of $\log \bar{Y}/(1 - Y)$ versus $\log p$ approach asymptotes with the slope, $n = 1$.

Wyman (1, 2) has shown that the perpendicular distance between the initial and the final asymptotes is related by the factor $RT\sqrt{2}$ to the overall free energy of interaction in the binding equilibrium *provided* that all the sites are intrinsically identical. Positively cooperative interactions mean that the work needed to saturate the molecule with ligand is lowered. The interaction energy is considered to be positive when the final asymptote lies above the initial one. If the sites or subunits are different and independent, the final asymptote will be below the initial one. Hence, in some region, the value of n would be less than unity and might suggest, erroneously, "negative cooperativity." In mammalian hemoglobins, therefore, the perpendicular distance gives only an apparent interaction energy. Wyman (2) emphasizes that this apparent energy can be equated with the true free energy of interaction only if the Hill plot is assumed to be linear with $n = 1$ in the absence of any interaction. The fact that a mixture of components or conformations *not* at equilibrium with one another will give rise to $n < 1$, and to *apparent* negative cooperativity, even though no interaction at all is present, deserves renewed emphasis.

Several recent reports have appeared of negative cooperativity in the binding of substrate by certain enzymes and in the binding of ligands by respiratory proteins. These include phosphoenol pyruvate carboxylase (3), glyceraldehyde-3-phosphate dehydrogenase (4, 5), cytidine triphosphate synthetase (6), and glutamate dehydrogenase (7). Indeed, Levitzki and Koshland (6) suggest that negative cooperativity may be characteristic of many regulatory enzymes. In view of the recent report by Susor et al. (8) that many "homogeneous" protein preparations appear to be quite heterogeneous, there is no certainty that negative cooperativity exists at all. Recent studies of hemoglobin by Gibson (9), to be described later, suggest that under certain circumstances the equilibrium with some organic phosphates may be reached with hemoglobin very slowly even though the O_2 binding equilibrium is reached rapidly. The importance of such slow equilibria in enzyme kinetics has been considered in detail by Frieden (10).

It should be recognized that Wyman's method of estimating the free energy of interaction is difficult to achieve with precision in practice because accurate measurements are required in the ranges 0–5% and 95–100% oxygenation. Interpretation of measurements of oxygenation in these regions is simple in principle but complex in practice. The first O_2 to go on at very low pO_2 will presumably follow noncooperative ligand-binding behavior. Similarly, only one free binding site is present above 95% oxygenation. Therefore, at the extreme ends of the binding curve the data will appear "noncooperative" with $n = 1$. However, we must recognize that even the very best preparations may be heterogeneous in some functional sense, and there is no *a priori* reason to assume that the properties of whatever is being measured in these regions are representative of the bulk of the hemoglobin; after 98% of the hemoglobin is oxygenated, the rest may be minor components, preparative artifacts, or junk. Weber (11) has considered ligand binding in detail from the standpoint of information theory. His analysis makes clear that information on equilibrium binding approaches zero in the regions which are close to zero binding or complete binding. Hence it is essential in analyzing data for the region 0–5% to be sure that one is not dealing just with a stoichiometric reaction in which significant *free* ligand is absent.

The observations whose basis we seek are that oxygen affinity can be *lowered* in the following ways: (*a*) decrease in pH (6.5 < pH < 8)—the "normal," positive, or *alkaline Bohr effect;* (*b*) increase in pH (pH < 6.2)—the "negative," reverse, or *acid Bohr effect;* (*c*) increase in pCO_2 at constant pH > 6.5—the *specific CO_2 effect;* (*d*) addition of low concentrations of 2,3-diphosphoglycerate or other organic phosphate—*the phosphate effect.*

Each of these linked effects has its reciprocal: oxygenation at physiological pH tends to drive off protons, CO_2, and 2,3-DPG. This means that the binding constants for H^+, CO_2, and DPG are lower for oxy- than for deoxyhemoglobin.

The general relationships of such reciprocal linkage effects have been analyzed in detail by Wyman (1, 2, 12). Some of the equivalent basic relationships are as follows:

$$\left(\frac{\partial \bar{X}}{\partial \ln y}\right)_{x,z} = \left(\frac{\partial \bar{X}}{\partial \ln x}\right)_{y,z}, \tag{2}$$

$$\left(\frac{\partial \bar{X}}{\partial \ln y}\right)_{\bar{Y},z} = \left(\frac{\partial \bar{Y}}{\partial \ln x}\right)_{\bar{X},z}, \tag{3}$$

$$\left(\frac{\partial \bar{X}}{\partial \bar{Y}}\right)_{x,z} = -\left(\frac{\partial \ln y}{\partial \ln x}\right)_{\bar{Y},z}. \tag{4}$$

In these equations $\bar{X}$ and $\bar{Y}$ are the quantities of substances X and Y bound per unit quantity of protein; x and y are their activities; z is the activity of a third ligand, here kept constant. These relationships give the basis for much of the experimentation so far carried out. Thus, for example, protons released during oxygenation can be measured directly with a pH-stat or differential acid-base titration, but equation 4 indicates that the same information can be obtained by measurement of the dependence of the oxygen equilibrium on pH. If $\bar{X}$ represents the protons released and $\bar{Y}$ the oxygen bound, and we let $y = pO_2$ and x be the proton activity, expression 4 becomes

$$\left(\frac{\partial \bar{X}}{\partial \bar{Y}}\right)_{x,z} = \left(\frac{\partial \log p}{\partial \text{pH}}\right)_{\bar{Y}}. \tag{5}$$

The right-hand side then gives the number of protons released per O_2 molecule bound at constant $\bar{Y}$ (usually this is given at $\bar{Y} = \frac{1}{2}$).* Similarly, the expression

$$\left(\frac{\partial \log p}{\partial \log [\text{DPG}]}\right)_{\bar{Y}} \tag{6}$$

should give the number of 2,3-DPG molecules released per O_2 molecule

---

* Perhaps it should be mentioned that the clinical literature occasionally utilizes $\partial p/\partial \text{pH}$ as an expression of the Bohr effect, but this is very misleading because it can cause one to see an apparent difference in Bohr effect when none is present. For example, Morpurgo et al. (13) describe a higher Bohr effect for the bloods of Peruvian Indians from high altitudes than for sea-level Europeans. However, if their data are given in terms of

$$\frac{\partial \log p}{\partial \text{pH}} \left(= 2.303 \frac{1}{p} \frac{\partial p}{\partial \text{pH}}\right)$$

rather than of $\partial p/\partial \text{pH}$ the apparent difference largely vanishes.

bound at the specified value of $\overline{Y}$, provided that [DPG] represents the *free* 2,3-DPG concentration. This expression, of course, says nothing about how DPG binding varies with $\overline{Y}$.

2. THE BOHR EFFECT

The physiological advantages of the alkaline Bohr effect have been recognized since its discovery by Bohr et al. (14) in 1904. The term originated in reference to the effect of carbon dioxide in lowering the oxygen affinity; CO_2 was thought to exert its effect entirely by its behavior as carbonic acid, H_2CO_3, but it is now recognized that the two effects must be distinguished. I consider here the effects of pH and in the next section the effects of CO_2.

Wyman (12) analyzed the existing data of the time and showed that the pH dependence of the oxygen equilibria of mammalian hemoglobins could be well described by assuming the presence of two oxygenation-linked acid groups per heme, one of which (group I) became a stronger acid (alkaline Bohr effect) whereas the other (group II) became a weaker acid (acid Bohr effect) upon oxygenation. Thus he could describe the pH dependence of the O_2 equilibrium by the equation

$$\log p_{50} = C + \log \frac{(H^+ + K_1')(H^+ + K_2')}{(H^+ + K_1)(H^+ + K_2)}, \tag{7}$$

where K_1 and K_1' are the dissociation constants for the acid groups of oxy- and deoxyhemoglobin, respectively, responsible for the alkaline Bohr effect, and K_2 and K_2' are the corresponding constants for the acid Bohr effect; H^+ represents the hydrogen-ion activity; and C is a constant which incorporates all other effects on the oxygen affinity. The data are fitted with $K_1 > K_1'$ and $K_2 < K_2'$ to account, respectively, for the alkaline and the acid Bohr effects. The acid Bohr effect is commonly thought to be of no physiological consequence because it lies outside the range of normal pH. Nevertheless the "nonphysiological" region of pH is important to examine because it tells something about the conformational changes which take place during oxygenation. At high pH, where K_2' and $K_2 \gg H^+$, the acid Bohr group is completely dissociated. Under these conditions equation 7 becomes

$$\log p_{50} = C + \log \frac{(H^+ + K_1')}{(H^+ + K_1)} + \Delta\, pK_2, \tag{8}$$

where $\Delta\, pK_2$ is the difference in the pK of the acid Bohr groups in oxy- and deoxyhemoglobins. Since $K_2' > K_2$, $\Delta\, pK_2 > 0$; hence the larger the value of $\Delta\, pK_2$, the lower will be the oxygen affinity in alkaline solution *even though no change in protonation of this group occurs during oxygenation: K_2 reflects* oxygenation-linked conformational changes in the vicinity of group II even though the group remains unprotonated.

According to equation 7 the oxygen affinity can be altered by changing either C or H^+, or by changing any one of four dissociation constants. Binding of other ligands at or in the neighborhood of either of these groups can therefore be expected to alter the affinity for hemoglobin.

Equation 7 describes the alkaline Bohr effect in terms of a single ionizable group per heme. However, recent structure studies and the analysis of modified hemoglobins, to be described below, show that at least six groups per tetramer are involved, two per α chain and one per β chain. Therefore, values of K_1 and K_1' represent the overall behavior of at least six ionizable groups. In the presence of 2,3-DPG, at least eight groups are involved.

Calculations of the *total* Bohr effect from expression 5 and deductions based on equation 7 assume that the value of n in the Hill equation is independent of pH; that is, the $\overline{Y}$ versus log pO_2 plots have the same shape at different pH values, and the magnitude of the right-hand side of equation 5 does not depend on the value of $\overline{Y}$ at which it was evaluated. Although this is essentially true for mammalian hemoglobins in the absence of complicating substances, the behavior of hemoglobins from lower vertebrates is often quite different: n is a function of pH. Although no satisfactory description of the mechanism of the dependence of n on pH in these hemoglobins has yet appeared, it is important to recognize here that, if n is pH dependent, expression 5 will give a different number of protons released per O_2 bound at each level of oxygenation, $\overline{Y}$. Under these circumstances numerical integration between $\overline{Y} = 0$ and $\overline{Y} = 1$ is needed to obtain the total number of protons released. This point has not always been recognized in the comparative physiological literature on fish hemoglobins, which often give "the" Bohr effect as that at 50% oxygenation.

The extensive X-ray studies of Perutz and his colleagues (15–17) have provided strong evidence for the identity of some of the groups involved in the alkaline Bohr effect. Their comparison of the electron-density maps for human oxy- and deoxyhemoglobins indicates differences in the relative positions of the α and β subunits. Each α chain makes contact with two β chains; the two contacts are designed $\alpha_1\beta_1$ and $\alpha_1\beta_2$. The greatest shifts which they observe between the structures of deoxy- and oxyhemoglobins occur at the points of contact designated $\alpha_1\beta_2$ and at the chain termini. The valyl α —NH_2 terminal group of each α chain appears linked to the COOH—terminus of the neighboring α chain in deoxyhemoglobin, but no such linkage is apparent in oxyhemoglobin: the terminal valyl α —NH_2 groups are free. These workers suggest that the proximity of the —COO^- group raises the pK of the α —NH_2 group of the α chains in deoxyhemoglobin, relative to that in oxyhemoglobin. This would account for part of the

Bohr effect. They find that the imidazole group of the COOH— terminal histidine of the β chain is linked to the γ —COOH group of aspartyl residue 94 in deoxyhemoglobin; this link breaks as a result of oxygenation. This appears to account for almost half of the Bohr effect. Histidyl residue α 122 also appears to participate in the Bohr effect (16). Thus the measured Bohr effect in normal phosphate-free mammalian hemoglobin appears to involve at least three different acid groups: the α —NH$_2$ terminal group of each α chain, α 122 histidine, and β 146 histidine.

Perutz (16) remarks that horse oxyhemoglobin crystals undergo a large lattice change at pH 5.9 and a further change below pH 5.4. Although the details have not been investigated, he speculates that these shifts may be responsible for the acid Bohr effect, which appears to involve a pK change from 5.7 to 4.9 upon deoxygenation. We can speculate that carboxyl groups at one of the α-β interfaces may be involved. The three carboxylic acid residues at the α_1-β_2 interface are α 94 Asp, β 101 Glu, and β 99 Asp. According-ing to Bolton and Perutz (18) the α 94 Asp residue is within 4 Å of the β 101 Glu residue in oxyhemoglobin but is further away in deoxyhemoglobin. Thus oxygenation would be expected to raise the pK of each of these groups. At present this seems a reasonable hypothesis for the acid Bohr effect.

It should be emphasized here that the explanation of the Bohr effect in different animal hemoglobins must involve a variety of different groups and that a single explanation cannot be invoked for all hemoglobins. Thus, for example, the hemoglobin of the lamprey, *Petromyzon marinus*, has a large alkaline, but no acid, Bohr effect (19–21). The molecule has none of the residues which have been invoked to explain the Bohr effect in mammalian hemoglobins (22). The imino terminal proline cannot be involved because the pK value is about 1.5 units higher for proline than for α —NH$_2$ groups. The COOH— terminal residue is tyrosine. Since the Bohr effects of *Petro-myzon marinus* hemoglobin and of *Lampetra fluviatilis* hemoglobin are the same (see review by Riggs in ref. 23), there seems to be little likelihood that the prolyl residue is involved because in the former hemoglobin it is free whereas in the latter hemoglobin the prolyl residue is formylated.

Andersen (24) and Andersen and Gibson (25) have carried out careful kinetic measurements of ligand binding and find that their data are completely explained on the basis that protonation of a single residue favors dimerization of the deoxygenated monomer, and that the dimer reacts slowly and the monomer rapidly with ligand. The apparent pK of this group is 6.05. Al-though such a pK is typical of histidyl residues, lamprey hemoglobin has only two histidines; these are internal in the heme pocket and therefore cannot be directly involved in aggregation. The heme pocket is quite similar to that in mammalian hemoglobins, where the heme-linked histidines are not involved in the Bohr effect. Furthermore, myoglobin has almost no Bohr effect but possesses both these residues. The most likely possibility is that carboxyl groups at the points of contact between the subunits of the dimer are responsible for the Bohr effect (23).

3. THE CARBON DIOXIDE EFFECT

A substantial amount of CO_2 is directly carried by hemoglobin. Although only a few animal species have been examined, it is probable that the extent of this direct transport differs substantially in different animals. In human hemoglobin about 40% of the total respiratory CO_2 is transported as bicarbonate, HCO_3^-, and about 60% by direct combination with hemoglobin (see discussion by Roughton in ref. 26). About 0.15 mole of CO_2 per heme is bound by oxyhemoglobin, and about 0.40 mole by deoxyhemoglobin. The direct binding of CO_2 occurs by the formation of carbamates with the uncharged amino group:

$$R—NH_3^+ \rightleftharpoons R—NH_2 + H^+,$$
$$R—NH_2 + CO_2 \rightleftharpoons R—NHCOO^- + H^+. \tag{9}$$

The most probable groups in hemoglobin for this reaction are the α —NH_2 groups; the ϵ —NH_2 groups of lysine can be excluded because their pK values appear always to lie far above the physiological range of pH. The bicarbonate ion, HCO_3^-, itself appears to have little or no effect on oxygenation and does not seem to be bound (27).

Kilmartin and Rossi-Bernardi (28) have provided strong evidence that the α —NH_2 groups of both α and β chains are the carriers of CO_2. They showed that the specific lowering of the oxygen affinity of horse hemoglobin by CO_2 was completely abolished by reaction of the α —NH_2 groups of both kinds of chain with cyanate:

$$R—NH_2 + HCNO \rightarrow R—NHCONH_2.$$

They also showed that the α —NH_2 groups of both kinds of chain were associated with the CO_2 effect by preparing hybrid hemoglobins in which either the α subunits or the β subunits were reacted with cyanate. Each of these hybrids showed partial inhibition of the CO_2 effect. Furthermore, reaction of cyanate with the α —NH_2 groups of the α chains caused a 25% reduction in the Bohr effect, but reaction with the α —NH_2 groups of the β chains did not alter the Bohr effect. This finding provides strong support for the conclusion of Perutz et al. (15, 16) that the α —NH_2 group of the α chain provides a significant part of the Bohr effect. The presence of CO_2 diminishes the Bohr effect because the binding of CO_2 by deoxyhemoglobin is associated with the release of protons. This release balances the uptake of protons so that the Bohr effect as such vanishes at pH 7.45 (29).

4. THE ORGANIC PHOSPHATE EFFECT

A. Metabolic and Historical Aspects

Since about 1909, dialysis has been known to increase the oxygen affinity of hemoglobin in hemolysates (see ref. 30). Remarkably, for more than 50 years this fact was not systematically investigated. This delay is most curious because one of the major activities of biochemists in the early part of this century concerned the extraction of enzymes from tissues and the discovery that enzymic activity was frequently lost or modified by dialysis. The first enzyme cofactors were discovered by restoring components that had been dialyzed out of the enzyme preparation, but no such investigation was undertaken by those studying hemoglobin although the possibility of the effects of other components was occasionally mentioned. For example, Krogh and Leitch (31) wrote in 1919:

We believe that the adaptation of fish blood must be brought about by some substance or substances present along with the haemoglobin within the corpuscles, and we wish to point out the general significance of the haemoglobin being enclosed in corpuscles surrounded by semipermeable membranes. By this arrangement just that chemical environment can be secured which is most suitable for the respiratory function of the haemoglobin in that particular organism, while at the same time the chemical composition of the blood plasma can be adapted, as it must needs be, to the general requirements of the body cells · · · the possession of semipermeable red corpuscles is therefore a necessary condition for utilizing to the full the wonderful respiratory properties of the haemoglobin.

It seems likely that this lag in investigation resulted in part from the conviction of biochemists that hemoglobin was something special and aberrant, and certainly not an enzyme. The physiologists studying hemoglobin did not look to enzyme studies for inspiration. Had this interchange occurred, the presence and importance of cooperativity in enzyme reactions and the existence of allosteric control mechanisms might very well have been discovered much earlier.

Sugita and Chanutin (32) first demonstrated an interaction between an organic phosphate and hemoglobin: a new electrophoretic component of hemoglobin could be demonstrated in the presence of organic phosphates. This interaction was shown by Benesch and Benesch (33) and simultaneously by Chanutin and Curnish (34) to lower the affinity of hemoglobin for oxygen. This finding has resulted in a massive effort to understand the role of organic phosphates and hence to a plethora of symposia. Much of the evidence for a central controlling role of organic phosphates can be found in the papers of

Benesch et al. and in several recent symposia and reviews; see, for example, refs. 35 and 36 and the papers which follow them, and the reviews edited by Brewer (37), de Verdier et al. (38), Brewer and Eaton (39).

The following discussion will be confined largely to the molecular basis for the effects of 2,3-diphosphoglycerate. No attempt will be made to examine metabolism. However, it should be recognized that a delicate balance exists and that alterations in erythrocyte metabolism can have substantial effects on oxygen transport. Thus inherited defects in the red cell enzymes, pyruvate kinase and hexokinase, have been shown to result in changes in the affinity of the blood cells for oxygen (40). Just as the level of 2,3-DPG alters the level of oxygenation, the reciprocal effect also occurs. Therefore a lowering of oxygen saturation frequently results in a stimulation of glycolysis and an accumulation of 2,3-DPG which can further decrease the oxygen saturation. Thus the 2,3-DPG concentration in red cells rises if oxygen saturation is decreased either by exposure to high altitudes (41) or by cardiac disease, which impairs oxygen transport (42), and it now appears that the thyroid hormone has a directly stimulating effect on the diphosphoglycerate mutase enzyme so that increased thyroxine results in an accumulation of 2,3-DPG (43). Ion transport in red cells appears to be closely linked in some way to changes in the 2,3-DPG concentration and in the level of oxygenation (44). Schaefer et al. (45) have made the interesting observation that chronic exposure of rats or guinea pigs to 15% CO_2 and 21% O_2 results in a transient drop in oxygen affinity which is paralleled by a similar drop in the red cell Na^+ and K^+ concentrations.

B. Stoichiometry of 2,3-Diphosphoglycerate

A complete understanding of the mechanism of action of organic phosphates requires that the number and nature of the binding sites be determined and that the kinetic and equilibrium constants be accurately measured for each site. The binding behavior can, in principle, be approached either directly by measuring the actual amount of organic phosphate bound or indirectly by measuring the effect of change in the *free* phosphate concentration on the oxygen equilibrium (see p. 5). Although several studies of the effect of phosphates on the oxygen equilibrium have been made, none has simultaneously measured the free phosphate concentration (particularly as a function of oxygenation). All other studies have involved direct measurement of the extent of 2,3-DPG (or other phosphate) binding. Since these other experiments, by Benesch et al. (46, 47), Chanutin and Hermann (48), Garby et al. (49), and Lo and Schimmel (50), utilized quite different conditions, the data obtained are difficult to compare: the protein concentrations differ more than tenfold, and the ionic strength, specific ions, and temperature

Table 1. The Binding of 2,3-Diphosphoglycerate by Human Hemoglobin: a Comparison of Results and Experimental Conditions

Experimental Conditions	Results		
	Benesch et al. (46, 47)	Garby et al. (49)	Chanutin and Hermann (48)
Hemoglobin concentration, mM	0.046–0.28	3.1	0.077
Maximum DPG/Hb ratio used	~5	~10	~5
Temperature	22° ± 2°	4°	"Cold room"
pH	7.30	7.20	6.5
Salt, buffer	0.1 M NaCl	3.0 mM Mg^{2+}	0.05 M cacodylic acid
Technique	Equilibrium dialysis	Sedimentation velocity	Equilibrium dialysis
K^a (association), M^{-1}			
Deoxy	4.8 × 10^4	0.013 × 10^4	13.5 × 10^4
Oxy	0	0.003 × 10^4	7.0 × 10^4
K_D/K_O	∞	4.3	1.9
Model (text)			
K_D/K_O	37.2	⋯	⋯

[a] Benesch et al. (46, 47) interpret their data on the basis of a single site, whereas Garby et al. (49) and Chanutin and Hermann (48) each analyze their data on the basis of at least three sites. Only the *first* association constant provided by each of the latter two sets of authors is presented in the table.

12

all differ (see Table 1). The Benesch data (46, 47) were interpreted in terms of a single site, whereas the data of Chanutin and Hermann (48) and of Garby et al. (49) were interpreted on the basis of multiple sites. The first association constant given by Chanutin and Hermann differs from the single Benesch constant by only a factor of 2, but the first constant given by Garby et al. is lower by a factor of 240. Part of this enormous difference must arise from the different experimental conditions, and part from the different techniques used to analyze the data.

The following factors may be responsible in part for the low binding of DPG observed by Garby et al. (49). The pentacyclohexylammonium salt of DPG was used; since no mention is made of the removal of this cation, we presume that it was present at the same concentration as DPG. Magnesium ions were present in nearly all of the experiments by Garby et al. Both cyclohexylammonium and Mg^{2+} ions must have been competing with hemoglobin for phosphate. Although Garby et al. provide some data suggesting that Mg^{2+} has no effect on their results, one would reasonably expect that these ions would lower DPG binding to hemoglobin because DPG binds Mg^{2+} with an association constant of $0.5-0.8 \times 10^3 \ M^{-1}$ (51). Collier and Lam (51) suggest that as much as 70% of the erythrocyte Mg^{2+} ions may be bound to DPG.

For purposes of illustration, let us suppose that a single site is present on hemoglobin which competes with Mg^{2+} ions for DPG and that the DPG-Mg complex does not bind to hemoglobin. Garby et al. (49) presumably measured as "free" DPG the total which was not bound to hemoglobin, $C_{DPG} =$ [DPG] + [DPG-Mg]. If K_{Mg} is the association constant for the reaction of DPG with magnesium, the free [DPG] would be expressed as

$$[DPG] = \frac{C_{DPG}}{1 + [Mg]K_{Mg}} \, .$$

In most of the experiments of Garby et al. (49), $[Mg] = 3 \times 10^{-3} \ M$. Therefore, [DPG] is only about one-fourth the value of C_{DPG}, and the calculated association constant would be one-fourth as great as that which would be determined in the absence of magnesium. Garby et al. state that up to 10 mM inorganic phosphate had no effect on DPG binding. Could it be that this lack of effect was due partly to the formation of a magnesium phosphate complex?

A further decrease in binding of 2,3-DPG to deoxyhemoglobin in Garby's experiments may have resulted from the inadvertent admission of oxygen to the solution of deoxyhemoglobin, which was transferred by thin polyethylene tubing to polycarbonate centrifuge tubes; there the solutions were covered with paraffin oil and capped with Scotch tape. Inasmuch as polyethylene, polycarbonate, Scotch tape, and paraffin oil are all highly permeable to oxygen, significant oxygenation might have occurred. Perhaps this is one reason why the difference between the binding ratio (DPG/Hb) for oxy- and deoxyhemoglobin is never greater than about 0.5, whereas Benesch et al. (46, 47) found a difference close to 1.0.

A lowering of the measured binding may also have resulted from the use of a very low salt concentration in the experimental technique. Garby et al. sedimented the deoxyhemoglobin solution for 36 hrs. They assumed that the concentration of DPG in the clear supernatant was nearly identical with that in the lower part of the tube, even though they observed an apparent sedimentation of 2,3-DPG itself. Such a centrifugal experiment is subject to a Donnan effect, just as in an osmotic pressure measurement, the only difference being the mechanism of the constraint by which the macromolecule is confined to a region. Above the isoelectric point of hemoglobin, a Donnan effect would result in a DPG gradient, with the highest DPG concentration at the top of the tube. This would result in an overestimate of the free DPG and hence an apparent lower binding ratio.

Garby et al. (49) studied the pH dependence of the binding of DPG to both oxy- and deoxyhemoglobin. Their results indicate that the *difference* in the extent of binding of DPG to oxy- and deoxyhemoglobin *increases* with pH. This is very puzzling because the effect of DPG on the oxygen equilibrium is known to *decrease* with rising pH; above pH 8, DPG has little or no effect. This fact implies that the binding of DPG above pH 8 is not linked to oxygenation, but the data of Garby et al. indicate that such a relationship exists. This peculiar result may arise because changes in the pH dependence of the binding of DPG to the other constituents; cyclohexylammonium and Mg^{2+} ions, may overshadow the pH dependence of the binding to hemoglobin. The binding curves ($\bar{v}$ versus [DPG], where $\bar{v}$ is the average number of moles of DPG bound) in these experiments are distinctly sigmoid, indicating that the binding may be cooperative. The Scatchard plot of the data ($\bar{v}/[DPG]$ versus $\bar{v}$) shows a conspicuous, unusual maximum reminiscent of the binding of guanosine triphosphate by glutamate dehydrogenase (52), where departures from linearity were related to the presence of an association-dissociation reaction of the protein. Such an explanation seems unlikely for $3\,mM$ hemoglobin; substantial dissociation can be expected only at concentrations more than 100-fold lower. A conformational transition could be invoked to explain the results, but it would have to be a transition among deoxy conformations. The existence of a sigmoid binding curve does suggest such a change, just as in the sigmoid O_2 binding by hemoglobin. The Scatchard plot, however, is capable of simple interpretation only when the ligand binding groups are assumed to be independent of one another. If the groups are not independent, the association constants derived by the curve fitting of segments of a Scatchard plot may be quite wide of the actual values.

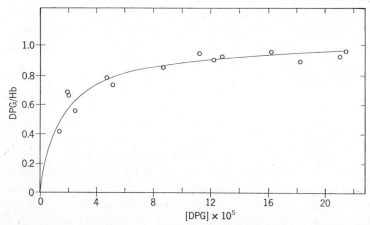

Fig. 1. The binding of 2,3-DPG by human deoxyhemoglobin. The experimental points are those of Benesch *et al.* (46), obtained in 0.1 *M* NaCl at pH 7.3 at 22° ± 2°. The smooth curve is computed on the basis of a model described by equation 10 in the text, which supposes that two binding sites are present that interact negatively with one another (see text).

For the many reasons outlined above, it seems that the data of Garby et al. do not provide adequate information to determine either the number of sites or the association constants. Nevertheless, their data *do indicate* that at least two classes of binding site are present under the experimental conditions that they used. Although similar difficulties are encountered in interpreting the experiments of Chanutin and Hermann (48), which were carried out in 0.05 *M* cacodylate buffer, pH 6.5, these experiments also indicate more than one binding site.

The binding data of Benesch et al. (46) appear to be much clearer, for these workers carried out experiments under conditions (0.1 *M* NaCl) in which nonspecific O_2-independent DPG binding is minimized. Their binding data (Fig. 1) appear to show that only one site is present. However, the maximum DPG/Hb ratios used by Garby et al. were greater than those of Benesch and his colleagues (46). Since the latter workers found evidence for more than one site, it may be instructive to ask whether the Benesch data are really inconsistent with a multiple-site model.

A theoretical binding curve is given in Fig. 1 for a molecule with two binding sites calculated essentially as described by Hill (53). The saturation function used is*

$$\bar{S} = \frac{k[L] + k^2[L]^2 e^{-w/RT}}{1 + 2k[L] + k^2[L]e^{-w/RT}}, \tag{10}$$

* I am indebted to Dr. Gary K. Ackers for this illustrative simulation.

where k is the intrinsic microscopic association constant; $[L]$, the DPG concentration; w, the interaction energy; R, the gas constant; and T, the absolute temperature. The curve was fitted with $w = 2.7$ kcal/mole, and $k = 3.33 \times 10^4$. In terms of the two macroscopic association constants, $K_1 = 6.67 \times 10^4$ and $K_2 = 1.83 \times 10^2$, or $K_1/K_2 = 360$. What this suggests is that the leveling off of the binding curve can be very deceptive and that accurate calculation of the association constant of the first site demands a knowledge of the second site even if the second site has only $1/360$ the affinity for DPG possessed by the first site when the latter is liganded. It is of some interest to note that this ratio, 360, is a reasonable number for two sites 5–10 Å apart which interact electrostatically. What this model does is to replace a single constant with two association constants, K_1 and K_2, one 4 times greater than that given by Benesch et al. (46) and the other $1/87$ as large. The value of K_2 is not very different from the value of the first association constant given by Garby et al. (49). The Benesch data shown in Fig. 1 are evidently of insufficient precision and cover too narrow a range of DPG concentrations to exclude a second site, but they are at least consistent with the presence of two sites. We have supposed the two sites to be identical but to interact negatively, that is, the binding of one DPG makes it less probable that a second will be bound. This is not consistent with the indication of positive cooperativity in Garby's data, but the experiments were carried out under very different conditions.

Renthal et al. (54) have recently found not only that pyridoxal phosphate competes with DPG for binding to deoxyhemoglobin but also that reduction with borohydride results in the formation of a covalent link between the pyridoxal group and the α —NH_2 terminal group of the β chain. This implicates one primary site under the conditions of the experiments. The α —NH_2 groups of each β chain are so situated that steric considerations make it most probable that DPG occupies the space between them on the diad axis of the molecule (16). However, it is not inconceivable that at high DPG concentrations two DPG anions might be bound, one to each β chain in deoxyhemoglobin. If so, negative cooperativity might be expected on both steric and electrostatic grounds.

The first studies of the effect of 2,3-DPG on the oxygen equilibrium of human hemoglobin (33, 34) showed clearly that this organic phosphate greatly lowered the affinity of hemoglobin for oxygen. One observation of Benesch and Benesch (33) deserves particular comment. The question is, Does DPG alter the cooperativity between subunits? These workers first reported not only that 2,3-DPG lowers the affinity for oxygen but also that the addition of a thirty-fold molar excess of 2,3-DPG to an *unbuffered* aqueous solution of hemoglobin increased the value of n in the Hill equation from 1.1 to 2.7. Their later measurements (46) suggested that this result was in error and was caused by the incomplete removal of 2,3-DPG. They demonstrated that essentially all the 2,3-DPG could be removed by passing the hemoglobin through a column of Sephadex G-25 equilibrated with 0.1 M NaCl; dialysis against distilled water removed little or no 2,3-DPG. Their data show that addition of 0.5 mole of 2,3-DPG to completely "stripped" hemoglobin resulted in biphasic oxygen equilibria (Fig. 2). Their data for completely stripped hemoglobin was adjusted by correcting the pO_2 values in

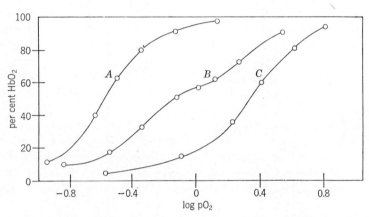

Fig. 2. The oxygen equilibrium of human hemoglobin (0.3%) in 0.01 M NaCl at 10° and pH 7.0 (before deoxygenation). A: phosphate-free hemoglobin; B: A + 0.5 mole of 2,3-DPG per mole of hemoglobin; C: A + 1.0 mole 2,3-DPG per mole of hemoglobin. Data from Benesch *et al.* (46).

the unbuffered solution for the change in pH with oxygenation; they assumed a normal Bohr effect. The Δ pH was about 0.3 during oxygenation, and hence this is a large correction. Since the Bohr effect is independent neither of salt concentration (55) nor of DPG concentration (56), this correction may be inadequate.

The existence of a biphasic curve implies that two molecular species (one with and one without phosphate) are present and that these are *not* at equilibrium with one another; if they were, the plateau evident in Fig. 2 should vanish. The presence of a plateau indicates either that a mixture of two intrinsically different kinds of molecule is present or that equilibrium with phosphate is extremely slow compared with the time required for the oxygen equilibrium measurements. These measurements were carried out at a low salt concentration, 0.01 M NaCl. Since much 2,3-DPG remains bound to hemoglobin even after dialysis for 24 hrs against 0.01 M NaCl, it seems likely that true equilibrium with 2,3-DPG was not present in these experiments. The absence of equilibrium under these conditions raises serious questions about the validity of the binding studies carried out at low salt concentration (48, 49).

Much evidence indicates that the binding of DPG to hemoglobin is electrostatic: positively charged groups on the protein appear to be involved in the binding of the negatively charged organic phosphates. Benesch et al. (47) have shown that sufficient NaCl can displace DPG entirely from hemoglobin. It appears that at a low salt concentration ($\sim$0.01 M NaCl) the DPG does not equilibrate with the hemoglobin rapidly. Under these conditions some molecules may have DPG and others not, so that a spurious kind of heterogeneity may be present, as discussed earlier. Gibson (9) has found that different organic phosphates react at quite different rates with human hemoglobin. Thus pyridoxal phosphate reacts with a half-time of minutes rather than milliseconds. This permits extensive kinetic analysis of the intermediates.

It must be emphasized here that most current measurements of ligand binding by hemoglobin are spectrophotometric; "equilibrium" curves are assumed to be at true equilibrium. The spectrophotometric assumption is that the absorbance change at any wavelength is linearly related to the fraction of the total number of ferroheme groups combined with ligand. Although this assumption appears to be valid within the errors of measurement (57, 58), it should be borne in mind that measurements of ligand binding of the hemes in the presence of substances which may perturb the protein conformation may elicit substantial departures from this linearity. Olson and Gibson (59) have made the important discovery that the binding of n-butyl isocyanide requires that at least two spectrally distinct components be considered. They believe that these components are the α and β subunits. They found not only that diphosphoglycerate and inositol hexaphosphate alter the overall affinity for the isocyanide but in addition that the functional properties of the α and β chains become quite distinct. Thus an apparent heterogeneity is introduced.

C. The pH Dependence of Diphosphoglycerate Binding

A detailed analysis of currently available data has recently been carried out by Riggs (60). This analysis and the resultant model depend largely on the data of Benesch et al. (47) and on the more recent experiments of Tomita and Riggs (56). The most important experimental facts upon which the analysis rests can be summarized as follows.

1. Diphosphoglycerate is bound to deoxyhemoglobin to a much greater extent than to oxyhemoglobin (46, 48–50). The stoichiometry (see Section

4.B) is approximately 1:1 at low DPG concentrations and in moderate salt concentrations (∼0.1 M NaCl) (46, 47).

2. The effect of DPG on O_2 binding is greatest at low pH (47, 56). Similarly, the binding of DPG is also greatest at low pH (47, 49).

3. The binding of DPG appears to be electrostatic; DPG can be completely displaced at a sufficiently high ionic strength (47).

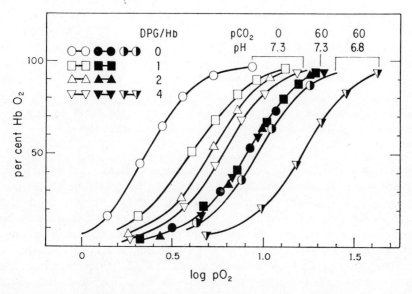

Fig. 3.   The oxygen equilibrium of mouse hemoglobin at 20°: influence of CO_2 and 2,3-DPG. Buffer: Tris-HCl, ionic strength, 0.025. Hemoglobin concentration, 0.2%. Data from Tomita and Riggs (56).

4. Carbon dioxide inhibits the effect of DPG on oxygen binding (Fig. 3) (56). This effect depends on the pH; the higher the pH, the lower is the pCO_2 required. Carbon dioxide is known to be carried by hemoglobin by combination with the α —NH$_2$ groups of all four chains (see Section 4.B). This reaction occurs only with the uncharged amino groups.

5. The α —NH$_2$ groups of the α chains contribute to the Bohr effect; those of the β chains appear not to participate in the Bohr effect in the absence of DPG (28).

These observations are completely explained by the assumption (16) that the primary binding site is between the two α —NH$_3^+$ groups of the β chains. The proposed arrangement is shown in Fig. 4. The pH dependence of binding suggests that DPG is bound *only* when both α —NH$_2$ groups are protonated.

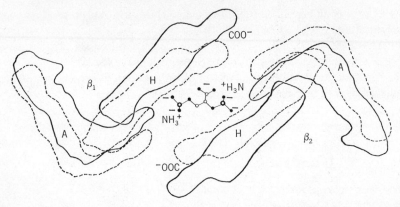

Fig. 4. Proposed site of binding of 2,3-DPG by hemoglobin. This is a projection of the electron density maps of oxy (- - - - -) and deoxy (——) hemoglobins at 5.5 Å resolution. From Perutz (16). The letters A and H refer to the A and H helical segments.

Since CO_2 combines only with the unprotonated form (equation 9), the scheme in Fig. 5 appears the most reasonable way to explain the pH-dependent competition between CO_2 and DPG. Kilmartin and Rossi-Bernardi (28) have shown that the release of protons associated with CO_2 binding by deoxyhemoglobin balances the number of protons normally taken up when the hemoglobin is deoxygenated, and hence the Bohr effect either vanishes or is greatly depressed. The greater the extent of DPG-Hb combination, the more will reaction 1 shift in favor of protonation. This means that the α —NH_2 groups of the β chains become oxygenation linked in the presence of DPG, although they have no such function in its absence. The arrangement in Figs. 4 and 5 suggests that a common mechanism may explain why both CO_2 and DPG lower the affinity of hemoglobin for oxygen. The carboxyl terminus of each β chain in deoxyhemoglobin is about 12 Å from the α

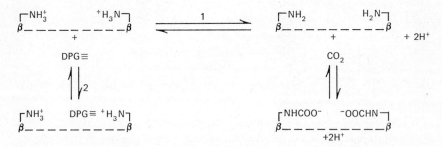

Fig. 5. Proposed reactions of 2,3-DPG and CO_2 with the α —NH_2 terminal groups of the β chains of hemoglobin.

—NH$_3^+$ group of the neighboring β chain. In contrast, this distance is only about 6 Å in oxyhemoglobin (61). Thus the approach of the —COO$^-$ group toward the —NH$_3^+$ group would be subject to substantial electrostatic repulsion by either the negative carbamino group, —NHCOO$^-$, or the DPG anion. As a result both CO$_2$ and DPG would stabilize and favor the deoxy conformation. Perhaps of equal importance is the fact that the space available is smaller in oxy- than in deoxyhemoglobin (Fig. 4).

The reactions shown in Fig. 5 can be analyzed quantitatively (60). Benesch et al. (47) give the association constant (K_B) at three pH values. Clearly their constant includes a protonation reaction. We seek to show that this reaction is in fact the protonation of the α —NH$_2$ groups of the β chains. Since these groups are ~16 Å apart in deoxyhemoglobin and ~20 Å apart in oxyhemoglobin (16), they cannot be influencing one another electrostatically to any significant extent. Hence we can describe their ionization in terms of two independent steps:

$$Hb^{2+} \xrightarrow{K_1} Hb^+ \xrightarrow{K_2} Hb.$$

Here K_1 and K_2 are the macroscopic dissociation constants. It is more convenient, however, to use the microscopic ionization constant, k, related to the macroscopic constants by $K_1 = 2k$ and $K_2 = k/2$. If Hb^{2+} — DPG is the only form of complex between hemoglobin and DPG, then Benesch et al. (47) measured the ratio of this form to all other forms not binding DPG: Hb, Hb$^+$, Hb^{2+}. Under these circumstances we can relate the pH-independent intrinsic association constant for DPG, K_D, with deoxyhemoglobin to the pH-dependent constant, K_B, by the expression

$$K_D = K_B\left(1 + \frac{k}{H^+}\right)^2. \tag{11}$$

If $1/\sqrt{K_B}$ is plotted against $1/H^+$, a straight line should result with a slope of $k/\sqrt{K_D}$ and an intercept $1/\sqrt{K_D}$. The Benesch data conform closely to this line; departures are less than 0.8%. The data yield a value of $K_D = 4.07 \times 10^5\ M^{-1}$ and $k = 7.7 \times 10^{-8}\ M$ at 25°, corresponding to an intrinsic pK of 7.11 at 25°. We can compare this value with that obtained by Forster et al. (62) for the groups associated with CO$_2$ binding. They found a pK value of 7.14 for the ionization of the α —NH$_2$ groups at 37°. Since CO$_2$ combines with the α —NH$_2$ groups of all four chains, this pK value represents an average for both chains. We can correct the value at 37° to 25° if we assume a value for the standard enthalpy change $\Delta H°$ for the ionization of an α —NH$_2$ group. These values appear to fall in the neighborhood of ~11 kcal/mole (63). If so, the pK obtained by Forster et al. (62) would be about 6.8 at 25°. We can consider this as only a gross approximation because their preparation was dialyzed just against distilled water. Consequently virtually all the erythrocyte DPG must still have been present, and so the extent of ionization would have been shifted. Nevertheless, the similarity of pK values and the fact that CO_2 combines only with the uncharged α —NH_2 groups and competes with DPG strengthen the validity of the scheme shown in Fig. 5.

It is possible that the β 143 histidyl residues may also contribute to the pH dependence of DPG binding, since they appear to be within hydrogen-bonding distance of the DPG (16). Such close proximity would be expected to modify the histidyl pK values. Nevertheless, the data presently available can be well described in terms of only two independent groups without resorting to any additional groups. It should be mentioned here that the Benesch data cannot be fitted at all if we assume that only a single ionizing group is present, and the fit is poor if we assume that the ionization is concerted, that is, that the two protons are either on or off together. Of course, introducing additional groups with additional terms and constants can do nothing but improve any fit.

Benesch et al. (47) also present data on the temperature dependence of K_B. They give the associated values of the thermodynamic functions: $\Delta H° = -13$ kcal/mole for the standard enthalpy change, and -22 entropy unit/mole for the standard entropy change. Unfortunately K_B, as we have shown, depends on pH. Equation 11 can be used to estimate the temperature dependence of the intrinsic constant, K_D, in terms of K_B and k. This estimate requires that we know the standard enthalpy change for the ionization reaction. We assume, as before, that $\Delta H°$ for α —NH_3^+ group ionization is about 11 kcal. The relationship

$$\frac{\partial\, pK}{\partial(1/T)} = \frac{\Delta H°}{2.3R}$$

then permits estimating pK and k for each temperature. Substitution of the Benesch values of K_B and these estimates for k in equation 11 yields K_D at each temperature. The value of the resulting K_D is found to be about 3.92 × $10^5\ M^{-1}$ and is *essentially independent of temperature*, so that $\Delta H°$ is zero or close to zero rather than -13 kcal. This means that the observed temperature dependence of K_B is due virtually entirely to the ionization reaction and *not* to DPG binding. Recalculation of $\Delta S°$ for DPG binding shows that $\Delta S°$ is not -22 eu but $+24$. These new estimates provide a more satisfactory picture of the actual binding process. A value of $\Delta H°$ close to zero must mean that there is an almost complete compensation of the enthalpy of DPG binding by a concomitant displacement of inorganic ions. The pocket in which DPG sits is a nest containing six positively charged groups, each of which must have one or more associated Cl^- ions in the absence of DPG

0.1 M NaCl. Clearly, as the ionic strength is greatly
 binding should become large and negative. This
sis against distilled water is ineffective in removing di-

nsider the changes in ionization of the β-chain α —NH$_2$ groups
associated with oxygenation *in the presence of DPG*. We first calculate the
number of protons bound by deoxyhemoglobin, *neglecting all protons other
than those associated with the β-chain α —NH$_2$ groups*. A similar calculation
is then made for oxyhemoglobin. The difference is the number of *additional*
protons released as a consequence of oxygenation in the presence of DPG.
The average number of additional protons bound per molecule of deoxyhemoglobin, $\bar{h}_D$, is given by

$$\bar{h}_D = \frac{1}{C_H}(2[\text{Hb}^{2+} - \text{DPG}] + 2[\text{Hb}^{2+}] + [\text{Hb}^+])$$

$$= \frac{2(H^+/k)[1 + (H^+/k)] + 2K_D(\text{DPG})_D[H^+/k]^2}{[1 + (H^+/k)]^2 + K_D(\text{DPG})_D[H^+/k]^2}. \tag{12}$$

The total molar concentration of hemoglobin is C_H; H^+ is the hydrogen-ion
activity; K_D is the association constant for DPG with *deoxyhemoglobin;* and
k is the microscopic constant for the α —NH$_3^+$ group dissociation reaction.
The free DPG concentration, $(\text{DPG})_D$, can be expressed in terms of the total
DPG concentration by the equation

$$(\text{DPG})_D{}^2 + \left[\frac{1}{K_D}\left(1 + \frac{k}{H^+}\right)^2 + C_H - C_D\right](\text{DPG})_D - \frac{C_D}{K_D}\left(\frac{k}{H^+} + 1\right)^2 = 0.$$

$$\tag{13}$$

We assume that the protons bound by oxyhemoglobin, $\bar{h}_O$, will be given by
equations identical with 12 and 13 except that K_O replaces K_D and $(\text{DPG})_O$
replaces $(\text{DPG})_D$, where the subscript letters D and O refer to the deoxy and
oxy states of the hemoglobin, respectively. The number of protons released
upon oxygenation will then be given by

$$\Delta\bar{h} = \bar{h}_D - \bar{h}_O.$$

Examination of equations 12 and 13 shows that at sufficiently high DPG
levels $\Delta\bar{h}$ approaches zero. Both Benesch et al. (47) and Tomita and Riggs
(56) provide data on $\Delta \log P_{50}/\Delta$ pH as a function of DPG concentration.
This is equivalent to a measurement of the number of protons released per
oxygen bound *at 50% oxygenation* (see p. 5). These data show clearly that
$\Delta\bar{h}$ does approach zero at high DPG concentrations. Although Benesch et al.
(47) assert that *no* DPG is bound by oxyhemoglobin under the conditions of

their experiments, this is clearly impossible, as the following consid
will show. Examine the overall reactions of the protonated species in
form of a "box":

$$DPG \cdot Hb^{2+} \xrightarrow[+O_2]{\Delta G_3} DPG \cdot Hb^{2+}O_2$$

$$\Delta G_1 \uparrow +DPG \qquad\qquad +DPG \uparrow \Delta G_2$$

$$Hb^{2+} \xrightarrow[\Delta G_4]{+O_2} Hb^{2+}O_2$$

Here the ΔG's are the associated free-energy changes. Transformation of
Hb^{2+} to $DPG \cdot Hb^{2+}O_2$ can occur along either of two pathways; hence

$$\Delta G_1^\circ + \Delta G_3^\circ = \Delta G_2^\circ + \Delta G_4^\circ.$$

The difference, $\Delta G_2^\circ - \Delta G_1^\circ$, is determined by K_D and K_O and is equal to
$RT \ln (K_D/K_O)$. If $K_O = 0$, $\Delta G_3^\circ - \Delta G_4^\circ$ evidently increases without limit
and the difference between the oxygen affinities of "stripped" and "DPG"
hemoglobins would also become indefinitely large; but we know that this
does not happen. This is but another way of looking at the linked functions
of DPG binding and oxygenation discussed already. The experimental data
for $\Delta \log P_{50}/\Delta$ pH versus [DPG] are completely inconsistent with the idea
that $K_O = 0$. If $K_O = 0$ in equations 12 and 13 and we make [DPG] large,
then $\Delta \bar{h}$ approaches a limiting value of $2k/(k + H^+)$ and never returns to
zero. This result indicates that the plateau level decreases with increasing H^+,
which is consistent with the Benesch data; $\Delta \bar{h}$ approaches a maximal value
of 2 in alkaline solution and of 1.0 where $k = H^+$, and is less than 1.0 where
$H^+ > k$.

Binding of organic phosphates by oxyhemoglobin has been found by
Garby et al. (49), Chanutin and Hermann (48), and Grisolia et al. (64).
Although Benesch et al. (47) found no binding by oxyhemoglobin, it seems
impossible that this is the case. Perutz (16) states that the atomic model of
oxyhemoglobin indicates that insufficient room exists between the α —NH$_2$
groups of the β chain to accommodate a DPG molecule. Nevertheless,
recent spin-label studies by Deal et al. (65), Moffat (66), and McConnell (67)
suggest that substantial shifts occur in the conformational equilibria of the
nearby COOH terminal residues in both oxy- and deoxyhemoglobin.
Such conformational shifts may account for the binding which does appear to
occur. Gibson and Gray (68) found that the rate constant for the dissociation
of oxygen from fully liganded oxyhemoglobin is increased by inositol
hexaphosphate from about 50/sec to 180/sec at 22°; this could happen only
if the phosphate is bound by oxyhemoglobin. They observed no such effect
with DPG or with pyridoxal phosphate. Surprisingly, they found that inositol
hexaphosphate did not alter the rate of release of CO from carboxyhemo-
globin. This suggested to them that carboxy- and oxyhemoglobins may have

574.8764   K962b

somewhat different conformations. It is worth emphasizing here, however, that the extraordinary sensitivity of organic phosphate binding to ionic strength and to pH warrants unusual care in ascertaining that the circumstances of two experiments are truly comparable.

It must be stressed that equations 12 and 13 contain only a single unknown constant, K_O, for the other constants, k and K_D, are independently determined by equation 11 and the pH-dependence data at 25°. However, the experimental data on $\Delta \log P_{50}/\Delta$ pH were obtained at 20°. If $\Delta H°$ for the ionization is 11 kcal, we estimate k to be $5.62 \times 10^{-8}\ M$ at 20°. Since K_D appears to be essentially independent of temperature, we can use the average of all values calculated from the Benesch data, $3.92 \times 10^5\ M^{-1}$. A least-squares search program has been used (60) to obtain a best-fit value of K_O, $1.05 \times 10^4\ M^{-1}$. This is about 2.7% of the value of K_D at 20°. The fit of these data is shown in Fig. 6. The model appears to account rather satisfactorily for the overall changes in $\Delta \log P_{50}/\Delta$ pH. Whether the deviations

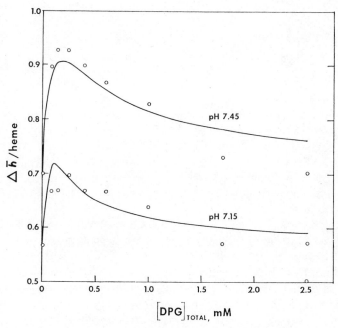

Fig. 6. The relationship between the Bohr effect and the total concentration of 2,3-DPG. The experimental points are those of Benesch *et al.* (47). The solid lines are computed on the basis of equations 11 and 12 in the text, with $K_D = 3.92 \times 10^5\ M^{-1}$, $K_O = 1.05 \times 10^4\ M^{-1}$, and $k = 5.62 \times 10^{-8}\ M$, and hemoglobin concentration $60\ \mu M$ on a tetramer basis. The value of $\Delta \bar{h}$ (text) has been added to the value of the Bohr effect in the absence of of diphosphoglycerate. From Riggs (60).

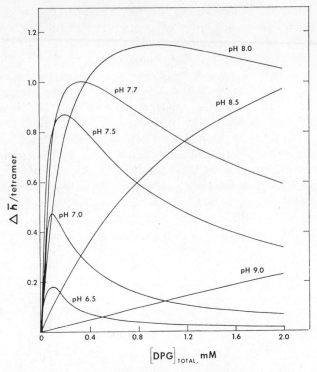

Fig. 7. Computed curves for the enhancement of the Bohr effect by diphosphoglycerate. From Riggs (23). Here $\Delta \bar{h}$ is the number of protons released from the α —NH$_2$ groups of the β chains as a result of oxygenation. The constants are the same as in Fig. 6.

which are present are real and reflect inadequacy in the model or whether they are due to experimental error can be determined only by more extensive and precise measurements.

The model we have discussed depends on several simplifying assumptions. The changes in the ionization of DPG with pH have been ignored. Only two ionizable groups are assumed to participate in DPG binding, and these are assumed not to ionize at all when DPG is bound. Presumably, at sufficiently high DPG levels, this would not hold true. Ionization of the β 143 histidyl residues could be involved but is not required to explain the data. Additional DPG binding sites have also been ignored. Error may be introduced by equating $\Delta \bar{h}$ with $\Delta \log P_{50}/\Delta$ pH, because $\Delta \bar{h}$ is the change resulting from complete oxygenation at constant pH, whereas $\Delta \log P_{50}/\Delta$ pH involves measurements made at 50% oxygenation over a pH interval. Thus $\Delta \bar{h}$ will equal $\Delta \log P_{50}/\Delta$ pH only if $\Delta \log P_{50}/\Delta$ pH is independent of degree of oxygenation and Δ pH is small. Since the shapes of the measured O$_2$-equilibrium curves are dependent on pH in the presence of DPG (56), this assumption may not be adequate.

We have assumed that no groups other than the α —NH_2 groups of the β chain change their pK as a result of DPG binding. Nevertheless, the argument above that both DPG and CO_2 may act in part by electrostatic repulsion of the COOH terminal histidine requires that these groups be stabilized in their deoxy conformation. It seems quite possible that this additional stabilization may be associated with a further elevation of the histidyl imidazole pK above its normal value in deoxyhemoglobin. The net result would be that oxygenation would be accompanied by a larger Δ pK for this group and hence more protons would be released, thus elevating the Bohr effect above the value predicted by the model.

Equations 12 and 13 predict how $\Delta \bar{h}$ should depend on pH and DPG concentration. The data in Fig. 6 show that the model is reasonable at two pH values. Computed curves (Figs. 7 and 8) indicate that a well-defined pH optimum should be present at pH 7.5. Thus, in the physiological range pH 7.0–7.4, an *increase in pH* should result in an enhancement of the Bohr effect from this source (the α —NH_2 groups of the β chains).

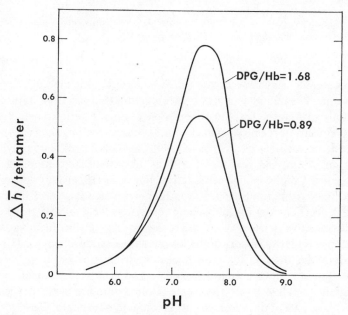

Fig. 8. The pH dependence of the enhancement of the Bohr effect, computed on the same basis as in Fig. 7, for two DPG/Hb ratios.

The model helps to explain the comparative physiological studies that have been carried out. Schmidt-Nielsen and Larimer (69) found, among a series of mammals, that the smaller the mammal and the higher its specific oxygen consumption, the higher is the oxygen pressure at which the blood releases oxygen. Riggs (70) showed that this depends on differences in the Bohr effect, and more recently Tomita and Riggs (56) reported that these variations in the Bohr effect result primarily from the presence of DPG. In the presence of low concentrations of DPG, mouse and human hemoglobins were shown to have different Bohr effects, even though they had very similar Bohr effects in the absence of DPG. Thus, at a DPG concentration of ~ 1 mole/mole of hemoglobin, the Bohr effect at 20° (protons released per O_2 bound) was found to be 0.9, 0.65, and 0.45 for mouse, human, and elephant hemoglobins under the same set of conditions; these differences presumably arise from variations in K_D, K_O, and k. Although these data are not sufficient for detailed analysis, they suggest that the K_D/K_O ratio is much larger in mouse than in human hemoglobin.

5. CONCLUSIONS

The heme proteins constitute perhaps the most varied and beautiful class of molecular machines known to the biochemist.

Allen et al.(71)

The intricate interplay between the linked functions of the binding of protons, CO_2, O_2, and organic phosphates by hemoglobin that has been unraveled by many studies is but the first-known example of an allosteric system. The sites and mechanisms of reaction of the major agents controlling oxygen transport now seem increasingly clear. The α —NH_2 groups of all four polypeptide subunits react with CO_2 and transport it as the carbamino compound. In the presence of 2,3-diphosphoglycerate all of the α —NH_2 groups appear to be involved in oxygenation-linked proton binding, which is part of the basis of the Bohr effect. The organic phosphate confers on the α —NH_2 groups of the β chains effective sensitivity to oxygenation. The β 146 histidyl residues account for a substantial part of the Bohr effect. In addition, α 122 His is also involved in the Bohr effect. The pocket in which the diphosphoglycerate anion is bound contains six positively charged groups. This space becomes substantially smaller when the hemoglobin combines with oxygen, and the result is a partial displacement of the organic phosphate. This "positive" pocket may be in part responsible for the large effect of changes in ionic strength on oxygen binding. Examination of the groups associated with this pocket in the hemoglobins from different species should prove particularly rewarding as the nature of the molecular adaptation of different mammalian hemoglobins gradually unfolds. The quantitative contributions of the different groups to DPG binding, to CO_2 transport, and to the Bohr effect may vary considerably in different hemoglobins.

ACKNOWLEDGMENTS

This review was written while on leave at the Department of Biochemistry, School of Medicine, University of Virginia, Charlottesville. I am indebted to Dr. Gary K. Ackers for many cheerful discussions. I have been supported by a Research Career Development Award from the National Institutes of Health.

REFERENCES

1. J. Wyman, Jr., *Advan. Protein Chem.*, **19**, 223 (1964).
2. J. Wyman, *J. Am. Chem. Soc.*, **89**, 2202 (1967).
3. L. Corwin and G. Fanning, *J. Biol. Chem.*, **243**, 3517 (1968).
4. A. Conway and D. E. Koshland, *Biochemistry*, **7**, 4011 (1968).
5. R. A. Cook and D. E. Koshland, *Biochemistry*, **9**, 3337 (1970).
6. A. Levitzki and D. E. Koshland, *Proc. Natl. Acad. Sci. U.S.*, **62**, 1121 (1969).
7. P. C. Engel and K. Dalziel, *Biochem. J.*, **115**, 621 (1969).
8. W. A. Susor, M. Kochman, and W. J. Rutter, *Science*, **165**, 1260 (1969).
9. Q. Gibson, *Biochem. Biophys. Res. Commun.*, **40**, 1319 (1971).
10. C. Frieden, *J. Biol. Chem.*, **245**, 5788 (1970).
11. G. Weber, "The Binding of Small Molecules to Proteins," in *Molecular Biophysics*, (Eds.: B. Pullman and M. Weissbluth), Academic Press, New York, 1965, p. 205.
12. J. Wyman, *Advan. Protein Chem.*, **4**, 407 (1948).
13. G. Morpurgo, P. Battaglia, L. Bernini, A. M. Paolucci, and G. Modiano, *Nature*, **227**, 387 (1970).
14. C. Bohr, K. Hasselbalch, and A. Krogh, *Skand. Arch. Physiol.*, **16**, 402 (1904).
15. M. F. Perutz, H. Muirhead, L. Mazarella, R. A. Crowther, J. Greer, and J. V. Kilmartin, *Nature*, **222**, 1240 (1969).
16. M. F. Perutz, *Nature*, **228**, 726 (1970).
17. H. Muirhead and J. Greer, *Nature*, **228**, 516 (1970).
18. W. Bolton and M. F. Perutz, *Nature*, **228**, 551 (1970).
19. G. Wald and A. Riggs, *J. Gen. Physiol.*, **35**, 45 (1951).
20. R. Briehl, *J. Biol. Chem.*, **238**, 2361 (1963).
21. E. Antonini, J. Wyman, L. Bellelle, N. Rumen, and M. Siniscalco, *Arch. Biochem. Biophys.*, **105**, 405 (1964).

22. S. L. Li and A. Riggs, *J. Biol. Chem.*, **245**, 6149 (1970).

23. A. Riggs, "Lamprey Hemoglobins," in *Biology of Lampreys* (Eds.: M. W. Hardisty and I. Potter), Academic Press, New York, in press.

24. M. E. Andersen, *Federation Proc.*, **29**, 855 Abs. (1970).

25. M. E. Andersen and Q. H. Gibson, *J. Biol, Chem.*, **246**, 4790 (1971).

26. F. J. W. Roughton, *Biochem. J.*, **117**, 801 (1970).

27. J. C. Kernohan, F. Kreuzer, L. Rossi-Bernardi, and F. J. W. Roughton, *Biochem. J.*, **100**, 49 (1966).

28. J. V. Kilmartin and L. Rossi-Bernardi, *Nature*, **222**, 1243 (1969).

29. L. Rossi-Bernardi and F. J. W. Roughton, *J. Physiol.*, **189**, 1 (1967).

30. J. Barcroft, *The Respiratory Function of the Blood. Part II: Haemoglobin*, Cambridge University Press, Cambridge, (1928).

31. A. Krogh and I. Leitch, *J. Physiol. (London)*, **52**, 288 (1919).

32. Y. Sugita and A. Chanutin, *Proc. Soc. Exptl. Biol. Med.*, **112**, 72 (1963).

33. R. Benesch and R. E. Benesch, *Biochem. Biophys. Res. Commun.*, **26**, 162 (1967).

34. A. Chanutin and R. R. Curnish, *Arch. Biochem. Biophys.*, **121**, 96 (1967).

35. R. Benesch and R. E. Benesch, *Nature*, **221**, 618 (1969).

36. R. E. Benesch and R. Benesch, *Federation Proc.*, **29**, 1101 (1970).

37. G. J. Brewer, (Ed.), *Red Cell Metabolism and Function*, Plenum Press. New York, 1970.

38. C. -H. de Verdier, L. Garby, C. F. Hogman, and O. Åkerblom (Eds.), *Försvarsmedicin*, **5**, 143 (1969).

39. G. J. Brewer and J. W. Eaton, *Science*, **171**, 1205 (1971).

40. M. Delivoria-Papadopoulos, F. Oski, and A. J. Gottlieb, *Science*, **165**, 601 (1969).

41. C. Lenfant, J. Torrance, E. English, C. A. Finch, C. Reynafarje, J. Ramos, and J. Faura, *J. Clin. Invest.*, **47**, 2652 (1968).

42. R. D. Woodson, J. D. Torrance, S. D. Shappell, and C. Lenfant, *J. Clin. Invest.*, **49**, 1349 (1970).

43. L. M. Snyder, W. J. Reddy, and L. Kurjan, *J. Clin. Invest.*, **49**, 1993 (1970).

44. R. Benesch and R. E. Benesch, *Science*, **160**, 83 (1968).

45. K. E. Schaeffer, A. A. Messier, and C. C. Morgan, *Respiration Physiol.*, **10**, 299 (1970).

46. R. Benesch, R. E. Benesch, and C. I. Yu, *Proc. Natl. Acad. Sci. U.S.*, **59**, 526 (1968).

47. R. E. Benesch, R. Benesch, and C. I. Yu, *Biochemistry*, **8**, 2567 (1969).

48. A. Chanutin and E. Hermann, *Arch. Biochem. Biophys.*, **131**, 180 (1969).

49. L. Garby, G. Gerber, and G.-H. de Verdier, *European J. Biochem.*, **10**, 110 (1969).

50. H. H. Lo and P. R. Schimmel, *J. Biol. Chem.*, **244**, 5084 (1969).

51. H. B. Collier and A. Lam, *Biochim. Biophys. Acta*, **222**, 299 (1970).

52. C. Frieden and R. F. Colman, *J. Biol. Chem.*, **242**, 1705 (1967).

53. T. L. Hill, *An Introduction to Statistical Thermodynamics*, Addison-Wesley, Reading, Mass., p. 144, 1960.

54. R. Renthal, R. E. Benesch, R. Benesch, and B. A. Bray, *Federation Proc.*, **29**, 723 Abs. (1970).

55. E. Antonini, J. Wyman, A. Rossi-Fanelli, and A. Caputo, *J. Biol. Chem.*, **237**, 2773 (1962).

56. S. Tomita and A. Riggs, *J. Biol. Chem.*, **246**, 547 (1971).

57. Y. Enoki and I. Tyuma, *Japan J. Physiol.*, **14**, 280 (1964).

58. S. R. Anderson and E. Antonini, *J. Biol. Chem.*, **243**, 2918 (1968).

59. J. S. Olson and Q. H. Gibson, *Biochem. Biophys. Res. Commun.*, **41**, 421 (1970).

60. A. Riggs, *Proc. Natl. Acad. Sci. U.S.*, **68**, 2062 (1971).

61. H. Muirhead, J. M. Cox, L. Mazzarella, and M. F. Perutz, *J. Mol. Biol.*, **28**, 117 (1967).

62. R. E. Forster, H. P. Constantine, M. R. Craw, H. H. Rotman, and R. A. Klocke, *J. Biol. Chem.*, **243**, 3317 (1968).

63. C. Tanford, *Advan. Protein Chem.*, **17**, 69 (1962).

64. S. Grisolia, J. Carreras, D. Diederich, and S. Charache, in *Red Cell Metabolism and Function*, Plenum Press, New York, 1970, p. 39.

65. W. J. Deal, S. G. Mohlman, and M. L. Sprang, *Science*, **171**, 1147 (1971).

66. J. K. Moffat, *J. Mol. Biol.*, **55**, 135 (1971).

67. H. M. McConnell, *Ann. Rev. Biochem.* **40**, 227 (1971).

68. Q. H. Gibson and R. D. Gray, *Biochem. Biophys. Res. Commun.*, **41**, 415 (1970).

69. K. Schmidt-Nielsen and J. L. Larimer, *Am. J. Physiol.*, **195**, 424 (1958).

70. A. F. Riggs, *J. Gen. Physiol.*, **43**, 737 (1960).

71. D. W. Allen, K. F. Guthe, and J. Wyman, *J. Biol. Chem.*, **187**, 393 (1950).

CHAPTER 2

Metabolic Regulations Involving Phosphoglycerates[*]

JOSEPH B. ALPERS

Department of Biological Chemistry, Harvard Medical School, Boston, Massachusetts

[*] The survey of literature pertaining to this chapter was concluded in July 1970.

33

1. INTRODUCTION

An abstract of work by Dische that appeared in 1941 in Travaux des Membres de la Société Chimie Biologique (1) reported:

In red corpuscles and their hemolyzates, added mono- and di-phosphoglyceric acids inhibit the spontaneous and coupled phosphorylation of glucose and the coupled phosphorylation of adenylic acid and adeninediphosphoric acid. The diphosphoglyceric acid normally present in the circulating red corpuscle is combined with colloidal substances which protect it from the action of the phosphatases of the corpuscles and also prevent it from interfering with the phosphorylation of glucose.

Now, after a lapse of nearly three decades, significant refinements have been made on these prophetic observations. In a period that has been increasingly concerned with biochemical regulations, the phosphoglycerate compounds have attained recognition and, in a certain sense, prominence. The various compounds in this group (3-phosphoglycerate, 2-phosphoglycerate, 2,3-diphosphoglycerate, and 1,3-diphosphoglycerate) have been shown to affect diverse processes, some of which pertain to the origin and the disposition of the phosphoglycerates themselves, while others are remote from what had been considered to be the normal precincts for these compounds. This chapter will deal with both of these aspects and also with the problem of how other specific cellular constituents in turn affect the metabolism of the phosphoglycerates. As might be expected, the considerations are necessarily intertwined.

The subjects we will consider in particular are the biosynthesis of 2,3-diphosphoglycerate (2,3-DPG), its hydrolysis to 3-phosphoglycerate, the interaction of 1,3-diphosphoglycerate (1,3-DPG) with mutase enzymes, the hydrolysis of 1,3-DPG, and the special role of 2,3-DPG in erythrocytes.

2. BIOSYNTHESIS OF 2,3-DIPHOSPHOGLYCERATE

Since the discovery in 1925 of high concentrations of 2,3-DPG in erythrocytes (2), there have been numerous reports of enzymes facilitating its biosynthesis (3, 4, 5). The reaction, in effect, is the transfer of a phosphoryl group from 1,3-DPG to 3-PGA, creating 2,3-DPG and regenerating the monophosphate (6). For red blood cells, the evidence has been most compelling that a specific diphosphoglycerate mutase is involved.

The fraction from red cells that performs 2,3-DPG biosynthesis exhibits certain kinetic features that seem restrictive but may be uniquely apt. 2,3-Diphosphoglycerate inhibits competitively with 1,3-DPG with a K_i of less than

1 μM. A second inhibitor, inorganic phosphate (P_i) exhibits a K_i of less than 1 mM (5). The former case, as will be emphasized later, may represent a major means for regulating biosynthesis, in conjunction with the binding of 2,3-DPG by the deoxy form of hemoglobin. The inhibition by P_i is more difficult to rationalize, since no physiological circumstances are presently recognized that would reduce its intracellular concentration to levels that could permit the reaction to proceed. The argument of intracellular compartmentation (the enzymologist's normal refuge under such circumstances) is out of the question in the case of the erythrocyte. Perhaps a likelier possibility is that some as yet undisclosed intracellular component may counteract the inhibition by P_i.

In a later section, we will review some alternative evidence that links the production of 2,3-DPG to the usual monophosphoglycerate mutase of glycolysis, at least in cells, unlike the red cell, where 2,3-DPG levels are not particularly elevated. To approach this problem in the red cell, monophosphoglycerate mutase from a sample of human blood was partially purified by the method of Rose (5), that is, by chromatography on DEAE-cellulose followed by ammonium sulfate fractionation. It was confirmed that fractions can be obtained with widely varying ratios of monophosphoglycerate mutase activity to the capability to generate 2,3-DPG. However, in the fractions richest in monophosphoglycerate mutase activity, the ability to generate 2,3-DPG survived exposure to 55° at pH 5, whereas the monophosphoglycerate mutase activity itself was destroyed (7). Two conclusions are possible. First, biosynthesis of 2,3-DPG in red cells may be accounted for by a protein that is distinct from monophosphoglycerate mutase. Alternatively, the monomutase may itself be the agent for 2,3-DPG biosynthesis and may be susceptible to a modification (*in vitro* and perhaps even in the cell) such that it loses the capability for the more complex overall reaction. Perhaps both agents contribute—a specialized enzyme and a more general one. Surely of all tissues, the red cell might be the one most likely to possess an enzyme specialized for 2,3-DPG biosynthesis. On the other hand, in muscle, and perhaps in most tissues where the level of 2,3-DPG is not extraordinarily high (because it is only casually produced), this function might reflect the monomutase action. The mechanism by which this could occur will be presented subsequently.

3. 2,3-DIPHOSPHOGLYCERATE PHOSPHATASE

If the reader has felt some surprise that the enzyme responsible for 2,3-DPG biosynthesis is a mutase inhibited by moderate concentrations of inorganic phosphate, and by very low concentrations of its own product, his astonishment may increase when he considers the enzyme that hydrolyzes 2,3-DPG— a phosphatase that is stimulated by P_i! An enzyme with this unusual property, and specific for cleaving 2,3-DPG to 3-PGA and P_i, was described more than 20 years ago by Rapoport and Luebering (8). Among its singular properties were stimulation by 2-PGA, inhibition by 3-PGA (taken as further evidence of its uniqueness for this path), and stimulation by certain ions that are normally recognized as inhibitors in enzyme systems: Ag^+, for example, caused a nearly twofold increase; Hg^{2+}, a fortyfold increase. An analogous enzyme from chicken breast muscle was described from the laboratory of Joyce and Grisolia (9), where the inhibition by 3-PGA was noted, but many of the other properties, such as the effects of P_i, Hg^{2+}, and 2-PGA could not be confirmed.

Recently an erythrocyte fraction was obtained with a strong capability for cleaving 2,3-DPG, although this activity amounted to only 5% of its monophosphoglycerate mutase activity (10). At moderate concentrations P_i powerfully stimulated the enzyme, when in the presence of high concentrations of chloride ion. Sulfite was seven times more potent than P_i under the same conditions, a fact which accords well with the stimulatory effects on 2,3-DPG degradation reported for sulfite and for other sulfur-containing anions, both in whole cells (11) and in hemolysates (12). An outstanding feature of the enzyme preparation was the stimulation by glycolate-2-P, a phosphoglycerate analog that may itself be present in red cells (13). This stimulation was 200 times greater than that achieved by P_i and was not dependent on chloride. The phosphoglycolate effect was inhibited by low concentrations of 3-PGA.

Some similar properties have been reported by others (14) in fractions from human hemolysates. One entity (clearly identified as monophosphoglycerate mutase) possessed phosphatase action on 2,3-DPG which was markedly increased by high concentrations of pyrophosphate. A second 2,3-DPG phosphatase isolated by the same workers was stimulated by bisulfite and was considered distinct from monophosphoglycerate mutase. The latter fraction could correspond to the phosphatase studied by Rose and Liebowitz (10).

It has been suggested (10) that the stimulation of 2,3-DPG phosphatase action by P_i could, through the ensuing pyruvate kinase step, support sagging ATP levels in the erythrocyte at the expense of 2,3-DPG; and that this is a physiologically appropriate control, since P_i and ATP levels are often inversely correlated. If this were the case (and it would be part of a "one-two punch," since P_i also inhibits further 2,3-DPG formation), then the increase of glycolysis in response to P_i must be distinguished from all other modes of glycolytic stimulation. In general, increased glycolysis is considered to create an increased flux of metabolites both in the Embden-Meyerhof scheme and in the 2,3-DPG path simultaneously (15). Indeed this notion was advanced by the same authors, when they rationalized the inhibition of 2,3-DPG phosphatase by 3-PGA: increased 3-PGA due to increased glycolytic flux would be reflected in a higher equilibrium concentration of 1,3-DPG and would promote 2,3-DPG production by mass action. Here, in another double-barreled action on the pathway, an increase of glycolysis would be reflected in an *increase* of 2,3-DPG accumulation; whereas an increase of P_i, which is a ready means for glycolytic stimulation (16, 17), restrains 2,3-DPG turnover. This discrepancy underlines the need for more data in order to bring the facts now available into harmony.

The stimulatory effect of phosphoglycolate is not unique for 2,3-DPG phosphatase and has been reported for both mono- and diphosphoglycerate mutase (5, 10). What seems involved is a site for positive allosteric control, and it is remarkable that the three enzymes behave similarly with respect to the same agent. The explanation proposed was along evolutionary lines, that is, that one of the enzymes has given rise to the other two. Although this is a possibility, it may also be imagined that one of the three enzymes could give rise to the others, not in an evolutionary sense, but by a direct transformation, either in the cell or in the test tube. (We will emphasize below that 2,3-DPG production is a normal partial reaction of the overall mutase enzyme.)

The most compelling evidence that these are distinct proteins is the demonstration that particular fractions of hemolysates obtained by chromatographic procedures are greatly enriched in the capability for production (5) or for degradation (10, 14) of 2,3-DPG. It must be borne in mind that these fractions are not, however, totally devoid of phosphoglycerate mutase activity. For example, the degradative enzyme isolated by Rose is still 18 times more active as a monophosphoglycerate mutase than as a 2,3-DPG phosphatase. To be sure, the relative capability of this fraction for 2,3-DPG cleavage has been improved more than 1000-fold over that of the usual erythrocyte monophosphoglycerate mutase.

Similarly, the fractions (obtained by DEAE-cellulose chromatography and subsequent ammonium sulfate precipitation procedures) that seem specialized for 2,3-DPG biosynthesis still contain detectable monophosphoglycerate mutase activity as well (5). In the writer's own laboratory, as indicated earlier, fractions prepared in this manner were found to be much more resistant to thermal inactivation with respect to 2,3-DPG synthesis than to overall mutase action (7). This again could indicate that completely separate proteins are involved; but in view of the uniform behavior of the various enzymes with phosphoglycolate, and the fact that the three functions under discussion have overlapped to such an extraordinary degree in the reports of the past three decades, we are inclined to reserve judgment. Our inclination at present is to give the notion of fully separate enzymes a qualified approval, particularly for red cells. Other explanations are less appealing. The situation could resemble that of triose phosphate dehydrogenase (18), or of other polycephalic enzymes whose separate functions can be individually measured and differentially inactivated. The isolation of protein fractions that "specialize" in one function over another might then be explained by a variety of influences, including the ionic environment. Since high concentrations of anions favor release of the diphosphate intermediate compared to the overall reaction (19), it is conceivable that the varying salt concentrations during chromatography may also favor the formation and elution of a particular form, perhaps one with a different state of aggregation of subunits. (This would, however, imply an equilibrium process, and hence a poor survival of the form when the salt concentration is subsequently lowered, which is not observed.) Another possibility, which would involve isozymes with differing ratios for the various functions, is presumably close to the position (of an evolutionary emergence) taken by others (10). It may be noted that chromatographically distinct isozymes have been observed in the case of phosphoglucomutase from many species (20).

Finally, there is accumulating evidence, which we will consider shortly, that phosphoglycerate mutase indeed involves alternating phospho-dephosphoenzyme forms. For phosphoglucomutase the phospho and dephospho forms are readily separated by ion-exchange chromatography (20, 21). Although phosphoglycerate mutase has thus far been "captured alive" only in the dephosphoenzyme form (22), conceivably some separation based on phosphoryl content or the ionic properties of the phosphoryl-binding site might also underlie the separations of the three enzymes that have been achieved.

4. MUTASE REACTIONS AND DIPHOSPHATE COMPOUNDS

The effect of 2,3-di-DPG on phosphoglycerate mutase is surely the earliest recognized example of "regulation" in this field. Sutherland et al. (23) proposed that the enzyme from muscle is a phosphotransferase in which 2,3-DPG acts by donating a phosphoryl group to a monophosphoglycerate, thus yielding a new molecule of 2,3-DPG and the other glyceryl monophosphate:

$$3\text{-PGA} + 2,3\text{-DPG} \rightleftharpoons 2,3\text{-DPG} + 2\text{-PGA}.$$

The word "cofactor" was introduced to describe this function because, although the diphosphate is regenerated, added 2,3-DPG greatly facilitates the *in vitro* reaction. Despite the many experiments that have been performed with muscle phosphoglycerate mutase using ^{32}P-labeled substrates, a direct demonstration of how 2,3-DPG functions to facilitate this reaction is still lacking. There is little serious doubt that the reaction involves a ping-pong mechanism, and that the enzyme alternates between phospho and dephospho forms, but only indirect, namely, kinetic, evidence is available to support this view. Our confidence that this mechanism is understood rests on the many parallels with the analogous phosphoglucomutase reaction, for which the mechanistic evidence is both direct and unambiguous. We will therefore examine some aspects of phosphoglucomutase and its interaction with its diphosphate, glucose-1,6-di-P, as a prelude to understanding the interactions of 2,3-DPG with phosphoglycerate mutase. Another reason for considering phosphoglucomutase in this chapter devoted to phosphoglycerates is that this enzyme has been discovered to interact with 1,3-DPG. Phosphoglycerate mutase indeed interacts with 1,3-DPG in a similar manner, and these interactions, as will be seen, strengthen the case for phosphoenzyme in the phosphoglycerate mutase reaction.

For phosphoglucomutase, the theory of an interaction between phospho and dephospho forms has been proved by isolation of the phosphoenzyme and the demonstration that it may be depleted of phosphate by repeated reaction with substrate (24). Phosphate on the enzyme can then be replenished by glucose-1,6-di-P, as demonstrated with ^{32}P. The requirement for added glucose-1,6-di-P has been explained by Najjar (25) on the basis of classical Michaelis kinetics, that is, that glucose-1,6-di-P is a dissociable intermediate. Ray and Roscelli (26) have presented evidence indicating that label from glucose-1-^{32}P appears faster in glucose-6-^{32}P than in glucose-1,6-di-^{32}P, and have concluded that free glucose-1,6-di-P is neither dissociable (in the sense of a reactant with reversible Michaelis-type binding) nor an intermediate. The data indicate specifically that glucose-1,6-di-P is released from the enzyme surface infrequently—about one time in twenty catalytic events. (An alternative statement might be that glucose-1,6-di-P does indeed behave like other dissociable reactants, but that in the immediate environment of the enzyme surface its affinity for the protein is so pronounced that equilibration with the medium is incomplete.) Whether one accepts the formulation of Ray or that of Najjar, added glucose-1,6-di-P is in any case required for the phosphoglucomutase reaction to proceed *in vitro*, because of the tendency for glucose-1,6-di-P, which may be considered one of the products of the reaction, to separate from its complex with dephosphoenzyme:

$$\text{Glucose-1,6-di-P} \cdot \text{E} \rightarrow \text{glucose-1,6-di-P} + \text{E}.$$

Added glucose-1,6-di-P restores phosphate on the dephosphoenzyme, thereby (see reaction 1, below) regenerating phosphoenzyme for reaction 2.

For muscle phosphoglucomutase, then, the concept of an alternating phos-phoenzyme-dephosphoenzyme system has rested comfortably on the follow-ing bases:

(*a*) Direct isolation of enzymatically active phosphoenzyme and dephos-phoenzyme forms and their interconversion by appropriate enzymatic means involving labeled substrates (24).

(*b*) Kinetic studies that indicate a ping-pong mechanism, wherein one of the products leaves the enzyme before all of the reactants have added (26), that is,

$$\text{Glucose-1,6-P} \cdot \text{E} \rightleftharpoons \text{glucose-6-P} + \text{PE}, \tag{1}$$

$$\text{Glucose-1-P} + \text{PE} \rightleftharpoons \text{glucose-1,6-P} \cdot \text{E}. \tag{2}$$

A novel contribution to the analysis of enzyme mechanisms was made recently by Britton and Clarke (27), using phosphoglucomutase as a model. These workers measured the flux ratio at equilibrium of [14]C- and ^{32}P-labeled substrates. Their measurements supported the idea that in the reaction glucose is transferred from reactant to product directly, but phosphate is transferred only indirectly (with a single intermediate phosphate entity). Also, using the "induced transport" test, with ^{14}C-labeled substrates, they obtained con-clusive evidence for a rapid isomerization of the free phosphoenzyme. These findings with phosphoglucomutase are entirely in accord with the mechanism (of phospho-dephosphoenzyme formation) described above, and they speci-fically exclude certain other possibilities, namely, a direct precursor to product phosphate transfer or an intermolecular phosphate transfer from glucose-1,6-di-P to the substrate (the latter two cases not involving phos-phoenzyme formation).

A separate corroboration of the fact that phosphoglucomutase is phos-phorylated during the course of reaction was made by the present author and his coworkers while investigating the origin of glucose-1,6-di-P (28). We assumed that, since the *in vitro* reaction is dependent on added glucose-1,6-di-P, the concentration or availability of this compound might be a limiting factor in cells. It had been reported much earlier by Paladini et al. from the laboratory of Leloir (29) that glucose-1,6-di-P in skeletal muscle and yeast is synthesized by an ATP-dependent kinase specific for glucose-1-P. An alternative path was later revealed by Sidbury et al. (30), in skeletal muscle and *E. coli*, in which 2 moles of glucose-1-P undergo dismutation to 1 mole of glucose-1,6,di-P and 1 mole of glucose. Our own studies indicated that a net production of glucose-1,6-di-P could result from the interaction of glucose-1-P and 1,3-DPG in the presence of phosphoglucomutase itself (31). These studies also revealed that 1,3-DPG exerted powerful kinetic effects on the phosphoglucomutase reaction. At low concentrations, in the absence of added glucose-1,6-di-P, 1,3-DPG stimulated the reaction. Concentrations above 1×10^{-5} M (or lower concentrations when glucose-1,6-di-P is present) were shown to be inhibitory. These findings were compatible with the following formulation:

$$\text{Glucose-1-P} + \text{PE} \rightleftharpoons \text{glucose-1,6-di-P} \cdot \text{E}, \tag{3}$$

$$\text{Glucose-1,6-di-P} \cdot \text{E} \rightleftharpoons \text{glucose-6-P} + \text{PE}, \tag{4a}$$

or, alternatively:

$$\text{Glucose-1,6-di-P} \cdot \text{E} + \text{1,3-DPG} \rightarrow \text{glucose-1,6-di-P} + \text{3-PGA} + \text{PE}. \tag{4b}$$

That is, the glucose-1,6-di-P · E complex formed by interaction with one reactant (here glucose-1-P) might proceed to the product (glucose-6-P), or it might react with 1,3-DPG, in the course of which glucose-1,6-di-P would be displaced, 3-PGA released, and phosphoenzyme regenerated. Thus, in the presence of glucose-1-P and stoichiometric quantities of 1,3-DPG (or of a 1,3-DPG-generating system such as 3-PGA, PGA-kinase, ATP, phosphocreatine, and creatine phosphokinase), glucose-1,6-di-P would accumulate. At low concentrations 1,3-DPG could cause stimulation of the overall reaction in the absence of added glucose-1,6-di-P, because regeneration of phosphoenzyme was evidently limiting in this circumstance. Higher concentrations of 1,3-DPG could ultimately be inhibitory to the overall reaction because displacing glucose-1,6-di-P from the dephosphoenzyme disrupted the only complex that could give rise to product.

Confirmation of this hypothesis was achieved by demonstrating with the use of 1-labeled ^{32}P-1,3-DPG that the dephosphoenzyme was readily phosphorylated (32). The labeled product was acid-stable and alkali-labile, as had been shown by Najjar and his coworkers and by Kennedy and Koshland (33) for the enzyme labeled from ^{32}P-labeled substrates. Proof that the label was at the active site lay, moreover, in the fact that unlabeled glucose-1-P interfered with the labeling by 1,3-DPG, and that glucose-1-P facilitated the release of ^{32}P from the labeled enzyme into a labeled organic compound (32).

These studies did not realize their initial goal: to disclose a source of glucose-1,6-di-P, "external" to the phosphoglucomutase reaction, which might regulate the activity of the enzyme. It is untenable that the phosphoglucomutase enzyme would release a supply of glucose-1,6-di-P into a reservoir for future reference. In this connection, the enzyme systems disclosed by Paladini et al. (29) and by Sidbury et al. (30) would be more appropriate to the problem of regulating phosphoglucomutase, since they described glucose-1,6-di-P production by mechanisms outside the reaction itself. However, these systems have recently been called into question by Passoneau et al. in Lowry's laboratory (34). These workers have extended the notion that we developed concerning phosphorylation by 1,3-DPG, and have shown that this can also be accomplished by fructose-1,6-di-P. They suggested that the finding of Paladini et al. concerning a specific ATP-dependent kinase for glucose-1-P could in fact result from a sequence of familiar activities, as follows:

$$\text{Glucose-1-P} \rightleftharpoons \text{glucose-6-P} \quad \text{(phosphoglucomutase)}, \tag{5}$$

$$\text{Glucose-6-P} \rightleftharpoons \text{fructose-6-P} \quad \text{(phosphohexoseisomerase)}, \tag{6}$$

$$\text{Fructose-6-P} + \text{ATP} \rightarrow \text{fructose-1,6-di-P} + \text{ADP}$$
$$\text{(phosphofructokinase)}; \tag{7}$$

then, on the phosphoglucomutase enzyme,

$$\text{Fructose-1,6-di-P} + \text{glucose-1,6-di-P} \cdot \text{E} \rightarrow \text{PE} + \text{fructose-6-P} +$$
$$\text{glucose-1,6-di-P}, \tag{8}$$

$$\text{PE} + \text{glucose-1-P} \rightleftharpoons \text{glucose-1,6-di-P} \cdot \text{E}. \tag{9}$$

Sum: 2 Glucose-1-P + ATP → glucose-1,6-di-P + fructose-6-P + ADP

Lowry and his collaborators have explained the early findings of Paladini et al. on the basis of their own discovery that phosphoglucomutase can interact with fructose-1,6-di-P to yield glucose-1,6-di-P. (Paladini's own system was not free of phosphoglucomutase.) Lowry's group also expressed some skepticism concerning the dismutase of Sidbury et al. (which was also not free of phosphoglucomutase), and our own laboratory (28) has reported difficulty in demonstrating this activity. The systems previously described were dismissed as probably based on artifacts, and the principal source of glucose-1,6-di-P in cells was attributed to such mechanisms as the interaction of phosphoglucomutase with fructose-1,6-di-P or with 1,3-DPG (34). Of these alternatives Lowry et al. favored fructose-1,6-di-P, which, although thermodynamically much less favorable as a phosphoryl donor, was present in cells at demonstrably greater concentrations.

If Lowry's hypothesis is true, the fundamental notion of phosphogluco-mutase regulation in cells has undergone a serious change. If glucose di-phosphate is recognized as a significant intracellular product of the reaction, it cannot be invoked as an agent for regulating the enzyme's activity. Viewed in this light, glucose-1,6-di-P accumulates by virtue of the intracellular conditions that affect phosphoglucomutase; these may include the available concentrations of fructose-1,6-di-P or 1,3-DPG (and possibly of other compounds). It may also (by analogy with phosphoglycerate mutase, as we will see below) reflect available concentrations of certain inorganic salts. Whereas the flux of 1,3-DPG or fructose-1,6-di-P might occur in cells in such a way as to "prime" phosphoglucomutase, that is, to effect a net phosphorylation, the same would ordinarily not be true of the flux of glucose-1,6-di-P. This does not gainsay the proven superiority of glucose-1,6-di-P as an *in vitro* phosphorylating agent for phosphoglucomutase. It merely states that glucose diphosphate occurs as a product of the reaction, and not as a reactant.

Phosphoglycerate mutase, as we have noted, has resisted all attempts at direct verification of an alternating phosphorylation-dephosphorylation mechanism. The positive points can best be stated in terms of the available analogies between phosphoglycerate mutase and phosphoglucomutase, an enzyme for which the alternating mechanism has been established beyond a reasonable doubt:

1. *Participation of the corresponding diphosphate.* Added 2,3-DPG acts to stimulate phosphoglycerate mutase *in vitro*, just as glucose-1,6-di-P acts with phosphoglucomutase. The apparent K_m's are about 1.4×10^{-6} M (35) and 6.2×10^{-8} M (26), respectively.

2. *Substrate kinetics.* Just as Ray and Roscelli (26) earlier showed ping-pong kinetics for phosphoglucomutase, Grisolia and Cleland (35) have demonstrated this for phosphoglycerate mutase. Specifically, in this study, the response to 3-PGA at a variety of 2,3-DPG concentrations generated a family of parallel lines in the Lineweaver-Burk plot. Likewise, 2,3-DPG kinetics at a series of 3-PGA concentrations showed a similar family of parallel lines. The fact that there is no extrapolation of these lines to a common point requires that 3-PGA and 2,3-DPG not be bound to the same form of the enzyme and hence that one product leave the enzyme surface before all the reactants have added. This requirement would be well satisfied by a phosphorylation-dephosphorylation sequence such as the following:

$$2,3\text{-DPG} \cdot \text{E} \rightleftharpoons 2\text{-PGA} + \text{PE}, \tag{10}$$

$$\text{PE} + 3\text{-PGA} \rightleftharpoons 2,3\text{-DPG} \cdot \text{E}. \tag{11}$$

3. *Flux ratios.* The unique kinetic studies of Britton and Clarke (27), which were introduced with outstanding success for the case of phosphoglucomutase, have been similarly applied to phosphoglycerate mutase with results of the same nature (36). The flux ratio experiments using [14]C- and ^{32}P-labeled substrates indicate that carbon is transferred directly and phosphorus indirectly, and with a single intermediate phosphate. Whereas such documentation was virtually superfluous for phosphoglucomutase, it is exceedingly useful in the case of phosphoglycerate mutase to be able to state that, according to these criteria, the mechanism is the same as that for phosphoglucomutase, which is known with certainty from direct approaches.

4. *Interaction with 1,3-DPG and the net production of 2,3-DPG.* The ability of 1,3-DPG to stimulate the phosphoglycerate mutase reaction and

to encourage the accumulation of 2,3-DPG in the presence of either 3-PGA or 2-PGA has been demonstrated (7). The results are entirely analogous to what has been observed with 1,3-DPG and the phosphoglucomutase system (31). For the latter case, it was proved (32) that these reactions were based on the ability of 1,3-DPG to displace glucose-1,6-di-P and to phosphorylate the resulting dephosphoenzyme. In the case of phosphoglycerate mutase, direct phosphorylation of the enzyme from ^{32}P-labeled 1,3-DPG could not be shown (32). However, the similarity of the other effects with 1,3-DPG in the two cases strengthens the presumption that a phosphoenzyme is formed which is too labile to be captured under the conditions used. As with the case of phosphoglucomutase and its cofactor, the supposition is raised that 2,3-DPG may occur in some cells as a *consequence* of the interaction of 1,3-DPG with phosphoglycerate mutase.

That a net production of 2,3-DPG could be accomplished with the mono-phosphoglycerate mutase enzyme was in a sense anticipated by the work of Grisolia and his collaborators. Using radioactively labeled 3-PGA, Cascales and Grisolia (19) demonstrated that the incorporation of label during the reaction could be diverted from the normal product, 2-PGA, in favor of 2,3-DPG, by the presence of high concentrations of various inorganic salts. Subsequently, Grisolia and Cleland (35) demonstrated that at 0.4 M KCl dissociation of the cofactor occurred faster than the overall catalytic rate. These experiments, although they did not involve net transformations, were consistent with the notion that, given a phosphorylating source, the phosphoglycerate mutase reaction would be able to generate 2,3-DPG.

5. *Isolation of phosphoenzyme forms.* Although various laboratories have at times reported the transfer of ^{32}P into rabbit muscle phosphoglycerate mutase (37, 38), from either labeled substrate or labeled 2,3-DPG, convincing proof has been lacking (*a*) that such phosphate is covalently bound, and (*b*) that it can be further transferred to unlabeled substrate, with the release of ^{32}P-2,3-DPG. A recent report, however, has satisfied the first of these criteria: ^{32}P can be affixed to rabbit muscle phosphoglycerate mutase, using ^{32}P-labeled 2,3-DPG (22). The innovation is that the short-lived phosphorylated enzyme is captured in alkali. This surprising discovery is of course consistent with phosphohistidine, and not with phosphoserine, as the critical residue, and hence is irreverently unlike the situation in phosphoglucomutase.

It remains to be shown that phosphate at this site can be transferred in the expected fashion. If we assume that what has been labeled is the active site, it may be necessary to propose, for the sake of a unitary hypothesis, that phosphoglucomutase also passes through a phosphohistidine on its way to creating a phosphoserine (25). Kennedy and Koshland (33) reported on the acid stability of phosphoglucomutase labeled from ^{32}P-labeled substrates. They were aware that the acid conditions under which the enzyme was tested were especially conducive to acyl and phosphoryl shifts. Accordingly they took steps to hydrolyze their labeled enzyme under conditions in which migration of phosphate from nitrogen to oxygen could not take place (i.e., by urea denaturation and digestion with proteolytic enzyme). The point is, however, that an N—O shift may well occur, not (as they carefully excluded) during the acid treatment, but during the actual labeling process itself.

5. ACYL PHOSPHATASE

An alternative mechanism for discharging the high energy contained in 1,3-DPG and in analogous compounds involves an acylphosphatase that has been detected in a variety of tissues. A regulatory role for such an enzyme was appreciated more than a decade ago by Harary (39), who noted that its presence in cells released phosphoglycerate formation from a dependence on ADP. When ADP is limiting in cells (as in refs. 40 and 41), this enzyme could permit a faster glycolytic rate at the expense of ATP formation. Actually, in this view, no net ATP need accrue from *either* of the energy-yielding kinases of glycolysis. 1,3-Diphosphoglycerate is cleaved to 3-PGA and P_i. Through the successive action of monophosphoglycerate mutase, enolase, and pyruvate kinase, ATP is produced and is then consumed in the PGA-kinase reaction, reacting with a second molecule of 3-PGA (obtained from the cleavage of yet another 1,3-DPG molecule) to regenerate 1,3-DPG for a second turn of the cycle. The net reaction is 1,3-DPG yielding pyruvate and 2 P_i.

This kind of enzyme has recently been purified from neural tissue (42) and found to have a molecular weight of less than 9000. Its substrate, 1,3-DPG, must be considered not only a glycolytic intermediate and a precursor of 2,3-DPG, but also an agent that can directly modify proteins. Ramponi and Grisolia (43) have reported the transfer of phosphoglycerate from 1,3-DPG to lysine-rich and arginine-rich histones. This process, like the phosphorylation of mutase enzymes mentioned previously, occurs without a separate enzyme requirement. The authors have suggested that the acylphosphatase, by affecting the 1,3-DPG concentration, may help to regulate histone acylation. The modification of histones is considered of relevance in the regulation of animal protein synthesis (44).

6. THE SPECIAL ROLE OF 2,3-DIPHOSPHOGLYCERATE IN ERYTHROCYTES

Two important problems have existed perenially side by side concerning the red cell. On the one hand, for four decades its major phosphorylated constituent, 2,3-DPG, has been in search of a role. On the other hand, there was the long-standing observation that the O_2 affinity of hemoglobin at intracellular conditions of pH and ionic strength was substantially greater than that of whole blood. At this writing, a few brief years since the observations of Benesch and of Chanutin and their collaborators (45, 46) established the link between these issues and sparked a massive confirmatory literature, it seems astonishing that this now obvious connection was not perceived sooner. Numerous review articles and symposia on this subject have appeared (47, 48, 49).

Benesch et al. demonstrated that oxyhemoglobin lacked the capability to bind with 2,3-DPG but that deoxyhemoglobin could do this to the apparent limit of 1 mole/mole of hemoglobin tetramer (50) and with a binding energy of 6.4 kcal/mole. Only the β chains were involved in 2,3-DPG binding, since the binding occurred with hemoglobin H (i.e., β_4^A) in both its oxy and deoxyforms, but not at all with α chains (51). Since the binding was mole for mole, it was assumed to occur on the diad axis of symmetry. The likeliest site was considered to be at the positively charged residues in the central cavity that runs along this axis (52) and whose width is known to change with oxygenation (53) and to become sufficiently wide in the deoxy form to admit 2,3-DPG.

Speculation about the locus of binding has extended to individual residues. Histidine 21 of the β chains was implicated on the basis that, when it is replaced, O_2 affinity becomes altered (54). (In hemoglobin Hiroshima, which has a high O_2 affinity, this residue is replaced with Asp. In hemoglobin F, in which the affinity for 2,3-DPG is much reduced, γ-chain His 21 is replaced by Ser.)

From a physiological point of view, the experiments and formulations of Benesch and of Chanutin have refined our understanding of oxygen transport. 2,3-Diphosphoglycerate (or inositol hexaphosphate in avian species) is seen as a "cofactor of oxygen unloading," shifting the oxygen dissociation curve of hemoglobin to the right. It operates in conjunction with other special influences in the periphery, such as reduced oxygen tension, increased pCO_2 (and other sources of acidity), and higher temperatures.

Benesch himself concluded that variations of ATP or 2,3-DPG in the physiological range would have little effect on the oxygen affinity; the ability to unload oxygen would be decreased only with a substantial reduction (more than one fourth) of the normal organic phosphates. The apparent basis for this inaccurate prediction (numerous reports now indicate a close link between 2,3-DPG and hemoglobin concentration (55, 56) was that the original measurements were of *equilibria* between hemoglobin and oxygen at particular oxygen pressures (50). The *in vivo* circumstances which confirm a relationship do not reflect equilibria, but rather the rapidly fluctuating circumstances of hemoglobin oxygenation in the individual cell. Since 2,3-DPG presumably affects the *rate* with which hemoglobin equilibrates with oxygen, minor variations of 2,3-DPG concentrations may be expected to influence the oxygen-unloading process. One study, involving 300 healthy adults (55), revealed a distinctly inverse correlation between 2,3-DPG and whole blood hemoglobin. Another extensive study of normal and anemic patients revealed that, for each gram of hemoglobin fall, there was a 2,3-DPG decrease of 0.23 mM (56). 2,3-Diphosphoglycerate was also found to be elevated in non-cyanotic cardiac patients (57). This increase may be reviewed as a mode of widening the arterial-venous O_2 difference by affecting hemoglobin-oxygen binding—an adaptation to the reduced supply of oxygen that is secondary to the diminished cardiac output.

It is consistent with this hypothesis that the blood of individuals living at high altitudes exhibits high levels of 2,3-DPG, and that an increase of this compound occurs quite promptly when individuals move to higher altitudes from sea level (58). The reverse adjustment in highlanders descending to sea level also occurs rapidly, with a half-time of about 6 hr (59).

A large number of other physiological and clinical observations are also in good accord with the formulation. Changes in 2,3-DPG can compensate for acid-base derangements that affect oxygen availability through the Bohr effect. Thus, as the O_2 dissociation curve shifts to the right in acidosis, 2,3-DPG shows a compensatory decline (60), and the reverse process is seen in alkalosis. The increase of 2,3-DPG observed at high altitudes is in fact a response not only to the increased need for efficient oxygen unloading, but also to the alkalosis that attends hyperventilation. These changes in 2,3-DPG are now assumed to be mediated ultimately, according to mechanisms that we have already discussed, by the state of oxygenation of venous blood. Indeed, the measurement of 2,3-DPG has been considered of possible value as a clinical index of the state of oxygenation.

A practical therapeutic problem is that bank blood stored in acid-citrate dextrose (ACD) for 10 days (49) loses nearly all of its 2,3-DPG. (The deficit is restored *in vivo* within about 24 hr after transfusion.) This depletion could be a severe disadvantage in recipients for whom the release of oxygen is especially critical. Storage agents superior to ACD in supporting 2,3-DPG are citrate-phosphate-dextrose and others containing inosine (61). Apart from storage, the age of the cells is itself a factor—older red cells have increased O_2 affinity and generally low 2,3-DPG levels (49).

Hypoxic states that promote increased 2,3-DPG may be secondary to a variety of causes. In response to exercise, a correlation was found in adult males between increases in blood 2,3-DPG and blood lactate (62). However, in an actual study of coronary sinus blood, no rise in 2,3-DPG was noted, despite increased oxygen dissociation induced by exercise in patients with ischemic heart disease (63). Diminished oxygen affinity and increased

2,3-DPG have been noted in thyrotoxic patients. The latter has been attributed to direct stimulation by the thyroid hormone of a diphosphoglycerate mutase (64), but could also be exerted indirectly by anoxia and the concentration of deoxyhemoglobin.

In certain cases of nonspherocytic congenital hemolytic anemia 2,3-DPG may be markedly elevated, to a degree out of proportion to the increases of other glycolytic intermediates, while ATP is diminished (65). Such cases are now recognized as involving a pyruvate kinase deficiency. Conversely, individuals with abnormally high pyruvate kinase demonstrate exactly the opposite changes (66).

The only state that can, by the mechanisms described, give rise to increased 2,3-DPG is an absolute increase of deoxyhemoglobin. In anemic states generally, a greater proportion of the available hemoglobin surrenders its oxygen, and there is an increase in the amount of deoxyhemoglobin in the venous blood when this is expressed (as we also express 2,3-DPG) per volume of cells. As we have noted, several reports have appeared concerning the correlation of anemia with increased 2,3-DPG (55, 56), but there can evidently be exceptions. For example, Charache and Grisolia (67) have shown that patients with sickle cell anemia have generally high 2,3-DPG, but their study of a large group of patients failed to demonstrate a correlation of the 2,3-DPG increase per milliliter of cells with the reduction of hematocrit. This may perhaps be related to the poor packing of sickle cells. In any case, the response of 2,3-DPG to deoxyhemoglobin must be resolved within each separate cell. A study that might therefore be especially revealing would involve an anemia (e.g., iron deficiency) in which what is diminished is the mean corpuscular hemoglobin.

Elevations of blood 2,3-DPG might also be expected in the presence of hemoglobin variants that exhibit high oxygen affinity. Grisolia et al. (68) have alluded to the fact that this may occur in the case of certain heterozygotes with hemoglobin Chesapeake, but not with hemoglobin Rainier. A third hemoglobinopathy exhibiting high oxygen affinity is hemoglobin Kansas, in which an actual decrease of 2,3-DPG blood concentrations has apparently been observed (68). The situation is evidently complex and dependent on factors besides the possible benefit that might accrue from possession of the cofactor. Particularly relevant is the affinity of the variant for 2,3-DPG. Variants that bind 2,3-DPG poorly (69) (e.g., hemoglobin F) have little capability to exert a pull on its biosynthesis, irrespective of how high the oxygen affinity or how low the hematocrit.

Several workers have published experiments that disagree with some particulars of the Benesch scheme. Grisolia and his collaborators have reported that 2,3-DPG does not bind uniquely to deoxyhemoglobin. Binding experiments based on ultrafiltration and on gel filtration (on columns preequilibrated with 2,3-DPG) (70) demonstrated that oxyhemoglobin readily binds 2,3-DPG, in amounts approaching 1 mole/mole of tetramer. These experiments were done with widely varying concentrations of hemoglobin that approached the concentrations occurring in cells, whereas the experiments of Benesch were performed at low hemoglobin concentrations and in 0.1 N NaCl.

Similarly, Garby et al. (71) have studied 2,3-DPG-hemoglobin binding under conditions of hemoglobin concentration and ionic composition resembling those in intact erythrocytes. Like Grisolia et al., they observed binding of 2,3-DPG with oxyhemoglobin, the affinity being about one-half that with deoxyhemoglobin. At least two binding sites for 2,3-DPG seemed to be involved, and these showed strong cooperation.

The formulation that deoxyhemoglobin promotes 2,3-DPG biosynthesis implies an unending fluctuation in 2,3-DPG production as the red cell is transported from the anaerobic peripheral environment to aeration in the lungs. A possible detraction from the scheme would be that the continuous turnover of the 2,3-DPG-hemoglobin bond might constitute a serious energy drain. Grisolia has calculated that this process in a resting adult may entail some 3–6% of normal dietary calories. His calculation utilized the energy of binding of 2,3-DPG by hemoglobin proposed by Benesch. However, it was acknowledged that no account was taken of the fraction of DPG in the cell that is hemoglobin-bound, a figure that is not accurately available. Moreover, in terms of the ΔH for oxygen binding, it can be calculated that the 2,3-DPG interaction in fact results in a lower energy expenditure for overall O_2 transport (48).

The evident resolution of the 2,3-DPG enigma does not leave the metabolism of erythrocytes or of the phosphoglycerates without unsolved problems. The 2,3-DPG pathway has, for the erythrocyte, been called an "uncoupling" path, that is, a glycolytic mode in which the dependence on phosphorylation has been circumvented. Indeed it has been suggested that the adult red cell utilizing glucose must require some such idling process in order to avoid a disadvantageously high ATP/ADP ratio, since its opportunities for expending ATP are limited. The validity of this hypothesis can be tested best by determining the flux through 2,3-DPG as compared to that through the conventional glycolytic scheme, which is not a simple task. The rate of disappearance of 2,3-DPG from whole cells incubated without substrate may give a first approximation of the rate of 2,3-DPG biosynthesis in the steady state (72), but the situation hardly reflects physiological conditions. Alternatively, one might compare the rates of accumulation of ^{32}P label in 2,3-DPG and in ATP, starting with 1-labeled 1,3-DPG. Such an experiment would also have serious drawbacks: for example, it could be performed only in hemolysates, where the normal regulatory influences on all of the enzymes involved are disrupted. Moreover, the chemical lability of 1,3-DPG would give rise to ^{32}P$_i$, which could participate in annoying exchange reactions.

Although the flux through 2,3-DPG is difficult to quantitate, it can to some extent be manipulated. 2,3-Diphosphoglycerate production from 1,3-DPG—whether through a special diphosphoglycerate mutase or by phosphorylation of the monophosphoglycerate mutase—may be viewed as an alternative to the PGA-kinase reaction. The concentration of ADP, as phosphate acceptor in the latter reaction, can then be expected to influence the concentration of 2,3-DPG. This has been demonstrated in red cells in which ADP has been varied by agents, such as ouabain or dipyridamole, that inhibit membrane ATPase (72).

A phosphoglycerate compound whose role is not apparent is the nucleotide derivative, adenosine diphospho-PGA (ADP-PGA), which may be viewed as the pyrophosphate condensation product of AMP with the 3-phosphate of 2,3-DPG. This compound has been identified in erythrocytes and other tissues (73), and an enzyme that facilitates its cleavage to P_i and ADP-glyceric acid has been described (74). There is to date no clue concerning the metabolic role of such nucleotide-glycerate derivatives.

7. EFFECTS OF PHOSPHOGLYCERATES ON OTHER SYSTEMS

Some three decades after the observations of Dische (1) to which we alluded at the beginning of this chapter, 2,3-DPG was found by Brewer (75) to inhibit hexokinase in human hemolysates. The inhibition approached 80% and was nearly eliminated by ATP and magnesium in millimolar concentrations. The implication is that the oxygenation of hemoglobin can release 2,3-DPG, which would diminish hexokinase action. This mechanism would encourage glucose utilization and metabolism where oxygen tension is low, that is, in the periphery, particularly in such regions as the spleen.

Something resembling a Pasteur effect, that is, a modest increase of glycolysis under anaerobic conditions, has in fact been an unexplained observation for a long time in mature erythrocytes (15). Whereas in other cells ATP of mitochondrial origin provides, through inhibition of phosphofructokinase, the main basis for the regulation of glycolysis as opposed to oxidative metabolism, in the mature erythrocyte there is no oxidative phosphorylation, no mitochrondrial ATP, and ostensibly no need for this particular kind of control. Still, a need for controlling the glycolytic flux challenges the wandering erythrocyte.

The work of Rose et al. (16, 76) showed that not only phosphofructokinase but also hexokinase can participate in regulating red cell metabolism. [It is now well established that pyruvate kinase, inhibited by ATP and stimulated by fructose diphosphate, is also an important part of the overall control (77, 78).] Rose demonstrated that the inhibition of hexokinase imposed by glucose-6-P can be relieved by P_i. The physiological significance of this formulation has since been questioned on several grounds. The effective range of P_i in hemolysates is 0.3–0.5 mM, but the effective external concentration in the whole cell experiments is very high, that is, 50 mM (which achieves an intracellular P_i concentration of about 12 mM). Moreover, other agents besides inorganic phosphates are also effective, such as ethyl or butyryl phosphate and inorganic sulfate. Others have attributed Rose's observation to indirect effects from the relief of phosphofructokinase inhibition (79). Therefore the enthusiasm that attended the discovery of this effect of P_i (which provided at the same time the first direct explanation of the Pasteur effect on glucose utilization) has subsided, although it has not evaporated. The more recent experiments of Brewer with 2,3-DPG relate to this same vital problem of a direct regulation of hexokinase. It would be most compelling if this observation were to prove unique for 2,3-DPG and for the hexokinase from erythrocytes.

Brewer's findings are of additional interest in that they involve ATP in relieving the 2,3-DPG-imposed inhibition of hexokinase, although the full implications of this are not yet evident. The work of Benesch and of Chanutin revealed that ATP behaves very much like 2,3-DPG with respect to hemoglobin binding: the arguments concerning the binding of 2,3-DPG specifically to the deoxy form of hemoglobin apply with equal force to ATP. Therefore ATP can be considered to be allied with 2,3-DPG in the unloading of oxygen. This is reasonable, because oxygen release from hemoglobin should occur under circumstances where the metabolic needs are greatest and where glycolytic inhibition (as by free ATP) should be minimized. In the regulation of red cell glycolysis, ATP would likewise be allied with 2,3-DPG— the former inhibiting phosphofructokinase; the latter, hexokinase. But ATP is, according to Brewer's study, also a means of relieving the 2,3-DPG inhibition of hexokinase, and the summation of these effects could result in a massive accumulation of hexose monophosphate. More likely, as Brewer has suggested, it is not ATP per se but a Mg-ATP complex that is the significant neutralizer of 2,3-DPG-inhibited hexokinase. It is fascinating and perhaps necessary to speculate that the flux of Mg^{2+} in and out of red cells may be the controlling factor here. This begs the question of ATP involvement or renders it neutral. The problem of magnesium flux has received little attention in erythrocytes or in any other animal cells, but from the point of view of a vast number of enzymes it clearly needs investigation.

From time to time other enzymes have been noted to be inhibited by phosphoglycerates. In particular, transketolase and transaldolase from red cells were found to be inhibited by 2,3-DPG (80). Although 3-PGA is among a variety of intermediates that stimulate mammalian glycogen synthetase activity, it does this less strikingly than the presumed physiological effector, glucose-6-P (81). 2,3-Diphosphoglycerate seems intimately involved in the metabolism of adenine nucleotides, not only, as we have specified, in the regulation of hexokinase, but also as an inhibitor of adenylic acid deaminase (82), and of the enzyme that produces 5-phosphoribosyl-1-pyrophosphate (83).

This chapter, which has devoted very little attention to nonmammalian systems, should not ignore the fact that 3-PGA is the presumptive first product in photosynthesis, in the carboxylation of ribulose-1,5-di-P. That 3-PGA is also evidently the primary activator of ADP-glucose pyrophosphorylase in leaves and green algae (84) is, therefore, a good example of "feed-forward" activation. Moreover, it suggests that ADP-glucose formation may significantly regulate starch deposition. Inhibitory effects of the three stable phosphoglycerates have been reported on pea-seed phosphofructokinase (85), the significance of which remains to be disclosed.

Recent information suggests that one might seek a possible involvement of phosphoglycerates in bacterial regulatory phenomena. In a study of sporulation and germination in various *Bacillus* species, Nelson and Kornberg (86) identified 3-PGA as the predominant acid-soluble phosphate compound in spores. Substantial accumulation of 3-PGA (up to 15 mM/g of spores) began about 1 hr before the appearance of refractile spores. During germination this 3-PGA was rapidly utilized (87, 88), and the authors suggested that it may serve as a ready endogenous source of ATP, which is virtually absent in spores. Although in early germination mainly hydrolytic processes are involved, some biosynthesis (as of protein and RNA) occurs.

It should be noted that in these experiments the mode of extraction would not have permitted the survival of extralabile materials, and hence could not have distinguished between 3-PGA and 1,3-DPG. One could speculate that any 1,3-DPG present might itself participate in activation processes. Phosphorylation is the basis for modifying the state of activation of numerous proteins, including the enzymes of the glycogen path as well as nucleoside diphosphate kinase, pyruvate dehydrogenase, and possibly phosphofructokinase (89). Similarly, the behavior of histones has been shown to be altered by phosphorylation (90). Since 1,3-DPG is a very favorable phosphoryl donor that is capable of "priming" enzymes such as phosphoglucomutase (31) and phosphoglycerate mutase (7), a possible role for this labile intermediate in the germination process should perhaps be considered.

ACKNOWLEDGMENT

Investigative work from the author's laboratory was performed under a grant from the National Institute of Arthritis and Metabolic Diseases, from which Dr. Alpers also held a Research Career Development Award.

REFERENCES

1. Z. Dische, *Trav. Membres Soc. Chim. Biol.*, **23,** 1140 (1941) [*C.A.*, **36,** 6550 (1942)].

2. I. Greenwald, *J. Biol. Chem.*, **63,** 339 (1925).

3. S. Rapoport and J. Luebering, *J. Biol. Chem.*, **183,** 507 (1950).

4. B. K. Joyce and S. Grisolia, *J. Biol. Chem.*, **234,** 1330 (1959).

5. Z. B. Rose, *J. Biol. Chem.*, **543,** 4810 (1968).

6. J. Luebering and S. Rapoport, *J. Biol. Chem.*, **196,** 583 (1952).

7. M. T. Laforet, J. B. Butterfield, and J. B. Alpers, article in preparation.

8. S. Rapoport and J. Luebering, *J. Biol. Chem.*, **189,** 683 (1951).

9. B. K. Joyce and S. Grisolia, *J. Biol. Chem.*, **233,** 350 (1958).

10. Z. B. Rose and J. Liebowitz, *J. Biol. Chem.*, **245,** 3232 (1970).

11. S. Mányai and Z. Várady, *Biochim. Biophys. Acta*, **20,** 594 (1956).

12. S. Mányai and Z. Várady, *Acta. Physiol. Acad. Sci. Hung.*, **14,** 103 (1958).

13. A. Örström, *Arch. Biochem. Biophys.*, **33,** 484 (1951).

14. D. R. Harkness, W. Thompson, S. Roth, and V. Grayson, *Arch. Biochem. Biophys.*, **138,** 208 (1970).

15. T. Asakura, Y. Sato, S. Minakami, and H. Yoshikawa, *J. Biochem.*, **59,** 524 (1966).

16. I. A. Rose, J. V. B. Warms, and E. L. O'Connell, *Biochem. Biophys. Res. Commun.* **15,** 33 (1964).

17. S. Minakami and H. Yoshikawa, *Biochim. Biophys. Acta*, **99,** 175 (1965).

18. I. Krimsky and E. Racker, *Biochemistry*, **2,** 512 (1963).

19. M. Cascales and S. Grisolia, *Biochemistry*, **5,** 3116 (1966).

20. J. G. Joshi, J. Hooper, T. Kuwaki, T. Sakurada, J. R. Swanson, and P. Handler, *Proc. Natl. Acad. Sci. U.S.*, **57,** 1482 (1967).

21. J. A. Yankeelov, H. R. Horton, and D. E. Koshland, *Biochemistry*, **3,** 349 (1964).

22. Z. Rose, personal communication.

23. E. W. Sutherland, T. Posternak, and C. F. Cori, *J. Biol. Chem.*, **181,** 153 (1949).

24. V. A. Najjar and M. E. Pullman, *Science*, **119,** 631 (1954).

25. V. A. Najjar, in *The Enzymes*, Vol. VI (Eds.: P. D. Boyer, H. Lardy, and K. Myrback), Academic Press, New York, 1962, p. 161.

26. W. J. Ray and G. A. Roscelli, *J. Biol. Chem.*, **239,** 1228 (1964).

27. H. G. Britton and J. B. Clarke, *Biochem. J.*, **110,** 161 (1968).

28. G. S. Levey and J. B. Alpers, *J. Biol. Chem.*, **240,** 4152 (1965).

29. A. C. Paladini, R. Caputto, L. F. Leloir, R. E. Trucco, and C. E. Cardini, *Arch. Biochem. Biophys.*, **23,** 55 (1949).

30. J. B. Sidbury, L. L. Rosenberg, and V. A. Najjar, *J. Biol. Chem.*, **222,** 89 (1956).

31. J. B. Alpers, *J. Biol. Chem.*, **243,** 1698 (1968).

32. J. B. Alpers and G. K. H. Lam, *J. Biol. Chem.*, **244,** 200 (1969).

33. E. P. Kennedy and D. E. Koshland, *J. Biol. Chem.*, **228,** 419 (1957).

34. J. V. Passoneau, O. H. Lowry, D. W. Schultz, and J. G. Brown, *J. Biol. Chem.*, **244,** 902 (1969).

35. S. Grisolia and W. W. Cleland, *Biochemistry*, **7**, 1115 (1968).
36. H. G. Britton and J. B. Clarke, *Biochem. J.*, **112**, 10 P (1969).
37. N. Zwaig and C. Milstein, *Biochem. J.*, **98**, 360 (1966).
38. R. J. Jacobs and S. Grisolia, *J. Biol. Chem.*, **241**, 5926 (1966).
39. I. Harary, *Biochim. Biophys. Acta*, **29**, 647 (1958).
40. K. Pye, in *Control of Energy Metabolism* (Eds.: B. Chance, R. W. Estabrook, and J. R. Williamson), Academic Press, New York, 1965, p. 193.
41. P. K. Maitra and B. Chance, in *Control of Energy Metabolism* (Eds.: B. Chance, R. W. Estabrook, and J. R. Williamson), Academic Press, New York, 1965, p. 157.
42. D. A. Diederich and S. Grisolia, *J. Biol. Chem.*, **244**, 2412 (1970).
43. G. Ramponi and S. Grisolia, *Biochem. Biophys. Res. Commun.*, **38**, 1056 (1970).
44. R. J. De Lange, D. M. Fambrough, E. L. Smith, and J. J. Bonner, *Biol. Chem.*, **244**, 5669 (1969).
45. R. Benesch and R. E. Benesch, *Biochem. Biophys. Res. Commun.*, **26**, 162 (1967).
46. A. Chanutin and R. R. Curnish, *Arch. Biochem. Biophys.*, **121**, 96 (1967).
47. R. Benesch and R. E. Benesch, *Nature*, **221**, 618 (1969).
48. Physiological Society Symposium, "Control Mechanisms for Oxygen Release by Hemoglobin," *Federation Proc.*, **29**, 1101–1117 (1970).
49. H. F. Bunn and J. H. Jandl, *New Engl. J. Med.*, **282**, 1414 (1970).
50. R. Benesch, R. E. Benesch, and C. I. Yu, *Proc. Natl. Acad. Sci. U.S.*, **59**, 526 (1968).
51. R. Benesch, R. E. Benesch, and Y. Enoki, *Proc. Natl. Acad. Sci. U.S.*, **61**, 1102 (1968).
52. M. F. Perutz, H. Muirhead, J. M. Cox, and L. C. G. Goaman, *Nature*, **219**, 131 (1968).
53. H. Muirhead, J. M. Cox, L. Mazzarella, and M. F. Perutz, *J. Mol. Biol.*, **28**, 117 (1967).
54. C-H. de Verdier and L. Garby, *Scand. J. Clin. Lab. Invest.*, **23**, 149 (1969).
55. J. W. Eaton and G. J. Brewer, *Proc. Natl. Acad. Sci. U.S.*, **61**, 756 (1968).
56. J. Torrance, P. Jacobs, A. Restrepo, J. Eschbach, and C. A. Finch, *New Engl. J. Med.*, **283**, 165 (1970).
57. R. D. Woodson, J. D. Torrance, S. D. Shappell, and C. Lenfant, *J. Clin. Invest.*, **49**, 1349 (1970).
58. C. Lenfant, J. Torrance, E. English, C. A. Finch, C. Reynafarge, J. Ramos, and J. Faura, *J. Clin. Invest.*, **47**, 2652 (1968).
59. C. Lenfant, J. D. Torrance, R. D. Woodson, P. Jacobs, and C. A. Finch, *Federation Proc.*, **29**, 1115 (1970).
60. G. M. Guest and S. Rapoport, *Blood, Heart and Circulation* (Ed.: F. R. Moulton), American Association for the Advancement of Science, Publication 13, Science Press, Lancaster, 1940, p. 55.
61. H. F. Bunn, M. H. May, W. F. Kocholaty, and C. E. Shields, *J. Clin. Invest.*, **48**, 311 (1969).
62. J. W. Eaton, J. A. Faulkner, and G. J. Brewer, *Proc. Soc. Exptl. Biol. Med.*, **132**, 886 (1969).
63. S. D. Shappell, J. A. Murray, M. G. Nasser, R. E. Wills, J. D. Torrance, and C. J. M. Lenfant, *New Engl. J. Med.*, **282**, 1219 (1970).
64. L. M. Snyder and W. J. Reddy, personal communication.

65. M. A. Robinson, P. B. Loder, and G. C. de Gruchy, *Brit. J. Haematol.*, **7**, 327 (1961).

66. C. Zürcher, J. A. Loos, and H. K. Prins, *Proc. 10th Congr. Intern. Soc. Blood Transf.*, Stockholm, 1964, p. 549 (1965).

67. S. Charache, S. Grisolia, A. J. Fiedler, and A. E. Helliger, *J. Clin. Invest.*, **49**, 806 (1970).

68. S. Grisolia, J. Carreras, D. Diederich, and S. Charache, in *Red Cell Metabolism and Function* (Ed.: G. J. Brewer), Plenum Press, New York, 1970, p. 39 (*Advan. Exptl. Biol. Med.*, **6**).

69. H. F. Bunn and R. W. Briehl, *J. Clin. Invest.*, **49**, 1088 (1970).

70. J. Luque, D. Diederich, and S. Grisolia, *Biochim. Biophys. Res. Commun.*, **36**, 1019 (1969).

71. L. Garby, G. Gerber, and C-H. de Verdier, *European J. Biochem.*, **10**, 110 (1969).

72. J. Duhm, B. Deuticke, and E. Gerlach, *Biochim. Biophys. Acta*, **170**, 452 (1968).

73. T. Hashimoto, M. Tatibana, and H. Yoshikawa, *J. Biochem.*, **53**, 219 (1963).

74. G. T. Zancan, E. F. Recondo, and L. F. Leloir, *Biochim. Biophys. Acta*, **92**, 125 (1964).

75. G. J. Brewer, *Biochim. Biophys. Acta*, **192**, 157 (1969).

76. I. A. Rose and E. L. O'Connell, *J. Biol. Chem.*, **239**, 12 (1964).

77. R. D. Koler and P. Vanbellinghen, *Advan. Enzyme Regulation*, **6**, 127 (1968).

78. W. C. Mentzer and J. B. Alpers, article in preparation.

79. S. Rapoport, *Bibl. Haematol.*, **29**, part 1, 133 (1968).

80. Z. Dische and D. Igals, *Arch. Biochem. Biophys.*, **101**, 489 (1963).

81. R. Piras, L. B. Rothman, and E. Cabib, *Biochemistry*, **7**, 56 (1968).

82. A. Askari and S. N. Rao, *Biochim. Biophys. Acta*, **151**, 198 (1968).

83. A. Hershko, A. Razin, and J. Mager, *Biochim. Biophys. Acta*, **184**, 64 (1969).

84. G. G. Sanwal and J. Preiss, *Arch. Biochem. Biophys.*, **119**, 454 (1967).

85. G. J. Kelly and J. F. Turner, *Biochim. Biophys. Acta*, **208**, 360 (1970).

86. D. L. Nelson and A. Kornberg, *J. Biol. Chem.*, **245**, 1137 (1970).

87. D. L. Nelson and A. Kornberg, *J. Biol. Chem.*, **245**, 1146 (1970).

88. P. Setlow and A. Kornberg, *J. Biol. Chem.*, **245**, 3637 (1970).

89. E. Viñuela, M. L. Salas, M. Salas, and A. Sols, *Biochem. Biophys. Res. Commun.*, **15**, 243 (1964).

90. M. G. Ord and L. A. Stocken, *Biochem. J.*, **107**, 403 (1968).

CHAPTER 3

Molecular Properties of Phosphofructokinase and Its Mechanisms of Regulation

TAG E. MANSOUR AND BARBARA SETLOW

Department of Pharmacology, Stanford University School of Medicine, Stanford, California

1. INTRODUCTION

Phosphofructokinase is the third enzyme involved in the series of reactions which catalyze the conversion of glucose to lactic acid. It catalyzes the phosphorylation of fructose-6-P to fructose-1,6-di-P:

REACTION 1

$$\text{Fructose-6-P} + \text{ATP} \xrightarrow{\text{Mg}^{2+}} \text{Fructose-1,6-di-P} + \text{ADP}$$

Since its discovery in yeast and mammalian muscle by Dische et al. (1), studies on phosphofructokinase have been handicapped by the absence of a highly purified enzyme. The purification of the enzyme from different sources during the last 10 years has resulted in a greater understanding of its molecular properties and the different mechanisms involved in its regulation. We will emphasize in this chapter the work which has been done on heart and muscle phosphofructokinase as a prototype for this system since these enzymes have been more extensively studied than enzymes from other sources. Furthermore, comparison will be drawn between these phosphofructokinases and those from other mammalian tissues and other sources. The most recent summary of the properties of phosphofructokinase can be found in Stadtman's (2) general review on allosteric enzymes.

2. ROLE OF PHOSPHOFRUCTOKINASE IN THE REGULATION OF GLYCOLYSIS

The role of phosphofructokinase as a rate-limiting enzyme in the Embden-Meyerhof pathway was first reported by Cori (3) in 1942. This function was indicated by the early findings of the Coris (4, 5, 6) that after incubation of frog muscle with epinephrine the endogenous concentration of the hexose monophosphate was increased to a greater extent than was lactic acid production. Studies by C. F. Cori (3) on the frog gastrocnemius muscle demonstrated that contraction of the muscle following electric stimulation caused a marked accumulation of hexose monophosphate esters and an increase in lactic acid production. Since in these experiments the sum of the hexose monophosphates and lactic acid formed can account for the amount of glycogen lost during contraction, the assumption was made that other intermediates of glycolysis do not accumulate to an appreciable extent. These experiments thus indicated that ". . . the reaction glycogen ⟶ hexosemonophosphate occurs more rapidly than the reaction hexosemonophosphate lactic acid, and points to phosphofructokinase as the rate-limiting step for lactic acid formation" (3).

Subsequently, several workers have suggested that phosphofructokinase plays a role in explaining the Pasteur effect. Engel'hardt and Sakov (7) found that the enzyme is highly sensitive to oxidative agents, an effect they thought could explain its role in the regulation of glycolysis during the shift from anaerobic to aerobic conditions. On the basis of the estimated steady-state levels of glucose, glucose-6-P, and fructose-1,6-di-P in the cell, several studies which followed the Cori work suggested that an increase in phosphofructokinase activity could be responsible for the Pasteur effect. For example, Lynen et al. (8) reported that in yeast cells anoxia, which causes an increase in the rate of glucose phosphorylation, also caused a reduction in the intracellular concentration of glucose and fructose-6-P and an increase in the intracellular levels of fructose-1,6-di-P. This suggested an increase in the activity of phosphofrucktokinase. Subsequently to these findings the use of substrate and product levels to determine enzyme activity following anaerobiosis was studied in the heart (9) and in the brain (10). Similarly an effect on the enzyme caused by antimonials was reported in schistosomes (11, 12). More direct evidence of phosphofructokinase activation was demonstrated in our laboratory in the liver fluke, *Fasciola hepatica* (13, 14). Stimulation of glycolysis by 5-hydroxytryptamine (serotonin) resulted in a marked increase in phosphofructokinase activity. Evidence based on our previous finding that serotonin increases the synthesis of cyclic 3′,5′-AMP (15) suggested that the effect of serotonin on glycolysis was mediated through the cyclic nucleotide. It was indeed found that cyclic 3′,5′-AMP can activate phosphofructokinase both in fluke cell-free extracts and in partially purified enzyme preparations.

These findings gave phosphofructokinase a physiological role analogous to that of glycogen phosphorylase in the regulation of carbohydrate metabolism. A gap existed, however, in our understanding of the molecular properties of phosphofructokinase. During the sixties, attempts were made in several laboratories to study the properties of this enzyme.

3. PURIFICATION AND CRYSTALLIZATION

Early attempts to purify phosphofructokinase were handicapped by its marked instability (16, 17). The enzyme was reported to be sensitive to mildly acidic pH's, dialysis, or low salt concentration. Investigations in our laboratory to isolate heart phosphofructokinase demonstrated that the enzyme becomes extremely labile after a mild degree of purification. Certain tissue components, later identified as hexose phosphate and adenine nucleotides, were shown to be essential for enzyme stability. Of these agents, fructose-1,6-di-P was the most active stabilizer.

The use of these stabilizers facilitated the purification of the enzyme from sheep heart to homogeneity (19) with good yield. Other purified enzymes were isolated from skeletal muscle (20, 21), erythrocytes (22), yeast (23), and *E. coli* (24). The enzyme was crystallized from both skeletal and heart muscle in the presence of ATP.

4. OLIGOMERIC STRUCTURE AND ASSOCIATION-DISSOCIATION SYSTEM

One of the main properties of purified phosphofructokinase is its tendency to form large aggregates. Molecular forms of the enzyme were demonstrated with sedimentation coefficients which varied from 50 S to 7 S (19, 25). High enzyme concentrations give highly aggregated forms. The most thorough estimate of the molecular weight of the skeletal muscle enzyme was reported by Paetkau and Lardy (25), who cited a value of 380,000 daltons. The relationship between the degree of aggregation of the enzyme and its catalytic activity has not been studied, mainly because phosphofructokinase is always assayed when it is highly diluted. Assay levels of the enzyme are presumably in the deaggregated form.

Information concerning the presence of subactive or inactive forms of phosphofructokinase (26) was obtained from studies on partially purified as well as on crystalline heart enzyme (19, 27). At a pH range of 5.8–6.5 the enzyme is extremely unstable and, when present in low concentration, irreversibly loses its activity. However, at moderately high protein concentrations the inactivated enzyme can be reactivated by incubation at pH 8.0. The sedimentation velocity of the inactive enzyme varies from 7 to 8. The reactivated enzyme has a sedimentation coefficient which is identical to that of the native enzyme (14.5–15 S). The residual activity of the dissociated form of the enzyme shows all the kinetic properties of the associated enzyme. The effect of ATP and of hexose phosphates on the sedimentation behavior of the dissociated enzyme was investigated at an acidic pH; ATP favors the dissociated form of the enzyme (7.5 S), while fructose-1,6-di-P favors the fully active polymerized form.

The kinetics of dissociation and association of phosphofructokinase was studied in our laboratory. Whereas enzyme dissociation, as determined by measuring enzyme activity, occurs almost instantaneously, the reassociation

process is slower (26). Furthermore, the process of dissociation and association depends on enzyme concentration. For example, at high concentration
reactivation of the dissociated enzyme is enhanced, while at lower enzyme
concentrations the rate of reassociation is lower. These findings were recently
confirmed by Frieden (28) on muscle phosphofructokinase. Reactivation of
the dissociated enzyme at an alkaline pH is facilitated by the following nucleotides and hexose phosphates: ATP, ADP, cyclic 3',5'-AMP, fructose-6-P,
and fructose-1,6-di-P. A combination of a nucleotide and a hexose phosphate
is much more effective than either one alone. This suggests that the optimal
conformation for reassociation of monomers is that with a hexose phosphate
and an adenine nucleotide bound to it.

Extracts from the heart and other tissues can also reactivate the dissociated
form of phosphofructokinase in the presence of an adenine nucleotide (26).
The effect of the tissue extract does not appear to be enzymatic, and the nature
of this component is yet to be determined.

A similar dissociation-association system in yeast phosphofructokinase
has been recently studied by Liebe et al. (29). The enzyme purified by these
authors has a molecular weight of $570,000 \pm 28,000$ at pH 7. When the
enzyme was dialyzed against phosphate buffer at pH 6.0, it was converted
to a molecular species with an average molecular weight of 370,000, which had
half of the original specific activity. This molecular conversion was reversible
when the enzyme was dialyzed against pH 7.0 phosphate buffer. The presence
of fructose-6-P at pH 6.0 hindered the conversion of the enzyme of molecular
weight 570,000 to the 370,000 species.

The relationships between different molecular forms of the enzyme and
its activity suggest the following molecular model for phosphofructokinase:

$$(EE)_n \overset{(1)}{\rightleftharpoons} EE \overset{(2)}{\rightleftharpoons} 2E \longrightarrow 16e.$$

The model represents phosphofructokinase which is fully active as a polymer
(EE) with an average molecular weight of about 400,000 and a sedimentation
coefficient of 15 S. This form could be present in an aggregated form, $(EE)_n$.
The equilibrium in reaction 1 is a function of enzyme concentration. At high
protein concentrations there is more of $(EE)_n$, and vice versa. The specific
activity of this form of the enzyme is as yet undetermined. Whereas ATP
shifts the equilibrium toward the nonaggregated form (EE), fructose-1,6-
di-P and fructose-6-P shift the equilibrium toward the aggregated form.
According to the model the fully active enzyme (EE) can be dissociated at a
mildly acidic pH into two inactive (or subactive) forms (E). The equilibrium
in reaction 2 is a function of pH, enzyme concentration, and ATP concentration. The enzyme in the cell is presumably present in all these forms at
equilibrium. At high ATP levels the equilibrium is shifted to low-molecular-
weight forms, which are less active and less stable. On the other hand, when

the concentration of fructose-6-P or fructose-1,6-di-P is high, the molecular species favored are those with high molecular weights, which are comparatively more stable. Thus an associating-dissociating system allows regulation of the concentrations of different molecular forms of the enzyme in the cell. The model also shows that the enzyme can be dissociated further to 16 small monomeric subunits (e) by 5 M guanidine HCl. Reversal of this reaction has not yet been established.

The teleology behind enzymes like phosphofructokinase, phosphorylase, pyruvate carboxylase, and acetyl-CoA carboxylase having a mechanism by which they undergo association-dissociation conversion was discussed recently by Frieden (28). Such a mechanism may ensure changes in the amount of active enzyme without altering the regulatory type of kinetic behavior. The slow transition from inactive to active forms, according to Frieden, may serve to buffer certain metabolic alterations. In the case of skeletal muscle fatigued by contraction a slow recovery of glycolysis was reported by Danforth and Helmreich (30). A relationship between the slow recovery of glycolysis and the possible dissociation of phosphofructokinase to inactive forms by the low pH is conceivable. The slowness of recovery might be a regulatory process to protect against the further depletion of the carbohydrate reserves of the muscle.

An activation system for phosphofructokinase which appears to involve association of inactive subunits of the enzyme may be present in the liver fluke (*Fasciola hepatica*), a trematode living in the bile ducts of several mammals. Enzyme activity, when tested in the presence of different substrate concentrations of both fructose-6-P and ATP, was always very low. However, when the flukes were preincubated with serotonin (5-hydroxytryptamine) both the activity of phosphofructokinase and that of glycolycis were increased (13, 14). This indolealkylamine was shown to be present in these organisms (31). The mechanism of activation of phosphofructokinase by serotonin was the subject of an extensive investigation in our laboratory, primarily because the effect resembled the activating action of epinephrine on glycogen phosphorylase in mammals. This similarity was even more suspected when we learned that serotonin is the hormone in the fluke that stimulates adenyl cyclase to synthesize more cyclic 3′,5′-AMP and that cyclic 3′,5′-AMP activates phosphofructokinase. The activation of the enzyme by cyclic 3′,5′-AMP was not a kinetic type of activation; it was due to an actual change in the amount of active enzyme.

Experiments were subsequently carried out to study the nature of activation of the enzyme by cyclic 3′,5′-AMP (32). A cell fraction containing inactive phosphofructokinase was isolated from the liver fluke, and we were able to detect the inactive enzyme by including in the regular phosphofructokinase assay 160 mM $(NH_4)_2SO_4$. We were also able to isolate another cell fraction (a heavy particulate fraction) which in the presence of cyclic 3′,5′-AMP, ATP, and Mg^{2+} can convert the inactive enzyme to the active form. The most logical development to these findings would have been to purify both cell fractions and to isolate, in a pure form, inactive phosphofructokinase as well as the protein in the particulate fraction necessary for activation. Unfortunately this was not possible because of the limited supply of the material and the very low activity of the enzyme in the organism.

We were, however, able to learn a little more about the problem using this impure system. Further experiments showed that the activation of the enzyme occurs in two steps. First, incubation of the cell particulate fraction with $ATP-Mg^{2+}$ yields a thermostable fraction. Second, incubation of the thermostable fraction with the inactive enzyme and with cyclic 3',5'-AMP results in activation of phosphofructokinase. Analysis of the thermostable fraction demonstrated that ADP and inorganic phosphate could account for almost all the activity in this fraction. We were able then to replace the thermostable fraction with ADP, P_i, and Mg^{2+} and obtain most of the activation effect on phosphofructokinase. Inorganic phosphate could be replaced by sulfate anions or hexose phosphates. Activation of the enzyme was found to be readily reversible by dialysis. The process of activation here also appears to involve polymerization of the enzyme. Sucrose gradient ultracentrifugal analysis of the enzyme before and after activation gave an S_{20w} of 5.5 for the inactive enzyme and 12.8 S for the activated enzyme. The possibility that this polymerization step is the outcome of a certain conformational change in the enzyme caused by the ligands above mentioned, which first leads to its activation and then ends with its polymerization, has yet to be elucidated.

5. KINETICS OF PHOSPHOFRUCTOKINASE

The first hint of a regulation mechanism for phosphofructokinase by one of its substrates was an observation by Lardy and Parks (33) that partially purified enzyme from skeletal muscle was inhibited by ATP. This effect was subsequently confirmed in phosphofructokinase from several other sources. In our studies in 1962 on phosphofructokinase from the liver fluke (14), besides confirming the ATP inhibitory effect on the enzyme, we found that the curve for fructose-6-P was sigmoidal and that in the presence of cyclic 3',5'-AMP this curve became hyperbolic (see Fig. 2 in reference 14). Furthermore, V_{max} in the presence of cyclic 3',5'-AMP was not significantly changed, while the apparent K_m for fructose-6-P was much lower than in the control assay. We concluded from these findings that cyclic 3',5'-AMP increases the affinity of phosphofructokinase for its hexose phosphate substrate.

Subsequently, similar results were observed in mammalian enzymes from

Table 1. Allosteric Effectors of Mammalian Phosphofructokinase

Inhibitor	pH	K, μM	Activator	pH	K, μM
ATP	6.65	2^a	AMP	6.95	1.8^c
Citrate	6.65	8^a	ADP	6.95	0.5^c
3-P-glycerate	7.1	8^a	Cyclic AMP	6.95	0.6^c
2-P-glycerate	7.1	110^b	P_i	7.0	4500^d
2,3-PP-glycerate	7.1	110^b	NH_4	7.0	2700^d
P-enolpyruvate	7.1	148^b	F-1,6-PP	7.0	1^e
P-creatine	7.1	550^b	G-1,6-PP	7.1	43^f
			F-6-P	7.1	256^f

[a] Dissociation constants measured by equilibrium binding to sheep heart phospho-fructokinase (53).
[b] Inhibitor constants determined from kinetic studies of phosphofructokinase from sheep brain. ATP = 0.26 mM; F-6-P = 0.075 mM. Inhibition by triose phosphates was observed only when ATP was present at inhibitory concentrations (37).
[c] Dissociation constants measured by equilibrium binding to rabbit skeletal muscle phosphofructokinase (52).
[d] Concentration of activator giving half-maximal activation of sheep brain phospho-fructokinase. ATP = 2.4 mM; F-6-P = 0.22 mM (63).
[e] Same as d, except that ATP = 3.8 mM; F-6-P = 0.22 mM (63).
[f] Concentration of activator giving full activation of sheep brain phosphofructo-kinase which was completely inhibited by a combination of 3-P-glycerate, 2-P-glycerate, P-enolpyruvate, and P-creatine (37).

several sources (27, 34, 35, 36). Detailed studies on phosphofructokinase from the heart and skeletal muscle showed that, while the enzyme displayed Michaelis-Menten kinetics at pH 8.2, it showed typical allosteric kinetics at pH 6.9. In addition to ATP and cyclic 3′,5′-AMP, several other metabolites were reported to modify enzyme activity at a pH of about 7 (Table 1). The discovery that citrate inhibits enzyme activity was particularly interesting since it offered a possible molecular basis for the regulation of glycolysis via a metabolite from the tricarboxylic acid cycle. It can also be seen from Table 1 that among the activators are all the metabolites that are increased during anoxia, such as AMP, ADP, and inorganic phosphate, as well as the substrate fructose-6-P, the product fructose-1,6-di-P, and its analog, glucose-1,6-di-P. A concomitant decrease in P-creatine would also serve to activate the enzyme (37).

The kinetics of the interactions between phosphofructokinase, its substrate, and the numerous modifiers are complex. An attempt is made here to describe the principles of these kinetics in the simplest form. The graphs in

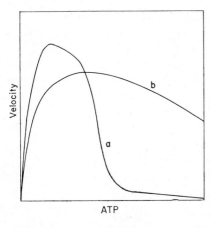

Fig. 1. Schematic representation of ATP saturation curves for phosphofructokinase. Curve *a* represents data typical of native enzyme at pH 6.9. Curve *b* represents data typical of enzyme which is not subject to ATP inhibition, such as native enzyme at a more alkaline pH, or native enzyme at pH 6.9 in the presence of an activator, or enzyme after photooxidation or ethoxyformylation.

Figs. 1 and 2 illustrate the types of curves obtained by increasing the concentration of either substrate at pH 6.9. Under these conditions V_{max} is approximately 40% of maximal activity at pH 8.2. A saturation curve for ATP in the presence of suboptimal concentrations of fructose-6-P gives a normal hyperbolic curve for the catalytic part, followed by a steep inhibition curve as the ATP concentration is increased (Fig. 1, curve *a*). The apparent K_i for ATP ranges from 0.08 to 0.3 mM in the presence of 0.2–0.4 mM fructose-6-P. In the presence of one of the activators listed in Table 1, the catalytic part of the curve is not changed, while the inhibitory curve reaches almost a plateau after maximal activity (Fig. 1, curve *b*). Thus the activators listed in Table 1 exert their effect by relieving ATP inhibition, that is, by "deinhibiting."

The saturation curve for fructose-6-P in the presence of inhibitory concentrations of ATP is shown in Fig. 2, curve *a*. The shape of the curve is

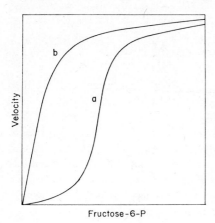

Fig. 2. Schematic representation of fructose-6-P saturation curves for phosphofructokinase. Curve *a* represents data typical of native enzyme at pH 6.9 at an inhibitory ATP concentration. Curve *b* represents data typical of enzyme which is not subject to ATP inhibition, such as native enzyme at a more alkaline pH, or native enzyme at pH 6.9 in the presence of an activator, or enzyme after photooxidation or ethoxyformylation.

sigmoidal. When this curve is replotted in terms of the Hill equation (38, 39) the apparent order of the reaction can be determined. At high concentrations ATP increases the apparent order of the reaction (increases the sigmoidicity); a maximum value for n was found to be 5.4. On the other hand, in the presence of one of the activators listed in Table 1 the sigmoidicity of the curve is decreased and the value of n could be reduced to 1, depending on the concentration of the activator (Fig. 2, curve b). The multimolecular nature of these kinetics has been interpreted on the basis of the model proposed by Monod et al. (39), although other interpretations are possible. A Hill coefficient of 5.4 indicates cooperative interaction between at least six fructose-6-P sites.

6. EVIDENCE FOR THE PRESENCE OF ALLOSTERIC SITES

According to the basic concepts formulated by Monod et al. (39), there are two requirements for an enzyme to be classified as an allosteric enzyme. The first is sigmoidal kinetics of the substrate and/or the modifier. The second is that enzyme activity be affected by ligands at a site other than the active site. The foregoing discussion of the kinetics of phosphofructokinase has shown that the enzyme fulfills the first requirement. Although the second requirement has not been met with by most of the "allosteric enzymes," we decided that it was imperative to show it in the case of phosphofructokinase because of the fact that ATP is both an inhibitor and a substrate. We wished to know whether the interaction of ATP with the enzyme occurs at the substrate site or at a separate allosteric site.

In the past several years a number of methods by which phosphofructokinase can be desensitized to the effects of its modifiers have been discovered (23, 40–50). Comparisons of the kinetic properties and the binding of substrates and modifiers to the native and desensitized forms of the enzyme have helped in verifying the presence of the allosteric sites as well as their nature. Experiments were carried out in our laboratory with the prime aim of obtaining a molecular form of phosphofructokinase which is insensitive to allosteric control.

Illumination of aerated solutions of proteins in the presence of a small amount of a light-sensitive dye (photooxidation) often results in destruction of histidine residues, although other side chains can also be affected (51). Photooxidation of phosphofructokinase from sheep heart in the presence of methylene blue resulted in an enzyme insensitive to ATP inhibition (Fig. 2, curve b), although there was very little effect on maximal catalytic activity (40). The cooperativity of the fructose-6-P kinetics was abolished as depicted in Fig. 2 and as shown by the decrease in the Hill coefficient to 1.0. Furthermore, the photooxidized enzyme was not inhibited by citrate or activated by

AMP. The kinetics of the enzyme at pH 8.2 was not significantly changed by photooxidation. Thus catalysis of the enzymatic reaction and regulation of the rate of this reaction by modifying ligands are separable functions of the phosphofructokinase molecule. This finding is consistent with the postulate that the regulatory site(s) is distinct from the catalytic site.

The availability of two forms of the enzyme, the native and the desensitized form, raised the question of which amino acid residues are affected by photooxidation. Amino acid analysis of native and photooxidized phosphofructokinase showed that the histidine residues were reduced from an average of 18.2 moles/10^5 g to 15.0, while the cysteine residues were reduced from 14.0 to 6.6 moles/10^5 g. The rest of the amino acid residues, including the photosensitive tryptophan, were not significantly changed (41). The results of previous work with sulfhydryl reagents (42) made it unlikely that cysteine was solely responsible for the desensitization (see below). It was therefore important to find a reagent which could interact specifically with histidine (the other amino acid oxidized) without affecting any other amino acid, particularly the cysteine residues. Although many amino acid reagents are not entirely specific, it was found that at pH 6.1 ethoxyformic anhydride can interact specifically with the histidine residues of phosphofructokinase (see reaction 2). Acylation of 3–4 histidine residues per 10^5 g of the enzyme with ethoxyformic anhydride resulted in the disappearance of both the homotropic interaction and the ATP inhibitory effect, just as occurs in photooxidation. The degree of loss of these properties was proportional to the number of histidine residues ethoxyformylated. As in photooxidation, the V_{max} was only slightly reduced and there was no significant change in the kinetics at pH 8.2.

Ethoxyformylation of the histidine residues can be reversed at pH 7 by hydroxylamine (reaction 3). The question was then asked whether such treatment would restore the kinetics of the native enzyme. This was found to be the case and constitutes evidence to support the view that the acylated side chain responsible for the desensitization is histidine.

The approach based on desensitization of the enzyme to allosteric control was also used in other laboratories. Desensitization of yeast phosphofructokinase to ATP inhibition has been accomplished by incubation of the enzyme with trypsin at pH 6.5 (29). Catalytic activity was unaffected. The densensitization was prevented by ATP, but not by ITP. The latter nucleotide is a good substrate for yeast phosphofructokinase but does not inhibit the enzyme. Although AMP, which counteracts the ATP inhibition of phosphofructokinase, did not protect against desensitization by trypsin, it prevented the protective effect of ATP.

A recent finding of Freyer et al. (47) showed that trypsin treatment results in the removal of a peptide which is 10,000 daltons per 160,000 daltons of

REACTION 2

$$\left(CH_3CH_2O\overset{O}{\underset{\|}{C}}\right)_2O + \underset{R\ \ H}{\boxed{N}} \xrightarrow{pH\,6.1} \underset{R}{\boxed{N}} + CO_2 + CH_3CH_2OH$$
$$\underset{\underset{O}{\overset{\|}{COCH_2CH_3}}}{}$$

REACTION 3

$$\underset{R\ \ \underset{\overset{\|}{O}}{COCH_2CH_3}}{\boxed{N}} + NH_2OH \xrightarrow{pH\,7.0} HON\overset{H\ \ O}{\underset{}{C}}OCH_2CH_3 + \underset{R}{\boxed{N}}$$

yeast phosphofructokinase. The fact that in the presence of ATP the enzyme still loses the same size peptide suggests that ATP may protect only a few residues of this peptide—those critical for allosteric inhibition by ATP— from trypsin action.

A search for an enzyme which could desensitize phosphofructokinase to ATP inhibition was made in many laboratories. In 1964 Viñuela et al. (48) described a system whereby phosphofructokinase in crude extracts of yeast could be converted to a form insensitive to ATP inhibition by incubation of the extracts with ATP, Mg^{2+}, and NaF. Cyclic 3′,5′-AMP greatly stimulated the conversion. Recently a similar desensitization system for pure yeast phosphofructokinase was reported by Atzpodian et al. (23, 49). It required a partially purified protein fraction from yeast, as well at ATP, Mg^{2+}, NaF, AMP, and fructose-6-P. Cyclic AMP had no effect. Phosphofructokinase incubated in this system remained insensitive to ATP inhibition after precipitation by ammonium sulfate or acetone provided the enzyme was redissolved in buffer containing Mg^{2+} and NaF. It was originally suggested by Viñuela et al. (48) that the desensitization of yeast phosphofructokinase might be caused by an enzyme-catalyzed phosphorylation of phosphofructokinase. The requirement for F^- was suggested to be necessary for inhibition of a protein-dephosphorylating enzyme which reconverted ATP-insensitive phosphofructokinase to the ATP-sensitive form. Such a system is analogous to that operating with glycogen phosphorylase. However, a recent report (50) shows that the desensitization is not due to direct enzymatic action on phosphofructokinase but rather is the result of conversion of ATP in the incubation mixture to ADP by an ATPase. Incubation of pure yeast phosphofructokinase with ADP, fructose-6-P, Mg^{2+}, and NH_4F is sufficient to desensitize the enzyme to ATP inhibition.

7. BINDING PROPERTIES OF PHOSPHOFRUCTOKINASE

Studies of the binding of various ligands to phosphofructokinase have helped to explain some of the kinetic properties of the enzyme, but have by no means clarified the entire picture. The first report of binding of ATP to skeletal muscle phosphofructokinase indicated that 3.3 moles of the nucleotide was bound per 10^5 g of enzyme at an acidic pH (52). A subsequent study using the heart enzyme, however, showed that a maximum of 3.6 moles of ATP was bound per 10^5 g of protein (53).The minor difference between the two results could probably be attributed to the presence of β-glycero-P as a buffer in studies with skeletal muscle enzyme. The triose-P was shown to decrease the maximum amount of ATP bound to the heart enzyme (53). A discrepancy in the stoichiometry of fructose-6-P binding to the muscle [1.1 sites/10^5 g (52)] and the heart [1.8 sites/10^5 g (53)] enzymes can probably also be explained by the inhibitory effect of β-glycero-P buffer on fructose-6-P binding (53). These studies suggest that phosphofructokinase contains two catalytic sites each for ATP and fructose-6-P, and two regulatory sites for ATP. This assumption is supported by studies on the binding properties of the enzyme under conditions where it does not respond to allosteric control, such as at pH 8.2 or after photooxidation. Under either condition the enzyme was shown to bind only 1.8–2 moles of both ATP and fructose-6-P per 10^5 g of enzyme (53). The binding properties of the heart enzyme are summarized in Table 2.

The stoichiometry of citrate binding also fits the kinetic properties of phosphofructokinase. Native heart enzyme binds 2 moles of citrate per 10^5 g at pH 6.65 and no citrate at pH 8.2, whereas the photooxidized enzyme does not bind citrate at either hydrogen-ion concentration (53). From these findings it is clear that ATP and citrate do not inhibit phosphofructokinase at pH 8.2 simply because they cannot bind to its regulatory sites at alkaline pH's. Exactly why they cannot bind, however, remains unknown. The effects of modifying histidine residues on the enzyme with ethoxyformic anhydride (41) could be interpreted to mean that these residues must be protonated before the two inhibitors can bind to phosphofructokinase.

Since the activators cyclic 3',5'-AMP and 5'-AMP also exert their effects only at acidic pH's, one might expect their binding to follow the same pattern as that shown by ATP and citrate. This is not, however, the case. Both mononucleotides have only 1 binding site/10^5 g of enzyme (52, 54).

Table 2.   Binding Properties of Sheep Heart Phosphofructokinase[a]

Ligand	Native PFK		Photooxidized PFK, pH 6.65 or pH 8.2
	pH 6.65	pH 8.2	
ATP	3.6	2	1.8
	$(K_D = 2 \mu M)$		$(K_D = 2.8 \mu M)$
F-6-P	1.8	1.8	1.6
	$(K_D = 21 \mu M)$	$(K_D = 17 \mu M)$	$(K_D = 20 \mu M)$
Citrate	2	0	0
	$(K_D = 0.3 \mu M)$		
Cyclic 3′,5′-AMP	1.0	0.9	1.1
	$(K_D = 1.5 \mu M)$	$(K_D = 2.0 \mu M)$	$(K_D = 1.7 \mu M)$
5′-AMP	1.0		
	$(K_D = 4.3 \mu M)$		

[a] Binding is expressed as moles of ligand bound per 10^5 g of enzyme.

For these activators, unlike ATP, fructose-6-P, and citrate, this value is not influenced by the hydrogen-ion concentration. Most surprising, however, is the fact that cyclic AMP binds to the enzyme at pH 8.2 with only a small increase in the dissociation constant (52, 54) and also binds to the photooxidized enzyme just as well as to the native enzyme (54). Since 5′-AMP binds to the same site as cyclic AMP (52, 54), presumably the nature of its binding is the same as that of the cyclic nucleotide. Competition between the binding of ATP and that of cyclic AMP to muscle phosphofructokinase indicated that the site for the cyclic nucleotide was identical with one of the ATP sites (53). It is somewhat difficult, however, to imagine that photooxidizing the enzyme or simply raising the pH to 8.2 can alter a binding site in such a way that the binding of one of two rather similar molecules is completely unaffected, whereas the other can no longer bind at all. If the sites for the two nucleotides are different, the kinetic activation by cyclic AMP must be due to an effect secondary to its binding and not to simple displacement of ATP from the inhibitory sites.

Two characteristics of the kinetics of phosphofructokinase have not been seen in the binding studies. One is the sigmoidal saturation curve for fructose-6-P observed at acidic pH's. All binding curves reported for the hexose-P have been linear, indicating lack of cooperativity (52, 53). However, sigmoidal kinetic curves are seen only in the presence of inhibitory concentrations of ATP and in the absence of activators (27). It is difficult to study binding under these conditions because the catalytic reaction can proceed. An analog of

ATP which cannot undergo P_i transfer is needed. Unfortunately the enzyme appears to show a very high specificity for the triphosphate moiety, both as substrate and as inhibitor (55).

The second kinetic characteristic of phosphofructokinase which has not been clarified is the fact that the presence of ATP enhances the inhibition caused by citrate and vice versa (56). High concentrations of citrate inhibited ATP binding to heart phosphofructokinase. Conversely, ATP competitively inhibited citrate binding (53). Thus it appears that ATP and citrate can bind near each other at a regulatory site on the enzyme. The effect of binding either inhibitor is to lower the affinity of the enzyme for fructose-6-P (52, 53). Although this observation might explain the inhibition of phosphofructokinase, it remains unclear why inhibition in the presence of both ATP and citrate is much greater than the sum of the inhibitions caused by each compound alone.

Perhaps surprisingly, it has been possible to correlate a number of the kinetic and binding properties of phosphofructokinase. The conditions for equilibrium binding are very different from those under which kinetic studies are conducted. For example, enzyme concentrations in the binding studies were higher by a factor of 10^3–10^4 than those used in assays of catalytic activity. This difference is particularly significant for an enzyme like phosphofructokinase, which undergoes concentration-dependent association (19, 25) and whose kinetic properties vary with protein concentration and, therefore, presumably with the state of aggregation of the enzyme molecule (57).

8. ROLE OF THIOL GROUPS IN THE REGULATION OF PHOSPHOFRUCTOKINASE

Several proteins are now known to have thiol groups which participate in their allosteric control. In the case of aspartic transcarbamylase, treatment of the enzyme with divalent or monovalent mercurials results in its dissociation into the regulatory and catalytic subunits with loss of its cooperative kinetics (58). The cooperative binding of oxygen by hemoglobin is changed by modifying the sulfhydryl group at position 93 of the β chains and depends on the sulfhydryl reagent used (59). Recent investigations on sheep heart phosphofructokinase in our laboratory showed that it contains approximately 15 sulfhydryl groups per 100,000 g of enzyme with no disulfide bonds (42). Approximately 6 thiol groups are accessible to titration with sulfhydryl reagents in the native enzyme (in the absence of denaturing agents). On the basis of the kinetics of reactivity of these thiol groups to different reagents, it was concluded that there are at least two types of thiol groups.

Enzyme activity is inhibited following titration of the enzyme with a sulfhydryl reagent. Such inhibition is greater at pH 6.9 than at pH 8.2 and was found to be dependent on the sulfhydryl reagent used. The monovalent mercurials were the least inhibitory, whereas the disulfide reagents caused a marked inhibition of activity. In the case of parachloromercuric benzoate (*p*-CMB) the maximum loss of enzyme activity at pH 8.2 was 50% when all the available thiol groups were titrated. The inhibition was reversible by 2-mercaptoethanol. Neither fructose-6-P nor ATP significantly changed the rate of titration of enzyme by *p*-CMB; nor did either change the number of groups titrated. On the other hand, both substrates prolonged the time for complete titration of the sulfhydryl groups by 5,5′-dithiobis(2-nitrobenzoic acid) (DTNB), while the number of thiol groups titrated with the same reagent was not significantly changed.

These results suggest that the thiol groups may not be present in the active site of the enzyme; rather they are important for enzyme activity at pH 6.9, which is optimal for allosteric kinetics. The role of sulfhydryl groups in the cooperative kinetics of the enzyme was shown when phosphofructokinase was reacted with *p*-CMB. A decrease in cooperativity between these binding sites was characterized by a decrease in the interaction coefficient of the fructose-6-P sites. Furthermore, positive effectors such as AMP or cyclic 3′,5′-AMP were capable of changing the cooperative kinetics of the *p*-CMB-treated enzyme completely to first-order kinetics, even in the presence of inhibitory concentrations of ATP. In contrast, neither positive effector could completely convert the cooperative kinetics of the native enzyme to first-order kinetics when inhibitory concentrations of ATP were present.

More confirmation for the idea that thiol groups are involved in allosteric control of phosphofructokinase comes from the work of Kemp and Forest (60), who showed that the muscle enzyme contains one sulfhydryl group per 90,000 g of enzyme which is at least 50 times more reactive toward DTNB than any other thiol in the protein. Disulfide formation between this group and DTNB resulted in almost complete abolition of the cooperative interactions manifested in the sigmoidal fructose-6-P saturation curves of the native enzyme (43, 44). However, there was no effect on the K_m's for ATP or fructose-6-P at the catalytic site. According to Kemp (45), MgATP greatly slowed the rate of reaction of DTNB with the reactive thiol group. The protective effect of MgATP was markedly reduced, however, by conditions which reverse the ATP inhibition of the enzyme, such as increased pH, increased fructose-6-P concentration, or the presence of AMP or 3′,5′-cyclic AMP. Kemp (44) proposed that MgATP can still bind to its regulatory site on the thiol-modified enzyme, but that the enzyme cannot respond normally to such binding by changing to its inhibited form. Evidence for such a hypothesis should be confirmed by studies of the binding of MgATP to the DTNB-modified enzyme. It would also be informative to determine whether the kinetic effects observed are due directly to formation of the disulfide or to the presence of the bulky, charged DTNB group in that particular region of the phosphofructokinase molecule.

9. IDENTIFICATION OF OTHER FUNCTIONAL GROUPS ON THE ENZYME

An investigation of the effects of lysine reagents on skeletal muscle phosphofructokinase indicated that these groups are important to the catalytic activity of the enzyme and possibly also to its regulation (61, 62). Reaction of either the heart or skeletal muscle enzyme with pyridoxal phosphate resulted in the formation of a Shiff base which could be reduced to a stable covalent linkage with sodium borohydride (reaction 4). Several different effects of this modification on phosphofructokinase were observed; these were dependent on the number of lysine residues titrated.

REACTION 4

Reaction of heart phosphofructokinase with up to 1 mole of pyridoxal phosphate per 100,000 daltons resulted in enzyme with an increased sensitivity to ATP inhibition at pH 6.9 (Table 3) (62). The increased sensitivity was found to be associated with a shift in the acidic portion of the pH-activity curve to higher pH's. Since pyridoxal-phosphofructokinase is inhibited by lower ATP concentrations as the pH is decreased, such a shift could account for the apparent increase in sensitivity to ATP inhibition. In contrast, a similar modification of the skeletal muscle enzyme was reported to result in decreased sensitivity to ATP inhibition (61). Considerably different conditions were used for assay of the ATP inhibition of modified heart and skeletal muscle phosphofructokinases, so that the discrepancy in the results may be only apparent.

Table 3.   Enhanced Sensitivity of Pyridoxal Phosphate-Treated Phosphofructokinase to ATP Inhibition

Lysine Modified, moles/10^5 g enzyme	Maximum Activity at pH 8.2, units/mg	ATP Concentration, mM, Giving	
		Maximum Activity at pH 6.9	50% Inhibition at pH 6.9
0	144	0.2	0.55
0.8	118	0.06	0.11
1.0	82	0.07	0.16
0	104	0.12	0.23
0.6	88	0.03	0.08

The ATP saturation curves were obtained at pH 6.9 with the enzyme from heart before and after small amounts of pyridoxal phosphate were covalently linked to lysine residues in the protein. Maximum activity at pH 8.2 and the ATP concentrations which gave maximal and 50% inhibited activities at pH 6.9 are shown. The fructose-6-P concentration was 0.4 mM.

More extensive modification of phosphofructokinase resulted in a progressive loss of catalytic activity, which reached 80% or more when about 4 lysine residues per 100,000 daltons had reacted with pyridoxal phosphate (61, 62). Both fructose-6-P and fructose-1,6-di-P, but not ATP, protected against the activity loss. This suggested that the reactive lysine residues might be involved in the binding of fructose phosphates to the catalytic site.

Another aspect of this finding—the fact that reaction of phosphofructokinase with high concentrations of lysine reagents causes the enzyme to dissociate into subunits of 80,000 daltons (61, 62)—was particularly interesting because dissociation of the native enzyme can occur only at an acidic pH, whereas dissociation by the lysine reagent, pyridoxal phosphate, can take place at an alkaline pH (pH 8.0). Dissociation of the heart enzyme (62) into 7 S subunits occurred readily at 3° when only 4 lysine residues per 100,000 daltons had reacted with pyridoxal phosphate, whereas the muscle enzyme required about 8–10 modified residues in order to dissociate at 20°. Warming the modified heart enzyme to room temperature caused it to reassociate. Since the dissociated form of phosphofructokinase is inactive (25, 27), the loss of catalytic activity observed after the modification of lysine residues may result at least in part from dissociation of the molecule to inactive subunits. The use of a lysine reagent for studies on the active site, therefore, must be regarded with reservation since it could affect enzyme activity through dissociation of the enzyme.

10. MECHANISM OF THE CATALYTIC REACTION

Allosteric regulation of phosphofructokinase is not observed when the enzyme is assayed at pH 8.2, where its catalytic activity is maximal (27). On the basis of kinetic studies performed under these conditions three different mechanisms have been proposed for the catalytic reaction. Lowry and Passonneau (63) reported kinetic data obtained with brain phosphofructokinase which are consistent with a reaction mechanism involving random order of substrate binding to the catalytic site (64). On the other hand, plots of kinetic data obtained with the enzyme from yeast (65), skeletal muscle (66, 67), and erythrocytes (68) indicate a mechanism in which release of one product occurs before addition of the second substrate (ping-pong mechanism) (64). A similar mechanism was suggested for the reverse reaction of the heart enzyme (69). The enzyme from the slime mold, *Dictyostelium discoideum*, which shows no evidence of allosteric regulation, also has the same kinetic pattern as that described for the muscle enzyme (70). In the case of the skeletal muscle enzyme (67) this mechanism is supported by the following observations: (*a*) the kinetic pattern of product inhibition; (*b*) enzyme-catalyzed hydrolysis of ATP and of fructose-1,6-di-P in the absence of the second substrate; and (*c*) enzyme-catalyzed phosphate exchange reactions between ATP and ADP and between fructose-6-P and fructose-1,6-di-P in the absence of the second substrate. A recent report of the exchange reactions catalyzed by ox heart phosphofructokinase contradicts the data above in that for this enzyme ADP was required for phosphate exchange between fructose-6-P and fructose-1,6-di-P (71). Accordingly a somewhat altered compulsory-order mechanism was proposed to accommodate this observation. The mechanisms proposed for both the skeletal muscle and ox heart enzymes involve a phosphorylated enzyme intermediate, but attempts to isolate it have so far been unsuccessful (67).

11. PHYSIOLOGICAL INTERPRETATIONS

The foregoing description of the molecular properties of phosphofructokinase draws attention to ATP as a critical ligand for allosteric control of the enzyme. An increase in the concentration of ATP in the environment where the enzyme is located can inhibit the catalytic activity of the phosphofructokinase, while a decrease in the ATP concentration has the opposite effect.

Furthermore, enzyme under the influence of ATP inhibition is subject to activation by 5'-AMP, cyclic 3',5'-AMP, and several hexose phosphate activators. There are two questions which should be considered in order to understand the possible mechanism of regulation of this enzyme. First, are changes in the level of ATP in the cell under different physiological conditions large enough by themselves to allow changes in the catalytic activity of the enzyme? Second, is the ATP level in the vicinity of the enzyme high enough at all times (acting or resting condition) to keep the enzyme in the ATP-inhibited form? Accordingly, changes in the levels of AMP and other

Table 4. Effect of Stimulation of Glycolysis on Nucleotide Levels in Different Tissues

Organ	Condition	ATP	ADP	AMP	Reference
Rat heart	Aerobic	6.7	1.28	0.096	72
Rat heart	Anaerobic	5.5	1.87	0.31	72
Frog sartorius	Control	7.2	2.3	0.4	73
Frog sartorius	Electric stimulation[a]	4.8	5.1	0.8	73
Rat heart	Control	8.3	1.07	0.116	74
Rat heart	Epinephrine[b]	7.25	2.0	0.32	74

[a] Muscles were stimulated by electric shock (36 shocks/min) for the same length of time as the control (30 min).
[b] Concentration of epinephrine in perfusate (5 min) was 0.2 μg/ml.

Data on the rat heart were recalculated from the papers indicated in order to determine the concentration (μ moles per milliliter) of intracellular H_2O. This was done on the basis that the amount of intracellular water is 2.5 times the dry weight reported in these papers.

activators would be of prime importance in modifying enzyme activity. Definite answers to these questions are not possible since any determination of ATP concentration in the cell is based on the assumption of uniform distribution of the nucleotide in the intracellular water. Indirect evidence of the presence of compartments for these substrates is now available.

The results summarized in Table 4 represent some published experiments from several laboratories to demonstrate the changes in the levels of adenine nucleotides accompanying activation of glycolysis by several means. Anoxia, electric stimulation, or epinephrine treatment results in only 15–30% decrease in ATP level. According to the kinetic data discussed above, the enzyme should still be inhibited at the level of ATP found after activation of glycolysis. What appears to be more significant is an increase in the levels of AMP, ADP, and inorganic phosphate (not shown in the table). These agents are all activators of the ATP-inhibited enzyme. Such activation is assumed to occur, as judged by a decrease in the levels of glucose-6-P and fructose-6-P, and an increase in the level of fructose-1,6-di-P, in these tissues. On the basis of an equilibrium state of 0.44 for adenylate kinase, Krebs (75) concluded that the tissue concentrations of AMP are subject to much greater percentage changes under physiological conditions than the concentration of ATP or ADP or the ATP/ADP ratio.

The problem is not as clear, however, as it appears. Helmreich et al. (73) reported an increase in glycolytic flux in the frog sartorius muscle after low-frequency electric stimulation with minimal changes in 5'-AMP levels. They concluded that phosphofructokinase must be controlled by a factor other than AMP or ADP, and Ca^{2+} was suggested as the link between activation of the contractile system and activation of both phosphofructokinase and glycogen phosphorylase b kinase. As yet, experiments in our laboratory have not shown such an effect with mammalian phosphofructokinase, although Ca^{2+} activation was demonstrated in the case of the liver fluke enzyme (14).

A unitarian view of the role of the adenine nucleotides in cell regulation has been suggested recently by Atkinson (76, 77). Kinetic results reported by him with nucleotides at physiological concentrations show that phosphofructokinase can respond to what he terms "adenylate charge," that is, small changes of AMP can affect enzyme activity. Citrate was shown to modulate this effect by increasing the steepness of the response curve to changes in the adenylate charge, while cyclic 3',5'-AMP acts directly on phosphofructokinase by releasing the enzyme from control through the adenylate system.

The idea of citrate as a metabolic regulator has been studied extensively by Randle and others (72, 78–81). Both glycolysis and phosphofructokinase activity are decreased in perfused rat hearts in alloxan diabetes, starvation, and fluoroacetate poisoning. All these conditions are accompanied by an increase in the intracellular levels of citrate. The idea was then proposed that citrate mediates the inhibitory effects of the conditions indicated above on phosphofructokinase and on glycolysis. The physiological importance of citrate as a regulator was tested by Randle and others and was found to be quantitatively feasible provided the citrate is in the vicinity of the enzyme. Since phosphofructokinase in the rat heart is found in the soluble fraction and citrate synthase is confined to the mitochondrion, a question arises about the source of cytoplasmic citrate. This would involve translocation either as citrate or as 2-oxoglutarate or glutamate with resynthesis in the cytoplasmic fraction by reversal of the NADP-linked isocitrate dehydrogenase (78). Another possible source which was considered is cytoplasmic synthesis of citrate from acetyl-CoA and oxaloacetate by citrate synthase, which is oriented in the mitochondrion in such a way as to allow cytoplasmic synthesis.

A pertinent question in connection with the effect of citrate on phospho-fructokinase is whether the enzyme from cells which have high glycolytic activity but no oxidative tricarboxylic cycle is sensitive to citrate inhibition. This question was investigated by Kühn et al. (22), who showed that phos-phofructokinase isolated from mammalian erythrocytes had low sensitivity to citrate inhibition.

A favorite procedure in many laboratories for studying the physiological role of phosphofructokinase involves studying oscillations of glycolysis in cell-free extracts as well as in intact yeast cells (82). In this method the re-duction-oxidation cycle of NADH can be observed during glycolysis over a long time range. These oscillations coincide with periodical acceleration and inhibition of glycolysis, as well as oscillations of concentrations and fluxes of glycolytic intermediates. Studies based on analysis of the control points during oscillation and on the crossover theorem (83) showed that these oscillations coincide with the activation and inhibition of phosphofructo-kinase. These results indicate that it is primarily this enzyme which generates the glycolytic oscillations (84, 85).

The question of how different hormones that stimulate glycolysis can change the activity of phosphofructokinase has yet to be answered. In this connection one major problem is finding the true levels of the active enzyme in resting tissues. Methods for the rapid arrest of enzymatic reactions which have been employed to study levels of glycogen phosphorylase have given no definite answer. Part of the explanation for this difficulty could lie in the fact that the optimal conditions used to assay the phosphofructokinase could very well result in enzyme association and thus enzyme activation. These conditions include an alkaline pH and the presence of comparatively high levels of fructose-6-P and ATP. On the other hand, assays under conditions closer to the physiological state in the cell, such as a slightly acidic pH and low substrate concentration, often result in enzyme dissociation and thus in-activation (26).

Consequently most of the information which concerns the effect of hor-mones on phosphofructokinase is based on extrapolation from experiments on tissue substrate levels before and after giving a particular hormone. Using this indirect method, Cori and his associates carried out an extensive number of experiments on frog muscles because of their ability to function for a long time under anaerobic conditions, thus limiting many interfering factors seen with exclusively aerobic tissues. In these experiments it was clearly shown that epinephrine, when added to resting frog sartorius muscle, increases the levels of hexose monophosphates, while the levels of fructose-1,6-di-P and of lactate remain relatively small (4, 86, 87). These results could be interpreted to mean that there is an activation of glycogen phosphorylase with no parallel activation of phosphofructokinase. The failure of epinephrine to increase lactic acid formation under these conditions is not in accord with the known physiological effect of this hormone in increasing energy production as a result of stimulation of glycogenolysis. Recent work appears to suggest that electric stimulation is the prime factor for activation of phosphofructokinase, either in the absence or in the presence of epinephrine. This is indicated by a rise in the levels of the hexose monophosphate pool as well as the levels of fructose-1,6-di-P and lactic acid. According to these experiments, an increase in phosphofructokinase by epinephrine cannot be seen until 30 min after addition of the hormone (87). In studies using the same indirect approach, insulin caused an apparent increase in phosphofructokinase activity in the heart, as shown by a decrease in the concentrations of glucose-6-P and of fructose-6-P, while the levels of fructose-1,6-di-P and lactic acid formation were increased (86).

The mechanism of activation of phosphofructokinase during electric stimulation and in connection with the effects of both insulin and epinephrine can only be speculated on. The magnitude of activation of phosphofructokinase during muscle contraction, particularly since it can occur under conditions in which the enzyme is saturated with fructose-6-P, suggests that an increase in V_{max} of the reaction is involved (86). Whether a conversion of an inactive form of the enzyme to an active form is also involved is not yet ascertained. In the heart, where the muscle is constantly contracting, an increase in the levels of cyclic 3′,5′-AMP, 5′-AMP, and ADP was shown to occur monotonically from the onset of the contractile force increase caused by epinephrine (74). Cyclic 3′,5′-AMP levels reached a peak value of 7 nM/g dry weight in 10 sec, while a maximum value was observed at 25 sec for ADP and AMP increase. This is equivalent to an intracellular concentration of cyclic 3′,5′-AMP of 2.8×10^{-6} M. This level of the nucleotide is within the concentration range that could affect phosphofructokinase.

12. COMPLEMENTARITY IN THE CONTROL OF CARBOHYDRATE ENZYMES

The activity of phosphofructokinase is linked to the activity of several other enzymes. An increase in its activity in the muscle results in a decrease in the level of fructose-6-P. Since there is equilibrium between glucose-6-P and fructose-6-P through the phosphoglucose isomerase reaction, the level of glucose-6-P will also fall, resulting in activation of hexokinase. Furthermore, activation of phosphofructokinase will result in an increase in the cellular levels of fructose-1,6-di-P, which is an allosteric activator of pyruvic kinase. Thus activation of the three enzymes is well coordinated to increase the rate of glucose degradation in order to meet the energy demand of the cell.

It appears that AMP plays a role in the regulation of a few other enzymes besides phosphofructokinase. An increase in the level of AMP in the cell, besides activating phosphofructokinase, will also increase the activity of glycogen phosphorylase *b* and will simultaneously inhibit the activity of fructose-1,6-diphosphatase. The first effect will favor an increase in glycogen breakdown, and the second will maintain high levels of fructose-1,6-di-P in the cell, which will activate both phosphofructokinase and pyruvic kinase. It will also keep fructose-1,6-di-P as a glycolytic substrate to be further degraded by aldolase.

The effect of sulfhydryl reagents on phosphofructokinase suggests a possible complementarity in the regulation of two opposing enzymes, fructose-1,6-diphosphatase and phosphofructokinase. Whereas disulfides such as cystamine activate fructose diphosphatase (88, 89), these reagents inactivate phosphofructokinase. It is possible, therefore, that, in addition to having the common effector AMP, these two enzymes could be controlled by the level of disulfides present in the cell, as has been suggested for fructose diphosphatase.

The role of fructose-1,6-diphosphatase in the regulation of glycolysis was discussed recently by Newsholme and Crabtree (90). In muscles with an extremely high demand for energy during contraction, such as in pheasant pectoral muscle, the activity of phosphofructokinase has to be stimulated to a level ninefold as great as the resting level. In such muscles, activation of phosphofructokinase through changes in AMP concentration may not be sufficient. Accordingly the sensitivity of the control mechanism can be markedly increased by the operation of a cycle between fructose-6-P and fructose-1,6-di-P catalyzed by the simultaneous activities of phosphofructokinase and fructose-1,6-diphosphatase in response to changes in AMP concentration. When the demand for energy is high, the increase in the level of AMP will assure simultaneous inhibition of the fructose diphosphatase and activation of phosphofructokinase, a complementary effect which ensures that loss of energy and restrictions of the rate of glycolysis at the stage of fructose-6-P are reduced to a minimum.

Citrate is another ligand that appears to occupy a central role in the regulation of energy metabolism. It inhibits phosphofructokinase activity at levels which may well exist in the cell (78, 91). Oxidation of fatty acids and ketone bodies in diabetes and starvation results in an increase in the levels of citrate in the cell muscle. Citrate has also been shown to activate acetyl-CoA carboxylase in liver and fat cells (92, 93). Physiologically speaking, the two complementary regulatory mechanisms are designed to provide energy to the muscle from fatty acids rather than carbohydrates and to stimulate fatty acid synthesis in the liver and in fat depots.

13. CONCLUSIONS

We have attempted to discuss in this chapter the present state of knowledge of the allosteric kinetics of phosphofructokinase, the oligomeric structure of the enzyme, and the various possible mechanisms of its regulation. The striking features of this enzyme are its multiple molecular forms and the fact that its activity is modified by a large number of cellular components. The reader who concludes, after reading this chapter, that phosphofructokinase is uncontrollable is mistaken because the overwhelming evidence, discussed above, emphasizes the involvement of the enzyme in the control of energy production in the cell. It is possible that because of the crucial position of phosphofructokinase in the control of cellular metabolism nature has provided this enzyme with a variety of ways to change its activity.

   In the last ten years there has been an upsurge in our understanding of the molecular properties of phosphofructokinase. Most of these studies were carried out on purified enzyme preparations. An adequate understanding, however, of the nature of the enzyme in the resting cell and of the changes which the enzyme undergoes in response to different physiological conditions is still lacking. Whatever the approach to these unanswered questions, it is wise to take into consideration all the mechanisms which are known to regulate enzyme activity.

ACKNOWLEDGMENTS

Some of the studies reported by the authors herein were supported by U.S. Public Health Service Grant AI04214 from the National Institutes of Allergy and Infectious Diseases and by a grant-in-aid from the American Heart Association.

REFERENCES

1. Z. Dische, *Biochem. Z.*, **280,** 248 (1935).
2. E. R. Stadtman, *Advan. Enzymol.*, **28,** 41 (1966).
3. C. F. Cori, in *A Symposium on Enzymes: Units of Biological Structure and Function* (Ed.: O. H. Gaebler), Academic Press, New York, 1956, p. 573.
4. A. H. Hegnauer and G. T. Cori, *J. Biol. Chem.*, **105,** 691 (1934).
5. G. T. Cori and C. F. Cori, *J. Biol. Chem.*, **116,** 119 (1936).
6. G. T. Cori and C. F. Cori, *J. Biol. Chem.*, **116,** 129 (1936).
7. V. A. Engel'hardt and N. E. Sakov, *Biokhimiya*, **8,** 9 (1943).

8. F. Lynen, G. Hartmann K. F. Netter and A. Schuegrat in *Ciba Foundation Symposium on Regulation of Cell Metabolism* (Ed.: G. E. W. Wolstenholme), Little, Brown, Boston, 1959, p. 256.

9. E. A. Newsholme and P. J. Randle, *Biochem. J.*, **80**, 655 (1961).

10. O. H. Lowry, J. V. Passonneau, F. X. Hasselberger, and D. W. Schulz, *J. Biol. Chem.*, **239**, 18 (1964).

11. T. E. Mansour and E. Bueding, *Brit. J. Pharmacol.*, **9**, 459 (1954).

12. E. Bueding and J. M. Mansour, *Brit. J. Pharmacol.*, **12**, 159 (1957).

13. T. E. Mansour, *J. Pharmacol. Exptl. Therap.*, **135**, 94 (1962).

14. T. E. Mansour and J. M. Mansour, *J. Biol. Chem.*, **237**, 629 (1962).

15. T. E. Mansour, E. W. Sutherland, T. W. Rall, and E. Bueding, *J. Biol. Chem.*, **235**, 466 (1960).

16. J. F. Taylor, *Phosphorus Metabolism Symposium*, Johns Hopkins University—McCollin-Pratt Institute Contribution 23, Vol. 1, Johns Hopkins Press, Baltimore, 1951, p. 104.

17. J. A. Muntz, *Arch. Biochem. Biophys.*, **42**, 435 (1953)

18. N. Wakid and T. E. Mansour, *Mol. Pharmacol.*, **1**, 53 (1965).

19. T. E. Mansour, N. Wakid, and H. M. Sprouse, *J. Biol. Chem.*, **241**, 1512 (1966).

20. K.-H. Ling, F. Marcus, and H. A. Lardy, *J. Biol. Chem.*, **240**, 1893 (1965).

21. A. Parmeggiani, J. H. Luft, D. S. Love, and E. G. Krebs, *J. Biol. Chem.*, **241**, 4625 (1966).

22. B. Kühn, G. Jacobasch, and S. Rapoport, *Acta Biol. Med. Ger.*, **23**, 1 (1969).

23. W. Atzpodian and H. Bode, *European J. Biochem.*, **12**, 126 (1970).

24. D. Blangy, H. Buc, and J. Monod, *J. Mol. Biol.*, **31**, 13 (1968).

25. V. Paetkau and H. A. Lardy, *J. Biol. Chem.*, **242**, 2035 (1967).

26. T. E. Mansour, *J. Biol. Chem.*, **240**, 2165 (1965).

27. T. E. Mansour and C. E. Ahlfors, *J. Biol. Chem.*, **243**, 2523 (1968).

28. C. Frieden, in *Regulation of Enzyme Activity and Allosteric Interactions* (Eds.: E. Kvamme and A. Pihl), Academic Press, New York, 1967, p. 59.

29. S. Liebe, G. Kopperschläger, W. Diezel, K. Nissler, J. Wolff, and E. Hofmann, FEBS *Letters*, **8**, 20 (1970).

30. W. H. Danforth and E. Helmreich, *J. Biol. Chem.*, **239**, 3133 (1964).

31. T. E. Mansour and D. B. Stone, *Biochem. Pharmacol.*, **19**, 1137 (1970).

32. D. B. Stone and T. E. Mansour, *Mol. Pharmacol.*, **3**, 161 (1967).

33. H. A. Lardy and R. E. Parks, Jr., in *Enzymes: Units of Biological Structure and Function*, Academic Press, New York, 1956, p. 584.

34. J. V. Passonneau and O. H. Lowry, *Biochem. Biophys. Res. Commun.*, **7**, 10 (1962).

35. T. E. Mansour, *J. Biol. Chem.*, **238**, 2285 (1963).

36. O. H. Lowry and J. V. Passonneau, *Arch. Exptl. Pathol. Pharmakol.*, **248**, 185 (1964).

37. J. Krzanowski and F. M. Matschinsky, *Biochem. Biophys. Res. Commun.*, **34**, 816 (1969).

38. A. V. Hill, *J. Physiol.* (London), **40**, 4P (1910).

39. J. Monod, J. Wyman, and J.-P. Changeux, *J. Mol. Biol.*, **12**, 88 (1965).

40. C. E. Ahlfors and T. E. Mansour, *J. Biol. Chem.*, **244**, 1247 (1969).

41. B. Setlow and T. E. Mansour, *J. Biol. Chem.*, **245**, 5524 (1970).

42. H. C. Froede, G. Geraci, and T. E. Mansour, *J. Biol. Chem.*, **243**, 6021 (1968).

43. P. B. Forest and R. G. Kemp, *Biochem. Biophys. Res. Commun.*, **33**, 763 (1968).

44. R. G. Kemp, *Biochemistry*, **8**, 4490 (1969).

45. R. G. Kemp, *Biochemistry*, **8**, 3162 (1969).

46. M. L. Salas, J. Salas, and J. Sols, *Biochem. Biophys. Res. Commun.*, **31**, 461 (1968).

47. R. Freyer, S. Liebe, G. Kopperschläger, and E. Hofmann, *European J. Biochem.*, **17**, 386 (1970).

48. E. Viñuela, M. L. Salas, and A. Sols, *Biochem. Biophys. Res. Commun.*, **15**, 243 (1964).

49. W. Atzpodian, J. M. Gancedo, V. Hogmaier, and H. Holzer, *European J. Biochem.*, **12**, 6 (1970).

50. E.-G. Afting, D. Ruppert, V. Hagmaier, and H. Holzer, *Arch. Biochem. Biophys.* in press. **143**, 587 (1971).

51. L. A. Cohn, *Ann. Rev. Biochem.*, **37**, 695 (1968).

52. R. G. Kemp and E. G. Krebs, *Biochemistry*, **6**, 423 (1967).

53. M. Y. Lorenson and T. E. Mansour, *J. Biol. Chem.*, **244**, 6420 (1969).

54. B. Setlow and T. E. Mansour, manuscript in preparation.

55. A. S. Otani and T. E. Mansour, unpublished observation.

56. J. V. Passonneau and O. H. Lowry, *Biochem. Biophys. Res. Commun.*, **13**, 372 (1963).

57. E. C. Hulme and K. F. Tipton, FEBS *Letters*, **12**, 197 (1971).

58. J. C. Gerhart and H. K. Schachman, *Biochemistry*, **4**, 1054 (1965).

59. J. F. Taylor, E. Antonini, M. Brunori, and J. Wyman, *J. Biol. Chem.*, **241**, 241 (1966).

60. R. G. Kemp and P. B. Forest, *Biochemistry*, **7**, 2596 (1968).

61. K. Uyeda, *Biochemistry*, **8**, 2366 (1969).

62. B. Setlow and T. E. Mansour, *Biochim. Biophys. Acta*, **258**, 789, (1972).

63. O. H. Lowry and J. V. Passonneau, *J. Biol. Chem.*, **241**, 2268 (1966).

64. W. W. Cleland, *Biochim. Biophys. Acta*, **67**, 104 (1963).

65. A. Sols and M. L. Salas, in *Methods in Enzymology*, Vol. IX (Eds.: N. O. Kaplan and W. A. Woods), Academic Press, New York, 1966, p. 440.

66. E. Viñuela, M. L. Salas, and A. Sols, *Biochem. Biophys. Res. Commun.*, **12**, 140 (1963).

67. K. Uyeda, *J. Biol. Chem.*, **245**, 2268 (1970).

68. R. B. Layzer, L. P. Rowland and W. J. Bank, *J. Biol. Chem.*, **244**, 3823 (1969).

69. M. Y. Lorenson and T. E. Mansour, *J. Biol. Chem.*, **243**, 4677 (1968).

70. P. Baumann and B. E. Wright, *Biochemistry*, **7**, 3653 (1968).

71. E. C. Hulme and K. F. Tipton, *Biochem. J.*, **122**, 181 (1971).

72. D. M. Regen, W. W. Davis, H. E. Morgan, and C. R. Park, *J. Biol. Chem.*, **239**, 43 (1964).

73. E. Helmreich, W. H. Danforth, S. Karpatkin, and C. F. Cori, in *Control of Energy Metabolism* (Eds.: B. Chance, R. W. Eastabrook, and J. R. Williamson), Academic Press, London, 1965, p. 299.

74. J. R. Williamson, *Mol. Pharmacol.*, **2**, 206 (1966).

75. H. A. Krebs, *Proc. Roy. Soc. B*, **159**, 545 (1964).

76. D. E. Atkinson, *Biochemistry*, **7**, 4030 (1968).

77. L. C. Shen, L. Fall, G. M. Walton, and D. E. Atkinson, *Biochemistry*, **7,** 4041 (1968).

78. P. J. Randle, T. M. Denton, and P. J. England, in *Metabolic Roles of Citrate* (Ed.:
     T. W. Goodwin), Academic Press, London, 1968, p. 87.

79. E. A. Newsholme and P. J. Randle, *Biochem. J.*, **80,** 655 (1961).

80. E. A. Newsholme and P. J. Randle, *Biochem. J.*, **93,** 641 (1964).

81. J. R. Williamson, E. A. Jones, and G. F. Azzone, *Biochem. Biophys. Res. Commun.*,
     17, 696 (1964).

82. B. Hess and A. Boitteux, *Ann. Rev. Biochem.* **40,** 237 (1971).

83. B. Chance, W. Holmes, J. Higgins, and C. F. Connelly, *Nature*, **182,** 190 (1958).

84. A. Ghash, and B. Chance, *Biochem. Biophys. Res. Commun.*, **16,** 174 (1964).

85. B. Hess, A. Boiteux, and J. Kruger, *Advan. Enzyme Regulation*, **7,** 149 (1969).

86. P. Özand and H. T. Narahara, *J. Biol. Chem.*, **239,** 3146 (1964).

87. E. Helmreich and C. F. Cori, *Pharmacol. Rev.*, **18,** 189 (1966).

88. S. Pontremoli, S. Traniello, M. Enser, S. Shapiro, and B. L. Horecker, *Proc. Natl.
     Acad. Sci. U.S.*, **58,** 286 (1967).

89. G. J. S. Rao, S. M. Rosen, and O. M. Rosen, *Biochemistry*, **8,** 4904 (1969).

90. E. A. Newsholme and B. Crabtree, FEBS *Letters*, **7,** 195 (1970).

91. E. A. Newsholme, in *Essays in Cell Metabolism* (Eds.: W. Bartley, H. L. Kornberg,
     and J. R. Quayl), John Wiley-Interscience, 1970, p. 189.

92. D. B. Martin and P. R. Vagelos, *J. Biol. Chem.*, **237,** 1787 (1962).

93. P. R. Vagelos, A. W. Alberts, and D. B. Martin, *J. Biol. Chem.*, **238,** 533 (1963).

CHAPTER 4

Primary Regulatory Enzymes and Related Proteins

ALBERTO SOLS* AND CARLOS GANCEDO*

*Instituto de Enzimologia, Centro de Investigaciones Biológicas,
Consejo Superior de Investigaciones Científicas, Madrid, Spain*

* Present address: Instituto de Enzimología del Consejo Superior de Investigaciones
Científicas, Departamento de Bioquímica, Facultad de Medicina de la Universidad
Autónoma, Madrid 20, Spain.

1.   INTRODUCTION

In recent years considerable evidence has accumulated that metabolic regulation involves enzymes that catalyze key reactions in metabolic pathways, particularly at the major metabolic crossroads. These "pacemaker" enzymes can control metabolic pathways because they are susceptible of being regulated in amount, activity, or both. Knowledge of these enzymes progressed very rapidly during the sixties, both in extension and in depth, and a rich variety of regulatory mechanisms is already known (2, 30, 58).

A specially sophisticated mechanism of regulation involves the action of enzymes whose physiological role is the regulation of other enzymes, without being themselves directly implicated in metabolic pathways. In this chapter we will consider some problems peculiar to these enzymes that do not appear for the generality of enzymes, or that tend to be overlooked in the usual treatments of mechanisms for the regulation of enzyme activity. Particular attention will be devoted to the kinetic problems that arise in systems involving specific macromolecular substrates, and the relative pool sizes of these enzymes and their substrates and cosubstrates.

A general discussion of the *convertible* enzyme systems, involving pairs of antagonistic *converter* enzymes, also includes a proposal for coherent nomenclature in this field. A brief reference to enzymes involved in irreversible changes of other enzymes is made, irrespective of their eventual involvement in regulation. Specific ligand proteins that modify the activity of certain enzymes are discussed as a group that shares with the above enzymes the problems of specific protein-protein recognition.

As a special group of primary regulatory enzymes that do not act on specific macromolecular substrates, the enzymes immediately involved in the formation or destruction of metabolic messengers are also briefly considered here.

Overall metabolic integration at the cellular level, where, as once pointed out by C. N. Hinshelwood, "everything depends on everything else," requires that the primary regulatory enzymes be themselves regulated. Allosteric mechanisms acting as chemical transducers offer much greater possibilities than controls through the availability of the corresponding micromolecular substrates or products of these enzymes. The fact that in the case of enzymes acting on specific proteins allosteric effects could involve conformational changes, either on the acting enzyme or on the macromolecular substrate, will be emphasized. Finally, it will be shown that efficient physiological control of interconvertible enzyme systems is likely to depend on control of the balance between the activities of the two members of each pair of antagonistic converter enzymes.

2. ENZYMES ACTING ON SPECIFIC PROTEINS; MODIFIER PROTEINS

A. Convertible Enzymes Involving Antagonistic Converter Enzymes

Regulation of enzyme synthesis and/or activity are the commonest mechanism for control of cellular metabolism. A more elaborate mechanism of regulation involves the existence of certain metabolic enzymes in two interconvertible forms with different activities (30).

Since the metabolic and regulatory properties of these systems are considered elsewhere in this volume (31), we will mention briefly here the systems in which conversion is well understood and consider some problems posed by the nomenclature used in referring to these systems.

Nomenclature: from Chaos to Order?

The rapidly increasing number of enzymes found to be able to alternate between two different forms suggests that there is a real need for a generally accepted standard nomenclature for this class of enzyme systems. Such a nomenclature would avoid the proliferation of a bewildering jargon by which the various laboratories working in the field designate the same process in different ways, or, even worse, apply the same designation to different processes.

Enzymes that can occur in two different forms, or "dimorphic" enzymes (56), could be referred to as *convertible enzymes. Dimorphic* seems to be a more static term than *convertible*, which forcefully conveys the idea of a dynamic entity "capable of being changed in form, condition, or qualities."*
What is critical is to clearly distinguish the conversion process, which results in the appearance of a different form by a chemical modification (30) of the initial enzyme, from mere conformational changes, microheterogeneity due

* In our definition we have tried to retain from the dictionary definition of *convert* the meaning of "change in character or function" (*The Shorter Oxford Dictionary*, 1967).

to autolytic processes, or genetically determined isoenzymes (with the eventual contribution of hybrid multimers).

For the designation of the two forms of convertible enzymes we suggest to follow the nomenclature of the phosphorylase system (23), that is, to use *a* to designate the physiologically more active form and *b* for the physiologically less active (or inactive) form. This criterion would apply independently of the chemical modification involved in the conversion, and of the chronological order of discovery. In three out of the five well-characterized convertible enzymes listed in Table 1 the more active form (*a*) is the chemically unmodified form. In our opinion, designations like I and D (for "independent" and "dependent" glycogen synthetase) or I and II (glutamine synthetase) should be abandoned. Proliferation of such particular terms will only superimpose confusion on complexity.

The *a* and *b* nomenclature presents another advantage when dealing with multimeric enzymes, a general case in convertible enzymes. Since in these enzymes hybrids are likely to occur, that is, enzymes having some converted subunits and some nonconverted ones, the hybrids can easily be designated by adding to the *a* or *b* a subindex indicating the number of such subunits in the multimer. Thus, in the case of glutamine synthetase, an enzyme with 12 subunits, there could be a_6b_6, a_3b_9, etc., depending on the number of adenylylated (*b* in this case)* subunits. However, this shorthand does not prejudge that the activity of the enzyme goes parallel to conversion; in other words, a functionally *b* tetrameric convertible enzyme could conceivably have a subunit composition of even a_3b_1.

The conversions mentioned above are carried out by specific enzymes, for which the term *converter enzymes* seems suitable and convenient These enzymes have as the only known (and presumable) physiological role the conversion of other enzymes, without acting directly in a metabolic pathway in the conventional sense. They are, therefore, a well-defined class of *primary regulatory enzymes*.

Each set of convertible and converter enzyme(s) is designated as a *convertible enzyme system*, with eventual specification of an *interconvertible enzyme system* whenever it is particularly relevant to stress the reversibility of the conversion, as far as the convertible enzyme is concerned.

The foregoing proposal for designating the forms of convertible enzymes is based on a physiological criterion (with historical background), although other alternatives—for instance, a nomenclature based on the fact of a chemical modification (and even its nature)—deserve consideration. What is really important is agreement on a given, coherent system.

---

* That the number of geometrically possible varieties is as high as 382 (58) seems to be a mathematical curiosity without significant functional counterpart.

Table 1. Designations of the Convertible Forms of Some Well-Defined Interconvertible Enzymes

Enzyme	Chemical Modification	Physiological Effect[a]	Nomenclature Currently Used		Proposed	
Glycogen phosphorylase	Phosphorylation	Inactive → active	*b*	*a*	*b*	*a*
Phosphorylase *b* kinase	Phosphorylation	Inactive → active			*b*	*a*
Glycogen synthetase	Phosphorylation	Active → inactive	{ I, *a*	*D*, *b*	*a*	*b*
Pyruvate dehydrogenase	Phosphorylation	Active → inactive	Dephospho form	Phosphorylated	*a*	*b*
Glutamine synthetase	Adenylylation	Active → inactive	{ *a*, I	*b*, II	*a*	*b*

[a] "Inactive" is used in the sense of less active in physiological conditions.

Well-Characterized Interconvertible Systems

The first enzyme known to appear in two different forms with different catalytic activities was glycogen phosphorylase (11, 23). At present the convertible system of this enzyme is one of the most thoroughly studied and best understood (Scheme I). At least one of the converter enzymes is a substrate for another converter enzyme, that is, it is also a convertible enzyme. Glycogen synthetase has also been shown to exist in two interconvertible forms (61), although as yet the system is not as well understood as that of glycogen phosphorylase.

The conversions of the enzymes mentioned above involve phosphorylation-dephosphorylation as a chemical process. The phosphoryl group is esterified with a particular serine residue of the protein. It is noteworthy that potato phosphorylase, which is not acted upon by rabbit phosphorylase kinase (37), does not contain this residue. In the recently elucidated conversion of pyruvate dehydrogenase (49, 64) phosphorylation-dephosphorylation also takes place, and it is precisely a serine residue that is phosphorylated (49). In this case a geometrical arrangement of the converter enzyme on the surface of its convertible substrate in the multienzyme complex has been suggested. Although this idea has not been clearly documented, it could be a solution to the problem posed when a large macromolecular substrate must be handled by an enzyme (see Section 4.A). Moreover, the possibility may exist that a castagnette or swinging arm mechanism is active; thus a molecule of a converter enzyme forming part of a multienzyme complex could be able to act on more than one subunit.

Glutamine synthetase from *Escherichia coli* exists also in two interconvertible forms (30). In this case the conversion involves adenylylation-deadenylylation. It has been shown (54) that the 5′-adenylyl group is bound in phosphodiester linkage to the hydroxyl group of a particular tyrosyl residue on each of the twelve subunits of glutamine synthetase. Subsequently, it was found (42) that the adenylyl-O-tyrosine bond is an energy-rich bond. To our knowledge no data are available on the energy content of the phosphoryl-serine bond in the enzymes mentioned before.

Recent evidence suggests that bacterial RNA polymerase can be controlled by adenylation. Chelala et al. (10) reported the existence of two interconvertible forms for the *E. coli* enzyme; one, adenylylated, was less active, and the other, deadenylylated, more active. The converter enzymes have not been characterized in this bacterium, but they have been studied in *Bacillus subtilis*. It has been found that in this case the adenylylating enzyme consists of a single chain of molecular weight 90,000, and the deadenylylating enzyme is a tetrameric protein of molecular weight 320,000. The subunit of this RNA polymerase that seems to be modified is ω, with a molecular weight of 12,000 (3).

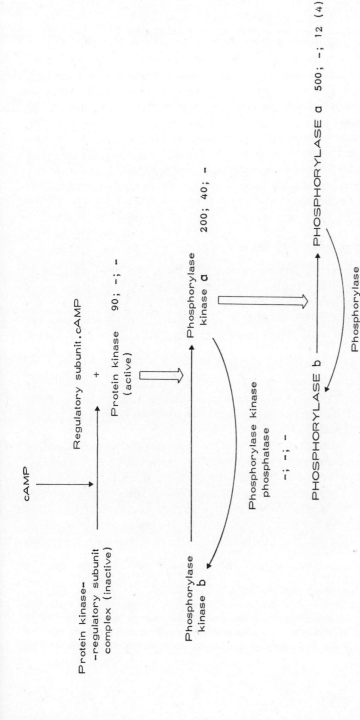

Scheme I. Convertible enzyme system of glycogen phosphorylase in muscle. Next to the more active form of each enzyme the following parameters are given: V_{max} (moles/min/40,000 daltons protein); molarity (μM); K_m ($\times 10^6 M$); molarity (μM/liter), followed by number of subunits in brackets. A dash in the appropriate place indicates lack of information. References to work from which the values shown were taken or calculated are, in the order listed above, protein kinase (62), phosphorylase kinase a (34, 15), phosphorylase a (12, 27, 19), and phosphorylase phosphatase (32).

B. Enzymes Involved in Irreversible Changes of Other Enzymes

Enzymically catalyzed irreversible changes of biologically active proteins constitute a broad field involving certain types of enzymes that share to various degrees some of the problems peculiar to the primary regulatory converter enzymes. In a way some of them can serve as models and may be precursors of converter enzymes in fully developed interconvertible enzyme systems. For these reasons it seems useful to consider briefly these processes here, even if most of the enzymes involved are not specifically regulatory and may not even be regulable at all, their activity *in vivo* being dependent mainly on substrate availability.

Completion after Translation

A well-defined group of enzymes whose completion, after the formation of the specific polypeptide chain(s), is dependent on another enzyme, consists of those requiring covalent attachment of a prosthetic group. This is the general case in the biotin-dependent carboxylases, the lipoate-involving dehydrogenases, and some others, including the heme in certain cytochromes. Very little is known of the specificity and kinetic properties of these "holo-enzyme synthetases" and related enzymes.

A second group of biologically active proteins that require certain covalent modification of the product of translation is that of the proteins initially synthesized as inactive precursors. The first known and probably the common-est type of enzyme-catalyzed irreversible transformation of a precursor protein into a biologically active protein (or peptide) involves cleavage(s) of the polypeptide chain. Another widespread phenomenon of covalent modification of proteins producing a stable change in some properties is the addition of glycosyl residues. These reactions are catalyzed by a variety of glycoprotein-glycosyl transferases, some of which seem to be regulated. A variety of other enzymically catalyzed covalent changes can also be involved. Among these possibilities, phosphorylation is in fact a key factor in several interconvertible enzyme systems.*

---

* A most peculiar type of completion involves the recently identified group of aminoacyl-tRNA protein transferases. These enzymes, which add an aminoacyl residue through peptide linkage to a polypeptide chain after translation, seem to be relatively specific for certain types of protein and could be concerned with the regulation of the function of some proteins they are particularly able to modify [R. L. Soffer, *J. Biol. Chem.*, **246**, 1602 (1971)].

Selective Degradation of Biologically Active Proteins

Inactivation of biologically active proteins is a very general process that has, in certain cases, a considerable degree of specificity. When *the key step* is enzymically catalyzed, the specificity of the inactivation process can depend on the potentially susceptible substrate as much as or even more than on the degradative enzyme. For this reason autolytic enzymes of low substrate specificity, such as most of the lysosomal enzymes, can be geared for relatively specific results. Thus hormonal induction of cellular autophagy through increased lysosomal fragility (16) can have relatively specific effects, like the markedly preferential inactivation of the glucokinase and pyruvate kinase in liver (46). The physiological significance of this preferential partial break-down of certain enzymes can have an important experimental counterpart in the covalent modification of certain enzymes during the process of extraction and purification, particularly when conditions favoring the activity of autolytic enzymes are provided. A dramatic illustration has been the recent identification of the "alkaline" fructose diphosphatase from liver composed of two chains of different size, as an autolytic artifact whose avoidance permits the isolation of a "neutral" enzyme with a single chain size (48).

In the case of inactivation of enzymes, saturation with their substrates (or cofactors) can be a major factor in their susceptibility to inactivation, either by conferring resistance to attack by some other enzyme or by labiliza-tion, as pointed out by Grisolia (24).

These possibilities are particularly important in relation to the control of the levels of certain enzymes in most cells of higher organisms. Here the level of intracellular enzymes provides a balance between synthesis and breakdown, as emphasized by Schimke (53), and from a functional point of view the key event in the "breakdown" process is any change that irreversibly inactivates the enzyme involved.

C. Specific Ligand Proteins That Modify the Activity of Certain Enzymes

In 1965 the identification by Gerhardt and Schachman (20) of two kinds of subunits in aspartyltranscarbamylase, one with catalytic activity and another apparently with only a regulatory function, acting as mediator between the allosteric effector CTP and the catalytic subunit, opened a new dimension in metabolic regulation at the molecular level. This has become the first well-characterized example of what by now can be considered as a very important family of specifically regulatory proteins that, acting as ligands, reversibly modify quantitatively or even qualitatively the activity of a variety of enzymes. They can do so either as chemical signals by themselves or as mediators of other chemical signals.*

Probably, most of these ligand regulatory proteins are made by the cell, not for any catalytic activity of their own, but solely for the control of the catalytic activity of certain enzymes. Nevertheless, it will be shown below that potential or even actual catalytic activity does not rule out the possibility that a protein can act as a regulatory ligand for another enzyme.

Gerhardt and Schachman (20) suggested that the intervention of regulatory subunits opens a way for the evolution of regulatory enzymes not constrained by the necessity of formation of both active and regulatory sites from the same polypeptide chain. Instead, the more critical mutations in such cases would be those which allowed the association and physiologically useful interactions between two proteins.

Quantitative Changes

The regulatory subunit of aspartyltranscarbamylase is the key to the quantitative modulation of this enzyme (pacemaker of the pyrimidine biosynthetic pathway), acting as mediator between an allosteric effector and the catalytic subunits in the multimeric native enzyme complex. Another type of quantitative regulation through a ligand protein that is inhibitory unless neutralized by an allosteric effector has recently been found in the field of cAMP protein kinases (50, 60). The cAMP-dependent protein kinase that converts inactive phosphorylase kinase into a phosphorylated active form (as well as glycogen synthetase a into the b form) is composed of two subunits, one catalytically active and the other regulatory and inhibiting the former. It is the second subunit that has high affinity for cAMP ($K_d =$ 0.1 μM). The regulatory subunit-cAMP complex dissociates from the catalytic subunit and activates it by removal of the inhibitor.

The contractile system of skeletal muscle, at the molecular level, is an architecturally elaborate complex built up of at least five different kinds of subunits, which include an enzyme (myosin ATPase), structural proteins (with F-actin as major component), and regulatory proteins (the troponin-tropomyosin β complex, or "relaxing protein," and the β-actinin). Troponin

---

* The discovery of two kinds of subunits in aspartyltranscarbamylase illustrates the importance of timing in the proper recognition of the significance of an experimental result. A decade earlier, the resolution of an apparently homogeneous preparation of the enzyme into approximately equal amounts of two kinds of proteins, with the activity restricted to one of them, would have been "obviously" interpreted in a very different way: it would have been concluded that the supposedly homogeneous preparation was in fact grossly contaminated, so that the really "pure" enzyme had about double the molecular activity formerly reported. Now, in the era of enzymatic regulation, such a really "pure" enzyme, although a product of gene expression, is physiologically an artifact lacking the regulatory character (or one of the regulatory properties) of the native enzyme.

itself has recently been resolved in up to four fractions, one of which binds Ca^{2+} but not tropomyosin (17). The complex system of regulatory proteins mediates the triggering effect of Ca^{2+} on the ATPase activity of the contractile actomyosin system. Moreover, preliminary observations suggest that one of the components of troponin is a convertible protein susceptible of being phosphorylated by a cAMP-dependent kinase to a form with greatly increased affinity for Ca^{2+} ions (4).*

The processes of gene expression probably involve a variety of specifically regulatory proteins. The key repressors, initially postulated entities of unknown nature, at the beginning of the process leading to the synthesis of inducible proteins have been identified as proteins. Isolation of the lac repressor from *E. coli* allowed its characterization as an allosteric protein with an exceedingly high affinity for the lac operator region ($K_d = ca.$ 10^{-13} M) which is greatly decreased by the binding of its specific allosteric effector, the inducer (7). The regulatory subunits involved in modulating the activity of RNA polymerases are mentioned below. Similar and perhaps other types of specifically regulatory proteins are likely to be part of the now-unfolding variety of ribosomal proteins (about 60) and of the initiation, elongation, and release "factors" (about 7, at least), as well as the peptide deformylase and equivalent enzymes involved in physiological peptide chain synthesis (39). Information on the precise role and physiological significance of these proteins that appear to be specifically involved in protein synthesis not only is very sparse but in addition comes mainly from preparations obtained from prokaryotic organisms, where already there are indications that one of the initiation factors (F_2) could be a convertible protein and participate in the regulation of translation (41). The development of this area with preparations from eukaryotic organisms will probably open the way to the identification at the molecular level of the mechanisms controlling the translation step, which, increased evidence indicates, is a major level of control in higher organisms.

An extreme case of positive change through the association of two proteins is that of ribonucleoside diphosphate reductase. The enzyme from *E. coli* consists of two nonidentical subunits without known catalytic activity. In the presence of Mg^{2+} these two proteins associate to form the active enzyme (8). There is not enough information to tell whether the apparently none-to-all change is induced by one of the proteins acting as activator or specifier ligand (presumably of the subunit that contains iron) or whether the two proteins contribute in an intrinsic way to the catalytic complex.

* Available information on the cilia and flagella of eukaryotic cells indicates that these organella also have contractile systems consisting of complexes of several kinds of subunits, which include an enzyme (ATPase) and allow for the possibility of both specifically structural and regulatory proteins.

One peculiar type of protein-protein interaction serving as a safety valve to prevent inconvenient activity of the primary product of translation is the mechanism recently described as "flip-flop" for the alkaline phosphatase of *E. coli* (36). This enzyme is composed of two identical subunits and has two active sites, but they cannot act independently of each other. In this way, cytoplasmic subunits of this digestive enzyme with a wide specificity are not active. Only upon a Zn^{2+}-dependent dimerization that takes place in the periplasm region does the enzyme becomes active, being then able to fulfill its role of digestive enzyme acting on a variety of phosphoric esters that could be available in the medium.*

The tryptophan synthetase in *E. coli* offers a curious example of how two enzymes can associate to a complex in which the two catalytic potentialities of both constituent proteins are markedly enhanced, and, when the two physiological substrates are available, a combined reaction takes place. The reactions catalyzed by the individual chains, alone or associated, are as follows:

A chain (or α subunit):
$$\text{Indoleglycerol-P} \rightarrow \text{indole} + \text{glyceraldehyde-3-P}$$
B chain (or β subunit):
$$\text{Indole} + \text{serine} \rightarrow \text{tryptophan} + H_2O$$
Tryptophan synthetase complex ($\alpha_2\beta_2$):
$$\text{Indoleglycerol-P} + \text{serine} \rightarrow \text{tryptophan} + \text{glyceraldehyde-3-P}$$

The evolutionary "freezing" of this useful association of two polypeptide chains by fusion of the two corresponding cistrons to give a longer polypeptide chain with the properties of the above complex, well characterized in *Neurospora crassa*, has been discussed by Smith (55).

Specific association between two enzymes can have a "negative" effect. Associations of this type, leading to marked decrease or even complete inhibition of one of a pair of *physiologically* antagonistic enzymes, can have a value for metabolic regulation that could justify its evolutionary selection. The first well-authenticated instance of this inhibition of one enzyme by another through specific association seems to be the rapid disappearance of ornithine transcarbamylase activity upon induction of the arginine catabolic pathway in yeast, discovered by Wiame (63) and characterized as involving a

---

* Protein-protein association is supposed to be essential for the catalytic activity of many multimeric enzymes composed of identical subunits. This conclusion is based on the observation that dissociation is accompanied by loss of activity. On first appearance these cases are similar to the dimerization of the alkaline phosphatase of *E. coli* described in the text. Nevertheless, the fact that in most instances substantial dissociation of multimeric proteins requires drastic environmental conditions makes it unlikely that activity could have been retained by the monomers.

complex of arginase and the ornithine transcarbamylase, in which the latter is inactive whereas arginase remains active. The fact that certain concentrations of small molecules that are specific ligands of the two enzymes (arginine and ornithine, respectively) are required for effective function of the inhibitory protein-protein complex is irrelevant in the broader context of this discussion. Before the characterization of the inhibitory protein as arginase, Wiame designated it as "epiprotein." However, the possibility of generalizing a term like *epiproteins* or *epienzymes* to specify protein-protein interactions is limited by the large variety of possibilities for interactions described in this chapter.

A similar case in which an enzyme inhibits another by complexing with it, with obvious regulatory value, involves the inhibition by phosphorylase a of liver glycogen synthetase phosphatase, demonstrated by Hers and his coworkers (28).

Qualitative Changes

The possibility that a ligand protein can determine an important qualitative change in the catalytic activity of an enzyme offers a more sophisticated way to exploit protein-protein interactions. A few cases already identified to various degrees illustrate different possibilities within this general mechanism.

α-Lactalbumin, probably a product of the evolution of a mammary gland lysozyme, binds specifically to a molecule of mammary gland galactosyltransferase and changes the specificity of the latter by modifying the affinity of the protein complex for acceptor sugars. The combination of a modifier protein unique to the mammary gland (the α-lactalbumin) and a functional enzyme (the transferase) results in the formation of a virtually new enzyme, the lactose synthetase.

Another example of qualitative change in catalytic activity is the interesting observation of Anderson and Stadtman (1) indicating that the deadenylylating system of glutamine synthetase b is composed of the adenylylating enzyme complexed with another protein. In this case the catalytic activity would be shifted from adenylylating to deadenylylating (involving phosphorolysis to ADP) by a protein component without known catalytic activity.

But perhaps the most impressive instance of qualitative change is that of RNA polymerase. The specific recognition of promoter sites on DNA needed for meaningful chain initiation requires the association to the "core enzyme" of a protein called σ (9). This specifier protein determines the specificity of the core enzyme; thus there are cases of phage infection in which genes not transcribed by the host polymerase system can be transcribed when new phase-specific σ factors are synthesized (66). Termination of RNA synthesis at adequate sites of the DNA template is also affected by another protein, which has been designated as the ρ factor (51).

3. ENZYMES IMMEDIATELY INVOLVED IN THE FORMATION OR DESTRUCTION OF METABOLIC MESSENGERS

The outstanding importance of the discovery and identification of 3′,5′-cyclic AMP (cAMP) as an intracellular (although occasionally also intercellular) metabolic messenger has become evident from the work of E. W. Sutherland, initiated a decade ago and vigorously pursued now in many laboratories in relation to an astonishing variety of regulatory problems [see Robinson et al. (52)]. We can define cAMP as the first well-authenticated case of a small molecule made by most types of cells solely for purposes of metabolic regulation; cAMP seems able to affect allosterically a variety of enzymes or otherwise biologically active proteins, as a *professional* messenger, instead of the variety of ordinary metabolites *secondarily* exploited as chemical signals. This contrast, as well as the fact that there are two types of specific enzymes, one for the formation and one for the removal of cAMP, is illustrated in Scheme II. Although ATP and AMP allosterically affect many enzymes, this regulatory function is secondary to their primary role as substrates or products for a variety of metabolic reactions. In contrast, cAMP seems to be neither a metabolic intermediate nor a building block, but solely a regulatory messenger.

Adenyl cyclase is a widespread enzyme which in a variety of tissues has been found in the cytoplasmic membrane, where it is responsive to certain hormones. By now, no less than about twelve tissues have been found in which the membrane-bound adenylcyclases respond to different hormones. Although the possibility of many regulatory isoenzymes (each with a binding site for a particular hormone) cannot be ruled out, it seems more likely that regulatory ligand proteins in the membranes will be involved as mediators between the hormonal signal in the outer surface of the membrane and the activity of the cyclase in the inner surface. Models of various degrees of complexity (6, 52) have been proposed, some including discriminator, transducer, and amplifier components built in the membrane.

There are a number of indications that cyclic GMP is likely to be also a regulatory messenger. A guanylcyclase, markedly different from the adenyl-cyclase, has been identified in a variety of tissues. It appears in the soluble, instead of in the particulate, fraction of tissue homogenates and seems to be hormone-insensitive (26). The latter results suggest that this presumably regulatory enzyme would in turn be regulated by some intracellular metabolite, rather than by an extracellular metabolic messenger as happens with the adenylcyclase.

Elimination of these cyclic nucleotides is usually accomplished by highly specific phosphodiesterases. High-affinity enzymes (K_m values *ca.* 5 μM)

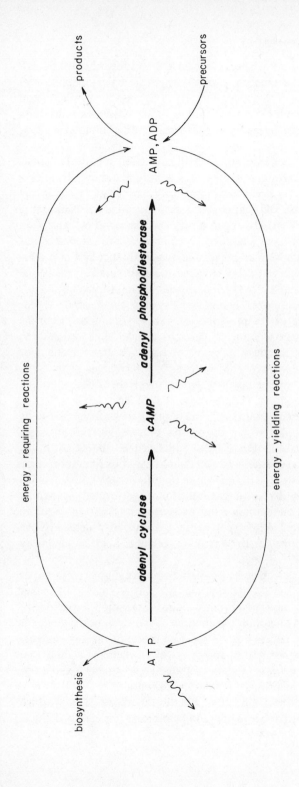

Scheme II

have been identified in brain, one specific for cAMP and the other for cGMP (22). There is no consistent evidence regarding the physiological regulation of these enzymes, which constitute the second part of the specific synthesis and degradation pair of enzymes for these cellular metabolic messengers, although they can be easily inhibited by caffeine and certain other biological compounds.

Several specific binding proteins for cAMP have been identified. Those related to protein kinase in animal tissues were mentioned above. In microorganisms, where cAMP has been recently identified as a key factor in the regulation of the synthesis of adaptive enzymes sensitive to "catabolite repression" (or "glucose effect") acting at a very early step in the translation process, Zubay et al. (67) have identified (in *E. coli*) a protein that binds cAMP with a K_d in the micromolar range, and whose absence in a mutant is accompanied by lack of ability of cAMP to act as derepressor. This microbial cAMP-binding protein could well play a specifically regulatory role.

Transmission at synapses frequently involves a specific chemical messenger released from the endings of the presynaptic neuron. Cholinergic neurons synthesize and liberate acetylcholine. Liberated acetylcholine crosses the synaptic cleft and acts on the membrane of the postsynaptic neurone. For efficient function the messenger should be rapidly removed as soon as its role has been fulfilled. Two specific enzymes are involved in acetylcholine-mediated transmission, choline acetylase and a specific acetylcholinesterase. At least the latter may be considered a primary regulatory enzyme since its activity is specifically coupled to the regulation of synaptic transmission. Timely activity of acetylcholinesterase is far more critical than that of the acetylase, since between the synthesis of acetylcholine and its liberation there is an intermediate stage of storage in vesicles, so that the onset of this chemical mediation depends immediately on the release of preformed messenger from the storage vesicles rather than from its synthesis. The occurrence in acetylcholinesterase of two kinds of polypeptide chains, of which only one has an active site (40), allows for the possibility of some kind of regulatory subunit in this enzyme.

Norepinephrine is a specific chemical signal of postganglionic sympathetic or "adrenergic" neurons in the autonomic nervous system. Still other chemical mediators in the nervous system include a transition from specific "professional" messengers to certain metabolites used as chemical signals for specific systems. For these reasons, some (or even most) of the key enzymes specifically involved in the formation and removal of chemical signals that act as mediators in the nervous system, although very likely regulatory, should not be considered as primary regulatory enzymes.

In the field of hormone liberation and elimination there are cases in which primary regulatory enzymes or proteins could be more or less clearly defined.

It would be too complex, however, to attempt to go into details here, although the subject deserves careful consideration.

In summary, enzymes specifically involved in the synthesis and degradation of cAMP—and, in general, enzymes involved in the formation (in the wide sense of making available for action, whether by synthesis or release) or destruction (by degradation or any other kind of transformation which suppresses the specific value as messenger) of metabolic messengers—can be considered to constitute a particular class of primary regulatory enzymes.

4. KINETIC PROBLEMS AND MOLAR RELATIONSHIPS

Usual kinetic treatments assume that the substrates of an enzymatic reaction are much smaller than the enzyme. New kinetic problems arise when dealing with systems in which two macromolecules are involved in the reaction, that is, in which the substrate is not much smaller (in some cases it can even be much bigger!) than the enzyme that acts on it. A complete understanding in quantitative terms of the behavior of these systems would require knowledge of definite values for several parameters, like the diffusion coefficients of macromolecules in the cell sap, the affinities between the reacting macromolecules, and the pool sizes of these macromolecules. Although these values are largely unknown, we think that even with the present state of knowledge some valid conclusions may be drawn which will clarify obscure points or bring to light areas of study to which little or no attention has been paid.

A. Diffusion of Macromolecules and Molecular Activities of Different Classes of Enzymes

In any chemical reaction, the rate of the total transformation depends on several factors:

1. Frequency of collisions between the reactants.
2. Efficacy of these collisions.
3. Rate of the chemical transformation.
4. Velocity of removal of products from the "reaction zone."

In a system at homogeneous temperature and without any external mechanical influence, factors 1 and 4 will be determined mainly by the efficiency of the diffusion process. Here we find already an obvious difference between the common enzymatic reactions in which the substrate is a micromolecule and the systems we are considering. Although the diffusibility of a molecule is related to its size, the diffusion coefficients of small molecules (or ions) and proteins do not differ very markedly in aqueous solutions [7×10^{-6} for Ca^{2+}, 3×10^{-6} for ATP^{3-} (35), and $4-11 \times 10^{-7}$ for proteins, corresponding to molecular weights in the range from 300,000 to 13,000 (2)]. However, although micromolecules seem to diffuse almost equally as well in the viscous medium of the cell as in aqueous solutions, except when strong binding is active (35), this is not true for macromolecules.* Lehman (38) has found that the diffusion coefficient of sucrose is reduced by only a third in *E. coli* cell sap, but that of a dextran fraction of molecular weight 15,000 decreases by a factor of 10. The influence of the diffusion of the products becomes of special importance when these are inhibitors of the reaction by an effect other than mass action. In addition, the existence of large hydrophobic regions in the "enzyme" and "substrate" considerably affects the diffusion of the substrate as well as that of the product. Moreover, steric and/or electrostatic interactions become important as the protein is surrounded by other macromolecules. As an example it can be mentioned that amylase action has been compared in aqueous solution and in a coacervate of starch, protamine, and gelatine; the activity was found to be lower in the coacervate than in the solution, the difference being ascribed to the higher viscosity of the coacervate (43).

We turn now to the second factor, the efficacy of collisions, which will be determined by the approach of both protein substrate and protein enzyme in an appropriate position with respect to the two corresponding recognition areas. Pollard (47) pointed out that, if only a part of a molecule is active in a process and this part must approach another active zone, collisions due only to random Brownian motion will give low reaction rates. He made the suggestion that a very temporary complex between both molecules permitting random rotational motion could be formed, thus allowing correct fitting of the two regions. In this way efficacy would be increased by two orders of magnitude. Very likely, induced fit (33) would play a significant role.

From the Smoluchowski equation

$$z = 2\pi(D_1 + D_2)N_1N_2(R_1 + R_2),$$

where D_1 and D_2 are the diffusion coefficients of the particles, N_1 and N_2 their numbers per cubic centimeter, and R_1 and R_2 their effective collision radii, one can calculate the collision frequency of two particles. Performing such a calculation for the collisions of glutamine synthetase and adenylyl-transferase, utilizing some data of the literature and reasonable assumptions where the former are lacking, we obtain an estimate of 10^{18} collisions/cm^3/sec. An estimate of the probability of two similar subunits coming into

---

* An entertaining analogy is a comparison of the movement of an elephant and an ant in a savanna with the progression of the same animals in a jungle. The "diffusion" of the ant will be similar in both terrains, but that of the elephant will be severely impaired by big trees in the jungle.

contact at the specific sites in a random collision gives a value of 0.03%.* When this value is introduced into the collision frequency calculated and the spatial arrangement of glutamine synthetase is taken into account, the number of collisions bringing active centers into contact will be of the order of 10^{14}, a figure which compares well with the rate of inactivation observed *in vivo* (44). In other words, it appears that a large proportion of the collisions bringing into contact a part of the active centers can be effective for reaction, a result that strongly supports Pollard's suggestion.

The fact that the probability of efficient collisions between complementary sites of two proteins is considerably smaller than that of such collisions between a binding site of a protein and a small molecule, plus the probability of hindrance of the diffusion of macromolecules in media with very high concentrations of macromolecules (like those prevailing in intracellular compartments), suggests that the molecular activities of enzymes acting on other enzymes (or other specific proteins) are likely to be rather small and could not be expected to reach the very high values familiar with many metabolic enzymes. A quantitative comparison of the few cases of converter enzymes for which information has been obtained with over 200 metabolic enzymes, as summarized in Table 2, has indeed fulfilled this expectation. Moreover, within the ligases class can be segregated the group of amino-acyl-tRNA synthetases which exhibit a range of "normalized" molecular activity values of the same order as the group of converter enzymes. The average values are *ca.* 80 and 100, respectively, as compared with 17,000 for metabolic enzymes in general. In this coincidence it seems more than fortuitous that in the reaction catalyzed by the aminoacyl-tRNA synthetases a specific macromolecular substrate is involved.

One important factor in controlling reaction rate will be the pH in the microenvironment formed by the two macromolecules, which will be different from the cell sap pH. Evidently the charged groups on the protein surface attract a thin film of oppositely charged ions, creating a microenvironment different in pH from the surrounding solution. If negative charges predominate on the surface of the enzyme, the microenvironment will be more acidic, and the contrary situation will prevail when the protein has positive charges. Therefore an enzyme acting in the cell sap will be subjected to a different kind of environment from that in a dilute solution.

* The assumptions made for this calculation are as follows: protein subunit of 40,000 daltons, radius (average) 40 Å; active center, taking as such the zone area determined by a secant circle (to the spherical subunit) of 4 Å radius (a middle value of the radius of a roughly spherical metabolite of 400 molecular weight). Thus the active center has an area of 28π Å^2. Now, since the total area of the subunit is 1600π Å^2, the active center represents only 1.7% of the complete surface. Suppose that this subunit collides with a micromolecular substrate; the probability of collision in which at least some point of each comes into contact is 0.8%. Suppose also that the collision takes place between two subunits of similar size. The probability now becomes $(1.7/100)^2 = 0.03\%$.

Table 2.   Molecular Activities of Different Classes of Enzymes

Enzymes Studied		"Normalized" Molecular Activities,[a] moles substrate/40,000 daltons/min	
		Average	Enzymes with Activity >500 %
Metabolic enzymes[b]			
Oxidoreductases (I)	67	18,000	85
Transferases (II)	61	13,000	82
Hydrolases (III)	35	36,000	86
Lyases (IV)	37	11,000	82
Isomerases (V)	10	8,000	90
Ligases (VI)			
Aminoacyl-tRNA synthetases	7	83	0
Others	9	3,700	89
Totals	226	17,000	81
Primary regulatory enzymes			
Phosphorylase kinase *a* (muscle)		200	
Protein kinase acting on phosphorylase kinase *b* (muscle)		90	
Glutamine synthetase adenylyltransferase (*E. coli*)		30	

[a] Since most proteins of moderate or large molecular weight are multimers frequently involving identical subunits, for many purposes it could be more useful to refer activities to subunits than to whole molecules (e.g., number of active sites per molecule of enzyme, or number of chemically modifiable sites). Since the average size of protein subunit is *ca.* 40,000 (57), a "normalized" molecular activity can be calculated on the basis of this protein weight. This value could be obtained even in cases where no information on the molecular weight of the enzyme is available, provided that the specific activity is known.

[b] Data on metabolic enzymes were taken from Barman's compilation (5), with the exclusion of six exceptionally active ones having "normalized" molecular activities greater than 250,000 (from *ca.* 400,000 to as much as 3,500,000 for a steroid Δ-isomerase, EC 5.3.3.1). When the rates of the two directions were known, the value corresponding to the faster one was used for the preparation of this table. Most of the values used for the present averages were measured at 25°. Roman numerals in brackets indicate class numbers of the Enzyme Commission.

B. Affinities Between Macromolecules

If it is assumed that other requirements of the reaction, such as metal ions and nucleotides, are present at saturating concentrations, the reaction of enzymes acting on specific proteins would be expected to be first order, at physiological concentrations of the macromolecular substrate and with dissociation constants not exceedingly small, as shown in Table 3. The physiological implications of these K_m values are discussed in Section 5 on physiological regulation. This kinetic behavior *in situ* would explain observations indicating a decrease in the affinity of the converter enzyme for the convertible substrate that parallels the extent of the conversion (21). Indeed the phenomenon observed is a decrease in the affinity for the *total* enzyme substrate; however, this behavior is easily explained when one thinks in terms of the normality of convertible subunits instead of the molarity of multimeric protein. As the reaction proceeds, the number of available subunits decreases; this means a decrease in actual substrate concentration. The effect on the reaction mimics a decrease in affinity for the whole protein, although the affinity for the corresponding subunits could remain virtually unmodified. Then, with a few possible exceptions, the activity of primary regulatory enzymes *in situ* will follow first-order kinetics. Extension of this conclusion to interconvertible enzyme systems will automatically provide a buffer between the two possible extremes: each of the two antagonistic converter enzymes will approach asymptotically a completion of its possible action that will tend to be more and more counteracted by the other.

C. Pool Sizes of Converter Enzymes and of Their Substrates and Products

A consequence of the low molecular activities of converter enzymes is that the molar ratios between converter and convertible enzymes in biological systems tend to be considerably greater than could otherwise have been expected. Indeed, from the few cases in which enough information is available it may be inferred that a value of about only 1 order of magnitude is a typical ratio between convertible and converter enzymes,* even in cases without any obvious need for very rapid metabolic changes.

---

* The molar ratios of phosphorylase *b* kinase/phosphorylase/phosphorylase *a* phosphatase seem to be approximately 10:100:4, from the data of Fischer, Krebs, and their coworkers (19, 32, 62). The ratio of the glutamine synthetase to its adenylyltransferase in *E. coli* is *ca.* 3, from the data of Ebner et al. (18) and Woolfolk et al. (65). And in the more elaborate "cascade" system known, the blood coagulation system, which involves a sequence of at least six proteins that can be sequentially activated from inactive precursors, mainly through limited, selective proteolysis (13), the relative proportion of each protein in the sequence increases by just about 1 order of magnitude.

Table 3. K_m (or K_d) Values for the Macromolecular Substrates (or Ligands) of Primary Regulatory Enzymes and Other Purely Regulatory Proteins

Regulatory Enzyme or Protein	Macromolecular Substrate or Regulated or Regulatory Ligand	K_m or K_d, μM	Reference
Protein kinase	Phosphorylase kinase b	0.1	62
	Glycogen synthetase a (I)	?	50
	Inhibitory subunit	Very low	
Phosphorylase kinase a	Phosphorylase b	40	32
Phosphorylase a phosphatase	Phosphorylase a	3	
	Glycogen synthetase b (D)	5	
Glycogen synthetase phosphatase	Inhibitory phosphorylase a	Very low	59
Glutamine synthetase adenylyltransferase	Glutamine synthetase	5	18
Aspartyltranscarbamylase	Regulatory subunit	1	20
Aminoacyl-tRNA synthetases	Corresponding tRNA	0.1–1	39
Lac repressor	Lac operator	10^{-7}	7
Myosin	Actin	No information available	
Actin	Tropomyosin B		
Tropomyosin B	Troponin		
Troponin	β-Actinin		

The interconvertible enzyme systems well characterized so far involve an adenosine nucleotide or inorganic phosphate or pyrophosphate as co-substrate or coproduct. This fact has invited repeated speculations on the probability of regulation of several of the known converter enzymes through the availability of ATP as substrate or through inhibition by the coproduct. The molarity or even the normality (57) of the convertible enzymes is so small with respect to that of the metabolites mentioned that the activity of any of the corresponding converter enzymes does not significantly alter the pools of cosubstrate or coproduct. This is particularly marked in the case of ATP, whose physiological concentrations in the millimolar range are at least 2–3 orders of magnitude greater than those of any of the enzymes (or subunits) whose conversion requires ATP. Some published observations suggestive of such controls can be accounted for by stoichiometric relationships mimicking kinetic constants in the experimental conditions used. It is also very unlikely that any of the conversions can markedly affect the inorganic phosphate or ADP pools. The same applies to the pyrophosphate pool, which, although smaller, affects very many important reactions that could not be neglected for the benefit of a single enzyme.

5. PHYSIOLOGICAL CONTROL OF PRIMARY REGULATORY ENZYMES

The ultimate question regarding the actual physiological value of the primary regulatory enzymes seems to be the classical one of "who controls the controllers" (56). In this regard allosteric systems acting as chemical transducers (45) offer the greatest range of possibilities and can be expected to be crucial factors in most cases.

Allosteric effects on systems involving two enzymes or, in general, one enzyme acting on a specific protein substrate raise the problem of ascertaining which of the two proteins is the one directly (without or with the involvement of regulatory subunits) affected by the modifier. Induction of a conformational change by the binding of a small molecule can as well affect the efficiency of a converter enzyme on a particular macromolecule as the susceptibility of the convertible enzyme (or protein) to be acted upon. The latter types of cases may be referred to as allosteric effects at the substrate level. In convertible enzyme systems, since the convertible enzymes themselves are regulatory enzymes, it is in general more likely that they are the real targets of effectors that influence the converter-convertible pair of enzymes. That is, the conformational change induced by an allosteric effector on a regulatory convertible enzyme can easily affect not only its activity but also its susceptibility as substrate of the corresponding converter enzyme. This is more economical, in evolutionary terms, than having an additional allosteric site in the converter enzyme. Enzyme conversion as a regulatory device is likely to be a relatively recent evolutionary acquisition, as compared to end-product inhibition. Therefore, it is tempting to speculate that end-product inhibition could have served as a good starting point for an end-product inactivation mechanism, followed by the selection of a converter enzyme particularly efficient in regard to the conformation of the convertible enzyme produced by the binding of the end-product.

In general, whenever the effector metabolite is known to be a specific ligand for the enzyme (or protein) acted upon by a primary regulatory enzyme, whether as allosteric modifier or as substrate or product (14, 33), it should be tentatively assumed that no extra site for this effector is required. In extreme cases, binding of any such ligand to the second protein could be a virtually essential requirement for the enzyme-catalyzed transformation, whether in absolute (V_{max}) or in relative terms (K_m versus concentration *in situ* of the potentially convertible protein). The following analyses of a few cases will illustrate these possibilities.

Hers and his coworkers (28) have reported a strong activation by glucose of the dephosphorylation of liver phosphorylase *a* by phosphorylase phosphatase. In this case, since glucose is an effector of liver phosphorylase *a*, it can be assumed in principle that the "stimulation of liver phosphorylase phosphatase" by glucose depends on allosteric modification of the protein substrate, rendering it more susceptible to the action of its specific converter phosphatase. The prior finding (29) that glucose, in the upper part of its physiological concentration range in liver, stimulates the glycogen synthetase phosphatase-glycogen synthetase system cannot yet be even tentatively ascribed to any one of the two proteins involved.

Reed (49) has reported that pyruvate dehydrogenase kinase is inhibited by pyruvate. Since the latter is a specific ligand as substrate of the convertible enzyme, it could be tentatively assumed that the observed effect of pyruvate on the pyruvate dehydrogenase kinase-pyruvate dehydrogenase system can probably be accounted for by a substrate-induced conformational change in the protein substrate, without the need for an additional allosteric site for pyruvate in the converter kinase.

The possible mechanisms involved in the allosteric regulation of the glutamine synthetase system are more complex. Stadtman (58) summarizes three antagonistic, apparently allosteric effects: glutamine seems to activate the adenylyltransferase and inhibit the deadenylylating enzyme system, whereas α-ketoglutarate and UTP have opposite effects. On the basis of the general considerations discussed above, the glutamine effect could be tentatively ascribed to the fact that glutamine is a specific ligand for the two convertible enzymes, as the main product of their metabolic activity. The immediate targets of the two other effects are more difficult to guess, since a binding site either in the glutamine synthetase (per subunit) or in the adenylyl-transferase (which in turn is part of the deadenylylating enzyme system) could serve as the basis for the observed effects.

In contrast to the generality of regulatory enzymes, for enzymes acting on specific proteins observation of a modifier effect by a small metabolite should not be taken as implying an allosteric site in the enzyme without evidence of specific binding. On the contrary, whenever an effector is known to be a specific ligand of the convertible protein, it should be tentatively assumed that no regulatory site in the converted enzyme is likely to be required, particularly if the apparent affinity of the ligand for the overall system is similar to that for the convertible protein.

The fact that, as pointed out above, the molarity of converter enzymes is in general definitely smaller than that of their corresponding convertible enzyme offers a mechanism for amplifying the effect of metabolic messengers occurring in biological systems in molar amounts much smaller than those typical of major metabolic regulatory enzymes. For instance, muscle phosphorylase b with a molarity of $ca.$ 24×10^{-6} (and a normality of $ca.$ 48×10^{-6}) could not be directly modulated by the metabolic messenger cAMP, whose maximum concentration is in the micromolar range. This subject has been amply discussed elsewhere (57). The possibility of amplification can be greatly multiplied in "cascade" systems. The intermediate catalytic effect(s) can allow a major metabolic change by even, theoretically, a single molecule of an initial metabolic messenger.

In Section 4.B and 4.C it was shown that enzymes acting on specific macromolecules typically have small molecular activities and that the K_m values for their macromolar substrates are usually low, but not very low. The data in Table 3 indicate a typical order of magnitude in the micromolar range, within 1 order of magnitude. This is well below the possibilities for specific complex formation between macromolecules. As indicated by the great stability of many well-characterized polymeric enzymes, and as illustrated in the table for the lac repressor-operator, K_d values in the nanomolar and even picomolar ranges can easily be reached. Enzymes acting on specific macromolecules, and most other regulatory proteins, in marked contrast to the genetic repressors, have as much need to dissociate readily from their products after accomplishing their effect as to associate effectively with their substrates.

Within the wide range of possibilities, the selective forces of evolution have ample opportunity for selection of a convenient degree of complementariness. The K_m values of these regulatory proteins, whether enzymes or not, can thus be expected to be balanced between the unfavorable extremes of too much above and too much below the physiological molar concentration of the more abundant [or available, as emphasized by Sols and Marco (57)] member of the pair. This situation offers the possibility of control through changes in the affinity between the two macromolecules, induced by specific ligands. At the same time it imposes an additional limitation on the molecular efficiency that can be expected *in situ* for converter enzymes.

We showed above that the relative pool sizes of macro- and micromolecular substrates and products in interconvertible enzyme systems make the latter inadequate in principle as potential feedback regulators at the substrate (or product) level. Moreover, in the case of the ATP-dependent converter enzymes the K_m values known for ATP in the vicinity of 10^{-4} M would ensure that these enzymes could be virtually saturated with respect to this cosubstrate within practically any conceivable metabolic situation in the cell. In other words, neither would the ATP-dependent conversion of an enzyme significantly affect the ATP pool, nor could changes in the latter within usual physiological ranges markedly decrease the efficiency of the corresponding converter enzyme. This general conclusion does not preclude the possibility that some of these enzymes could be allosterically regulated by one of the above metabolites. Convertible enzymes of considerable importance in energy metabolism could thus be linked for regulation to the "energy charge" of the cell (2).

The fact that in cell-free extracts phosphorylation-dephosphorylation systems are easily weighed in favor of dephosphorylation by the addition of Mg^{2+} to concentrations in the 10^{-2} M range should not be construed as indicative of physiological regulation of those convertible enzyme systems by changes in Mg^{2+} concentration *in situ*. For a protein with only moderate affinity for Mg^{2+}, its availability *in situ* in the generality of cells is smaller by an order of magnitude than the amount of total cellular Mg^{2+} (57).

In summary, regulation of primary regulatory enzymes is most likely to be allosteric, either by a direct mechanism or through a conformational change in a protein substrate.

Physiological control of interconvertible enzyme systems is probably based on changes in the balance of the activities of the two antagonistic converter enzymes, rather than in alternative strong inhibition (or de-activation) of one of them. If in different circumstances the activity of one or the other of the enzymes of an antagonistic pair were very effectively blocked, the system would be rather sluggish. For instance, a change from 99% to 10% inhibition would require a change in the concentration of the corresponding effector of up to 3 orders of magnitude. On the other hand, if converter enzymes in general maintain always an important activity, say at least some 10–20% of the maximal one, a relatively small change in the concentration of an effector could markedly invert the balance between the two enzymes of an antagonistic pair. The negative counterpart of this increased facility of response would be the continued operation of a futile cycle. However, the maximum amount of ATP that could be wasted in such a cycle can be calculated to be some 2 orders of magnitude smaller than the ATP (plus GTP) involved in the synthesis of the corresponding convertible enzyme, and therefore of relatively little significance for the overall metabolism of the cell. Efficient control of the antagonistic pairs of primary regulatory enzymes in interconvertible enzyme systems may well deserve the small energy expenditure of such futile cycles.

6. PERSPECTIVES

The subject of primary regulatory enzymes in general, and in particular the problems peculiar to enzymes acting on specific macromolecular substrates, can be expected to be clarified greatly in the near future. Within the framework of the general considerations discussed in this chapter, any significant progress in the knowledge of any of these enzymes could have potential value for an increase in the understanding of individual problems of other enzymes of this general type.

The macromolecular substrate specificity of *converter* enzymes is a matter that should be carefully evaluated, both from the point of view of a proper understanding of metabolic integrations and for the avoidance of unnecessary multiplication of a number of nonentities. As an illustration, it seems useful to point here to the very rapid recent evolution from "the phosphorylase kinase kinase" and "the glycogen synthetase kinase," to "a cAMP-dependent protein kinase," to "one" of an undefined number of cAMP-dependent protein kinases. An obvious next step should be the ascertaining of how many functionally different cAMP-dependent kinases exist in a cell. Identification of additional physiological substrates of converter enzymes should have an adequate quantitative basis (in terms of both V_{max} and K_m) and a plausible metabolic value. Changes in the activity of proteins caused by enzyme-catalyzed modification presently constitute one of the subjects in which it is easy to make physiological nonsense out of observations *in vitro* on the *possibility* of enzymatic modification of certain proteins.

The lack of fundamental knowledge concerning many physicochemical properties of substances in cellular or cell-like environments makes highly desirable further research directed toward gaining information in this field. The possible limiting effect of intracellular diffusion, particularly for macromolecules, in relation to a variety of processes deserves serious consideration. Although the old view, championed by Willstäter, that "enzymatic activity is a natural force not based upon substance" has been entirely superseded, sometimes there is little or no awareness of the actual physical existence of a great concentration of proteins in the cell sap, with concomitant changes in physicochemical properties, as compared with the dilute solutions that enzymologists normally use *in vitro*. In this context, the low molecular activities shown in this chapter to be typical of enzymes acting on specific macromolecules could easily become even lower under physiological conditions.

Progress in this field would benefit greatly from the prompt acquisition of much more definite information concerning affinities between proteins that form specific complexes, whether occasional (enzymes-substrates), permanent (specific multimers), or intermediate in physiological stability (regulatory proteins-regulated proteins).

ACKNOWLEDGMENTS

We are greatly indebted to the following colleagues who helped us in the preparation of this chapter by providing preprints, data, and/or useful comments: A. Albert, G. DelaFuente, H. De Wulf, J. M. Gancedo, E. Helmreich, C. F. Heredia, H. G. Hers, H. Holzer, E. G. Krebs, J. Llopis, S. P. Mistry, A. Sillero, and J. M. Wiame. Thanks are due also to Miss Clotilde Estévez for very efficient assistance in the preparation of the manuscript.

REFERENCES

1. W. B. Anderson and E. R. Stadtman, *Biochem. Biophys. Res. Commun.*, **41**, 704 (1970).
2. D. E. Atkinson, in *The Enzymes*, Vol. 1 (Ed.: P. D. Boyer), Academic Press, New York, 1970, p. 461.
3. J. Avila, J. M. Hermoso, E. Viñuela, and M. Salas, *European J. Biochem.*, **21**, 526 (1971).
4. C. Bailey and C. Villar-Palasí, *Federation Proc.*, **30**, 1147 Abs. (1971).
5. Th. Barman, *Enzyme Handbook*, Springer-Verlag, Berlin, 1969.
6. L. Birnbaumer, S. L. Pohl, H. Michiel, J. Kraus, and M. Rodbell, in *Role of Cyclic AMP in Cell Function* (Eds.: P. Greengard and E. Costa), Raven Press, New York, 1970, p. 185.
7. S. Bourgeois and J. Monod, in *Control Processes in Multicellular Organisms* (Eds.: G. E. W. Wolstenholme and J. Knight), Churchill, London, 1970, p. 3.
8. N. C. Brown, Z. N. Canellakis, B. Lundin, P. Reichard, and L. Thelander, *European J. Biochem.*, **9**, 561 (1969).
9. R. R. Burgess, A. A. Travers, J. J. Dunn, and E. K. F. Bautz, *Nature*, **221**, 43 (1969).
10. C. A. Chelala, L. Hirschbein, and H. N. Torres, *Proc. Natl. Acad. Sci. U.S.*, **68**, 152 (1971).
11. G. T. Cori and A. A. Grenn, *J. Biol. Chem.*, **151**, 31 (1943).

12. G. T. Cori, D. Illingworth, and P. Keller, in *Methods in Enzymology* (Eds.: S. P. Colowick and N. O. Kaplan), Vol. 1, Academic Press, New York, 1955, p. 200.

13. O. E. W. Davie, C. Hougie, and R. L. Lundblad, in *Recent Advances in Blood Coagulation* (Ed.: L. Pollen), Churchill, London, 1969

14. G. DelaFuente, R. Lagunas, and A. Sols, *European J. Biochem.*, **16**, 226 (1970).

15. R. J. DeLange, R. G. Kemp, W. Dixon Riley, R. A. Cooper, and E. G. Krebs, *J. Biol. Chem.*, **243**, 2200 (1968).

16. R. L. Deter and Ch. DeDuve, *J. Cell Biol.*, **33**, 437 (1967).

17. W. Drabikowski, R. Dabrowska, and B. Barylko, *FEBS Letters*, **12**, 148 (1971).

18. E. Ebner, D. Wolff, C. Gancedo, S. Elsässer, and H. Holzer, *European J. Biochem.*, **14**, 535 (1970).

19. E. H. Fischer and E. G. Krebs, *J. Biol. Chem.*, **231**, 65 (1958).

20. J. C. Gerhardt and H. K. Schachman, *Biochemistry*, **4**, 1054 (1965).

21. A. Ginsburg, in *Outlines of the Lectures Given at the First International Symposium on Metabolic Interconversion of Enzymes*, S. Margherita Ligure, Genova, 1970, p. 12.

22. N. D. Goldberg, W. D. Lust, R. F. O'Dea, J. Wei, and A. G. O'Toole, in *Role of Cyclic AMP in Cell Function* (Eds.: P. Greengard and E. Costa), Raven Press, New York, 1970, p. 67.

23. A. A. Green and G. T. Cori, *J. Biol. Chem.*, **151**, 21 (1943).

24. S. Grisolia, *Physiol. Revs.*, **44**, 657 (1964).

25. H. B. Halsall and V. N. Schumaker, *Biochem. Biophys. Res. Commun.*, **39**, 479 (1970).

26. J. G. Hardman and E. W. Sutherland, *J. Biol. Chem.*, **244**, 6363 (1969).

27. E. Helmreich, personal communication, 1971.

28. H. G. Hers, H. De Wulf, and W. Stalmans, *FEBS Letters*, **12**, 73 (1970).

29. H. G. Hers, H. De Wulf, W. Stalmans, and G. Van den Berghe, in *Advan. Enzyme Regulation*, **8**, 171 (1970).

30. H. Holzer, *Advan. Enzymol.*, **32**, 297 (1969).

31. H. Holzer and W. Duntze, this volume, p. 115.

32. S. S. Hurd, W. B. Novoa, J. P. Kickenbotton, and E. H. Fischer, in *Methods in Enzymology*, Vol. 8 (Eds.: E. F. Neufeld and V. Ginsburg), Academic Press, New York, 1966, p. 546.

33. D. E. Koshland, Jr., *Proc. Natl. Acad. Sci. U.S.*, **44**, 98 (1958).

34. E. G. Krebs, D. S. Love, G. E. Bratvold, K. A. Trayser, W. L. Meyer, and E. H. Fischer, *Biochemistry*, **3**, 1022 (1964).

35. M. J. Kushmerick and R. J. Podolsky, *Science*, **166**, 1297 (1969).

36. M. Lazdunski, C. Petitclerc, D. Chappelet, and C. Lazdunski, *European J. Biochem.*, **20**, 124 (1971).

37. Y. P. Lee, *Biochim. Biophys. Acta*, **43**, 25 (1960).

38. R. Lehman, cited in E. Pollard, *J. Theoret. Biol.*, **4**, 98 (1963).

39. P. Lengyel and D. Söll, *Bacteriol. Rev.*, **33**, 264 (1969).

40. W. Lenzinger, *Biochem. J.*, **123**, 139 (1971).

41. R. Mazumder, *Biochem. Biophys. Res. Commun.* (in press).

42. M. Mantel and H. Holzer, *Proc. Natl. Acad. Sci. U.S.*, **65**, 660 (1970).

43. A. D. McLaren and L. Packer, *Advan. Enzymol.*, **33**, 245 (1970).

44. D. Mecke and H. Holzer, *Biochim. Biophys. Acta*, **122**, 341 (1966).

45. J. Monod, J.-P. Changeux, and F. Jacob, *J. Mol. Biol.*, **6**, 306 (1963).

46. K. Otto and P. Schepers, *Z. Physiol. Chem.*, **348**, 482 (1967).

47. E. Pollard, *J. Theoret. Biol.*, **1**, 328 (1961).

48. S. Pontremoli, personal communication, 1971.

49. L. J. Reed, in *Current Topics in Enzyme Regulation*, Vol. 1 (Eds.: B. L. Horecker and E. R. Stadtman), Academic Press, New York, 1969, p. 233.

50. E. M. Reinmann, C. O. Brostrom, J. D. Corbin, C. A. Kind, and E. G. Krebs, *Biochem. Biophys. Res. Commun.*, **42**, 187 (1971).

51. J. W. Roberts, *Nature*, **227**, 1168 (1969).

52. G. A. Robinson, R. W. Butcher, and E. W. Sutherland, *Cyclic AMP*, Academic Press, New York, 1971.

53. R. T. Schimke, in *Current Topics in Cellular Regulation*, Vol. 1 (Eds.: B. L. Horecker and E. R. Stadtman), Academic Press, New York, 1969, p. 77.

54. B. M. Shapiro and E. R. Stadtman, *J. Biol. Chem.*, **243**, 3769 (1968).

55. E. L. Smith, in *The Enzymes*, Vol. 1 (Ed.: P. D. Boyer), Academic Press, New York, 1970, p. 267.

56. A. Sols, in *Outlines of the Lectures Given at the First International Symposium on Metabolic Interconversion of Enzymes*, S. Margherita Ligure, Genova, 1970, p. 1.

57. A. Sols and R. Marco, in *Current Topics in Cellular Regulation*, Vol. 2 (Eds.: B. L. Horecker and E. R. Stadtman), Academic Press, New York, 1970, p. 227.

58. E. R. Stadtman, in *The Enzymes*, Vol. 1 (Ed.: P. D. Boyer), Academic Press, New York, 1970, p. 397.

59. W. Stalmans, H. De Wulf, and H. G. Hers, *European J. Biochem.*, **18**, 582 (1971).

60. M. Tao, M. L. Salas, and F. Lipmann, *Proc. Natl. Acad. Sci. U.S.*, **67**, 408 (1970).

61. C. Villar-Palasí and J. Larner, *Arch. Biochem. Biophys.*, **94**, 436 (1961).

62. D. A. Walsh, E. G. Krebs, E. M. Reimann, M. A. Brostrom, J. D. Corbin, J. P. Kickenbotton, T. R. Soderling, and J. P. Perkins, in *Role of Cyclic AMP in Cell Function* (Eds.: P. Greengard and E. Costa), Raven Press, New York, 1970, p. 265.

63. J. M. Wiame, in *Current Topics in Cellular Regulation* (Eds.: B. L. Horecker and E. R. Stadtman), Academic Press, New York, Vol. 4, 1971.

64. O. Wieland and B. Jagow-Westermann, *FEBS Letters*, **3**, 271 (1969).

65. C. A. Woolfolk, B. Shapiro, and E. R. Stadtman, *Arch. Biochem. Biophys.*, **116**, 177 (1966).

66. W. Zillig, A. Heil, P. Palm, K. Zechel, W. Seifert, and D. Rabussay, in *Symposium on Reaction Mechanisms and Control Properties of Phosphotransferases*, Rheinhardsbrunn, GDR, in press.

67. G. Zubay, D. Schwartz, and J. Beckwith, *Proc. Natl. Acad. Sci. U.S.*, **66**, 104 (1970).

CHAPTER 5

Chemical Modification of Enzymes by ATP

HELMUT HOLZER AND WOLFGANG DUNTZE

Biochemisches Institut der Universität Freiburg im Breisgau, Germany

1. INTRODUCTION

There are four different important mechanisms by which metabolites control enzyme-catalyzed reactions: (1) limitation of an enzymatic reaction by stoichiometric participation of the controlling metabolite in the reaction; (2) control of enzyme activity by allosteric interaction of metabolites with the enzyme; (3) regulation of the rate of synthesis of an enzyme by control of transcription or translation; (4) control of enzyme activity by chemical modification of an enzyme.

The control of ATP production in glycolysis by the stoichiometric participation of phosphate and ADP in the oxidation of glyceraldehyde-3-phosphate

Table 1. Enzymes Regulated by ATP-Dependent, Enzyme-Catalyzed, Chemical Modification

Enzyme	Mechanism of Modification	Organism or Tissue	Reference
Glycogen phosphorylase	Phosphorylation/ dephosphorylation	Many organisms	3, 4
Phosphorylase *b* kinase	Phosphorylation/ dephosphorylation	Muscle	5, 6, 7
Phosphorylase *a* phosphatase (?)	Phosphorylation/ dephosphorylation	Liver, adrenal gland, avian skeletal muscle	14, 15, 32
Glycogen synthetase	Phosphorylation/ dephosphorylation	Many organisms	8
Glycogen synthetase phosphatase (?)	Phosphorylation/ dephosphorylation	Liver	16, 17
Fructose diphosphatase	Phosphorylation/ dephosphorylation	Kidney	9
Pyruvate dehydrogenase complex	Phosphorylation/ dephosphorylation	Heart muscle, liver, kidney	10, 11
Lipase (?)	Phosphorylation/ dephosphorylation	Adipose tissue	18, 50, 51
Palmityl-CoA synthetase (?)	Phosphorylation/ dephosphorylation	Liver	19
Glutamine synthetase	Adenylylation/ deadenylylation	*E. coli*	12, 13, 54
RNA polymerase (?)	Adenylylation/ deadenylylation	*E. coli*	62, 72

(1, 2) represents an example of mechanism (1) in which ATP is involved. In this reaction sequence glycolytic ATP production is coupled by a "stoichiometric feedback mechanism" to processes which consume ATP and thus increase the levels of the substrates ADP and phosphate.

The allosteric inhibition of phosphofructokinase by ATP provides an example of mechanism (2). The consequence of this regulatory mechanism is similar to that of the "stoichiometric feedback control" of oxidation of glyceraldehyde-3-phosphate mentioned above: changes in the ATP level caused by the consumption of this metabolite control the rate of its production. In this case, however, the regulatory effect of ATP is due to a "catalytic feedback control," since it acts by allosteric modification of the catalyst phosphofructokinase and not by stoichiometric participation in the controlled reaction.

So far, no example of mechanism (3), that is, control of transcription or translation by ATP has been observed, whereas several examples of the modification of enzymes by ATP-dependent phosphorylation or adenylylation of the protein molecule (mechanism 4) have been found. These enzymes are listed in Table 1. This chapter will deal with the different mechanisms of these chemical modifications by ATP. In addition, the general significance of metabolic control by ATP-dependent chemical modification of enzymes will be discussed.

2. SYNTHESIS AND DEGRADATION OF GLYCOGEN

Figure 1 shows the reactions involved in the synthesis and degradation of glycogen and the regulation of the participating enzymes by chemical modification. Glycogen synthetase (EC 2.4.1.11: UDP-glucose:glycogen α-4-glucosyltransferase) catalyzes the synthesis of glycogen by an irreversible reaction. The degradation of glycogen to glucose-1-phosphate catalyzed by glycogen phosphorylase (EC 2.4.1.1: α-1,4-glucan:orthophosphate glucosyltransferase) is reversible *in vitro*. Under *in vivo* conditions, however, the phosphorolysis of glycogen is favored. Both enzymes can be regulated either through allosteric control by different effectors or through chemical modification by phosphorylation and dephosphorylation. The kinases and phosphatases involved in the chemical modifications are shown in Fig. 1.

3. GLYCOGEN PHOSPHORYLASE

Glycogen phosphorylase has been purified from a variety of sources. The most extensive studies have been carried out with the enzymes from skeletal

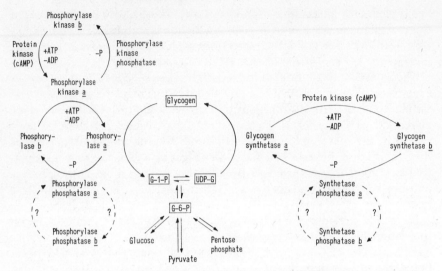

Fig. 1. Chemical modification of enzymes involved in glycogen synthesis and degradation.

muscle and liver. Glycogen phosphorylase can exist in two different molecular forms, phosphorylase *a* and phosphorylase *b*. Both forms are oligomers containing two or four presumably identical monomers with a molecular weight of 92,500. Accordingly, the dimers and tetramers have molecular weights of 185,000 and 370,000, respectively. Muscle phosphorylase *b* exists predominantly as a dimer which can be phosphorylated in the presence of ATP and Mg^{2+}, according to the following equation:

$$2 \text{ Muscle phosphorylase } b \text{ dimer} + 4 \text{ ATP} \rightarrow \text{muscle}$$

$$\text{phosphorylase } a \text{ tetramer} + 4 \text{ ADP.} \quad (1)$$

The reaction is catalyzed by a specific phosphorylase kinase which has been found in skeletal muscle. The reverse reaction requires the catalytic activity of another enzyme, phosphorylase phosphatase, which has been found in liver and several other tissues.

Experimental evidence for the existence of hybrids containing phosphorylated and nonphosphorylated monomers has been provided by Hurd et al. (20). The nonphosphorylated *b* form appears to have only little activity *in vivo*. *In vitro* the enzyme has an obligatory requirement for 5'-AMP for activity. In contrast, the phosphorylated form *a*, which is thought to be the active form *in vivo*, can be fully active without 5'-AMP at high substrate concentrations. At very low substrate concentrations 5'-AMP has a stimulatory effect.

A variety of allosteric effectors influence the conformation and the enzymatic activity of isolated phosphorylase *a*. Under appropriate conditions the tetrameric form can undergo reversible conformational transitions between an inactive T state and an active R state. In addition, both tetrameric forms may dissociate in the presence of the substrates glycogen and glucose-1-phosphate or under the influence of allosteric ligands such as glucose and 5′-AMP to form active (R-state) or inactive (T-state) dimers (21). These conformational transitions of phosphorylase *a* are summarized in Fig. 2. Similar conformational changes which influence catalytic activity have also been discussed for phosphorylase *b* (22, 23). However, since the allosteric properties of glycogen phosphorylase are profoundly changed by phosphorylation and dephosphorylation, respectively, the primary mechanism of regulation appears to be chemical modification of the enzyme.

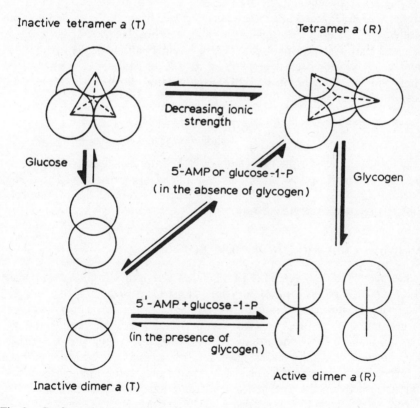

Fig. 2. Conformational transitions of muscle phosphorylase *a*. From Helmreich, *Comp. Biochem.*, **17**, 17 (1969).

Phosphorylation occurs at the alcoholic hydroxyl group of a seryl residue. The sequence of amino acids in the vicinity of the phosphorylated serine has been determined. Nolan et al. (24) isolated a tetradecapeptide containing the phosphorylated seryl residue and determined the amino acid sequence:

H_2N-Ser-Asp-Glu(NH_2)-Glu-Lys-Arg-Lys-Glu(NH_2)-Ileu-Ser(P)-Val-Arg-
Gly-Leu-COOH.

This tetradecapeptide can serve as a substrate for phosphorylase *a* phosphatase and, after dephosphorylation, for phosphorylase *b* kinase. The rate of phosphorylation, however, is only 1% of that found with phosphorylase *b* as the substrate. Likewise, phosphate is released from the phosphopeptide at a rate less than 3% of that for the dephosphorylation of phosphorylase *a* (24).

Glycogen phosphorylase from liver is modified in the same way as the enzyme from skeletal muscle by enzyme-catalyzed phosphorylation in the presence of ATP and Mg^{2+}. The molecular weight of this enzyme has been found to be 185,000. By analogy to muscle phosphorylase it is generally assumed that liver phosphorylase is also a dimer, although the existence of monomers has not been demonstrated as yet. In contrast to muscle phosphorylase, no aggregation of the phosphorylated dimers to tetramers has been found (see equation 2):

Liver phosphorylase *b* (dimer) + 2 ATP → liver

phosphorylase *a* (dimer) + 2 ADP.   (2)

In this case, the phosphorylated dimer is the catalytically active form of the enzyme. As with muscle phosphorylase a seryl residue is phosphorylated. There are, however, minor differences in the structure of the peptide containing the phosphoserine (25).

4. GLYCOGEN PHOSPHORYLASE KINASE

It has been found by Krebs et al. (26) that glycogen phosphorylase kinase (E.C. 2.7.1.38: ATP:phosphorylase phosphotransferase), the enzyme catalyzing the phosphorylation of phosphorylase *b*, also exists in two different forms that can be interconverted by enzymatic phosphorylation and dephosphorylation. Both forms have been isolated from rabbit skeletal muscle. The enzyme phosphorylating glycogen phosphorylase kinase has been enriched from the same tissue. The activity of this phosphorylase-kinase kinase depends on the presence of small concentrations of cyclic 3',5'-AMP. A concentration of $10^{-7} M$, which is in the physiological range, is sufficient for half-maximal activity. Since it is well established that epinephrine causes an increase in the concentration of cyclic AMP, one can visualize the activation of glycogen degradation by the "cascade" shown in Fig. 3.

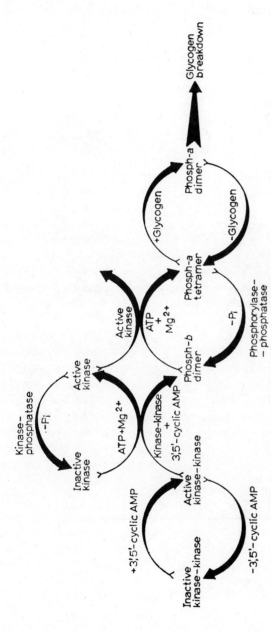

Fig. 3. Reaction cascade for the activation of muscle phosphorylase. From Helmreich, *Comp. Biochem.*, **17**, 17 (1969).

The nonphosphorylated phosphorylase kinase b can also be activated by trypsin or by high concentrations of Ca^{2+} ions. The latter activation needs a protein factor which possesses proteolytic activity (27). The biological significance of this Ca^{2+}-dependent irreversible activation is not clear. In addition, a reversible activation by low concentrations of Ca^{2+}, which is likely to be related to the control of muscle activity, has also been demonstrated (28). With phosphorylase kinase from locust flight muscle an additional activation by phosphate ions has recently been shown (29).

The dephosphorylation of phosphorylase kinase a by a specific enzyme, phosphorylase-kinase phosphatase, has been demonstrated in extracts from rabbit skeletal muscle (5). Thus a reversible interconversion of phosphorylase kinase has also been established (cf. Fig. 1).

5. GLYCOGEN PHOSPHORYLASE PHOSPHATASE

Phosphorylase a phosphatase (E.C. 3.1.3.17: phosphorylase phospho-hydrolase) has been partially purified from rabbit skeletal muscle (30). Dephosphorylation of phosphorylase a is completely inhibited by 10^{-5} M 5′-AMP. Probably this inhibition is not caused by the influence of 5′-AMP on the phosphatase itself, but rather is due to the fact that AMP induces a conformational change of phosphorylase a which renders the enzyme inaccessible to the phosphatase. Caffeine and glucose probably affect the activity of phosphatase also by changing the conformation of the substrate (24). The phosphorylase phosphatase from rabbit skeletal muscle does not appear to be modified itself by phosphorylation and dephosphorylation. In contrast, Chelala and Torres (31), working with homogenates from pigeon breast muscle, and Merlevede et al. (32), working with preparations from dog liver, have provided indirect evidence that the phosphatase from these tissues may be interconverted between an active and an inactive form by phosphorylation and dephosphorylation. However, further experiments, especially a direct demonstration of the covalent incorporation of phosphate into the enzyme protein and the purification of the participating enzyme(s), will be necessary to establish this mechanism.

6. GLYCOGEN SYNTHETASE

Glycogen synthetase (E.C. 2.4.1.11: UDP-glucose:glycogen α-4-glucosyl-transferase) has been shown by Larner and his coworkers to exist in two different molecular forms (for summaries see refs. 33 and 34). One form requires high concentrations of glucose-6-phosphate for full activity *in vitro* and has been named the "dependent" or D form. The other form does not require glucose-6-phosphate for activity and, accordingly, has been designated as the "independent" or I form. By analogy to the nomenclature of glycogen phosphorylase the I form, which under physiological conditions appears to be the only active form, has been termed glycogen synthetase *a* in contrast to the inactive D form or glycogen synthetase *b* (35, 36). However, synthetase *a* is not entirely independent of glucose-6-phosphate since at very low substrate concentrations the affinity of the enzyme to UDP-glucose is increased appreciably by glucose-6-phosphate (35).

The two forms of glycogen synthetase, which have been partially purified from a variety of mammalian tissues, can be interconverted by a phosphorylation-dephosphorylation cycle. In contrast to glycogen phosphorylase *a*, glycogen synthetase *a* is the nonphosphorylated form. Phosphorylation occurs in the presence of ATP, Mg^{2+}, and a cyclic AMP-dependent protein kinase which has been shown to be different from glycogen phosphorylase kinase (33, 37). The enzyme catalyzes the incorporation of one phosphate group per subunit of glycogen synthetase *a* (39). Reactivation of synthetase *b* involves the removal of the phosphoryl moiety by a phosphatase. In addition, a Ca^{2+}-dependent inactivation of glycogen synthetase *a* to *b*, which, in contrast to the inactivation by phosphorylation, is irreversible, has been observed (38).

There is evidence that glycogen synthetase phosphatase may also exist in an active phosphorylated and an inactive dephosphorylated form and that the two may be interconverted by specific enzymes (16, 17). However, additional experimental evidence is necessary to establish the chemical modification of this enzyme. Therefore glycogen synthetase phosphatase is listed with a question mark in Table 1.

7. FRUCTOSE-1,6-DIPHOSPHATASE

In 1966 Mendicino et al. (9) reported that purified fructose-1,6-diphosphatase (E.C. 3.1.3.11: D-fructose-1,6-diphosphate 1-phosphohydrolase) from kidney was inactivated by incubation with a crude kidney extract in the presence of ATP and Mg^{2+}. Upon addition of cyclic 3',5'-AMP the rate of inactivation was increased. Using ^{32}P-ATP, the authors observed that ^{32}P was incorporated into the protein simultaneously with the inactivation of the enzyme. Upon chromatography on cellulose phosphate a peak of radioactivity firmly bound to protein coincided with the peak of the enzymatic activity of fructose diphosphatase. Reactivation could be achieved by incubation of the inactive enzyme with a crude kidney extract. Simultaneously with the reactivation, incorporated ^{32}P was liberated (J. Mendicino, personal communication). These results suggest a reversible interconversion of fructose diphosphatase, as indicated in Fig. 4.

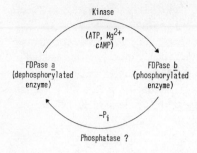

Fig. 4. Enzyme-catalyzed interconversion of kidney fructose-1,6-diphosphatase (FDPase) from an active (*a*) to an inactive (*b*) form and vice versa.

Using purified fructose-1,6-diphosphatase from swine kidney, Mendicino et al. (40) have shown that the inactivating system is tightly bound to mitochondria from kidney and other organs. In this system the rate of inactivation depends on the concentrations of ATP, ADP, and AMP and is influenced by fatty acids, tricarboxylic acids, and other oxidizable substrates, as well as by uncouplers of oxidative phosphorylation (41).

8. PYRUVATE DEHYDROGENASE COMPLEX

The pyruvate dehydrogenase complex (PDC) from mitochondria catalyzes a series of consecutive reactions by which pyruvate is oxidized according to the following overall reaction:

$$CH_3COCO_2H + CoA—SH + NAD^+ \rightarrow CH_3CO—S—CoA + CO_2 +$$
$$NADH + H^+. \quad (3)$$

The complex consists of three different types of component enzymes: pyruvate dehydrogenase [E.C. 1.2.4.1: pyruvate: lipoate oxidoreductase (acceptor acetylating)], dihydrolipoyl transacetylase (E.C. 2.3.1.12: acetyl-CoA: dihydrolipoate *S*-acetyltransferase), and dihydrolipoyl dehydrogenase (E.C. 1.6.4.3: reduced-NAD: lipoamide oxidoreductase). The structure of the enzyme complex and the function of the single component proteins and their cooperation within the complex have been studied extensively (for summaries see refs. 42 and 43). Recently Linn et al. (10) have shown that PDC from beef kidney, beef heart, and pork liver mitochondria is regulated by a phosphorylation-dephosphorylation reaction cycle (cf. Fig. 5). Similar results have been obtained by Wieland and his coworkers (11, 44) for the PDC complex from pig heart mitochondria. With the preparation from beef kidney mitochondria Linn et al. (10) could show that the pyruvate dehydrogenase component of the complex is phosphorylated by ATP in the presence of a specific kinase, that is, a pyruvate-dehydrogenase kinase. This kinase was strongly bound to the transacetylase component of the PDC complex (45). Of the three components of the complex only pyruvate dehydrogenase was inactivated in the presence of ATP. Moreover, radioactivity from $[\gamma\text{-}^{32}\text{P}]\text{ATP}$ was incorporated exclusively into the pyruvate dehydrogenase protein. The radioactive phosphoryl moiety is bound to the protein in acid- and alkali-labile linkage, thus indicating that it may be attached by an ester bond to the alcoholic hydroxyl group of a seryl or threonyl residue.

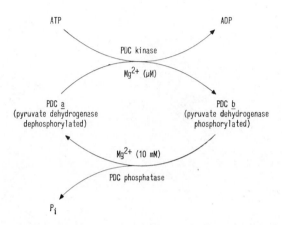

Fig. 5. Enzyme-catalyzed interconversion of the pig heart pyruvate dehydrogenase complex (PDC) containing an active (*a*) and an inactive (*b*) form of the component pyruvate dehydrogenase.

The pyruvate-dehydrogenase kinase is active at micromolar concentrations of Mg^{2+}. In the presence of higher concentrations of Mg^{2+}, the optimal concentration being 10 mM, phosphorylated PDC can be reactivated. The reactivation is accompanied by a release of bound ^{32}P as inorganic phosphate. The reactivation by dephosphorylation is catalyzed by a phosphatase. This PDC phosphatase, as well as the PDC kinase, appears to be an integral part of the PDC complex, although the phosphatase seems to be bound less tightly.

The pyruvate-dehydrogenase phosphatase from pig heart mitochondria has been separated from the other components of the pyruvate dehydrogenase complex by gradient centrifugation (46). Linn et al. (45) achieved resolution of the phosphatase from the PDC complexes from pork liver, beef kidney, and beef heart mitochondria by isoelectric precipitation and ultracentrifugation. These authors were able to show that the phosphatases from kidney, heart, and liver are functionally interchangeable.

Interestingly, dephosphorylation of PDC from pig heart muscle can be stimulated by cyclic $3',5'$-AMP, whereas the phosphorylation of PDC is not influenced by this nucleotide (11, 46). In this connection, the observation of Skosey (47) is of interest that the stimulation of α-ketoglutarate oxidation by adrenocorticotropin (ACTH) is mediated by cyclic $3',5'$-AMP. In isolated rat epididymal adipose tissue an increase of cyclic AMP stimulates the production of $^{14}CO_2$ from α-ketoglutarate-1-^{14}C. Since cyclic AMP stimulates the PDC phosphatase (11, 46), thus causing an activation of PDC, one could suggest that a similar mechanism operates in the activation of α-ketoglutarate oxidase of adipose tissue. However, the detailed mechanism of the stimulation of α-ketoglutarate oxidation by cyclic AMP remains to be established.

The content of active (i.e., dephosphorylated) and inactive (i.e., phosphorylated) PDC in rat kidney and heart mitochondria depends on the nutritional conditions (48). In normally fed animals the PDC seems to be in the active form to an extent of about 70%. By contrast, the amount of active PDC in fasting rats is less than 5%. Addition of carbohydrates to the diet increases the percentage of active PDC. Likewise, the ratio of active to inactive PDC is increased by insulin in alloxan diabetic rats. Finally, measurements of the concentration of free fatty acids in the plasma indicate that a correlation exists between low PDC activity and high concentrations of free fatty acids, and vice versa.

These results suggest that the inactivation and reactivation of PDC by a phosphorylation-dephosphorylation cycle plays an essential role in the regulation of the carbohydrate and lipid metabolism in mammals. This view is further emphasized by the observation of Linn et al. (45) that ADP is a competitive inhibitor of ATP in the inactivation reaction and that pyruvate protects the PDC against ATP-mediated inactivation. Since the inactivation of PDC depends on ATP and the reactivation is enhanced by high concentrations of Mg^{2+} ions, an increase in the ATP/ADP ratio will stimulate the inactivation in two additional ways (49): (a) the inactivating kinase reaction is favored at high concentrations of the substrate ATP, and (b) the reactivating phosphatase reaction can be inhibited as a consequence of a decreased Mg^{2+} concentration. Since the affinity of ATP for Mg^{2+} is much greater than that of ADP, an increased ATP/ADP ratio could decrease the Mg^{2+} concentration enough to prevent the reactivation of the PDC. Thus the inactivation of the PDC would be favored over the reactivation.

Another enzyme recently shown to be converted from an inactive to an active form is lipase (E.C. 3.1.1.3: glycerol-ester hydrolase) from adipose tissue (50, 51). The inactive enzyme is activated in the presence of ATP, Mg^{2+}, cyclic AMP, and a cyclic AMP-dependent protein kinase derived from skeletal muscle (51). Under these conditions, the γ-phosphate group from

^{32}P-ATP is covalently bound to the lipase and the time course of phosphorylation is paralleled by the activation of the enzyme. The chemical modification of lipase appears to be an essential mechanism in the hormonal control of the lipolytic activity in adipose tissues (50). A similar interconversion of two different molecular forms by phosphorylation and dephosphorylation has also been discussed for palmityl CoA-synthetase [E.C. 6.2.1.3: palmitate: CoA ligase (AMP)] from rat liver (19).

9. GLUTAMINE SYNTHETASE FROM *E. COLI*

In 1966 Mecke et al. (52) described a rapid inactivation of glutamine synthetase [E.C. 6.3.1.2: L-glutamate:ammonia ligase (ADP)] in intact cells of *Escherichia coli* after addition of 10^{-4} to 10^{-6} M NH_4^+ to the medium. Further experiments led to the preparation of a cell-free system in which glutamine synthetase was inactivated in the presence of ATP, Mg^{2+}, and glutamine (53). It was shown that a specific "inactivating enzyme" catalyzed this reaction (12). The reaction mechanism was found to involve an adenylylation of the enzyme protein according to the following equation (13, 54):

12 ATP + glutamine synthetase → glutamine synthetase

$$(-AMP)_{12} + 12 \text{ PP.} \quad (4)$$

In this reaction adenylic acid is attached to the phenolic hydroxyl group of a tyrosyl residue in phosphodiester linkage (55).

After removal of NH_4^+ from the growth medium a rapid reactivation of glutamine synthetase could be observed in intact *E. coli* cells (56). Studies in a cell-free system provided evidence that the reactivation is due to an enzyme-catalyzed deadenylylation (57, 58, 59). The reactivating (i.e., deadenylylating) system was studied in detail by Anderson et al. (60), who found that two different proteins are involved. One protein is likely to be identical with the inactivating (i.e., adenylylating) enzyme. The other protein appears to be required to modify the system in such a way that in the presence of specific effectors the deadenylylation reaction is catalyzed (60). In addition these workers found that the deadenylylation is caused, not by hydrolysis, but by phosphorolysis according to the following equation:

Glutamine synthetase$(-AMP)_{12} + 12 \text{ P} \rightarrow$ glutamine synthetase

$$+ 12 \text{ ADP.} \quad (5)$$

The reactions by which glutamine synthetase is modified are summarized in Fig. 6. In analogy to the nomenclature of glycogen phosphorylase, the adenylylated inactive form and the deadenylylated active form of the enzyme are designated as glutamine synthetase b and glutamine synthetase a, respectively. The main effectors controlling the adenylylating and deadenylylating enzyme systems are also included in Fig. 6.

It is obvious that the above mechanism of regulation is advantageous for the economy of the cells. The product of the glutamine synthetase reaction favors the formation of the inactive glutamine synthetase b by stimulation of inactivation and inhibition of activation; conversely, the precursor of the

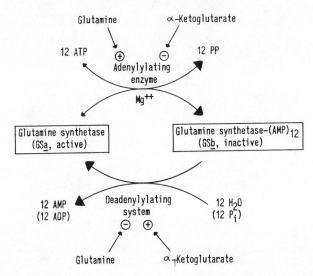

Fig. 6. Enzyme-catalyzed interconversion of glutamine synthetase (GS) from an active (a) to an inactive (b) form and vice versa.

glutamine synthetase reaction, α-ketoglutarate, favors the active form of glutamine synthetase by stimulation of activation and inhibition of inactivation.

According to Mantel and Holzer (61), the free energy of hydrolysis of the adenylyl-O-tyrosine bond in glutamine synthetase b is about 9 kcal/mole, which is comparable to that of the pyrophosphate bond of ATP (10 kcal/mole). The energy of the α-β-pyrophosphate bond of ATP, therefore, is conserved to a great extent in the adenylyl-O-tyrosine bond. In the phosphorolytic cleavage of this bond, which has been found by Anderson et al. (60), this energy can be utilized for the synthesis of ADP from orthophosphate and enzyme-bound AMP.

From the foregoing thermodynamic data one would expect that the equilibrium constant of the phosphorylase reaction (equation 5) is not very different from unity. Probably, the reaction is shifted toward the formation of glutamine synthetase a and ADP by an as-yet-unknown mechanism. Such a unidirectional course of the reaction must be anticipated if the controlling effectors, glutamine and α-ketoglutarate, are to exert their regulatory influence on the reactivation of glutamine synthetase b in a biologically useful way. In the adenylylation of glutamine synthetase a the reaction proceeds in one direction, since the liberated pyrophosphate is removed from the equilibrium by hydrolysis.

10. DISCUSSION

For a long time glycogen phosphorylase from mammalian tissues was the only enzyme known to be regulated by ATP-dependent enzyme-catalyzed chemical modification. During the last 10 years, however, four additional enzymes have been found to exist in two forms of different catalytic activity which are interconvertible by enzymatic phosphorylation and dephosphorylation, respectively (Fig. 7). At present, enzymatic modification by ATP-dependent adenylylation has been demonstrated only in the case of glutamine synthetase from E. coli. However, recent results by Goff and Weber (62) and by Chelala et al. (68) with RNA polymerase from E. coli indicate that this enzyme may also be modified by an ATP-dependent adenylylation.

Most enzymes from mammalian tissues, for which a regulation by phosphorylation and dephosphorylation has been observed (cf. Table 1), are key enzymes of carbohydrate metabolism. Cyclic 3′,5′-AMP serves as an important effector in the enzymatic interconversions between the different forms of these enzymes (for a summary see ref. 63). The activity of phosphofructokinase is also increased by cyclic AMP (64), although a modification of the enzyme by phosphorylation has not been established.

Since synthesis and degradation and, therefore, the level of cyclic AMP are controlled by several hormones, chemical modification of enzymes by phosphorylation and dephosphorylation represents an important step in the hormonal control of carbohydrate metabolism in mammalian tissues. Probably, cyclic AMP-mediated hormonal control of several other metabolic processes is also due to enzymatic phosphorylation of proteins. For example, it should be mentioned that cyclic AMP stimulates not only the kinase reactions shown in Fig. 1 but also the phosphorylation of other proteins (63). The phosphorylation of histones is especially dependent on the presence of cyclic AMP (65). This observation suggests that enzyme-catalyzed chemical modification of proteins may also be involved in the hormonal regulation of gene activity in eukaryotic cells.

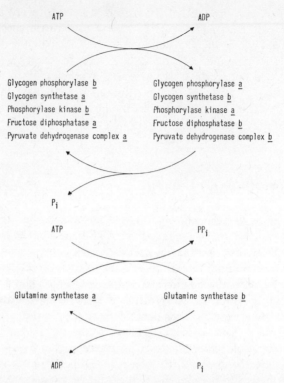

Fig. 7. ATP-dependent interconversion of enzymes between molecular forms of different catalytic activity by chemical modification: (*a*) catalytically active form, (*b*) catalytically inactive or less active form.

Control by enzymatic modification is distinguished by several characteristic properties from other mechanisms of enzyme regulation.

A. Reversible Production of Different Stable Forms of an Enzyme

In contrast to physical modification by allosteric effectors or regulatory proteins, chemical modification leads to the production of stable forms of the controlled enzymes exhibiting different catalytic and regulatory properties. Thus the effect of a modifying agent is maintained even after the effector has been removed or inactivated. Reversion of the modification is not initiated unless an effector for the back reaction is present.

The biological significance of this regulatory effect is illustrated in the case of glutamine synthetase from *E. coli*, where glutamine causes an inactivation of the biosynthetic activity by adenylylation of the enzyme. Even after glutamine has been exhausted, the enzyme remains inactive until the precursor of glutamine synthesis, α-ketoglutarate, which stimulates the deadenylylation, has accumulated. Reactivation of the enzyme immediately after the depletion of the glutamine pool in the absence of α-ketoglutarate would lead to a considerable decrease in the glutamate concentration since no glutamate could be formed from α-ketoglutarate. This would be disadvantageous for the cells because glutamate is required not only for the synthesis of glutamine but also for other important reactions (e.g., transaminations). Thus it seems useful for the economy of the cells that the *presence* of α-ketoglutarate and not the *absence* of glutamine releases the inhibition of glutamine biosynthesis.

B. Rate of Control

Chemical modification results in a change in the number of active enzyme molecules. The number of enzymatically active protein molecules may also be regulated by the control of *de novo* synthesis by induction or repression. There is, however, a striking difference in the rates of these two mechanisms. For instance, in mammalian tissues the adaptation to metabolic requirements by induction or repression is rather slow and usually takes several hours before becoming effective. By contrast, phosphorylation and dephosphorylation of interconvertible enzyme forms can be completed within a few minutes. In *E. coli*, glutamine synthetase is more than 90% inactivated by adenylylation within 120 sec (52). On the other hand, even under conditions of full repression of enzyme synthesis, many cell divisions are required to dilute the enzyme to negligible levels. Complete reactivation of inactive glutamine synthetase by deadenylylation is achieved within a few minutes, whereas under comparable experimental conditions it takes about 1 hr to reach maximal activity by *de novo* synthesis (52, 56).

C. Energy Requirements

In comparison to regulation by induction or repression of enzyme synthesis, little energy is required for the control of activity by chemical modification. On the assumption that a subunit which is phosphorylated or adenylylated once in the course of the modifying reaction consists of 200–800 amino acids, the amount of energy required for *de novo* synthesis of an enzyme will be several orders of magnitude higher than that consumed in the modification reaction. Also, inactivation by proteolytic degradation represents a much higher waste of energy than inactivation by chemical modification.

In principle, energy could be wasted by a continuous cycle of phosphorylation and dephosphorylation or adenylylation and deadenylylation of an enzyme. However, such "futile cycles" do not appear to be significant *in vivo* (66). The activities of the modifying enzymes which could catalyze such cycles are regulated in a manner that prevents simultaneous antagonistic function.

D.   Persistence of Partial Activities of Multifunctional Enzymes

Another aspect of chemical modification is the possibility that in multifunctional enzymes one partial activity of the modified enzyme may remain unimpaired while another is completely abolished. This is the case with glutamine synthetase from *E. coli*. Under certain conditions adenylylation of this enzyme leads to the inactivation of its biosynthetic activity, whereas its γ-glutamyltransferase activity is only slightly decreased. The transferase activity can be determined with hydroxylamine as an acceptor according to the following equation:

$$\text{Glutamine} + \text{hydroxylamine} \rightarrow \gamma\text{-glutamyl hydroxamic acid} + NH_3. \quad (6)$$

Although no biological acceptor for the γ-glutamyl group has been found as yet, the existence of a differential effect of adenylylation on the biosynthetic activity and the transferase activity, respectively, suggests that the latter is of biological significance. If the intracellular concentration of glutamine is sufficiently high, it is advantageous for the economy of the cells to inhibit further synthesis. However, if the transfer of the γ-glutamyl group is essential in a biosynthetic reaction, it is also important to maintain the glutamine transferase activity. Obviously, both objectives are achieved by the adenylylation of glutamine synthetase.

E.   Increase in the Number of Effector Sites by Augmentation of the Available Amount of Protein

In the three-dimensional structure determining a catalytic site or an effector binding site of a protein, several functional groups are arranged at specific distances and angles. To achieve the correct structure a framework composed of a considerable number of amino acids is required. It can be reasonably estimated that between 40 and 200 amino acids, representing a molecular weight of 5000–25,000, are involved in the formation of a single functional site. Clearly, involvement of several different proteins in one and the same regulatory mechanism constitutes a considerable increase in the amount of protein available for the realization of additional functional sites which will permit a more sensitive and refined control. This consideration could explain why the enzymes regulated by chemical modification generally are key enzymes of metabolic pathways. Such key enzymes have to respond to a great number of different controlling effectors, and therefore their regulation requires a multitude of different controlling sites.

F. Amplification

Finally, it should be noted that enzyme-catalyzed chemical modification provides a possibility for the amplification of regulatory signals. Bowness (67) has tried to calculate the extent of the amplification of an epinephrine stimulus in the activation cascade of glycogenolysis, comparing the concentrations of the hormone administered and of glucose-1-phosphate produced on a molar basis. From experimental data of Krebs et al. (26, 69) and Illingworth and Cori (70) it was estimated that in response to 1 molecule of epinephrine 10–20 molecules of cyclic AMP are formed within 1 min and that 1 mole of cyclic AMP leads in turn to the formation of 1600 moles of glucose-1-phosphate per minute. For the overall amplification of the epinephrine signal Helmreich et al. (71) found a somewhat lower factor of about 200.

11. SUMMARY

The ATP-dependent chemical modification of enzymes represents a regulatory mechanism for which an increasing number of examples have been detected recently. Modification by phosphorylation and dephosphorylation has been demonstrated for the following enzymes from mammalian tissues: glycogen phosphorylase, phosphorylase *b* kinase, glycogen synthetase, fructose diphosphatase, and the pyruvate dehydrogenase complex. There is also experimental evidence that phosphorylase *a* phosphatase, glycogen synthetase phosphatase, lipase, and palmityl-CoA synthetase are regulated by a phosphorylation-dephosphorylation mechanism. In *E. coli* regulation of glutamine synthetase has been shown to involve modification of the enzyme by adenylylation and deadenylylation. A similar regulatory mechanism probably operates in the case of RNA polymerase from *E. coli*.

The characteristic properties and the biological significance of metabolic control by enzyme-catalyzed chemical modification were discussed in the preceding chapter.

REFERENCES

1. F. Lynen, *Liebigs Ann. Chem.*, **546,** 120 (1941).
2. M. J. Johnson, *Science*, **94,** 200 (1941).

3. G. Cori and A. A. Green, *J. Biol. Chem.*, **151**, 31 (1943).

4. E. H. Fischer and E. G. Krebs, *J. Biol. Chem.*, **216**, 121 (1955).

5. R. J. DeLange, R. G. Kemp, W. D. Riley, R. A. Cooper, and E. G. Krebs, *J. Biol. Chem.*, **243**, 2200 (1968).

6. D. A. Walsh, J. P. Perkins, and E. G. Krebs, *J. Biol. Chem.*, **243**, 3763 (1968).

7. W. D. Riley, R. J. DeLange, G. E. Bratvold, and E. G. Krebs, *J. Biol. Chem.*, **243**, 2209 (1968).

8. C. Villar-Palasi and J. Larner, *Arch. Biochem. Biophys.*, **94**, 436 (1961).

9. J. Mendicino, C. Beaudreau, and R. N. Bhattacharyya, *Arch. Biochem. Biophys.*, **116**, 436 (1966).

10. T. C. Linn, F. H. Pettit, and L. J. Reed, *Proc. Natl. Acad. Sci. U.S.*, **62**, 234 (1969).

11. O. Wieland and E. Siess, *Hoppe-Seylers Z. Physiol. Chem.*, **350**, 1160 (1969).

12. D. Mecke, K. Wulff, K. Liess, and H. Holzer, *Biochem. Biophys. Res. Commun.*, **24**, 452 (1966).

13. H. S. Kingdon, B. M. Shapiro, and E. R. Stadtman, *Proc. Natl. Acad. Sci. U.S.*, **58**, 1703 (1967).

14. W. Merlevede and G. A. Riley, *J. Biol. Chem.*, **241**, 3517 (1966).

15. C. A. Chelala and H. N. Torres, *Biochim. Biophys. Acta*, **198**, 504 (1970).

16. W. Stalmans, H. DeWulf, B. Lederer, and H.-G. Hers, *European J. Biochem.*, **15**, 9 (1970).

17. H. DeWulf, W. Stalmans, and H.-G. Hers, *European J. Biochem.*, **15**, 1 (1970).

18. J. D. Corbin, E. M. Reimann, D. A. Walsh, and E. G. Krebs, *J. Biol. Chem.*, **245**, 4849 (1970).

19. M. Farstad, *Biochim. Biophys. Acta*, **146**, 272 (1967).

20. S. S. Hurd, D. Teller, and E. H. Fischer, *Biochem. Biophys. Res. Commun.*, **24**, 79 (1966).

21. E. Helmreich, M. C. Michaelides, and C. F. Cori, *Biochemistry*, **6**, 3695 (1967).

22. L. L. Kastenschmidt, J. Kastenschmidt, and E. Helmreich, *Biochemistry*, **7**, 3590 (1968).

23. L. L. Kastenschmidt, J. Kastenschmidt, and E. Helmreich, *Biochemistry*, **7**, 4543 (1968).

24. Chr. Nolan, W. B. Novoa, E. G. Krebs, and E. H. Fischer, *Biochemistry*, **3**, 542 (1964).

25. D. P. Wolf, E. H. Fischer, and E. G. Krebs, *Biochemistry*, **9**, 1923 (1970).

26. E. G. Krebs, D. J. Graves, and E. H. Fischer, *J. Biol. Chem.*, **234**, 2867 (1959).

27. R. B. Huston and E. G. Krebs, *Biochemistry*, **7**, 2116 (1968).

28. E. Ozawa, K. Hosoi, and S. Ebashi, *J. Biochem.* (*Japan*), **61**, 531 (1967).

29. R. G. Hansford and B. Sacktor, *FEBS Letters*, **7**, 183 (1970).

30. S. S. Hurd, W. B. Novoa, J. P. Hickenbottom, and E. H. Fischer, in *Methods in Enzymology*, Vol. 8 (Eds.: S. P. Colowick and N. O. Kaplan), Academic Press, New York, 1966, p. 546.

31. C. A. Chelala and H. N. Torres, *Biochim. Biophys. Acta*, **178**, 423 (1969).

32. W. Merlevede, J. Goris, and C. DeBrandt, *European J. Biochem.*, **11**, 499 (1969).

33. J. Larner, C. Villar-Palasi, N. D. Goldberg, J. S. Bishop, F. Huijing, J. I. Wenger, H. Sasko, and N. B. Brown, *Advan. Enzyme Regulation*, **6**, 409 (1968).

34. J. Larner, *Trans. N.Y. Acad. Sci.*, **29**, 192 (1966).

35. H. J. Mersmann and H. L. Segal, *Proc. Natl. Acad. Sci. U.S.*, **58**, 1688 (1967).

36. H. DeWulf, W. Stalmans, and H.-G. Hers, *European J. Biochem.*, **6**, 545 (1968).

37. D. L. Friedmann and J. Larner, *Biochemistry*, **4**, 2261 (1965).

38. E. Belocopitow, M. Del Carmen Garcia Fernandez, L. Birnbaumer, and H. N. Torres, *J. Biol. Chem.*, **242**, 1227 (1967).

39. R. T. Soderling, J. P. Hickenbottom, E. M. Reimann, F. Hunkeller, D. A. Walsh, and E. G. Krebs, *J. Biol. Chem.*, **245**, 6317 (1970).

40. J. Mendicino, H. S. Prihar, and F. M. Salama, *J. Biol. Chem.*, **243**, 2710 (1968).

41. N. Kratovich and J. Mendicino, *J. Biol. Chem.*, **245**, 2483 (1970).

42. L. J. Reed and D. J. Cox, *Ann. Rev. Biochem.*, **35**, 57 (1966).

43. L. J. Reed and R. M. Oliver, *Brookhaven Symp. Biol.*, **21**, 397 (1969).

44. O. Wieland and B. von Jagow-Westermann, *FEBS Letters*, **3**, 271 (1969).

45. T. C. Linn, F. H. Pettit, F. Hucho, and L. J. Reed, *Proc. Natl. Acad. Sci. U.S.*, **64**, 227 (1969).

46. O. Wieland and E. Siess, *Proc. Natl. Acad. Sci. U.S.*, **65**, 947 (1970).

47. J. L. Skosey, *J. Biol. Chem.*, **241**, 5108 (1966).

48. O. Wieland, E. Siess, and F. H. Schulze-Wethmar, *First International Symposium on Metabolic Interconversion of Enzymes*, St. Margherita, 1970, p. 52.

49. L. J. Reed, T. C. Linn, F. H. Pettit, and F. Hucho, *First International Symposium on Metabolic Interconversion of Enzymes*, St. Margherita, 1970, p. 48.

50. J. K. Huttunen, D. Steinberg, and St. E. Mayer, *Proc. Natl. Acad. Sci. U.S.*, **67**, 290 (1970).

51. J. K. Huttunen, D. Steinberg, and St. E. Mayer, *Biochem. Biophys. Res. Commun.*, **41**, 1350 (1970).

52. D. Mecke and H. Holzer, *Biochim. Biophys. Acta*, **122**, 341 (1966).

53. D. Mecke, K. Wulff, and H. Holzer, *Biochim. Biophys. Acta*, **128**, 559 (1966).

54. K. Wulff, D. Mecke, and H. Holzer, *Biochem. Biophys. Res. Commun.*, **28**, 740 (1967).

55. B. M. Shapiro and E. R. Stadtman, *J. Biol. Chem.*, **243**, 3769 (1968).

56. L. Heilmeyer, Jr., D. Mecke, and H. Holzer, *European J. Biochem.*, **2**, 399 (1967).

57. F. Battig, L. Heilmeyer, Jr., and H. Holzer, *Fifth Meeting of the Federation of European Biochemical Societies (FEBS)*, Prague, 1968, Abstr. 775, p. 194.

58. B. M. Shapiro and E. R. Stadtman, *Biochem. Biophys. Res. Commun.*, **30**, 32 (1968).

59. L. Heilmeyer, Jr., F. Battig, and H. Holzer, *European J. Biochem.*, **9**, 259 (1969).

60. W. B. Anderson, S. B. Hennig, A. Ginsburg, and E. R. Stadtman, *Proc. Natl. Acad. Sci. U.S.*, **67**, 1417 (1970).

61. M. Mantel and H. Holzer, *Proc. Natl. Acad. Sci. U.S.*, **65**, 660 (1970).

62. C. G. Goff and K. K. Weber, *Cold Spring Harbor Symp. Quant. Biol.*, **35**, 101 (1970).

63. G. A. Robison, R. W. Butcher, and E. W. Sutherland, *Ann. Rev. Biochem.*, **37**, 149 (1968).

64. T. E. Mansour and C. E. Ahlfors, *J. Biol. Chem.*, **243**, 2523 (1968).

65. R. H. Stellwagen and R. D. Cole, *Ann. Rev. Biochem.*, **38**, 951 (1969).

66. H.-G. Hers, *Advan. Enzyme Regulation*, **8**, 171 (1970).

67. J. M. Bowness, *Science*, **152**, 1370 (1966).

68. C. A. Chelala, L. Hirschbein, and H. N. Torres, *Proc. Natl. Acad. Sci. U.S.*, **68,** 152 (1971).

69. J. B. Posner, R. Stern, and E. G. Krebs, *J. Biol. Chem.*, **240,** 982 (1965).

70. B. Illingworth and G. T. Cori, *Biochem. Preparations*, **3,** 1 (1953).

71. E. Helmreich, S. Karpatkin, and C. F. Cori, in *Control of Glycogen Metabolism* (Ciba Foundation Symposium), (Ed.: W. J. Whelan), Churchill, London, 1964, p. 223.

CHAPTER 6

Chemotrophic Basis for and Regulation of Protein Turnover: An Irreversible Type of Elastoplastic Enzyme Modification

SANTIAGO GRISOLIA AND W. HOOD

Department of Biochemistry and Molecular Biology, University of Kansas Medical Center, Kansas City, Kansas

137

1. INTRODUCTION

To explain the specificity and mechanism of enzyme action Fisher (1) suggested in 1894 the lock-key and template concepts; they required that the substrate and the enzyme have complementary surfaces. In chemical language, this means that, because of steric hindrance, substrate molecules that are very large, small, or geometrically different from the enzymic active site will not react or will do so with difficulty. The induced fit theory of some 60 years later (2) suggests that some conformational changes are induced by the substrate on the enzyme. An excellent discussion of strain, distortion, and conformational changes of both enzyme and substrate can be found in the book of Jencks (3).

It is by now a classical concept in enzyme chemistry that an enzyme may be protected by substrates or cofactors against denaturation, and many examples of this can be found in the literature. Only relatively recently, however, has the *phenomenon of inactivation of an enzyme by substrate or substrate-analog been recognized* (4). Many investigators, particularly in the last few years, have come to realize the close relationship of protection and inactivation of enzymes by substrates. Moreover, others have pointed out the fact that the introduction of a covalently bonded group or pseudosubstrate may enhance *or* decrease the stability of enzymes. For example, Martin and Bhatnager (5) have illustrated this extensively. The elastoplasticity theory (4) explains how the stress produced by the conformational change induced by substrate(s)* can facilitate inactivation of enzymes by physical, chemical, or biological agents. In cases where there is enzyme inactivation, the conformational change induced by the substrate when the reaction does not proceed to completion will probably result in slow inactivation (unless an additional effector molecule is present and/or rapid dissociation into subunits occurs). Thus this type of regulation was not recognized as early as others which rely on kinetic effects or alterations of pathways. An admittedly inaccurate but possibly pedagogically useful scheme of these points is shown in Fig. 1.

A few years ago it was found that substrates apparently can produce modification of enzymes in other ways than via the active site with covalent bond formation (6) and can, in fact, modify enzymes which are not related to either their synthesis or their utilization. Examples are the inactivation of triosephosphate dehydrogenase or of ribonuclease by carbamyl phosphate, as illustrated in Table 1. These findings clearly indicate a new modality of protein modification due to substrates and clarify several effects previously observed. Moreover, in this context, substrates can then exert a regulatory action on closely connected enzymes or on a closely connected sequence, and thus exert a feedback type of effect. An example is the inactivation of glutamate dehydrogenase by carbamyl phosphate (6), which will be discussed in detail in Section 2.D. Although these effects could be considered, in essence, allosteric or quasiallosteric, they are to a greater degree chemical effects, since they result in covalent bond formation between reactive amino

---

* This also includes allosteric modifiers.

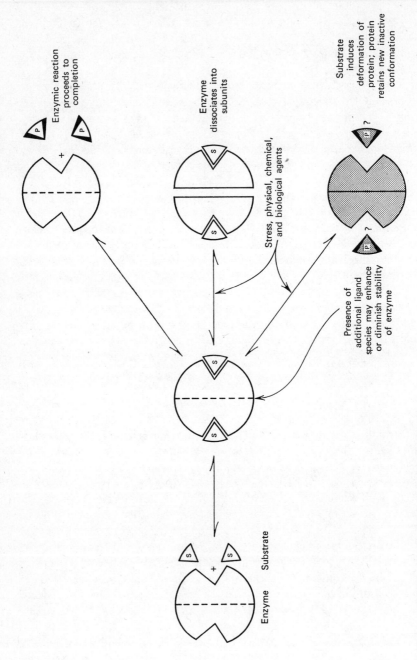

Fig. 1. Substrate-induced inactivation of enzyme.

Enzymic reaction proceeds to completion

Enzyme dissociates into subunits

Substrate induces deformation of protein; protein retains new inactive conformation

Stress, physical, chemical, and biological agents

Presence of additional ligand species may enhance or diminish stability of enzyme

Enzyme   Substrate

140

acid residues of the enzyme and the substrate. The term *chemotrophic* was, therefore, proposed to describe these effects. In the particular cases where they affect regulation at the beginning of a chain they will be designated as chemotrophic feedback, as illustrated in Fig. 2.

Elastoplasticity is a specific phenomenon since it is related to the changes induced by substrate(s) at the active site(s) of an enzyme.* On the other hand,

Table 1. Effect of Carbamyl Phosphate on Activity and Carbamylation of Several Enzymes

Unpublished Data of Murdock, Chabas, Bettis, and Grisolia.

Enzyme	A, % Remaining Activity	B, nM Carbamyl Protein
Beef liver glutamic dehydrogenase	2	190, 250
Beef brain glutamic dehydrogenase	0	N.D.[a]
Yeast glutamic dehydrogenase	0	N.D.
Rabbit muscle glycerophosphate dehydrogenase	29	Traces
Enolase	100	N.D.
Alcohol dehydrogenase	100	N.D.
Ribonuclease	0	235, 78
Trypsin	N.D.	191
Catalase	N.D.	0
Urease	100	0

NOTE TO TABLE 1

From 1 to 5 mg of enzyme was incubated with 0.08 M carbamyl phosphate at pH 7.0 for 1 hr at 38°. Enzyme activities were measured by standard procedures.
[a] ND: not determined.

chemotrophic effects, in which substrates (or possibly other compounds not related to enzymes directly connected with their synthesis or utilization) can covalently attach to an enzyme, may be largely unspecific. Indeed, the chemotrophic agents thus far studied may be attached either at the active site of the enzyme or at sites remote from the active site. Although, as mentioned previously, many of these effects may be unrelated to the active site, there is evidence of some specificity and/or affinity, as shown by the higher reactivity of some susceptible groups that on a chemical basis should

* Includes also allosteric sites.

have been equally reactive, for example, one lysine out of some twenty reactive lysines per subunit of glutamic dehydrogenase (see below).

The nonspecific nature of these effects could indicate a more widespread role and possibly a regulatory mechanism in cellular metabolism other than the control of some aspects of protein turnover. As will be illustrated with many examples, it is rapidly becoming apparent that reactive molecules such as carbamyl phosphate and ATP modify many proteins by covalent

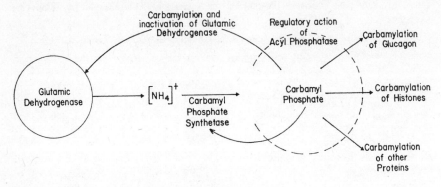

Fig. 2.   Chemotrophic regulation by carbamyl phosphate.

attachment of either the carbamyl, phosphoryl, and adenylyl or the phosphoryl portions of these reagents. Moreover, it can be postulated that other highly reactive compounds such as CoA derivatives may be found to be chemotrophic effectors.

Further work on the clarification and physiological significance of the phenomenon of inactivation of enzymes by substrates in the elastoplasticity sense remains to be done. Nevertheless, this has been an exciting and rapidly developing area which has, indeed, expanded so much that it can no longer be covered thoroughly in a single chapter. Therefore the reader is referred to a number of recent reviews and papers (7–14). In this presentation we shall concentrate only on chemotrophic effects or closely related phenomena. For convenience some examples of chemical modifications of proteins by reagents that are not substrates (but that yield similar products) will be presented. Of course, the number of chemotrophic cases which have been studied thus far is limited. For simplification these effects have been divided somewhat arbitrarily according to the main modification reactions, namely, carbamylation, acylation, phosphorylation, and adenylylation; some of the biological implications are briefly discussed, particularly at the end of the chapter.

2. CARBAMYLATION

A. Biological Basis for Postulating the Importance of Carbamyl Phosphate in Protein Modification

The main site of carbamyl phosphate synthesis in ureotelic animals is the liver mitochondria (other pathways for carbamyl phosphate synthesis in animal tissues are quantitatively less important). Man can make at least 0.5 mole of urea a day; and, since 1 mole of carbamyl phosphate is necessary per mole of urea synthesized in the liver, it is self-evident that at least some 0.5 mole or more of carbamyl phosphate can be made per day in the 300–500 ml occupied by the mitochondria! Taking into consideration the high reactivity of carbamyl phosphate, it is not surprising that, if a small portion of this effector escapes interaction with ornithine to form citrulline, it may react with a number of acceptors, including proteins.

Although Stark (15), who has studied in detail the carbamylation of proteins by cyanate (see Section 2.C), considered protein carbamylation via carbamyl phosphate unlikely, carbamylation of proteins by carbamyl phosphate has recently been shown to occur in a number of cases. In fact, in unpublished comparative studies with cyanate (Silverstein, Chabas, Bettis, and Grisolia) carbamyl phosphate has been found to be a very good carbamylating agent; although in general cyanate is superior for this purpose at low pH's, carbamyl phosphate can be just as good. Of course nonenzymatic carbamylation, in the chemotrophic sense, appears to proceed in all cases thus far tested via cyanate.

B. Relationships between Carbamyl Phosphate, Cyanate, and Urea

In view of the close relationship between cyanate, urea, and carbamyl phosphate, carbamylation of proteins (and other acceptors) by cyanate is of interest and should be briefly considered. The decomposition of carbamyl phosphate, particularly at neutral or alkaline pH's, involves the intermediate formation of cyanate; therefore the possibility that cyanate is an intermediate in carbamylation reactions carried out by carbamyl phosphate, including transcarbamylation reactions, has not been excluded. Indeed, even in the enzymic hydrolysis of urea, the intermediate formation of carbamyl phosphate has been considered by Varner (16). However, recent experiments (17) have clearly excluded the intermediate formation of cyanate during enzymic hydrolysis of carbamyl phosphate and during transcarbamylation to ornithine

(Carreras, Chabas, and Grisolia), thus making even more unlikely the intermediate formation of cyanate during other enzyme reactions involving carbamyl phosphate. On the other hand, as further discussed below, the possibility exists that urea may act as a carbamylating agent via carbamyl phosphate.

C. Chemical Basis for Carbamylation and Examples of Direct Carbamylation by Cyanate

On extension of studies leading to the use of cyanate for the determination of NH_2-terminal residues in proteins, Stark and Smyth (18) showed that in addition to reacting with amino groups cyanate reacts reversibly with sulfhydryl groups of proteins (19) and with carboxyl groups (15). Stark (15) predicted that the carbamylation of α-amino groups in proteins by cyanate at pH 7 or below occurred 100 times faster than that of ϵ-amino groups, and he related the rate constant for carbamylation to the pK_a of several amino acids and peptides. This is shown in Table 2. Cejka et al. (20) claimed similar preferential reactivity of the α-amino groups when studying the carbamylation of globin by urea. It appears from these studies that hydroxyl or sulfhydryl

Table 2. Apparent Specific Rate Constants for Reactions of Amino Acids and Peptides with Potassium Cyanate

Data from Stark (15).

R-NH$_2$	pK_a	[KNCO], M	$k_1 \times 10^3$, M^{-1}/min
Tetraglycine	7.75	0.2	275, 279
Triglycine	7.91	0.2	213, 200
Glycylglycine	8.17	0.2	129
		0.4	150
Threonine	9.12	0.2	33.5
		0.4	31.0
Glycine	9.60	0.2	20.2, 20.2
		0.4	21.6
Alanine	9.69	0.2	8.42
		0.8	8.13
		1.0	8.56
β-Alanine	10.19	0.8	5.59
		1.0	5.74
ϵ-Aminocaproic acid	10.75	1.0	2.04
		2.0	1.98

Table 3.   Carbamyl Content of Pepsin on Carbamylation with Potassium Cyanate and After Treatment with Hydroxylamine

Data from Rimon and Perlmann (21).

Protein	Carbamyl Groups per Molecule		Relative Specific Activity	
	Ultraviolet Absorption	^{14}CNO Incorporated	Hemoglobin	Synthetic Substrate
Pepsin, control	None	None	100	100
NH$_2$OH-treated				
Carbamylpepsin I	5.8		42	278
NH$_2$OH-treated	0		80	120
Carbamylpepsin II	5.6	4.5		
NH$_2$OH-treated	0			

groups which on carbamylation do not increase the negative charge of the molecule are decomposed under slightly alkaline conditions and thus should be considered in terms of cell physiology only under acidotic conditions. Indeed, at pH 6 cyanate reacts with carboxyl groups (15) as follows:

$$R_1{-}COOH + HCNO + H_2O \longrightarrow R_1{-}\overset{\overset{\displaystyle O}{\|}}{C}{-}O{-}\overset{\overset{\displaystyle O}{\|}}{C}NH_2,$$

which can then react with an amino group as follows:

$$R_1{-}\overset{\overset{\displaystyle O}{\|}}{C}{-}O{-}\overset{\overset{\displaystyle O}{\|}}{C}{-}NH_2 + R_2{-}NH_2 \longrightarrow R_1{-}\overset{\overset{\displaystyle O}{\|}}{C}{-}NH{-}R_2 + NH_3 + CO_2.$$

Therefore a cross-link between the ϵ-amino group of lysine and glutamic or aspartic carboxyl residues could be mediated by cyanate, *without* carbamylation of the end product.

Results different at first glance from those of Stark (15) were obtained by Rimon and Perlmann (21), who found that cyanate reacted with pepsin and with pepsinogen. With pepsin, the activity decreased on using hemoglobin as a substrate and, interestingly, the activity with a synthetic substrate increased. Moreover, these changes were reversed by decarbamylation with hydroxylamine (Table 3). With pepsinogen, nine of the ten ϵ-NH$_2$ groups of lysine were carbamylated (to homocitrulline) but not the α-NH$_2$-terminal of

Table 4. Specific Optical Rotation ($[\alpha]_{366}$) and Potential Pepsin Activity of Pepsinogen and Carbamylated Pepsinogens

Data from Rimon and Perlmann (21).

Substance	Number of Lysine Residues Modified	$-[\alpha]_{366}$	Change, %	Loss of Potential Pepsin Activity, %
Pepsinogen	0	212	None	0
Carbamylpepsinogen	3	212	None	0
Carbamylpepsinogen	4	259	33.0	17.0
Carbamylpepsinogen	7	275	74.0	68.0
Carbamylpepsinogen	8	293	95.5	74.0
Carbamylpepsinogen	9	297	100	86.0

leucine. These divergent results may possibly be due to different conditions. As already indicated, Stark (15) carbamylated proteins at pH 7, whereas Rimon and Perlmann (21) used pH's in the range of 8–9, where reaction also occurs with four tyrosine residues. Progressive carbamylation of pepsinogen results in a change in the molecular configuration of the protein, which is accompanied by a decreased susceptibility to activation to pepsin (Table 4). A more recent paper by Grizzuti and Perlmann (22) presents extensive data on optical rotatory dispersion and circular dichroism of pepsinogen, carbamylpepsinogen, and also succinylpepsinogen. Table 5 illustrates some of the

Table 5. Properties of Pepsinogen, Carbamylpepsinogen, and Succinylpepsinogen

Data from Grizzuti and Perlmann (22).

Property	Pepsinogen	Carbamyl-pepsinogen	Succinyl-pepsinogen
Number of lysine residues modified per molecule of protein	None	9	10
$[\eta]$, $(g/ml)^{-1}$	3.0	8.7	24.0
$-[\alpha]_{366}$	210	297	320
λc	236	220	224
Potential pepsin activity, %	100	16	<10

properties of pepsinogen and of modified pepsinogen. Thus, according to the results of Perlmann and his coworkers (22), modification of an enzyme by a carbamylating agent may result in altered enzymic properties either by causing a conformational change of the protein or by action at the active site.

As shown by Avramovic and Madsen (23), inactivation of phosphorylase b by cyanate also involves the carbamylation of the ϵ-amino groups of lysine to form homocitrulline. The rate of inactivation of the enzyme is affected by various ligands. For example, with increasing additions of the activator AMP the rate of enzyme inactivation decreases. Inorganic phosphate, glucose-1-phosphate, and the competitive inhibitor UDP-glucose have no effect on the rate of inactivation by cyanate, but *in the presence of AMP* they provide further protection. The allosteric inhibitors ATP, ADP, and glucose-6-phosphate do not affect inactivation but antagonize the protection offered by AMP. The interrelationships among these ligands in the stability context are analogous to those seen in kinetic studies. They have been interpreted on the basis of the proposed model for allosteric transitions between two conformational states. This interpretation assumes that the enzyme in these different conformations is inactivated at different rates by cyanate. Moreover, carbamylation may result in a conformational change.

The inactivation of subtilisin type Novo by cyanate also involves carbamylation of the lysine residues, converting them into unchanged homocitrulline residues, as reported by Svendsen (24). Sedimentation and optical rotatory dispersion studies revealed that both the ϵ-amino groups and the α-amino group of the enzyme were freely exposed to react with cyanate; in contrast to the case of carbamylpepsinogen, however, the carbamylation of these amino groups did not significantly disrupt the native conformation of the molecule and apparently produced only minor changes in enzymic activity. It was suggested that cyanate could inactivate by reaction with serine at the active center.

Keech and Farrant (25) have shown that cyanate, as well as other amine reagents, inactivates sheep kidney pyruvate carboxylase and that a lysyl residue is reactive. Of particular interest is the demonstration that the allosteric effector acetyl-CoA, which is an absolute requirement for catalytic activity, protects against inactivation.

Carbamylation has helped to clarify the nature of the binding of CO_2 to hemoglobin. This involves a carbamino reaction between CO_2 and an α-amino-group hemoglobin molecule as follows: $\alpha\text{-}NH_2 + CO_2 \rightleftharpoons \alpha\text{-}NHCOO^- + H^+$. That the α-amino group is involved in the binding of CO_2 has been demonstrated by Kilmartin and Rossi-Bernardi (26). When the α-amino group of the α chain of horse hemoglobin was carbamylated with cyanate,* the Bohr effect was reduced by 30%. When the α-amino groups of both the α and β chains were carbamylated, there was no oxygen-linked CO_2 binding.

D.   Carbamylation with Carbamyl Phosphate

At the physiological concentrations of glutamic dehydrogenase and carbamyl phosphate present in liver this reagent rapidly inactivates the dehydrogenase, resulting in carbamylation of the protein (6). At the physiological concentrations which may occur in mitochondria, there was essentially a complete loss of enzyme activity in 1 hr of incubation at 38° and at pH 7.4. This is illustrated in Fig. 3. Thin-layer chromatography revealed that carbamylation involved lysine exclusively since homocitrulline was the only carbamyl amino acid detected. On the basis of the chromogenecity and with the use of a sensitive carbamyl method it appeared that carbamylation of approximately 4% of the lysine (one lysine per subunit) of bovine glutamic dehydrogenase results in complete inactivation. This finding seems of particular importance since recently Smith et al. (27) have shown that the lysine in position 97 of bovine liver glutamic dehydrogenase (subunit approximately 60,000 molecular weight) is an essential residue which can be modified with pyridoxal-5'-phosphate (28); inactivation of this enzyme by carbamyl phosphate also involves this lysine residue.

Although an exhaustive study has not been carried out as yet, experiments with some twenty proteins indicate that their susceptibility to carbamylation at or near neutral pH varies a great deal. As shown previously, glutamic dehydrogenase is very readily carbamylated, whereas ornithine transcarbamylase and phosphoglycerate mutase, for example, are not. It appeared of interest, then, to determine whether carbamyl phosphate would react with histones, proteins known to be easily acylated. Carbamylation of histones both with added carbamyl phosphate and with a generating system for

---

* It is possible also to carbamylate hemoglobin extensively, that is, 1.5 moles/mole, with carbamyl phosphate (Carreras, Diederich, and Grisolia, unpublished). Moreover, the carbamylation of hemoglobin with either KCNO or carbamyl phosphate can be carried out with red cells *in vitro* and without hemolysis. These observations are of interest since hemoglobin can then be subject to chemotrophic modification. Moreover, they may have practical significance in several ways—for example, to follow the half-life of red cells (labeled with ^{14}C-KCNO or carbamyl phosphate), to investigate the site and mechanism of degradation of the globin portion of hemoglobin, and possibly to investigate the so-called effects of urea on sickle cell anemia. As pointed out many times in this chapter, urea, cyanate, and carbamyl phosphate are in equilibrium so that the reported protection of sickling by high concentrations of urea may be due to and/or influenced by carbamylation. Indeed, experiments from this and other laboratories have demonstrated that carbamylation reduces sickling.

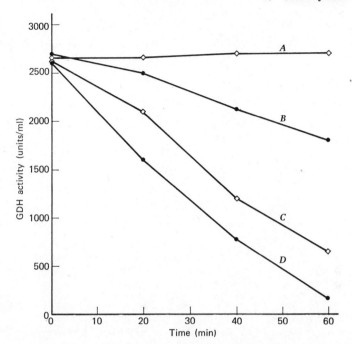

Fig. 3. Effect of carbamyl phosphate on glutamic dehydrogenase activity. *A*: enzyme incubated without carbamyl phosphate; *B*: enzyme incubated with 12.7 m*M* carbamyl phosphate; *C*: enzyme incubated with 32 m*M* carbamyl phosphate; *D*: enzyme incubated with 114 m*M* carbamyl phosphate.

carbamyl phosphate was tested (29). It was clearly demonstrated, both by colorimetric and by radioactivity measurements using ^{14}C-carbamyl phosphate, that under equal conditions there was negligible carbamylation of albumin, whereas arginine-rich and particularly lysine-rich histones were readily carbamylated (Fig. 4). Carbamylation of lysine-rich histone increased with increasing carbamyl phosphate. To determine whether carbamylation from carbamyl phosphate proceeds via cyanate, carbamylation of histones with KCNO was tested and found to proceed more readily than with carbamyl phosphate. Thus carbamylation with cyanate may also be of practical interest in that carbamyl histones could be formed as artifacts when high concentrations of urea are used to solubilize nuclear protein.

Table 1 presents a resume of some of the enzymes which are susceptible to inactivation by carbamyl phosphate. A comparison of the results shown in the table reveals that there is no good agreement between the inactivation studies and the extent of carbamylation as judged colorimetrically (30). This discrepancy is due to the fact that the chromogenic equivalence of the carbamyl amino acid varies greatly, and that, as indicated above, the inactivation

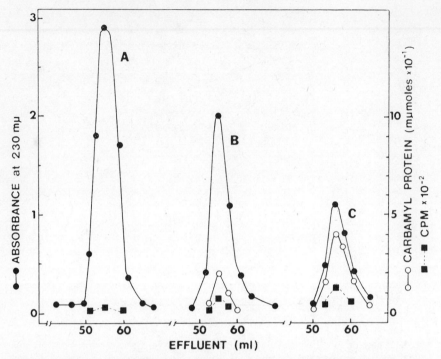

Fig. 4. Profile of incorporation of carbamyl phosphate into several proteins. After incubation with carbamyl phosphate, the proteins were isolated by gel chromatography as described by Ramponi and Grisolia (29). *A*: albumin; *B*: arginine-rich histone; *C*: lysine-rich histone.

due to carbamylation may depend on many factors, including pH, the position of the susceptible group, and the protein. Indeed, it is of interest that, even with a small peptide such as glucagon, carbamylation with either cyanate or carbamyl phosphate yields complete carbamylation of the terminal histidine, while lysine is not affected (Grande, Grisolia, and Diederich, in press).

In view of the above findings, it appeared of interest to determine the relative susceptibility of the enzymes mainly concerned with ammonia utilization, namely, glutamic dehydrogenase, glutamine synthetase, carbamyl phosphate synthetase, ornithine transcarbamylase, aspartic transcarbamylase, and carbamyl phosphate phosphatase. These studies have revealed preferential inactivation of glutamic dehydrogenase (Chabas and Grisolia, unpublished studies). The susceptibility of these enzymes in subcellular fractions of liver and under conditions leading to the biosynthesis of carbamyl phosphate (e.g., high ammonia and bicarbonate concentrations) was tested. Again glutamic dehydrogenase was inactivated extensively. The initial success in demonstrating preferential inactivation of glutamic dehydrogenase has led to the conducting of experiments at the organ level. Indeed, in preliminary experiments carried out in collaboration with Dr. Anthony Barak (Medical Research Laboratories, Veterans Administration Hospital, Omaha, Nebraska), rat livers were perfused with different concentrations of urea, ammonia, and carbamyl phosphate and were then processed. Although the identity of the modified proteins remains to be clarified, extensive protein carbamylation occurs in both the supernatant and the mitochondrial fractions; livers perfused with carbamyl phosphate contain four times as much carbamyl protein as do controls.

These facts prompted us to test the extent of "*in vivo*" carbamylation following administration of carbamyl phosphate and cyanate. (The effect of chronic administration is discussed in Section 6.) As indicated above, both the red-cell membranes and Hb are rapidly carbamylated with either cyanate or carbamyl phosphate. As will be illustrated later, some serum proteins are very extensively carbamylated whereas others are not. Moreover, there is extensive carbamylation of proteins from liver and brain following the injection of carbamyl phosphate or cyanate.

Male Holtzman rats of 250–350 g were injected either through the tail vein or intraperitoneally with 2.5 ml of 0.15 M ^{14}C carbamyl phosphate or ^{14}C-NaCNO. After 6 hr the rats were anesthetized and bled from the thoracic aorta. The extensive incorporation into brain, liver, and blood proteins is illustrated in Table 6.

Table 6. The Effect of Carbamyl Phosphate and NaCNO Administration on Protein Carbamylation

Unpublished Data of Grisolia, Bettis, and Diederich.

Experiment	Tissue	Carbamyl Phosphate	NaCNO
		mμmoles/mg protein	
1[a]	Liver	0.41	0.75
	Brain	0.32	0.44
	Serum	1.16	1.8
	RBC	0.68	0.80
2[b]	Liver	1.1	0.79
	Brain	0.58	0.43
	Serum	3.1	2.0
	RBC	0.82	1.31

[a] Animal injected intravenously.
[b] Animal injected intraperitoneally.

Since in all cases tested carbamylation with carbamyl phosphate seems to proceed through cyanate, the time-dependent carbamylation of the proteins of blood was tested with cyanate under the conditions described above. We found a rapid decrease in nonprotein-bound radioactivity and a progressive increase to a steady value, in about 4 hr, of carbamylated protein.

To determine if cyanate decomposed rapidly or was metabolized, a rat was injected with cyanate, the CO_2 and urine collected at intervals, and then was sacrificed at 24 hr. The major part, $\sim70\%$, of the cyanate injected becomes bound to the tissue proteins, $\sim7\%$ to blood proteins, $\sim13\%$ appears as CO_2, and $\sim3\%$ of the injected cyanate appears in the urine.

E. Inactivation by Urea

The formation of carbamylated products during electrophoresis or chromatography in the presence of urea has been observed with insulin by Cole (31), with casein by Manson (32), and with some wheat proteins by Cole and Mecham (33), but the carbamylating agent in these cases is cyanate, present

Table 7. Effect of TPNH, Trypsin, and Urea on the Stability of Glutamic Dehydrogenase

Data from Grisolia et al. (34).

Reagents Added to Standard Reaction Mixture	Residual Activity, %	
	Time of Incubation, min	
	15	30
None	83	60
0.2 M urea	60	50
0.4 M urea	60	30
1.0 M urea	26	10
0.1 mM TPNH	40	26
0.1 mM TPNH and 0.2 M urea	16	10
0.1 mM TPNH and 0.4 M urea	6	0
0.1 mM TPNH and 1.0 M urea	0	. . .
5 μM trypsin	76	40
10 μM trypsin	43	20
100 μM trypsin	0	. . .
0.1 mM TPNH and 5 μM trypsin	6	0
0.1 mM TPNH and 10 μM trypsin	0	. . .

Tubes contained 0.5 mg glutamic dehydrogenase, pH 7.2 Tris in a total volume of 1 ml, and the additions listed above.

in trace amounts in urea solutions. A combination of 0.5–1 M urea with 1 mM TPNH potentiates the cofactor-induced inactivation of glutamic dehydrogenase with apparent concomitant protein carbamylation (34). This is further illustrated in Table 7 which illustrates also the higher susceptibility of the liganded protein to trypsin. On the other hand, no such effect was demonstrated with triosephosphate dehydrogenase (35). The relatively higher susceptibility of glutamic dehydrogenase to urea is, then, in agreement with the data on carbamylation presented previously.

F. Physiological Implications

There has been speculation that a close connection exists between glutamic dehydrogenase and urea synthesis, that is, via NH_4^+ as a common intermediate, but the possibility that they are connected in the fashion indicated in Fig. 5 has never been considered. Thus the present findings are of further interest in suggesting a regulatory role for the very active and widely distributed acyl phosphatase extensively studied in our laboratory (36). According to the scheme presented in Fig. 5, when the main pathway for ammonia detoxication in ureotelic animals, that is, urea production from carbamyl phosphate is not operating at full capacity because, for example, of insufficient ornithine and/or ornithine transcarbamylase, as has been suggested in hepatic comma or cirrhosis, carbamyl phosphate may accumulate to such high levels that it inactivates glutamic dehydrogenase. It is at this point that the acyl phosphatase may perhaps exert a physiological function in preventing the accumulation of deleterious levels of carbamyl phosphate, particularly in sensitive areas (most tissues, except liver and kidney, lack most if not all the enzymes for urea synthesis). Some evidence for this has already been found by Ramponi and Grisolia (29).

Whether there are enzymes catalyzing the formation of the carbamyl histones or of other carbamyl proteins as discussed above remains to be investigated. Carbamyl proteins may be formed readily in biological systems simply by chemical interaction, and therefore regulation at this level is more likely to be carried out by regulation of the substrate concentrations and/or related enzymes, as discussed below.

G. Miscellaneous

Carbamylation of cholinesterase has also been reported by O'Brien (37). This does not appear to involve covalent bond formation, however, and therefore is not discussed further.

Argyroudi-Akoyunoglou and Akoyunoglou (38) have shown that carbamyl phosphate protects spinach chloroplast ribulose diphosphate carboxylase

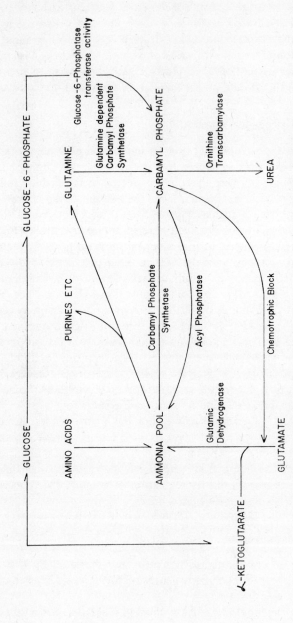

Fig. 5. Metabolic relationship between glutamate and carbamyl phosphate.

154

against iodoacetamide inhibition. This is not due to the binding of the protector on the active site alkylated by iodoacetamide.

The finding that carbamyloxyurea degrades DNA, as reported by Rosenkranz and Rosenkranz (39), seems, at first glance, of remote interest. In solution, hydroxyurea has no effect on the properties of DNA. Upon protracted incubation, however, it degraded the polynucleotide. This is of much interest since hydroxyurea is known to be a specific inhibitor of DNA synthesis. It appears that in solution hydroxyurea is converted into carbamyloxyurea and that the latter is the reagent responsible for the degradation of DNA.

As reported elsewhere, urea and cyanate, as well as carbamyl phosphate, are in equilibrium. Moreover, an enzyme has recently been described which, in the presence of TPNH, converts hydroxyurea into urea:

$$\underset{\text{NHOH}}{\overset{\text{NH}_2}{\text{C}=\text{O}}} + \text{TPNH} + \text{H}^+ \rightleftharpoons \underset{\text{NH}_2}{\overset{\text{NH}_2}{\text{C}=\text{O}}} + \text{TPN} + \text{H}_2\text{O}.$$

It is conceivable that hydroxyurea may be formed or degraded by the above reaction and that urea and hydroxyurea may be converted into carbamylurea (biuret) or carbamyloxyurea as follows:

$$\text{NH}_2-\overset{\text{O}}{\overset{\|}{\text{C}}}-\text{NH}_2 + \text{P}\sim\overset{\text{O}}{\overset{\|}{\text{C}}}-\text{NH}_2 \longrightarrow \text{NH}_2-\overset{\text{O}}{\overset{\|}{\text{C}}}-\text{NH}-\overset{\text{O}}{\overset{\|}{\text{C}}}-\text{NH}_2,$$

$$\text{NH}_2-\underset{\text{O}}{\overset{\|}{\text{C}}}-\text{NHOH} + \text{P}\sim\overset{\text{O}}{\overset{\|}{\text{C}}}-\text{NH}_2 \longrightarrow \text{NH}_2-\underset{\text{O}}{\overset{\|}{\text{C}}}-\text{NHO}-\underset{\text{O}}{\overset{\|}{\text{C}}}-\text{NH}_2.$$

Several drugs are known which exert their pharmacological effects by their ability to form a covalent linkage with certain enzymes. For instance, acetylcholinesterase is inhibited by carbamates such as physostigmine and neostigmine. These compounds transfer their R_2N—CO group to a serine residue, thus forming carbamyl derivatives of the enzyme (see, e.g., ref. 40).

Although a carbamylation type of reaction is not involved, it is interesting nevertheless to note that arginine and bicarbonate inactivate kidney transamidinase (41). A more accurate description is that of a CO_2 stimulation of substrate (arginine, canavanine, or guanidoacetate) inactivation of the transamidinase. In analogy to the chemistry of amino isothiuronium salts Grazi et al. (41) have proposed that, since the substrate probably binds to a

SH group of the enzyme in the presence of bicarbonate, there is an intra-molecular transfer of the amidine group to yield mercaptoethyl guanidine and 2-aminothiazoline derivatives as in the model:

$$
\begin{array}{ccc}
\begin{array}{cc}
CH_2 & CH_2 \\
| & | \\
NH_2 & S \\
& \diagdown \\
& C \\
\diagup & \diagdown \\
NH & NH_2
\end{array}
\rightleftharpoons
\begin{array}{cc}
CH_2 & CH_2 \\
| & | \\
NH & S \\
\diagdown & \diagup \\
& C \\
\diagup & \diagdown \\
NH_2 & NH_2
\end{array}
\rightleftharpoons
\begin{array}{cc}
CH_2 & CH_2 \\
| & | \\
NH & SH \\
& \diagdown \\
& C \\
\diagup & \diagdown \\
NH & NH_2
\end{array}
\end{array}
$$

$$\downarrow HCO_3^-$$

$$
\begin{array}{cc}
CH_2 & CH_2 \\
| & | \\
NH & S \\
\diagdown & \diagup \\
& C \\
& | \\
& NH_2
\end{array}
\quad + NH_4^+.
$$

Interestingly, ornithine is a good protector when arginine is the inactivator. Of course, since the mechanism of the transamidinase, particularly the role of the SH group, has not been proved, the role of CO_2 should be investigated in further detail. This matter is of special interest since little is known about the regulation of the metabolic paths connected with biosynthesis, and particularly about the utilization of arginine.

In summary, it has been illustrated, although admittedly with only a few examples, as shown in Table 8, that carbamyl phosphate, cyanate, and derivatives thereof are good carbamylating agents with a number of proteins and that some of the carbamylated proteins* can be decarbamylated with hydroxylamine. In this regard it is of interest that acetyl groups can be easily removed from lysine residues by hydroxylamine, as will be shown in the next section. It has been demonstrated that in a number of cases carb-amylation results in extensive changes in the properties of the proteins, ranging from loss of activity to conformational effects, and that the sus-ceptibility to carbamylation in turn may depend on the presence of ligands and on the conformation as well as the type of protein. It appears that no generalization can be made thus far as to the susceptibility of proteins. In contrast to Stark's attempt with simple model systems, in some cases there is preferential reactivity with lysine, although in others, for example, with

---

* However, homocitrulline, either in free form or as part of a protein, is very slowly decarbamylated (if at all) with hydroxylamine. Moreover, the carbamyl group cannot be converted to hydroxyurea when reacted with hydroxylamine. In addition, as will be reported below, the acetylated lysines of trypsin are also resistant.

glucagon, the terminal amino group is the reactive one. It appears safe to predict that any protein or amino acid susceptible to carbamylation with cyanate may also be susceptible to carbamylation with carbamyl phosphate (in a few cases carbamylation by carbamyl phosphate may appear to be greater than that by cyanate). Although carbamyl phosphate acts via cyanate in nonenzymic carbamylation reactions, *the physiological carbamylating agent is carbamyl phosphate, since it is produced in large quantity by mammals whereas cyanate is not.* Also carbamylation with carbamyl phosphate can be demonstrated at different levels ranging from amino acids to proteins, cells and cell fractions, perfused organs, and the whole animal. Finally, it should be noted that carbamyl phosphate, in addition to participating in carbamylation reactions, may also take part in protein phosphorylation, as will be illustrated later.

3. ACYLATION

A. Existence of Acyl Proteins

It has been recognized, particularly in recent years, that there are proteins containing acyl groups either at the amino-terminal end of the molecule (e.g., cytochromes c and a, fibrinogen, tobacco mosaic virus, ovalbumin, myosin, frog hemoglobin, and collagen) or on the ϵ-amino group of lysine. Furthermore, O-acylation and S-acylation have also been described and are receiving increased attention. The mechanism by which acyl groups are added to proteins is far from clear. Evidence has been presented, in analogy with formylmethionine as an initiator of protein synthesis in bacterial systems, suggesting that acyl N-terminal groups are added during the initiation of protein synthesis. However, data from several laboratories do not support this as a universal concept. With the possible exception of the participation of N-acetylvalyl-tRNA, no other N-acetyl amino acid is known to be involved in protein synthesis. However, a number of acylation-deacylation reactions must occur after the proteins and/or subunits have been made, as, for example, in the case of the acetylated chain of fetal hemoglobin (42). Protein acylation is of interest, then, in relation to the general subject of this chapter. Furthermore, the finding that histones are very rapidly acylated and deacylated is particularly important and will be discussed in some detail.

B. Some Models of Acylation of Proteins

Acylations of proteins have been studied extensively as models, particularly since chemical modification of biologically active proteins is often employed

Table 8.   Examples of Protein and Peptide Carbamylation

Protein	Carbamylating Agent	Effect	Reference
Pepsinogen	Cyanate	Carbamylation of 9 of the 10 ϵ-amino groups of lysine and 4 tyrosine residues, resulting in a conformational change and a decreased susceptibility to activation to pepsin (loss of potential pepsin activity parallels degree of carbamylation).	21, 22
Pepsin	Cyanate	Altered proteolytic activity.	21, 22
Phosphorylase	Cyanate	Carbamylation of lysine residues, resulting in inactivation. Possibly a conformational change occurs. Susceptibility to carbamylation markedly affected by ligands.	23
Subtilisin	Cyanate	Inactivation, possibly by reaction with serine at active center. Carbamylation of α and ϵ-amino groups occurs but does not disrupt conformation.	24
Pyruvate carboxylase	Cyanate	Inactivation. Lysine group at allosteric site carbamylated.	25

to characterize functional groups which may be involved in the active site. However, interpretation of such studies may be difficult because conformational changes, rather than actual modification of functional groups, may influence the biological activity.

In a model study, Habeeb (43) succinylated bovine serum albumin, human γ-globulin, and lysozyme and studied changes in conformation. The Stokes radius was used to quantitate conformational changes which affected the overall shapes of the macromolecules. In addition, the susceptibility of the disulfide groups to either sulfitolysis or reduction, as well as the ability of the succinylated proteins to precipitate with antibodies against the respective native protein, was evaluated. Another related study is that of Bezkorovainy

Table 8 (*continued*)

Protein	Carbamylating Agent	Effect	Reference
Hemoglobin	Cyanate, carbamyl phosphate	Carbamylation of terminal amino group cf both chains. Inability to bind CO_2.	26, Carreras and Grisolia, unpublished data
Insulin	Cyanate present in urea solutions	Carbamylation of amino terminal glycine and phenylalanine.	31
Casein	Cyanate present in urea solutions	Carbamylation of α- and ε-amino groups.	32
Wheat proteins	Cyanate present in urea solutions	Carbamylation, resulting in altered electrophoretic mobility.	33
Glucagon	Cyanate and carbamyl phosphate	Glycolytic and lypolytic properties almost lost.	Grande and Grisolia, and Diederich, unpublished data
Arginine- and lysine-rich histones	Cyanate and carbamyl phosphate	Carbamylation of lysine-rich histones, yielding homocitrulline.	29, Ramponi, Leaver, and Grisolia, unpublished data
Glutamic dehydrogenase	Carbamyl phosphate	Inactivation. Carbamylation of 4% of lysine residues per mole of enzyme.	6
Oxytocin and oxytocin analogues	Cyanate	Inactivation and/or inhibitors	179
Eye lens α-crystallin	Cyanate; cyanate present in urea solutions	Altered electrophoretic mobility	180

et al. (44) on the properties of succinylated transferrin, conalbumin, and orosomucoid. Reduced-alkylated-succinylated transferrin and conalbumin had random coil conformations in dilute buffer solutions and molecular weights near those of native proteins, indicating a lack of subunit structure.

Iron protected transferrin against denaturation by succinylation. Approximately ten fewer amino groups were susceptible to succinylation in the iron-saturated protein than in iron-free transferrin, indicating either the participation of these groups in the iron-binding phenomenon or the fact that iron-saturated transferrin assumes a more compact conformation than iron-free transferrin. Reduced-alkylated-succinylated orosomucoid has a molecular weight close to that of the native protein; however, to judge from its β value, it failed to assume a random coil shape in dilute aqueous buffers.

The succinylation of 19 S thyroglobulin induces its dissociation into half-sized 12 S subunits, the degree of dissociation depending on the degree of iodination of the molecule (45). This is thought to involve alterations in electrostatic interactions, since succinylation results in an increase of the carboxyl groups at the expense of the amino groups, causing an increase in the net charge of the molecules at neutral pH. The result is an increase in the electrostatic repulsion of the subunits. A further example of modification reactions and of aggregation due to acylation and of the effect of ligands is that of Hoagland (46), who prepared a series of acylated β-caseins. He showed that the nature of the alkyl group affects both the Ca^{2+} sensitivity and the aggregation of modified β-casein. The increase in net negative charge resulting from acylation apparently reduces Ca^{2+} sensitivity and aggregation through electrostatic repulsion.

C.   Deacylation

As shown by Marzotto et al. (47), amino groups of proteins such as RNAase and lysozyme can be acylated and the acetyl, like the carbamyl groups discussed previously, can be removed at 25° from the enzymes by treatment with hydroxylamine at pH 7. The hydroxylamine-treated product resembles native proteins in its properties and recovers full enzymic activity. Additional examples of deacylation are discussed below.

D.   Use of Acetylimidazole

This reagent is being used extensively at present to acylate a number of proteins because it seems to be relatively specific for the acylation of tyrosyl groups. Thus, although acetylimidazole is an unlikely physiological intermediate, it will be considered as a model and a number of examples given to typify both the modification reactions and the fact that control of acylation may occur at several points, and in turn it is also affected by ligands and other environmental factors (4).

For example, the effect of a competitive inhibitor on the acylation by acetylimidazole of both the tyrosyl and the lysyl residues of nucleases from a staphyloccus has been studied by Cuatrecasas et al. (48), who showed acylation of 5 tyrosyl and 9–10 lysyl residues, with almost complete loss of activity of DNAase and of RNAase. Complete activity could be restored by alkaline treatment after removal of the O-acetyl groups. Moreover, deoxythymidine-3,5-diphosphate offered some protection against acylation, and this compound also protected the activity of DNAase.

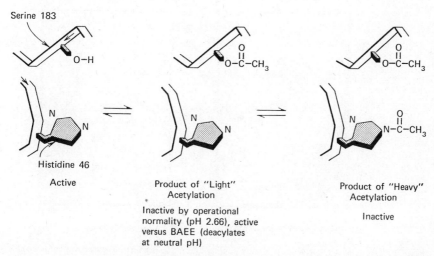

Serine 183

Histidine 46

Active

Product of "Light" Acetylation

Inactive by operational normality (pH 2.66), active versus BAEE (deacylates at neutral pH)

Product of "Heavy" Acetylation

Inactive

Fig. 6. A schematic representation of the special effects of acetylation upon the catalytically functional groups of trypsin. Data from Houston and Walsh (50).

The inactivation of the α-amylase of *Bacillus subtilis* by N-acetylimidazole has been described by Connellan and Shaw (49). Again (see above) full activity was restored when the modified protein was treated with hydroxylamine. All the evidence indicates that the changes in activity are due to reactivity with tyrosine residues. Moreover, when the acetylation was carried out in the presence of substrate, there was excellent protection both of activity and of tyrosyl residues. Interestingly, free amino groups were acetylated under the conditions used, and hydroxylamine was then ineffective. This contrasts with previous cases discussed in which hydroxylamine can reverse carbamylation and acetylation. In agreement with this finding is the fact that modification with maleic anhydride produced only limited loss of activity. Although this enzyme is probably composed of two subunits, no attempts to separate them have been carried out, nor is it known whether acylation separates the subunits as is the case for glutamine synthetase.

Many studies have been made on the acetylation of trypsin. Although extensive acetylation (as much as 85–100% acetylation of the ε-amino groups) yields largely active trypsin, acetylation of the tyrosyl residues, in contrast, results in complete inactivation. It should also be noted that the α-amino group is not acylated in this case. Coverage of these points, as well as an imaginative way to study trypsin acetylation and inactivation with acetyl-imidazole, has been presented by Houston and Walsh (50), who show that at an early stage the transient inactivation of trypsin is due to acetylation of

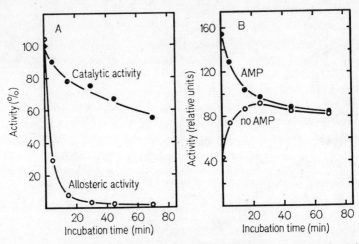

Fig. 7. Effect of acetylation on catalytic and allosteric activity of phosphofructokinase. *A*: remaining catalytic and allosteric activity as a function of incubation time; *B*: activity in presence and absence of AMP. Data from Chapman et al. (51).

residue 183 and that the inactivation is reversible at neutral pH (Fig. 6). The so-called heavy acylation produced in a second stage probably involves the histidine 46 residue. These two acetyl groups can be removed with imidazole in contrast to the lysyl residue, which cannot be deacetylated even with 1 *M* hydroxylamine.*

Chapman et al (51) have studied the effect of acetylation and of sulfhydryl blocking agents on phosphofructokinase from rabbit muscle. Treatment of phosphofructokinase with *N*-acetylimidazole led to rapid loss of allosteric function, as measured by the ability of AMP to stimulate the ATP-inhibited enzyme (Fig. 7). The catalytic activity was far less sensitive to acetylation. The loss of allosteric function on acetylation was due to the fact that the modified enzyme was insensitive to ATP inhibition. The presence of ATP during acetylation prevented the loss of allosteric activity. The results

* See the footnote on p. 148.

indicate that tyrosine residues may play an important role in the inhibitory binding sites of ATP.

Vul'fson and Kozlova (52) have shown that phosphorylase a can be inactivated by acetylimidazole, the degree of inactivation depending on the concentration of acetylimidazole. Glucose-1-phosphate and AMP or glycogen plus AMP protected the enzyme. During acetylation, the number of free amino groups decreased 25–30%. In the presence of AMP and the substrate, both of which protect the enzyme, the amino groups were not acetylated. Ultracentrifugation data showed that the disruption of structure occurred only with acetylimidazole at $2 \times 10^{-2}\ M$, where the tetramer dissociates into monomer units.

By far the most sophisticated study along these lines may be that of Wilk et al. (53). These workers showed that the native octameric glutamine synthetase from sheep brain was partially acetylated when treated with N-acetylimidazole and that the enzymatic activity declined with progressive acetylation. A two-step dissociation process was observed in which a 9 S component was formed first, followed by formation of a 3.8 S component. The 9 S component could be reassociated to the native 15 S form by increasing the ionic strength. Approximately 12–16 acetyl groups per subunit were introduced, as determined with $[1\text{-}^{14}C]N$-acetylimidazole. Incubation of the acetylated enzyme with hydroxylamine led to reassociation of the 9 S species to yield the octamer, and to the removal of about half of the acetyl groups. Acetylation in the presence of ATP and Mg^{2+} did not lead to dissociation; however, subsequent removal of ATP and Mg^{2+} by gel filtration induced dissociation, although no 3.8 S form appeared under these conditions. Addition of ATP and Mg^{2+} at concentrations approximately equal to the K_m for ATP in the catalytic reaction led to reassociation. The sedimentation coefficients of the products formed were in excellent agreement with the expected values for tetramer (9) and monomer (3.7). The two-step dissociation process is in accord with a model possessing D_4 symmetry. The results suggest the participation of tyrosyl residues in the maintenance of the structural integrity of the enzyme.

E.   Studies with Triosephosphate Dehydrogenase

In an extension of previous studies, Park's group (54, 55) has shown that acetyl phosphate acetylates triosephosphate dehydrogenase and that this may result in enzyme inactivation (Table 9). Possibly acetyl phosphate is a model substrate for the enzyme, since there is little evidence for its extensive synthesis or utilization and since it is used at a much lower rate than the "physiological" substrate, 1,3-P-glycerate. One of the more interesting aspects of this investigation is that the reaction occurs more easily with the

Table 9. Effect of Substrate Concentration on the Acetylation of Glyceraldehyde-3-phosphate Dehydrogenase

Data from Park et al. (54).

[14]C-Acetyl Phosphate, μmoles	^{14}C-Acetyl Groups Bound per Mole DPN-Free Enzyme, Moles		
	pH 4.6	pH 7.0	pH 8.5
2		1.8	
4	1.5	2.8	2.9
10	2.0	3.5	3.2
20	2.1	4.0	3.7

animal than with the yeast enzyme. Furthermore, under the conditions used, other proteins are much less susceptible to acetylation, although there is some reaction with them. It seems quite clear that the cysteine residue, which can be acetylated very easily at neutral pH and at low temperatures, is intermediate to the acetylation of the lysine residues. In other words, at higher temperatures or higher pH the acetyl group migrates,* resulting in the acetylation of the lysine groups with concomitant enzyme inactivation. Of course, the transfer of acyl groups through the reaction

$$RCO—SR_1 + R_2SH \longrightarrow RCO—SR_2 + R_1SH$$

has been known for many years. For example, Wieland and Bukelmann (56) reacted S-acetylthiophenyl with glutathione to yield S-acetylglutathione. The reaction was carried out at low pH (approximately 3) to prevent acetylation of the free amino groups in glutathione.

Since 3–4 cysteine residues can be acylated per mole of enzyme, and since substrates, cofactors, and time of incubation, as well as temperature and pH, influence the modification of the enzyme, together with the fact that acetyl phosphate is more effective than *p*-nitrophenyl acetate in acetylating the enzyme, triosephosphate dehydrogenase is clearly an excellent example of an enzyme that may be subject to chemotrophic inactivation and/or regulation. Many details regarding binding sites, including changes in DPN binding, protection by DPN binding, and protection by DPN against acylation, have been elegantly studied by Park and her coworkers (54, 55). Therefore it will be of great interest to check whether 1,3-diphosphoglycerate, the natural substrate, affects the enzyme in the same way as acetyl phosphate. Also, since carbamyl phosphate readily inactivates this enzyme (see above),

---

* Such migrations may be more common than is presently realized and may be misleading, particularly in studies designed to identify binding to an amino acid at active centers.

comparative experiments which thus far have not been carried out with enzyme free of the DPN may be very relevant. Possibly the enzyme is modified by carbamyl phosphate along the lines shown by Park et al. with acetyl phosphate.

F. Studies with Insulin

The acetylation of insulin has also been reported (57). Acetylation of free amino groups or the lysine B29 does not affect its biological activity. On the other hand, modification of the phenylalanine B1 amino group resulted in a large decrease in the affinity of insulin for anti-insulin antibodies, thus enabling a correlation of these observations with the tertiary structure of insulin to be carried out. As indicated previously, insulin is susceptible to carbamylation, and it would be interesting, therefore, to test carbamylated insulin for biological activity.*

G. Studies with Histones

The presently extreme interest of biochemists in histones stems from the suggestion, as yet unproved, that these proteins may regulate or play a role in certain aspects of nucleic acid synthesis and regulation, including gene transcription. Thus modified histones, including acyl histones, may also be important in this regard. Phillips (58) demonstrated, in an extension of his previous work, the presence of N-acetylserine as the N-terminal residue of approximately 40% of the calf thymus histones. This work has been extended and confirmed, and the complete amino acid sequence of calf thymus histone IV has been published (59). Interestingly, these workers (59) found also ϵ-N-methyllysine and ϵ-N-acetyllysine in residue 16. This fact led to the suggestion that it is in residue 16 that rapid incorporation of acetyl groups occurs. In other words, it was shown earlier that acetyl group turnover is more rapid than histone turnover even in the adult liver, where there is little histone synthesis. Thus it has been concluded that acetylation involves the complete protein. However, other workers believe that the acetyl N-terminal is the residue that turns over rapidly. It is possible that both groups do.

A variety of studies dealing with enzymatic acetylation of histones has appeared. For example, Gallwitz (60) has reported studies with a kinase isolated from rat liver nuclei. This enzyme uses acetyl-CoA and is not Mg^{2+} dependent; in fact, Mg^{2+} inhibits the kinase. Gallwitz observed also an approximately 10% nonenzymatic incorporation. Pigeon liver has also been used to test acylation of histones, since acetylation of proteins was shown

---

* Particularly since, as previously indicated, carbamylation of the terminal histidine inactivates glucagon.

earlier to occur both in this tissue and in reticulocytes. For example, Takashashi (61) demonstrated the incorporation of acetate with a preparation derived from pigeon liver. Activity was noted mostly with arginine-rich histones; the lysine-rich histones were poor substrates. Interestingly, a portion of the acetate incorporated could be removed with hydroxylamine, indicating some O-acylation. The studies of Ozawa et al. (62), who have presented evidence for the existence of acyl proteins in rat liver mitochondria, should be extended and may be most important, since they show that the observed effects depend on cellular respiration. Very recently Inoue and Fujimoto (63) studied in some detail the deacetylation of histones in preparations from calf thymus. The properties of the enzyme activity were well outlined, and evidence for several species of the enzyme was obtained by gel filtration. Interestingly, the activity is predominantly in the cytoplasm, as is also true for histone phosphatase, although no evidence for a decarbamylase enzyme has been found as yet for carbamylated histones. The only decarbamylase of carbamyl amino acids thus far known is also a cytoplasmic enzyme (64).

A comparison of the acetylation of proteins and nucleic acids has been carried out by Paik and Kim (65), who conclude that, although active acetylation occurs in rat and frog liver homogenates (where there is a dependence on ATP, Mg^{2+}, and CoA), there is no measurable acetylation of nucleic acids. The article of MacGillivary and Monjardino (66) is interesting since it exemplifies the fact that acylation of histones may be nonspecific. Gallwitz (67), however, has demonstrated the existence of multiple histone acetyltransferases obtained from the nuclei of rat liver, thymus, and kidney, and it is suggested that the different chromatographic behaviors of these enzymes from different organs indicate a certain organ specificity. Unfortunately, most of the work done with the acylation and deacylation reactions of histones has been carried out on very crude preparations. [See, for example, Wilhelm and McCarty (68) and Mori et al. (69).]

In our laboratory we have demonstrated (29) very fast reaction of carbamyl phosphate and 1,3-diphosphoglycerate with histones. The incorporation of 1,3-diphosphoglycerate by arginine-rich histones, and particularly by lysine-rich histones, was demonstrated with two generating systems for 1,3-diphosphoglyceric acid as follows: (a) from triosephosphate dehydrogenase and (b) with the thermodynamically unfavorable phosphoglycerate kinase system. Again, under these conditions there was very little acylation of albumin. This work raises two important points: (a) the occurrence of acyl derivatives of histones may often be due to nonenzymatic reactions, and thus (b) regulation of histone acylation may occur via variations in the substrate concentration, in addition to enzymic regulation. In fact, Paik et al. (70) have shown that considerable nonenzymic acetylation of histones by acetyl-CoA takes place. The incorporation of label from [1-^{14}C]acetyl-CoA into calf thymus histones does not appear to be enzymatic, since boiling of the histone for 1 hr resulted in only a reduction of 60% of the incorporation. The different rates of acetylation between native and boiled histone may be due to changes in the tertiary structure brought about by boiling. Despite its nonenzymic nature, the acetylation reaction depends on pH, period of incubation, ionic strength, and ionic species. Furthermore, the nonenzymic acetylation reaction shows a certain degree of specificity in that slightly lysine-rich histone is most susceptible to acetylation. It is apparent, therefore, that variations in the chemical environment *in vivo* can cause variations in the amount of acetylation of the histones and thus influence the physiology of the cell at the gene level. This nonenzymic acetylation by acetyl-CoA thus resembles the nonenzymic modification, already described, of histones with carbamyl phosphate and 1,3-diphosphoglycerate.

As indicated above, Ramponi and Grisolia (29) demonstrated fast acylation of lysine and of arginine-rich histones with 1,3-diphosphoglycerate. Like carbamyl phosphate, 1,3-diphosphoglycerate is formed in very large quantities by mammals; for example, an average-sized man metabolizes 1–2 moles of carbohydrate daily, thereby making 2–4 moles of 1,3-diphosphoglycerate per day. Thus it is not surprising that, if this very highly reactive anhydride is not consumed rapidly in further metabolic reactions, mostly via the Embden-Meyerhof pathway, it may be used to modify other substances including proteins.

The fast modification of histones by acyl phosphates and the demonstration of marked inhibition of histone acylation by acylphosphatase suggest that histone modification by 1,3-diphosphoglyceric acid may be regulated by acylphosphatase, as is the case for carbamyl phosphate. Such a role, as well as the formation and presence of phosphoglyceryl histones in tissues, merits further investigation. It should be noted that Phillips (79) detected approximately 1 mole of carbohydrate positive material in lysine-rich histone isolated from calf thymus. In view of the above findings, it is not surprising that regulation through acylphosphatase on the main pathway of carbohydrate metabolism may be of much importance. This is illustrated in Fig. 8.

As indicated above, the problem of whether the *N*-acetyl groups in proteins are added after completion of the peptide chain, or whether the acetyl amino acids are the starting points for the propagation process, remains to be solved. Furthermore, it is necessary to clarify whether the biosynthesis of some proteins which do not have amino-terminal acyl groups involves initiation of chain growth by activated *N*-acyl amino acids, followed by deacylation of the completed peptide chains. This has motivated additional interest in modified histones and in acylated tRNAs as chain initiators (71, 72, 73). The possibility that an acyl tRNA other than formyl or valyl tRNA is involved has not been

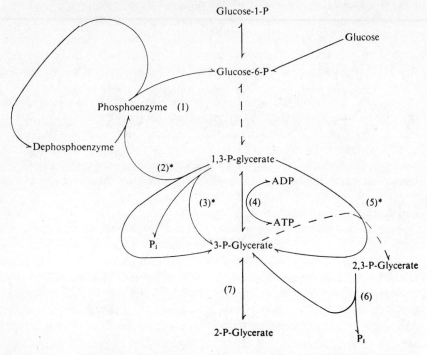

Fig. 8. The 1,3-diphosphoglycerate trident shunt. (1) Phosphoglucomutase; (2) Alper's pathway; (3) acyl phosphatase; (4) phosphoglycerate kinase; (5) diphosphoglycerate mutase; (6) diphosphoglycerate phosphatase; (7) 2,3-diphosphoglycerate-dependent phosphoglyceromutase. An asterisk (*) indicates the steps at which 1,3-diphosphoglycerate may shunt away from the "normal" pathway catalyzed by phosphoglycerate kinase.

tested. In view of our findings relating to histone acylation, the possibility that carbamyl phosphate and 1,3-diphosphoglycerate could acylate amino acid-tRNAs has been tested and preliminary experiments are encouraging (Escarmis and Grisolia, unpublished). A resumé of some proteins susceptible to acylation is shown in Table 10.

H.   Pharmacological Implications

As discussed in Section 2 on carbamylation, a number of drugs are able to form covalent linkages with certain enzymes. Thus penicillin, which is a metabolite antagonist for D-alanyl-D-alanine, is thought to combine irreversibly with a transpeptidase involved in the biosynthesis of cell wall components by forming a covalent bond. This involves an irreversible acylation at the active site of the enzyme by the β-lactam ring of the antibiotic (see, e.g., ref. 40).

Table 10. Examples of Acylated Proteins

Protein	Modification	Result	Reference
Bovine serum albumin, human γ-globulin lysozome	Succinylation	Conformational change.	43
Transferrin conalbumin orosomucoid	Succinylation	Conformational change.	44
Thyroglobulin	Succinylation	Dissociation into subunits.	45
Casein	General acylation (acetylation, propylation, etc.)	Nature of alkyl group affects both the Ca^{2+} sensitivity and the aggregation properties.	46
Ribonuclease, lysozyme	Acetoacetylation	Inactivation.	47
DNAase, RNAase	Acetylation	Acetylation of tyrosyl and lysyl residues, resulting in a reversible loss of activity.	48
α-Amylase	Acetylation	Tyrosine residues modified, resulting in loss of activity.	49
Trypsin	Acetylation	Acetylation of tyrosyl residues, resulting in inactivation.	50
Phosphofructo-kinase	Acetylation	Loss of allosteric properties.	51
Phosphorylase *a*	Acetylation	Acetylation of free amino groups, resulting in inactivation.	52
Glutamine synthetase	Acetylation	Tyrosyl residues may be modified, resulting in loss of activity.	53
Triose phosphate dehydrogenase	Acetylation	Inactivation. Lysyl group acetylated.	54, 55
Insulin	Acetylation	Biological activity unaffected, but molecule has modified response to antibody.	57
Histones	Acetylation		58–62, 65, 66
Histones	Phosphoglycerylation		29

Of much interest is the finding that human serum albumin can be acetylated by aspirin under physiological conditions. On the basis of competition with binding properties and by the use of isotopes it was shown that the protein is irreversibly modified by acetylation. These observations are of much interest since there is an increasing awareness that aspirin intolerance may be due to acetylation of protein (74). Studies in our laboratory have shown that, in addition to acetylating human serum albumin, both salicylate and acetyl salicylate can extensively inactivate a number of enzymes *in vitro*. For instance, it was shown that triosephosphate dehydrogenase and glutamic dehydrogenase became unstable in the presence of salicylate at a wide variety of concentrations and conditions, including therapeutic, and at physiological pH and temperature (75, 76). Studies involving a large number of enzymes have indicated that only those with a requirement for metal ions either as a cofactor for activity or as a part of the integral structure of the enzyme were inactivated by salicylate (77). Furthermore, the enzymes that were inactivated by salicylate were also inactivated by imidazole, a good metal-binding agent. It is possible, therefore, that salicylate inactivates these enzymes by a metal chelation mechanism. The possibility that salicylates act therapeutically by forming chelates was considered several years ago (78). In fact, salicylate forms stable chelates with cobalt, iron, and copper with binding constants of 10^5–10^7.

4. PHOSPHORYLATION

A. Occurrence of Phosphorylated Proteins and Some Considerations Regarding Chemotrophic Phosphoryl Effectors

The identification of phosphoproteins in tissues from a wide variety of species, and the subsequent discovery of phosphokinases and phosphorylases which catalyze the transfer and the removal, respectively, of phosphate groups on proteins, suggest a role for this type of phosphorylation mechanism in enzyme regulation. Numerous examples of enzyme-catalyzed phosphorylation and dephosphorylation reactions have been studied in recent years. In most cases the phosphate group undergoing transfer is supplied by ATP. This is not unexpected in view of the large amount of ATP produced and used in metabolism. For example, the average man uses approximately 2000–3500 calories per day, which is equivalent to the production and utilization of approximately 100–150 moles of ATP as high-energy phosphate. This is approximately equivalent to the utilization and production of approximately 0.1 mole of ATP per minute (80).

Like carbamyl phosphate and other chemotrophic effectors, ATP may play a dual role since it may be involved in both the phosphorylation and the adenylylation of enzymes. (Adenylylation is discussed in Section 5.) A phosphate group(s) may be covalently attached to the enzyme, resulting in modified catalytic activities. Removal of the phosphate group(s) causes a return to the original catalytic activity. The ATP may cause structural changes in the enzyme, such as its dissociation into subunits, resulting in either a reversible inactivation, inhibition of enzyme activity, or a modified response to effector molecules. The binding of ATP to an enzyme molecule may also cause an alteration in the rate at which it is degraded by proteolytic enzymes, as has been shown in the case of rabbit muscle glyceraldehyde-3-phosphate dehydrogenase, where AMP increases the rate of chymotryptic degradation (81). This subject will be discussed in detail later.

B.   Enzyme-Dependent Protein Phosphorylation

There is presently much interest in the cyclic-AMP-dependent protein phosphokinases. These enzymes catalyze the transfer of the terminal phosphate group of ATP to a protein acceptor and may be ubiquitous in the animal kingdom (82). Figure 9 illustrates some of the enzymes which are subject to regulation in this manner. Cyclic-AMP-dependent protein kinases purified from adipose tissue (83), skeletal muscle (84), brain (85), liver (86), and *Escherichia coli* (87) have been studied. The apparently ubiquitous occurrence of cyclic-AMP-dependent protein kinases in the animal kingdom has led to the suggestion that all the actions of cyclic AMP are mediated through such protein kinases (87). Interestingly, histones are more effective than casein as substrates for mammalian enzymes, whereas the reverse is true for the protein kinase prepared from the nonmammalian species. This may reflect an important (possibly evolutionary) difference between the protein kinase of mammals and that of other species. The mechanism by which cyclic AMP stimulates protein kinases is unknown, but the suggestion has been made that it may adenylate tissue-specific protein kinases.

Much attention has been focused on muscle phosphorylase kinase, which catalyzes the conversion of phosphorylase *b* (the inactive form) to phosphorylase *a* (the active form), a reaction in which the terminal phosphate group of ATP is transferred to a specific serine residue (88–91). Although this reaction is irreversible, phosphorylase *b* can be reformed through the action of phosphorylase phosphatase, which catalyzes the hydrolytic cleavage of the bound phosphate in phosphorylase *a* (92, 93). In muscle, as well as in other tissues, phosphorylase *a* is also under hormonal control (94, 95). Phosphorylase kinase, like phosphorylase, can exist in an active and an inactive form. Incubation of the kinase with ATP in the presence of magnesium ions

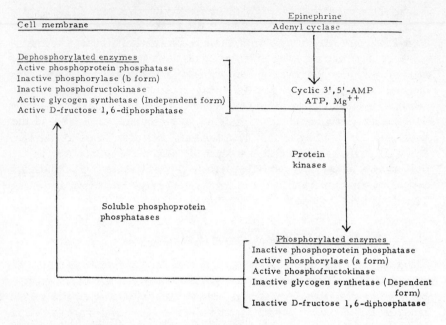

Cell membrane

Epinephrine
Adenyl cyclase

Dephosphorylated enzymes
Active phosphoprotein phosphatase
Inactive phosphorylase (b form)
Inactive phosphofructokinase
Active glycogen synthetase (Independent form)
Active D-fructose 1,6-diphosphatase

Cyclic 3',5'-AMP
ATP, Mg^{++}

Protein
kinases

Soluble phosphoprotein
phosphatases

Phosphorylated enzymes
Inactive phosphoprotein phosphatase
Active phosphorylase (a form)
Active phosphofructokinase
Inactive glycogen synthetase (Dependent
form)
Inactive D-fructose 1,6-diphosphatase

Fig. 9. Proposed mechanism and plausible interrelationships for the regulation of the activity of the renal and hepatic enzymes involved in glycolysis, gluconeogenesis, and glycogenesis at the two stages in which the breakdown or synthesis of carbohydrate is catalyzed by phosphorylase and phosphofructokinase or glycogen synthetase and D-fructose-1,6-diphosphatase, respectively. From Mendicino et al. (113).

activates the enzyme, and this process is accompanied by phosphorylation of the protein (96). Glucose markedly activates phosphorylase phosphatase, thus decreasing the level of phosphorylase *a* (97).

Similarly the activity of the multienzyme pyruvate dehydrogenase complex from beef kidney mitochondria is regulated by a phosphorylation-dephosphorylation reaction sequence (98). Phosphorylation and concomitant inactivation of pyruvate dehydrogenase are catalyzed by an ATP-specific kinase; dephosphorylation and concomitant reactivation, by a phosphatase. Figure 10 depicts the kinetics of the phosphorylation-dephosphorylation reaction. The kinase and the phosphatase appear to be regulatory subunits of the pyruvate dehydrogenase complex.

Recently Lueck and Nordlie (99) reported that glucose-6-phosphatase possesses potent carbamyl phosphate-glucose phosphotransferase activity, and a physiologically significant synthetic role for this phosphotransferase was proposed with both carbamyl and phosphoryl donors.

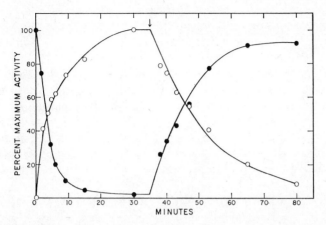

Fig. 10. Relationship between degree of phosphorylation and activity of the pyruvate dehydrogenase complex. Pyruvate dehydrogenase was incubated with γ-P^{32}-ATP. At the time indicated by the arrow, $MgCl_2$ was added to give a final concentration of 10 mM. Data from Linn et al. (98). Symbols: ●—●, enzyme activity; ○—○, protein-bound radioactivity.

C. Protein Phosphorylation During Gene Expression and Protein Synthesis

There are many reports of regulatory phosphorylation of chromosomal proteins, including protamines (see, e.g., refs. 100 and 101). Insulin and prolactin stimulate a protein kinase involved in the phosphorylation of histone and nonhistone nuclear proteins, and this can induce the synthesis of specific milk proteins (102). Glucagon and insulin have also been shown to stimulate histone phosphorylation in rat liver cells (86). In addition, a histone phosphatase specific for phosphorylated histones and protamine has been described (103).

Kabat (104, 105) has shown that rabbit reticulocyte suspensions incorporate ^{32}P-phosphate extensively into serine and threonine residues of ribosomal structural proteins, thus providing evidence that eukaryotic ribosomes are heterogeneous and are subject to chemical modifications after their assembly into functional units. Furthermore, it was postulated that this chemical modification performs a regulatory role in controlling protein biosynthesis.

D. Other Examples of Nonenzymic Protein Phosphorylation

Although not susceptible to the protein phosphokinase system, a number of enzymes can be phosphorylated by a variety of agents. For example, alkaline phosphatase from *E. coli* or bovine liver has been shown to be phosphorylated when incubated with orthophosphate (106). The serine at the active center of phosphoglucomutase may be phosphorylated by glucose-1,6-diphosphate (see ref. 107). Alpers and Lam (108) have reported that [1-^{32}P]-labeled 1,3-diphosphoglyceric acid can phosphorylate rabbit muscle phosphoglucomutase, as previously postulated on the basis of kinetic evidence (109). Bovine liver and human erythrocytic nucleoside diphosphokinase are susceptible to phosphorylation by ATP (110). Alkaline hydrolysis of the product isolated after incubation of either kinase yielded [1-^{32}P]phosphohistidine, [3-^{32}P]phosphohistidine, *N*-[ϵ-^{32}P]phospholysine, and other radioactive products. The radioactive phosphoenzyme was shown to transfer its label to ADP (111).

Particulate preparations from a variety of tissues inactivate renal fructose-1,6-diphosphatase in the presence of ATP, Mg^{2+}, and cysteine, and it has been postulated that a protein kinase and a phosphoprotein phosphatase may be involved in regulating the activity of this enzyme. Moreover, it has been suggested that phosphofructokinase is also regulated in this way (112, 113).

Bonsignore et al. (123) have demonstrated in a rat liver preparation the presence of three enzymes that inactivate glucose-6-phosphate dehydrogenase, and it seems possible that several proteins may regulate the activity of glucose-6-phosphate dehydrogenase *in vivo*. Glucose-6-phosphate dehydrogenase activity declines during starvation or diabetes and rises after refeeding without any new synthesis of the enzyme, suggesting that this enzyme is controlled not only at the level of synthesis but also by different factors that may involve (*a*) interconversion between catalytically different forms of glucose-6-phosphate dehydrogenase and (*b*) the three proteins mentioned above. The glucose-6-phosphate dehydrogenase-inactivating enzymes could be involved in controlling the turnover of this enzyme. Moreover, Kenney, as will be discussed later, has suggested the existence of a specific tyrosine aminotransferase-degrading enzyme.

E.   Inactivation of Enzymes by Phosphoryl Substrates

A number of enzymes known to be directly susceptible to chemotrophic inactivation by phosphoryl compounds are listed in Table 11. Some of these will be discussed in more detail.

Anderson et al. (28) discovered the inactivation of glutamic dehydrogenase by pyridoxal 5'-phosphate. Incubation of glutamic dehydrogenase at 34° with pyridoxal phosphate causes a time-dependent loss in activity, as illustrated in Fig. 11. Dilution of the enzyme-inhibitor mixture does not result in reversal of the inactivation process. Inactivation does not, therefore, arise from the formation of a freely dissociable enzyme-inhibitor complex. The loss of glutamic dehydrogenase activity was accompanied by a concomitant loss in the alanine dehydrogenase activity of the enzyme. In this respect, inactivation by pyridoxal phosphate differs from the *inhibition* of glutamic dehydrogenase by many effector molecules, since the latter is generally accompanied by an enhancement of the alanine dehydrogenase activity. However, it resembles the TPNH-induced inactivation of glutamic dehydrogenase, in which (34) there is parallel decrease of both alanine and glutamic dehydrogenase activities.

Considerable inhibition of the enzyme was observed at very low concentrations of pyridoxal phosphate, and the degree of inactivation during a given period of incubation increases as the ratio of pyridoxal phosphate to the enzyme increases. The inhibition of both glutamic dehydrogenase and alanine dehydrogenase activities of the enzyme is accompanied by a profound effect on the ability of the enzyme to aggregate. Incubation of the enzyme with pyridoxal phosphate at a pyridoxal phosphate/glutamic dehydrogenase ratio of 50 resulted in 90% inactivation and in the prevention of aggregation at higher protein concentrations. Over the protein concentration range where aggregation of native glutamic dehydrogenase occurs, the sedimentation coefficient of the pyridoxal phosphate-inactivated enzyme remained essentially constant at approximately 12.5. This was taken to indicate that pyridoxal phosphate interferes with the polymerization process responsible for the formation of higher-molecular-weight material for the monomeric form but does not promote further dissociation of the monomeric form to subunits of lower molecular weight.

The formation of a Schiff base between pyridoxal phosphate and an amino group of the protein, in most cases an ϵ-amino group of a lysine residue, has often been demonstrated. The spectral changes observed on the addition of glutamic dehydrogenase to pyridoxal phosphate are consistent with those found on the formation of pyridoxal phosphate Schiff bases; the fact that larger changes were not observed indicates that relatively few pyridoxal phosphate molecules are needed to inactivate the enzyme. From the 325 nm maximum of the absorption spectrum of the pyridoxyl-enzyme produced at a pyridoxal phosphate/glutamic dehydrogenase ratio of 50, it was calculated that 4–5 moles of pyridoxal phosphate was bound per 250,000 molecular weight of the enzyme. This estimated value is based on a molar extinction coefficient of 9710 for ϵ-pyridoxyllysine and on the assumption that the pyridoxyl-enzyme should exhibit an absorption comparable to that of free pyridoxyllysine.

These observations are consistent with the recent demonstration by Smith and his coworkers (27) that lysine residue 97 is the site for the attachment of one pyridoxal phosphate per subunit.

Table 11.   Examples of Enzyme Inactivation by Phosphoryl Compounds

Enzyme	Modifier	Effect	Reference
Rabbit muscle aldolase	Fructose diphosphate	Dissociation of enzyme into active subunits by a general reaction involving fructose diphosphate and more than half the ϵ-amino groups of lysine residues in the protein.	124
Rabbit muscle aldolase	Glyceraldehyde 3-phosphate, erythrose 4-phosphate	Inactivation by modification of residues at active site.	125
Human muscle aldolase, human liver "fructose-intolerance" aldolase	ATP	Inhibition.	129
Triose phosphate isomerase	1-Hydroxy-3-iodo-2-propanone phosphate (analog of the substrate, dihydroxy-acetone phosphate)	Inactivation by modification of active site of enzyme.	130

The sulfhydryl groups of glutamic dehydrogenase do not appear to be affected by the reaction of the enzyme with pyridoxal phosphate, since the same number of sulfhydryl groups was present in both the native and the pyridoxyl-enzyme. Since mercurials enhance the glutamic dehydrogenase activity and inhibit the alanine dehydrogenase activity of the enzyme, an involvement of sulfhydryl groups would not be expected in view of the fact that both activities are affected by pyridoxal phosphate to the same extent. Moreover, it should be remembered that in the TPNH-induced inactivation of glutamic dehydrogenase (34, 115) there is no decrease—in fact, there is some increase—in titratable SH groups.

Pyridoxal also inactivates glutamic dehydrogenase, although much higher concentrations are required. It is possible that the phosphate group of pyridoxal phosphate confers a greater specificity on the interaction of this inactivator with the enzyme; however, it has been pointed out that the existence of pyridoxal in the unreactive internal hemiacetal form could help to

Table 11 (*continued*)

Enzyme	Modifier	Effect	Reference
Yeast glyceral-dehyde-3-phosphate dehydrogenase	ATP	(*a*) Inhibition caused by competition of ATP or NAD binding sites and (*b*) inactivation caused by dissociation of enzyme into subunits.	126
Glyceraldehyde-3-phosphate dehydrogenase	Deoxyribose 5-phosphate	Inactivation.	131
Sheep heart phospho-fructokinase	ATP	Dissociation of enzyme into subunits, resulting in inhibition.	127
Kidney fructose-1,6-diphosphatase	Pyridoxal 5'-phosphate	Enzyme rendered insensitive to AMP or substrate inhibition.	116
Bovine liver glutamic dehydrogenase	Pyridoxal 5'-phosphate	Inactivation by modification of essential lysine residue at active site.	28
Yeast hexokinase	Xylose in the presence of ATP and Mg^{2+}	Inactivation by structural change in the protein.	121
Serine dehydratase	Serine, threonine	Inactivation by formation of oxazolidine ring by reaction of serine with enzyme-bound pyridoxal phosphate.	117, 118
D-Glycogen synthetase	ATP-Mg^{2+}	Inactivation by possible phosphorylation of enzyme.	128

explain the greater reactivity of pyridoxal phosphate with nucleophilic reagents.

Many compounds, such as purine nucleotides and pyridine nucleotides, affect both glutamic dehydrogenase and the oxidation of glutamic acid in intact mitochondria. The importance of these effects may be related to the control mechanisms of the mitochondrial dehydrogenase. In this respect, it is of interest that the mitochondrial oxidation of glutamic acid can occur through a pyridoxal phosphate-dependent transaminase pathway, as well as by the glutamic dehydrogenase-catalyzed reaction. Although the importance of the pyridoxal phosphate inactivation of glutamic dehydrogenase to the normal functioning of the enzyme is not immediately obvious, this finding clearly indicates another example of chemotrophic feedback or regulation.

Pyridoxal phosphate also modifies the characteristics of fructose-1,6-diphosphatase, which is normally subject to inhibition by high substrate

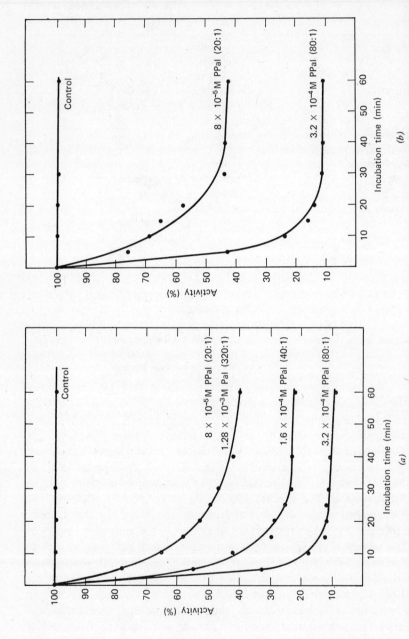

Fig. 11. (a) Inactivation of glutamic dehydrogenase activity by pyridoxal 5′-phosphate and pyridoxal. (b) Inactivation of alanine dehydrogenase activity by pyridoxal 5′-phosphate. Ratios shown in parentheses are molar ratios of inhibitor to enzyme. Data from Anderson et al. (28).

concentrations. However, treatment of pig kidney fructose-1,6-diphosphatase, followed by borohydride reduction, leads to the formation of an active pyridoxamine phosphate derivative of the enzyme which is no longer sensitive to either the allosteric AMP inhibition or the high substrate inhibition (116). Both of these effects are due to the modification of different ϵ-aminolysyl residues of the enzyme, as in the case of glutamic dehydrogenase and other enzymes.

Pyridoxal phosphate has also been implicated in the inactivation of threonine dehydratase by its substrates threonine and serine (117, 118). It is postulated that an oxazolidine ring is formed between serine and bound pyridoxal phosphate, thus inactivating the enzyme, as shown in Fig. 12. It has further been shown that this enzyme can be reversibly inactivated by elemental sulfur, and this may be a physiological device for the control of methionine metabolism (119).

A tyrosine aminotransferase inactivating system, which probably consists of two protein components, has been characterized in rat liver homogenates (120). Tyrosine aminotransferase has been shown to undergo very rapid turnover in rat liver, and thus its cofactor, pyridoxal phosphate, conceivably could, as in the case of glutamic dehydrogenase and fructose-1,6-diphosphatase, exert a regulatory action in this process.

One of the best examples of substrate-induced inactivation, including *in vivo* effects, is that of DelaFuente (121), who studied the inactivation of yeast hexokinase by xylose in the presence of a phosphoryl donor substrate. This is shown in Fig. 13. It has been concluded that lyxose and xylose, nonphosphorylated analogs of the sugar substrate, induce a conformational change in the protein, and that as a consequence the very weak ATPase activity of hexokinase becomes greatly increased and the apparent affinities for the nucleotide substrates are markedly changed. With xylose, there is concomitant progressive and apparently irreversible inactivation of the enzyme, in both its transferase and its hydrolase activities. Furthermore, this xylose-dependent inactivation of yeast hexokinase has also been obtained in whole cells. The inactivating effect is very specific for xylose and requires a metal-nucleotide complex. The specificity with respect to the latter is similar to that of the hexokinase as well as the hydrolase activities, since only ITP can substitute for ATP and only Mn^{2+} for Mg^{2+}.

To test whether the efficiency of xylose as an inactivating agent may be related to its binding to the enzyme, as revealed by its ability to competitively inhibit the hexokinase reaction, the effect of the concentration of xylose on the rate of inactivation at a fixed concentration of magnesium and ATP was measured. Since the inactivation rate is assumed to be directly related to the binding of the inactivating agent, it may be admitted that the inactivation constant is coincident with the dissociation constant of the complex formed

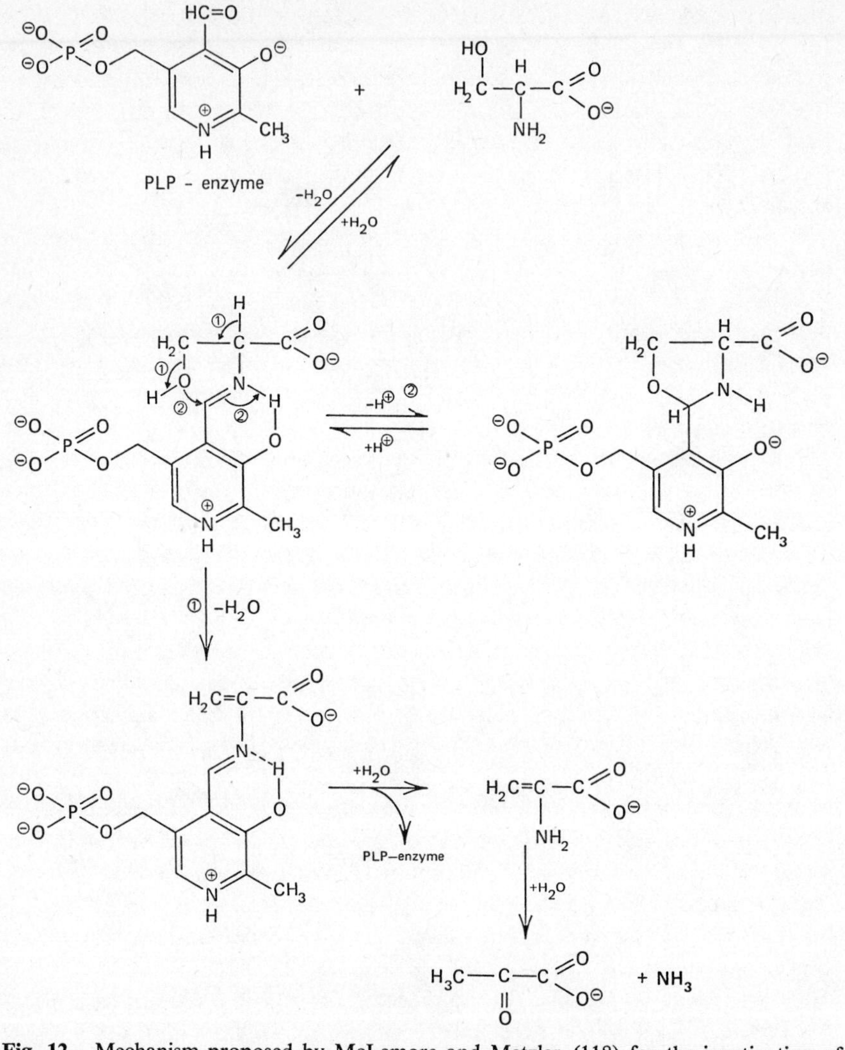

Fig. 12. Mechanism proposed by McLemore and Metzler (118) for the inactivation of L-threonine dehydratase by L-serine. PLP-pyridoxal phosphate.

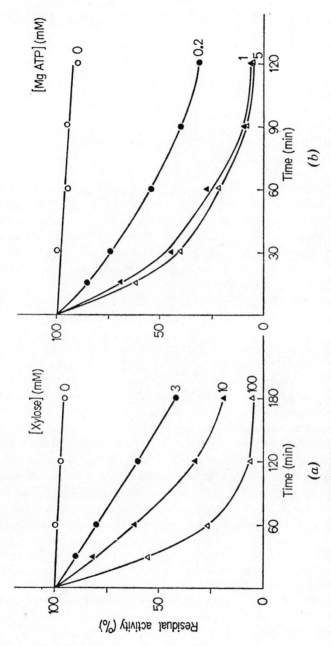

Fig. 13. Effect of the concentration of xylose and MgATP on the rate of inactivation of hexokinase. (a) Variable concentrations of xylose at a fixed concentration of MgATP. (b) Variable concentrations of MgATP at a fixed concentration of xylose. Data from DelaFuente (121).

by the inactivating agent and the enzyme. The value found for xylose was approximately 10 mM, which is the K_i value for xylose as a competitive inhibitor of the hexokinase reaction.

A similar study was carried out with variable concentrations of MgATP at a fixed concentration of xylose, and an apparent dissociation constant near 0.1 mM was obtained. It is important to note that this value is roughly coincident with the K_m found for MgATP in the ATPase reaction in the presence of lyxose, but not in its absence, and also with the K_m of MgATP in the hexokinase reaction.

Lyxose and N-acetylglucosamine can act as protective agents against inactivation by xylose. Their respective efficiencies are closely related to the corresponding K_i values of these compounds as competitive inhibitors of the hexokinase reaction. It is of much interest, particularly medical (see below), that glucose, as a protecting agent, proved markedly better than could be expected from the K_m in the hexokinase reaction; 0.4 mM glucose counteracted approximately 50% of the inactivating effect of xylose at a concentration of the latter of about 10 times its dissociation constant. It is well known that diets rich in xylose produce cataracts and that increasing the amount of glucose in the diet prevents their development. Possibly the mechanism of xylose-produced cataract formation is related to the inactivating effects of xylose on hexokinase in addition to modification of the phosphogluconate pathway.

Attempts to recover the activity of the xylose-induced inactivated enzyme by prolonged dialysis, incubation with an excess of glucose, treatment with —SH compounds, a change to pH 11.5 for a short time, and treatment with 4 M urea followed by dialysis were unsuccessful.

Resting cell populations, when incubated aerobically with xylose in the presence of ethanol as a source of energy to provide for intracellular ATP, lose a large proportion of their hexokinase activity without impairment of other enzymatic activities tested. At periods of 3 or 4 hr, the values of hexokinase activity reached a plateau, so that no marked changes occurred in the following 2 hr. The fact that in most cases no more than 80% of the hexokinase became lost could be adscribed either to heterogeneity of the yeast population or to the existence of two isoenzymes of yeast hexokinase, one of which could be resistant to inactivation by xylose. The latter possibility was considered improbable since preparations from several sources showed the same rate of inactivation. On the other hand, in our own laboratory we have demonstrated (Santos and Grisolia, unpublished) that preparations of TPD inactivated by NADH, when added to active TPD in the presence of NADH, offered much greater protection against inactivation than other nonspecific proteins such as albumin.

Inactivation of the enzyme with [γ-^{32}P]ATP or with ^{14}C-labeled xylose did not give detectable labeling of the protein. According to DelaFuente, no change either in the ultraviolet spectrum or in the molecular weight of the purified enzyme, as judged by the pattern of elution from gel filtration columns, could be detected. The mechanism responsible for the xylose-dependent inactivation cannot be ascertained from the information available. Possibly the inactivation process is related to the formation of a potentially active ternary complex. Moreover, it seems to be related to the effect which induces labilization of the terminal phosphoryl of the nucleotide triphosphates, since the relative efficiencies of ATP, ITP, and GTP as inactivating agents are similar to those of these compounds as substrates in the hexokinase as well as in the hydrolase activities of the enzyme. Since other results from DelaFuente's laboratory clearly indicated that sugar substrates can induce a conformational change in the protein, it was assumed that the same flexibility of the active site which allows an "induced fit" can, in special circumstances, give rise to an inactive conformation.

DelaFuente pointed out that, among the known examples of inactivation of enzymes by their substrates or compounds related to them, probably the one showing the greatest similarity to the case described here is that reported by Bertland and Kalckar (122). In this instance D-galactose, D-fucose, and L-arabinose inactivate the enzyme UDPG-4-epimerase in the presence of UMP.

The findings of DelaFuente may provide a good model to test many aspects of the elastoplasticity theory.

Woodfin (124) has attributed the dissociation of rabbit muscle aldolase into subunits upon incubation with its substrate, fructose diphosphate, to a reaction involving fructose diphosphate and more than half the ε-amino groups of lysine residues in the protein. However, Lai et al. (125) have reported the specific inactivation of rabbit muscle aldolase by its substrates, glyceraldehyde-3-phosphate and erythrose-4-phosphate, as shown in Fig. 14. The loss of activity is irreversible and is obtained upon incubation of the enzymes with stoichiometric amounts of substrate; four equivalents of substrate are incorporated per mole of enzyme. Dihydroxyacetone phosphate and its substrate analog, hexitol diphosphate, can protect against the inactivation, suggesting, in contrast to Woodfin, that modification of the residues at the active site is involved. The nature of the covalent bonds formed and the site of the attachment are not known, but it is conceivable that a type of mechanism discussed previously in regard to the inactivation of triosephosphate dehydrogenase by acetyl phosphate (i.e., the acylation of lysine residues with the intermediate formation of acylated cysteine groups) may be involved (54, 55).

With yeast glyceraldehyde-3-phosphate dehydrogenase, ATP produces two separate effects. One effect is due to ATP competition with NAD for binding sites and leads to instantaneous inhibition of activity (126). The

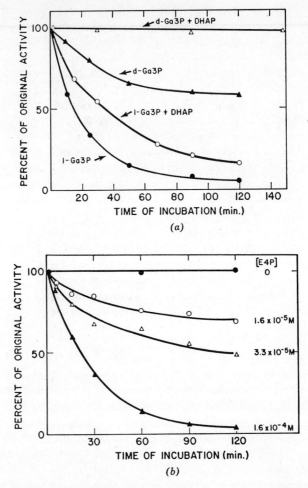

Fig. 14. Inactivation of aldolase by substrate. (a) By D- or L-glyceraldehyde 3-phosphate (Ga3P) and dihydroxyacetone phosphate (DHAP). (b) By D-erythrose 4-phosphate (E4P). Data from Lai et al. (125).

other effect involves dissociation of the enzyme into subunits by ATP and also results in loss of activity (126). These effects, which are illustrated in Fig. 15, can occur at physiological concentrations of ATP. The enzyme can also be stabilized by NAD against the ATP-induced dissociation, thus providing a mechanism involving NAD in control of degradation of this enzyme.

Mansour and Ahlfors (127) have demonstrated the inhibition of heart phosphofructokinase by its substrate ATP through a mechanism believed to involve dissociation of the enzyme into subunits. The kinetics of this inhibition is affected markedly by other ligands such as cyclic AMP. In the presence of Mg^{2+}, ATP causes an inactivation of the glucose-6-phosphate-dependent activity of glycogen synthetase (128). It is thought that this results in the production of an inactive molecular form of the enzyme, which is more phosphorylated than the glucose-6-phosphate-dependent enzymes.

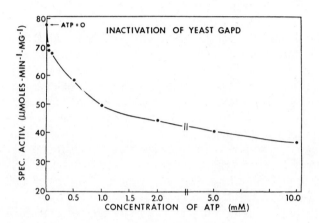

Fig. 15. Activity of yeast glyceraldehyde-3-phosphate dehydrogenase after $4\frac{1}{2}$ hr incubation with ATP. Data from Stancel and Deal (126).

5. ADENYLYLATION

There are two ways in which ATP can be used for protein modification reactions: (*a*) as a phosphoryl donor, as exemplified elsewhere in this chapter, and (*b*) by transfer of the adenylyl moiety. Interest in this area of protein modification was started by the initial findings of Holzer and his coworkers; it appears that adenylylation may have extensive biological significance and that many proteins may be modified in this way. Several examples of adenylylation are discussed below, and a more extensive coverage is presented in Chapter 5.

A. Glutamine Synthetase

Because of the complex effects of adenylylation on glutamine synthetase, a brief background of this enzyme is presented. The overall reaction catalyzed by glutamine synthetase can be summarized as follows:

$$\text{Glutamate} + \text{ATP} + \text{NH}_3 \overset{\text{Mg}^{2+}}{\longleftrightarrow} \text{glutamine} + \text{ADP} + \text{P}_i.$$

The reaction is generally believed to proceed in two stages:

Synthetase reaction   Transferase reaction

$$\text{Glutamate} \underset{\text{ADP}}{\overset{\text{ATP}}{\rightleftharpoons}} \text{glutamyl phosphate} \underset{\text{NH}_2}{\overset{P_i}{\rightleftharpoons}} \text{glutamine.}$$

The enzyme from *Escherichia coli* can exist in several forms. Removal of Mn^{2+} from *E. coli* glutamine synthetase converts the enzyme from a "taut" to a "relaxed" form which is catalytically inactive (132) and is susceptible to deaggregation by alkaline pH or mild denaturants (133). In addition the sulfydryl groups of the relaxed form react readily with iodoacetate or organic mercurials with resulting deaggregation of the subunits (134). The metabolic effectors of glutamine synthetase can cause alterations in the rate and the extent of inactivation by mercurials. Whereas CTP enhances the degree of inactivation, AMP and histidine decrease the rate of inactivation (134). When divalent cations are added to the relaxed enzyme, it is converted into a tightened form that is indistinguishable from the original taut enzyme on the basis of its catalytic properties and resistance to denaturants or to reaction with sulfydryl group reagents. However, under certain conditions the tightened form differs from the taut form in being much less soluble in dilute buffer solution (135).

Holzer's group has shown (136, 137) that purified glutamine synthetase from *E. coli* can be enzymatically inactivated according to the equation:

$$\text{Glutamine synthetase } a \underset{\text{ATP, Mg}^{2+}, \text{ glutamine}}{\overset{\text{inactivating enyzme}}{\rightleftharpoons}} \text{Glutamine synthetase } b.$$

The inactivating enzyme, as well as the active and inactive forms of glutamine synthetase, has been purified. In contrast to the altered activity in the glutamine synthetase reaction, other properties are not changed (e.g., γ-glutamyltransferase activity, sedimentation in the ultracentrifuge, behavior in gel filtration, ion-exchange chromatography, and electrophoresis on polyacrylamide and agar-gel). It was proposed that the inactivating reaction involved glutamylation, amidation, phosphorylation or adenylylation of the enzyme, proteolytic action, and conformational changes. Eventually evidence was presented for the binding of the adenine part of ATP to glutamine synthetase (138, 139). This binding is catalyzed by the inactivating enzyme in the presence of ATP, Mg^{2+}, and glutamine.

It would perhaps be more correct in this case, as with others, to call the phenomenon a "modification" of glutamine synthetase rather than an "inactivation." The stability of glutamine synthetase is also susceptible to substrates, and this enzyme can be protected or made less stable to heat in the presence of ATP and other reagents (4).

The adenylylated glutamine synthetase from *E. coli* has markedly different properties from the unadenylylated; for example, the adenylylated enzyme is less active in catalyzing glutamine biosynthesis than is the unadenylylated (138, 139). In addition, the unadenylylated enzyme is specifically activated by Mg^{2+}, whereas the adenylylated enzyme is dependent on Mn^{2+} for activity (139). The adenylylated enzyme is also much more susceptible to inhibition by some of the products of glutamine metabolism than is the unadenylylated enzyme, when either the biosynthetic activity or the glutamyl transfer activity of glutamine synthetase is examined (139). The pH optimum for catalysis of α-glutamyl transfer differs in the two enzyme forms, being 6.8 for the adenylylated preparation and 7.9 for the unadenylylated one; both have the same activity when studied at pH 7.15 (140).

The extent of adenylylation is dependent on the growth conditions of *E. coli* (141, 142). Deadenylylation of glutamine synthetase occurs *in vivo* upon switching from a nitrogen-rich to a nitrogen-poor medium (141, 143). Cell extracts of *E. coli* contain another enzyme which catalyzes the removal of adenylyl residues from glutamine synthetase (144, 145). The glutamine synthetase deadenylylating enzyme is activated by α-ketoglutarate and inhibited by glutamine. The complex deadenylylating system has been shown to consist of at least two protein components, designated P_1 and P_2 (146), and Anderson et al (147) and Hennig et al. (148) have shown that the adenylyltransferase activity can be associated with the P_1 protein fraction of the deadenylylating enzyme system. The deadenylylation reaction involves phosphorolysis rather than hydrolysis of the phosphodiester linkage, since the reaction is almost completely dependent on inorganic phosphate and ADP is produced as a result of deadenylylation.

Thus glutamine can exert two effects on the regulation of glutamine synthetase; it can activate the adenylyltransferase, but on the other hand it inhibits the activity of the deadenylylating enzyme system. The substrate glutamate inhibits both the adenylylase and the deadenylylase, and ATP, also a substrate in the overall reaction, inhibits the adenylyltransferase. Kapoor and Bray (149) have demonstrated that glutamine synthetase from *Neurospora crassa* differs from the *E. coli* enzyme in that it is inhibited by GTP and DPN. Whereas DPN was shown to enhance heat inactivation, glutamate and Mg^{2+} together diminish inactivation, suggesting that there are different binding sites on the enzyme for glutamate and DPN. Somewhat similar results have been obtained with brain glutamine synthetase, which has been shown to be less stable to heat in the presence of ATP (114). It is possible that a parallelism exists between the enzymic inactivation of the *E. coli* and the brain enzyme.

B.   RNA Polymerase

A further example of enzyme adenylylation of great biological interest has been reported (154). *Escherichia coli* DNA-dependent RNA polymerase can be reversibly inactivated by incubation with ATP, Mg^{2+}, and a cellular fraction. The γ-phosphorus on the labeled ATP is not incorporated into the protein. However, labels from the α-phosphorus and the carbon of ATP appears in the enzyme, indicating an adenylylation type of mechanism.

C. DNA Repair

Other examples of adenylylation reactions are provided by the enzymes capable of repairing single-strand breaks in DNA, restoring the ester bond between a 3' hydroxyl group and a 5' phosphomonoester (150, 151). The DNA ligase from uninfected E. coli cells has a requirement for DPN, which is broken down during the course of the reaction into AMP and NMN. Little et al. (152) have examined the first step of the reaction catalyzed by DNA ligase of uninfected E. coli. With DPN in the absence of DNA, the enzyme forms a stable enzyme-adenylylate complex, releasing NMN. The enzyme-adenylylate complex can be isolated by gel filtration and is capable of sealing single-strand breaks in DNA in the absence of DPN; during this reaction AMP is released. These reactions may be summarized as follows:

$$DPN + enzyme \rightleftharpoons enzyme-AMP + NMN,$$

$$Enzyme-AMP + nicked \ DNA \rightleftharpoons enzyme + AMP + joined \ DNA.$$

Thus the adenylylated enzyme formed in this case, though stable and produced in the absence of DNA, should be regarded as an intermediate in the overall reaction.

D. Histones

The adenylylation of histones by an enzyme associated with mammalian chromatin has been reported by Nishizuka et al. (153). The enzyme catalyzed the transfer of the ADP-ribose moiety of NAD with the simultaneous release of nicotinamide. The reaction product showed a variety of molecular size distribution from a monomer up to polymers composed of several ADP units. The monomer, as well as the polymers of ADP-ribose, is thought to be linked covalently to histones, and DNA appeared to be required for elongation of the polymer.

6. SOME BIOLOGICAL IMPLICATIONS

The substrate-induced inactivation of enzymes and the elastoplasticity theory reflecting stress at the molecular level have many biological implications, as indicated in a previous review (4). Some of these implications, as briefly discussed above, together with some new ones, may also apply to chemotrophic modifications.

Although, in general, substrate-induced enzyme inactivation may be a relatively slow process, chemotrophic modifications may be very fast. A protein modified by a chemotrophic effector (as well as by elastoplastic modifiers) may be rendered either more or less susceptible to proteolytic attack. Furthermore, as the protein loses its original enzymatic activity, it may acquire other properties, for example, immunological specificity and/or immunological power. As discussed below, protein turnover, at least in many instances, is probably related to or controlled by this type of effect. Another highly speculative implication is memory (4). Although this area remains speculative, chemotrophic modifications may be more closely related to memory than general elastoplastic effects; therefore, because of its importance, memory will be briefly considered.

A. Memory

If it proves to be that some chemotrophically modified proteins turn over slower, they may be ideal candidates to explain memory. As indicated in the earlier review (4) and discussed by many other investigators, memory must leave an imprint on either the proteins or the nucleic acids of brain. As already discussed, carbamylation and protein modification thereof can occur not only with proteins but also with nucleic acids or nucleoproteins, particularly at the histone level, for example, with carbamylurea (39). Nevertheless, we favor the idea that chemotrophic effectors yield modified proteins which, in some cases, do not turn over or do so slowly and may, with or without association to other molecules or structures, retain the memory of the event. This could explain how a relatively small number of chemotrophic reagents possibly, via similar mechanisms, explain the many modified proteins of necessity involved in memory, since it seems unlikely that every memory event will yield a different type of protein modification. Thus, if by a finite number of general mechanisms the neuron can fix imprints and reinforce them in a repetitious manner, at least certain types of memory would be explained fairly well. Moreover, it should be taken into consideration that the very effective chemotrophic reagents thus far described (e.g., 1,3-diphosphoglycerate and ATP) can be regulated by enzymes which are present at extremely high concentrations in the brain, such as acylphosphatase and ATPase.

It should be kept in mind also that much of the energy required by the brain is used for protein synthesis and that a large part of this protein leaves the neuron body and is degraded in the axon. However, part of the protein could be retained after chemotrophic modifications with or without cross linking or in association to other macromolecules for memory effects. Whether protein modification occurs at the Nissl bodies, which seem to be very active in protein synthesis, or in the other parts of the neuron remains to be determined. This suggestion has the advantage that it probably could be put to test by changing the environmental conditions, for example, in ageing and learning situations involving a number of animals, followed by checking for the presence of modified proteins in the neurons of the animals subjected to these processes.

Another area in which chemotrophic effects could have implications is that of mental defects. In defects related to the urea cycle, that is, ammonemia, citrullinuria, and possibly argininosuccinuria, there is a possible common denominator in the highly reactive carbamyl phosphate, which may, as in other cases, interact with brain proteins. Animals could be subjected to chronic ammonia and urea intoxication, and their learning abilities could be correlated with possible changes in carbamylated proteins. As indicated above, there has been a brief report (155) on the existence of an enzyme catalyzing the interconversion of urea and hydroxyurea, and these may be converted to carbamyloxyurea. This may be an important effect in uremia. These considerations led us to test (Crist and Grisolia) the effects of chronic administration of carbamyl phosphate and cyanate on brain function. 75 μmoles NaCNO were injected IP daily to 30–35 gm weanling littermate mice for 16 days. 50, 100, and 150 μmoles NaCNO were injected daily to 65–70 gm littermate Holtzman rats for 16 days. 150, 300, and 450 μmoles carbamyl phosphate were injected IP daily for 14 weeks to other groups of 65–70 gm littermate rats. Learning (Lashley water maze) was tested immediately after ceasing CNO^- injections and retention checked 6 weeks later. Growth curves for all animals were recorded. Control littermates receiving saline were used in all cases. The animals receiving either carbamyl phosphate or cyanate were retarded (errors multiplied by time to criterion) in a dose-related degree, $p < 0.001$. Also, retention of prior task learning was reduced by administration of either cyanate or carbamyl phosphate. Thus a biochemical model for memory involving chemotrophic modification of proteins is now available.

B. Protein Turnover

The subject of protein turnover has been considered for nearly 100 years. For instance, in 1881 Voit (156) stressed the differences between "inert protein" and circulating protein and posed an unanswered question: What is the mechanism which tends to keep the nitrogen content almost constant? In the early 1900's, Folin (157) proposed that because of the continuous use of structures they had to be replenished. More recently, Schoenheimer (158) postulated that nearly all proteins in the body are in a state of flux, being continually synthesized and degraded. The fundamental question is, of course, Why do living organisms require a protein turnover? It seems likely that by and large this is the penalty that organisms must pay to maintain their capacity for growth and regeneration.

There is now overwhelming evidence, not only that rapid protein turnover occurs, but indeed that it is much more rapid than at first realized and that a great deal of the basal energy needs is used to carry out this function. In 1953, Simpson (159) demonstrated the need for energy during protein catabolism, and since then others have confirmed his observations. Interestingly, Brostrom and Jeffay (160) have recently re-evaluated the energy requirements for protein catabolism and have suggested that, although there may be a structural component for this process which requires energy, the enzymes involved in the actual degradation of proteins probably do not have energy requirements.

C. Some Quantitative Aspects of Protein Turnover and Its Regulation

It has been postulated that 400 g of protein is synthesized and degraded per day by a 70-kg man. However, this estimate may be too conservative. Consider, for example, that in this average man some 30 ml of red cells (or about 10 g of hemoglobin) is destroyed per day in the spleen, an organ that weighs about 100 g and contains approximately 20 g of protein. If this organ is assumed to be composed only of macrophages responsible for hemoglobin degradation, it can destroy at least half of its own weight of protein pér day! In other words, the normal daily protein requirement does not reflect the extensive reutilization of amino acids, and as indicated before (181), at least 5 moles of ATP are needed for synthesis of each peptide bond. Therefore at least 30–40% of basal metabolic needs of man are expended in this process!

It should be noted that man's body is able to change the amount of nitrogen secretion by a factor of about 15, that is to say, from about 2 to 30 g of nitrogen per day. The mechanisms involved in this change and/or adaptation are not known, particularly in cases where there is a shortage of dietary protein. In such situations, the body maintains a higher protein supply in some tissues but not in others. For example, in times of starvation, the brain and heart lose only 3% of their bulk, whereas muscle may lose 31%, liver 54%, and spleen 67%. Even in acute starvation, blood albumin is maintained at a fairly constant level, although other protein stores are depleted, because albumin is most important to maintain the oncotic pressure in the blood. A decrease in the albumin level would result in edema. When the supply of protein is low, there is competition between liver and muscle activities, the former being maintained at high levels at the expense of muscle. The same applies to enzymes within the cell, as indicated by Schimke (12).

D.   Methods and Possible Pitfalls in the Measurement of Protein Turnover *in Vivo*

The rate of turnover of many proteins in a variety of animals and tissues has been determined, and Rechcigl (161) has summarized the half-lives of many of the enzymes so far studied. These range from about 1 hr for δ-amino-laevulinate synthetase in rat liver to 100 days for glyceraldehyde-3-phosphate dehydrogenase in rabbit skeletal muscle. Indeed, it has been shown (182) that the half-life of ornithine decarboxylase is 11 min!

The techniques used to determine the rate of protein turnover fall into two basic categories. One involves measuring time-dependent alterations in the enzyme level, caused by changes in diet or the administration of enzymic activators and inhibitors. Isotopic techniques, which constitute the other type, have also been used extensively, particularly the pulse-labeling method in which a single dose of labeled amino acid is administered to a series of animals and the specific activity decay of the labeled enzyme is followed. It is essential with this method, as well as with other isotopic techniques, that there be little or no reutilization of the labeled amino acid once the enzyme has been labeled. Thus the size of the amino acid pool at the time that the label is administered is critical.

The recent paper of Kemp and Sutton (162) exemplifies these difficulties. It raises the question again of whether or not the amino pool is truly in equilibrium, and speculates that there may be a two- to threefold difference in the specific radioactivity of the soluble amino acid pool. Further dangers in using isotopic techniques for determining average protein half-life are discussed by Klevecz (163), who demonstrated the large reutilization of amino acids in mammalian cells. An additional factor which must be considered is the variation in rhythmicity of the plasma amino acid pool and its relation to the dietary intake (164). Thus the ribosomes, as well as any other systems which may affect protein synthesis, may be influenced by periodicity or circadian rhythms, as reflected by the changes in free amino acids in the blood. Millward (165) pointed out the difficulties of isotopic techniques when measuring protein turnover in skeletal muscle. Since labeled aspartate and glutamate are reutilized to a very low extent, the rates of protein synthesis and catabolism may be measured accurately using these two amino acids.

E. Importance of Control of Protein Degradation in Cell Metabolism

Under steady-state conditions, where the level of a given enzyme in a given tissue remains constant, the rate at which that enzyme is synthesized is equal to the rate at which it is degraded. Whereas much work has been done in recent years on the regulation of protein synthesis, little is known about the opposite process of protein degradation. Yet clearly the latter process will be as important a factor in regulation as protein synthesis.

A number of considerations make it appear likely that protein degradation is not a completely random process and that many factors may influence the overall rate of turnover. The heterogeneity of the half-lives of many proteins, for instance, indicates that some specificity is involved. Furthermore, the half-lives of the cells themselves may change in response to various external factors. For instance, during periods when the external supply of protein is limiting, certain types of cells or proteins are able to maintain their nitrogen equilibrium at the expense of other types. Thus the cell can recognize altered external conditions and correspondingly adjust its turnover, that is, an organism or even a cell has certain priorities in respect to the type of protein it wishes to conserve.

As already discussed, the factors which are influential in controlling protein synthesis have been extensively elucidated in recent years; on the other hand, attention has been focused only relatively recently on the other feature of protein turnover, namely, degradation. It is apparent that control over protein degradation may be achieved largely by one of three means.

A. The activity of proteases may be controlled by the intrinsic properties of the enzymes and by many other factors, including hormones.

B. There may be a specific protease for each enzyme, but this seems unlikely as a specific protease would be required for a specific enzyme *ad infinitum* (12). However, there may be specific proteases for a certain number of enzymes and for certain enzyme groups.

C. Protein degradation may be controlled by variations in the properties of the protein molecules as substrates.

Each of these possibilities will be discussed in more detail.

A. The mechanism by which denervation results in atrophy of skeletal muscle is not known, although Goldberg (166) has shown that after denervation and the resultant atrophy muscles show indications of increased protein degradation. Nevertheless, the specific activity on the muscle proteins remains unchanged, indicating that protein synthesis must have also decreased following denervation. Cortisone also produces marked atrophy of the plantaris muscle, and it appears that this increases protein degradation and decreases protein synthesis. Thus this hormone can affect both the synthesis and the degradation of protein. Goldberg has also presented evidence that in *E. coli* changes in the level of aminoacyl-tRNAs regulate protein breakdown by mechanisms similar to those controlling the synthesis of ribosomal RNA, and has further suggested that protein synthesis is not essential for protein degradation (167). Rechcigl and Heston (168) have demonstrated that there are genetic differences between the rates of catalase destruction in different substrains of mice, thus indicating gene control over this process.

Although most work on protein turnover has been done with whole animals, another type of approach has involved *in vitro* studies using proteolytic enzymes. Bond (169), for example, studied the stability of lactate dehydrogenase, arginase, catalase, serine dehydratase, and tyrosine aminotransferase to a number of proteolytic enzymes and tried to relate their susceptibility to proteolytic enzymes to their half-lives *in vivo*. Enzymes with short *in vivo* half-lives were found to be especially vulnerable to proteolytic attack, whereas long-lived enzymes were more resistant. Furthermore, there was good correlation between the relative rates of inactivation *in vivo* and *in vitro* only with specific proteases such as trypsin and chymotrypsin, and not with nonspecific proteases (e.g., pronase and subtilisin). These results suggest that proteolytic enzymes with some degree of specificity are involved in the intracellular degradation of enzymes. Pyridoxal phosphate had no effect on the inactivation of lactate dehydrogenase, arginase, or catalase. However, serine dehydratase and tyrosine aminotransferase, which require pyridoxal phosphate for activity, were rendered less vulnerable to inactivation in the presence of this compound. It is suggested, therefore, that ligands may be expected to modify the sensitivity of some enzymes to proteases and thus have a regulatory function *in vivo* (see *C*). Segal et al. (170) studied the relationship between the thermostability of alanine aminotransferase *in vitro* and its half-life *in vivo* and concluded that the two are not related. A liver lysosomal preparation inactivated the enzyme under conditions in which it was otherwise stable. These workers concluded that the turnover of this enzyme *in vivo* is a reflection of an active (i.e., enzymatic) process which does not depend on prior thermal denaturation.

The interest in proteases, particularly those from lysosomes and similar particles, is exemplified by Gray and his coworkers (171). They have recently shown that a neutral protease able to degrade protein resides in the peroxisomal fraction of rat liver. Nevertheless, it appears that this neutral protease activity is too low to account for the turnover of much of the cytoplasmic protein. On the other hand, it is self-evident that there are other proteases that may affect the cytoplasmic protein.

Brain tissue, which has a high rate of protein turnover, contains cathespin, which operates at an acid pH, and also a highly active neutral proteinase system. The latter enzyme is more active in the white matter than in the gray and appears to be associated particularly with axon material, since it is present also in peripheral nerve [see Richter (172)].

Kuehl and Sumsion (173) have also found that there is no correlation between the thermal stabilities of aldolase and lactate dehydrogenase and their rate of turnover *in vivo*. It is suggested that degradation involves a random enzymatic attack of cellular proteins.

B. There is evidence that some enzymes may be inactivated by specific proteases. Kenney (174), for example, proposed that the inactivation of tyrosine transaminase is due to a specific protein. Indeed, as already indicated, such specific proteases do exist, as demonstrated by Bonsignore et al. (123) for the glucose-6-phosphatase-degrading enzymes of liver and by Mendicino (112) for fructose-1,6-diphosphatase. However, it seems unlikely that this is a general mechanism since it would necessitate the presence of one enzyme to degrade each enzyme *ad infinitum* (12).

C. As suggested a number of years ago (4), it is now becoming increasingly apparent that the rate of proteolytic degradation is directly related to the conformation of the substrate protein. Since the protein may exist in a number of conformational states, a shift in the equilibrium between these states caused by addition of a ligand will produce an alteration of the overall rate of proteolysis of the protein. There are numerous examples of ligand-protein interactions that affect the rate of proteolysis, and some recent ones have been mentioned; others may be cited, such as the previously discussed case in which pyridoxal phosphate rendered serine dehydratase and tyrosine aminotransferase less stable to proteolytic degradation. Studies of the susceptibility of ferritin and apoferritin to proteolysis have also provided some evidence for the dependence of proteolytic degradation on the conformation of the protein substrate. The binding of a 70-Å-diameter micelle of iron to apoferritin results in a lowered susceptibility to tryptic degradation (175). Furthermore, digests of apoferritin contained peptides that were absent from ferritin digests. Chymotrypsin and subtilisin digests yielded similar findings, but with nonspecific pepsin and a cathepsin little difference was found, suggesting that these two enzymes could not distinguish between the liganded and the nonliganded protein. McClintock and Marcus (176) have demonstrated that the proteolytic inactivation of aspartate transcarbamylase of *E. coli* is increased in the presence of CTP but inhibited in the presence of aspartate. Polgar (177) has shown that AMP renders glyceraldehyde-3-phosphate more sensitive to attack by proteolytic enzymes. This work has been extended by Cantau et al. (81), who have shown that the arsenolytic activity of this enzyme disappears rapidly whereas the oxidative activity remains.

In vivo studies have also indicated that ligands may exert regulatory actions during protein degradation. Thus Schimke et al. (178) have shown that the substrate tryptophan prevents the proteolytic breakdown of tryptophan pyrrolase *in vivo*. Administration of tryptophan to animals whose tryptophan pyrrolase had previously been increased to high levels prevented the decay of the enzyme (as shown in Table 12). Evidence that tryptophan prevents the degradation of the enzyme was also found by isotopic procedures.

The fact that the cell can recognize when to degrade its proteins suggests that the proteins may become modified in some way. One way in which this may happen is through the modification of cell protein by means of ligands, as already mentioned. A chemotrophic modification of the protein, involving the formation of a covalent bond between it and a ligand, results in a complete change in the molecule, in effect creating a new type of protein and thus rendering it more (or less) susceptible to proteolytic breakdown, including recognition by lysozymes.

Thus, as has been illustrated in this chapter, modification of proteins by chemotrophic agents introduces a new dimension to protein degradation since it results in essentially irreversible changes in the enzyme. These may, of course, result in altered catalytic properties of the enzyme and possibly in protein degradation. Changes in the structure of the protein can cause an altered susceptibility to breakdown by proteolytic enzymes and/or recognition by lysozymes.

Table 12.   Effect of Various Agents on the Loss of Tryptophan Pyrrolase Activity Data from Schimke (178).

Treatment and Agent	Tryptophan Pyrrolase Activity at $4\frac{1}{2}$ hr, units/g liver
No treatment	3.0 ± 0.5
Pretreatment with hydrocortison + tryptophan:	
Zero time assay	97 ± 5
NaCl	25 ± 3
Histidine	25 ± 4
Hydrocortisone	35 ± 3
Tryptophan	96 ± 5

NOTE TO TABLE 12

Tryptophan pyrrolase activity was increased to high levels by treatment with hydrocortisone and tryptophan. Animals were then injected with one of the following: NaCl, histidine, hydrocortisone, or tryptophan, and after $4\frac{1}{2}$ hr the liver tryptophan pyrrolase activity was measured.

F. Hemoglobin Carbamylation and the Sickle Cell Crisis

There have been many studies and much interest in sickle cell hemoglobin, in the sickling phenomenon, and particularly in its treatment. Sickling and sickle cell crises occur at relatively high concentration of deoxy S hemoglobin. Many therapeutic attempts to alter *in vivo* sickling have been largely unsatisfactory and often are more dangerous than helpful. In view of the recent interest to modify sickling with urea and related carbamylating agents (cyanate and carbamyl phosphate) and since some of the implications of protein carbamylation and behavior of carbamylated hemoglobin may not be fully appreciated, we should like to emphasize a number of points: (*a*) Urea has been reported to prevent sickling by affecting intermolecular hydrophobic bonding (183). It is important to note that urea has been reported to increase the oxygen affinity of the erythrocyte (184). (*b*) Cerami and Manning have shown (185) that cyanate inhibits *in vitro* sickling. They find "a normal capacity to bind and release O_2" for carbamylated hemoglobin S. On these bases, they suggested that cyanate may be used *in vivo* to prevent sickling. (*c*) Kraus and Kraus reported (186) that carbamyl phosphate is more effective than cyanate in preventing sickling. (*d*) Diederich, Carreras, and Grisolia found diminished sickling after carbamylation; there is *also an increase in* O_2 *affinity after carbamylation of adult, fetal, and sickle cell blood.* As already discussed, carbamylated horse hemoglobin was shown to have increased O_2 affinity (26). At equal concentrations, carbamylation of hemoglobin is always more rapid with cyanate than with carbamyl phosphate. Carbamylation prevents sickling in proportion to the induced increase in affinity of hemoglobin for O_2 (up to 2 moles carbamyl incorporation/mole hemoglobin, D. Diederich, unpublished).

As discussed above, there is a marked variability and susceptibility of proteins and even residues thereof to carbamylation. While acute toxicity is low for cyanate and lower for carbamyl phosphate, as shown here the intravenous injection of cyanate results in extensive and relative indiscriminate carbamylation of protein (in addition to that of hemoglobin). Moreover, in our experience with carbamylation of the whole animal (over 6-month observation after extensive carbamylation), there are changes in immunological reactivity of some enzymes, and as briefly illustrated, extensive carbamylation *in vivo* as an experimental model for memory with mice and rats yields (Crist and Grisolia) a marked impairment of learning and memory.

The half-life of cyanate is $\sim$10 times larger than carbamyl phosphate, but carbamyl phosphate *decomposes, at physiological* pH, *to cyanate* (187). Therefore, at first glance there are serious drawbacks in the suggested use

of cyanate for sickle cell treatment and little advantage in the use of carbamyl phosphate over cyanate. However, there is a very large amount of
carbamyl phosphate phosphatase in tissues.

The possibility and advantages of manipulating hemoglobin carbamylation in sickle cell treatment (D. Diederich, personal communication) are to
diminish sickling to manageable levels by judicious modification of hemoglobin S. To that effect conditions should be worked out *in vitro* for both
rapid and limited carbamylation of erythrocytes; extensive carbamylation
(i.e., 4 residues) will shift the O_2 dissociation curve so much that the hemoglobin will become essentially ineffective for O_2 transport (D. Diederich,
personal communication).

ACKNOWLEDGMENTS

The work presented in this paper was supported by National Institutes of
Health Grants AM13119 and AM01855 and American Heart Association
Grant 70-678. We wish to thank the National Academy of Sciences, American
Chemical Society, *Journal of Biological Chemistry*, *European Journal of
Biochemistry*, *Biochemical Journal*, Academic Press Inc., Elsevier Publishing
Company, and the authors cited in the chapter for permission to reproduce
tables and figures. We also thank Mrs. Gwynne Streit for her expert help in
the preparation of the manuscript.

REFERENCES

1. E. Fischer, *Ber. Deut. Chem. Ges.*, **27**, 2985 (1894).
2. D. E. Koshland, Jr., *Advan. Enzymol.*, **22**, 45 (1960).
3. W. P. Jencks, *Catalysis in Chemistry and Enzymology*, McGraw-Hill, New York, 1969.
4. S. Grisolia, *Physiol. Rev.*, **44**, 657 (1964).
5. C. J. Martin and G. M. Bhatnager, *Biochemistry*, **5**, 1230 (1966).
6. S. Grisolia, *Biochem. Biophys. Res. Commun.*, **32**, 56 (1968).
7. J. Monod, J. P. Changeaux, and F. Jacob, *J. Mol. Biol.*, **6**, 306 (1963).
8. D. E. Atkinson, *Ann. Rev. Biochem.*, **35**, 85 (1966).
9. E. R. Stadtman, *Advan. Enzymol.*, **28**, 41 (1966).
10. H. Holzer, *Advan. Enzymol.*, **32**, 297 (1969).
11. A. Sols and S. Grisolia, *Metabolic Regulation and Enzyme Action*, Academic Press, London–New York, 1970.
12. R. T. Schimke, *Ann. Rev. Biochem.*, **39**, 929 (1970).
13. H. Holzer, D. Mecke, K. Wulff, K. Liess, and L. Heilmeyer, Jr., *Advan. Enzyme Regulation*, **5**, 211 (1967).

14. H. M. Kalckar, H. Klenow, A. Munch-Petersen, and J. H. Thaysen, *The Role of Nucleotides for the Function and Conformation of Enzymes* (Alfred Benzon Symposium I), Academic Press, New York, 1969.

15. G. R. Stark, *Biochemistry*, **4**, 1030 (1965).

16. J. E. Varner, in *The Enzymes*, Vol. 4 (Eds.: P. D. Boyer, H. Lardy, and K. Myrback), Academic Press, New York, 1960, p. 247.

17. D. Diederich, S. Grisolia, and G. Ramponi, *FEBS Letters*, **15**, 30 (1971).

18. G. R. Stark and D. G. Smyth, *J. Biol. Chem.*, **238**, 214 (1963).

19. G. R. Stark, *J. Biol. Chem.*, **239**, 1411 (1964).

20. J. Cejka, Z. Vodrazka, and J. Salak, *Biochim. Biophys. Acta*, **154**, 589 (1968).

21. S. Rimon and G. E. Perlmann, *J. Biol. Chem.*, **243**, 3566 (1968).

22. K. Grizzuti and G. E. Perlmann, *J. Biol. Chem.*, **244**, 1764 (1969).

23. O. Avramovic and N. B. Madsen, *J. Biol. Chem.*, **243**, 1656 (1968).

24. I. Svendsen, *Compt. Rend. Trav. Lab. Carlsberg*, **36**, 235 (1967).

25. D. B. Keech and R. K. Farrant, *Biochim. Biophys. Acta*, **151**, 493 (1968).

26. J. V. Kilmartin and L. Rossi-Bernardi, *Nature*, **222**, 1243 (1969).

27. E. L. Smith, M. Landon, D. Piszkiewicz, W. J. Brattin, T. J. Langley, and M. D. Melamed, *Proc. Natl. Acad. Sci. U.S.*, **67**, 724 (1970).

28. B. M. Anderson, C. D. Anderson, and J. E. Churchich, *Biochemistry*, **5**, 2893 (1966).

29. G. Ramponi and S. Grisolia, *Biochem. Biophys. Res. Commun.*, **38**, 1056 (1970).

30. D. Hunninghake and S. Grisolia, *Anal. Biochem.*, **16**, 200 (1966).

31. R. D. Cole, *J. Biol. Chem.*, **236**, 2670 (1961).

32. W. Manson, *Biochim. Biophys. Acta*, **63**, 515 (1962).

33. E. G. Cole and D. K. Mecham, *Anal. Biochem.*, **14**, 215 (1966).

34. S. Grisolia, M. Fernandez, R. Amelunxen, and C. L. Quijada, *Biochem. J.*, **85**, 568 (1962).

35. D. Tucker and S. Grisolia, *J. Biol. Chem.*, **237**, 1068 (1962).

36. D. Diederich and S. Grisolia, *J. Biol. Chem.*, **244**, 2412 (1969).

37. R. D. O'Brien, *Ann. N.Y. Acad. Sci.*, **160**, 204 (1969).

38. J. H. Argyroudi-Akoyunoglou and G. Akoyunoglou, *Biochem. Biophys. Res. Commun.*, **32**, 15 (1968).

39. H. S. Rosenkranz and S. Rosenkranz, *Biochim. Biophys. Acta*, **195**, 266 (1969).

40. A. Albert, *Selective Toxicity*, Methuen, London, 1968.

41. E. Grazi, G. Ronca, and V. Vigi, *J. Biol. Chem.*, **240**, 4267 (1965).

42. W. A. Schroeder, J. T. Cua, G. Matsuda, and W. D. Fenninger, *Biochim. Biophys. Acta*, **63**, 532 (1962).

43. A. F. S. A. Habeeb, *Arch. Biochem. Biophys.*, **121**, 652 (1967).

44. A. Bezkorovainy, R. Zschocke, and D. Grohlich, *Biochim. Biophys. Acta*, **181**, 295 (1969).

45. M. Rolland and S. Lissitzky, *Biochim. Biophys. Acta*, **214**, 282 (1970).

46. P. D. Hoagland, *Biochemistry*, **7**, 2542 (1968).

47. A. Marzotto, P. Pajetta, L. Galzigna, and E. Scoffone, *Biochim. Biophys. Acta*, **154**, 450 (1968).

48. P. Cuatrecasas, S. Fuchs, and C. B. Anfinsen, *Biochim. Biophys. Acta*, **159,** 417 (1968).

49. J. M. Connellan and D. C. Shaw, *J. Biol. Chem.*, **245,** 2845 (1970).

50. L. L. Houston and K. A. Walsh, *Biochemistry*, **9,** 156 (1970).

51. A. Chapman, T. Sanner, and A. Pihl, *European J. Biochem.*, **7,** 588 (1969).

52. P. L. Vul'fson and N. B. Kozlova, *Biokhimiya*, **33,** 658 (1968).

53. S. Wilk, A. Meister, and R. H. Haschemeyer, *Biochemistry*, **9,** 2039 (1970).

54. J. H. Park, D. C. Shaw, E. Mathew, and B. P. Meriwether, *J. Biol. Chem.*, **245,** 2946 (1970).

55. E. Mathew, B. P. Meriwether, and J. H. Park, *J. Biol. Chem.*, **242,** 5024 (1967).

56. T. Wieland and F. Bukelmann, *Ann.*, **576,** 20 (1952).

57. D. G. Lindsay and S. Shall, *Biochem. J.*, **121,** 737 (1971).

58. D. M. P. Phillips, *Biochem. J.*, **107,** 135 (1968).

59. R. J. DeLange, D. H. Fambrough, E. L. Smith, and J. Bonner, *J. Biol. Chem.*, **244,** 319 (1969).

60. D. Gallwitz, *Biochem. Biophys. Res. Commun.*, **32,** 117 (1968).

61. T. Takashashi, *Niigata Igakkai Zasshi*, **82,** 169 (1968).

62. T. Ozawa, N. Yamanaka, and K. Yagi, *J. Biochem. (Tokyo)*, **66,** 285 (1969).

63. A. Inoue and D. Fujimoto, *Biochim. Biophys. Acta*, **220,** 307 (1970).

64. J. Caravaca and S. Grisolia, *J. Biol. Chem.*, **235,** 684 (1960).

65. W. K. Paik and S. Kim, *Biochem. J.*, **116,** 611 (1970).

66. A. J. MacGillivary and J. P. P. V. Monjardino, *Biochem. J.*, **108,** 22P (1968).

67. D. Gallwitz, *FEBS Letters*, **13,** 306 (1971).

68. J. A. Wilhelm and K. S. McCarty, *Cancer Res.*, **30,** 418 (1970).

69. A. Mori, K. Tanaka, T. Tomita, K. Nakamura, and T. Hayashi, *Biochim. Biophys. Acta*, **192,** 555 (1969).

70. W. K. Paik, D. Pearson, H. W. Lee, and S. Kim, *Biochim. Biophys. Acta*, **213,** 513 (1970).

71. G. C. Liew, G. W. Haslett, and V. G. Allfrey, *Nature*, **226,** 414 (1970).

72. A. E. Smith and K. A. Marcker, *Nature*, **226,** 607 (1970).

73. G. Polz and G. Kriel, *Biochem. Biophys. Res. Commun.*, **39,** 516 (1970).

74. D. Hawkins, R. N. Pinckard, and R. S. Farr, *Science*, **160,** 780 (1968).

75. S. Grisolia, I. Santos, and J. Mendelson, *Nature*, **219,** 1252 (1968).

76. S. Grisolia, J. Mendelson, and D. Diederich, *Nature*, **223,** 79 (1969).

77. S. Grisolia, J. Mendelson, and D. Diederich, *FEBS Letters*, **11,** 140 (1970).

78. R. J. Henry, *Clinical Chemistry: Principals and Techniques*, Hoeber Medical Division, Harper and Row, New York, 1964, p. 265.

79. D. M. P. Phillips, *Biochem. J.*, **87,** 258 (1963).

80. S. Grisolia, L. Mokrasch, and H. Grady, *J. Kansas Med. Soc.*, **60,** 111 (1959).

81. B. N. Cantau, G. J. Jaureguiberry, and J. Pudles, *European J. Biochem.*, **16,** 208 (1970).

82. J. F. Kuo and P. Greengard, *Proc. Natl. Acad. Sci. U.S.*, **64,** 1349 (1969).

83. J. D. Corbin and E. G. Krebs, *Biochem. Biophys. Res. Commun.*, **36,** 328 (1969).

84. D. A. Walsh, J. P. Perkins, and E. G. Krebs, *J. Biol. Chem.*, **243**, 3763 (1968).

85. E. Miyamoto, J. F. Kuo, and P. Greengard, *Science*, **165**, 63 (1969).

86. T. A. Langan, *Science*, **162**, 579 (1968).

87. J. F. Kuo and P. Greengard, *J. Biol. Chem.*, **244**, 3417 (1969).

88. E. H. Fischer and E. G. Krebs, *J. Biol. Chem.*, **216**, 121 (1955).

89. E. G. Krebs and E. H. Fischer, *Biochim. Biophys. Acta*, **20**, 150 (1956).

90. E. G. Krebs, A. B. Kent, and E. H. Fischer, *J. Biol. Chem.*, **231**, 73 (1958).

91. C. Nolan, W. B. Novoa, E. G. Krebs, and E. H. Fischer, *Biochemistry*, **3**, 542 (1964).

92. G. T. Cori and A. A. Green, *J. Biol. Chem.*, **151**, 31 (1943).

93. D. J. Graves, E. H. Fischer, and E. G. Krebs, *J. Biol. Chem.*, **235**, 805 (1960).

94. E. W. Sutherland, in *Phosphorus Metabolism*, Vol. I (Eds.: W. D. McElroy and B. Glass), John Hopkins Press, Baltimore, 1951, p. 53.

95. T. W. Rall, E. W. Sutherland, and J. Berthet, *J. Biol. Chem.*, **224**, 463 (1957).

96. R. J. DeLange, R. G. Kemp, W. D. Riley, R. A. Cooper, and E. G. Krebs, *J. Biol. Chem.*, **243**, 2200 (1968).

97. P. A. Holmes and T. E. Mansour, *Biochim. Biophys. Acta*, **156**, 275 (1968).

98. T. C. Linn, F. H. Pettit, and L. J. Reed, *Proc. Natl. Acad. Sci. U.S.*, **62**, 234 (1969).

99. J. D. Lueck and R. C. Nordlie, *Biochem. Biophys. Res. Commun.*, **39**, 190 (1970).

100. W. B. Benjamin and R. M. Goodman, *Science*, **166**, 629 (1969).

101. L. J. Kleinsmith, V. G. Allfrey, and A. E. Mirsky, *Proc. Natl. Acad. Sci. U.S.*, **55**, 1182 (1966).

102. R. W. Turkington and M. Riddle, *J. Biol. Chem.*, **244**, 6040 (1969).

103. M. H. Meisler and T. A. Langan, *J. Biol. Chem.*, **244**, 4961 (1969).

104. D. Kabat, *Biochemistry*, **9**, 4160 (1970).

105. D. Kabat, *Biochemistry*, **10**, 197 (1971).

106. L. Engstrom, *Biochim. Biophys. Acta*, **52**, 47 (1961).

107. V. A. Najjar, in *The Enzymes*, Vol. 6 (Eds.: P. D. Boyer, H. Lardy, and K. Myrback), Academic Press, New York, 1962, p. 161.

108. J. B. Alpers and G. K. H. Lam, *J. Biol. Chem.*, **244**, 200 (1969).

109. J. B. Alpers, *J. Biol. Chem.*, **243**, 1698 (1968).

110. O. Walinder, *J. Biol. Chem.*, **243**, 3947 (1968).

111. P. L. Pedersen, *J. Biol. Chem.*, **243**, 4305 (1968).

112. J. Mendicino, C. Beaudreau, and R. N. Bhattacharyya, *Arch. Biochem.*, **116**, 436 (1966).

113. J. Mendicino, H. S. Prihar, and F. M. Salama, *J. Biol. Chem.*, **243**, 2710 (1968).

114. S. Grisolia and B. K. Joyce, *Biochem. Pharmacol.*, **3**, 167 (1960).

115. J. C. Warren, D. O. Carr, and S. Grisolia, *Biochem. J.*, **93**, 409 (1964).

116. F. Marcus and E. Hubert, *J. Biol. Chem.*, **243**, 4923 (1968).

117. A. Pestaña and A. Sols, *FEBS Letters*, **7**, 29 (1970).

118. W. O. McLemore and D. E. Metzler, *J. Biol. Chem.*, **243**, 441 (1968).

119. A. Pestaña and A. Sols, *Biochem. Biophys. Res. Commun.*, **39**, 522 (1970).

120. F. Auricchio and A. Liguori, *FEBS Letters*, **12**, 329 (1971).

121. G. DelaFuente, *European J. Biochem.*, **16**, 240 (1970).

122. A. V. Bertland and H. M. Kalckar, *Proc. Natl. Acad. Sci. U.S.*, **61**, 629 (1968).

123. A. Bonsignore, A. DeFlora, M. A. Mangiarotti, I. Lorenzoni, and S. Alema, *Biochem. J.*, **106**, 147 (1968).

124. B. M. Woodfin, *Biochem. Biophys. Res. Commun.*, **29**, 288 (1967).

125. C. Y. Lai, G. Martinez-deDretz, M. Bacila, E. Marinello, and B. L. Horecker, *Biochem. Biophys. Res. Commun.*, **30**, 665 (1968).

126. G. M. Stancel and W. C. Deal, *Biochem. Biophys. Res. Commun.*, **31**, 398 (1968).

127. T. E. Mansour and C. E. Ahlfors, *J. Biol. Chem.*, **243**, 2523 (1968).

128. M. Rosell-Perez and P. Morey, *FEBS Abstr.*, p. 213 (1969).

129. Y. Nordmann, F. Schapira, and J. C. Dreyfus, *Biochem. Biophys. Res. Commun.*, **31**, 884 (1968).

130. F. C. Hartman, *Biochem. Biophys. Res. Commun.*, **33**, 888 (1968).

131. N. K. Nagradova and A. E. Lisauskaite, *Biokhimiya*, **32**, 624 (1967).

132. H. S. Kingdon, J. Hubbard, and E. R. Stadtman, *Biochemistry*, **7**, 2136 (1968).

133. C. A. Woolfolk and E. R. Stadtman, *Arch. Biochem. Biophys.*, **118**, 736 (1967).

134. B. M. Shapiro and E. R. Stadtman, *J. Biol. Chem.*, **242**, 5069 (1967).

135. R. C. Valentine, B. M. Shapiro, and E. R. Stadtman, *Biochemistry*, **7**, 2143 (1968).

136. D. Mecke, K. Wulff, K. Liess, and H. Holzer, *Biochem. Biophys. Res. Commun.*, **24**, 452 (1966).

137. D. Mecke, K. Wulff, and H. Holzer, *Biochim. Biophys. Acta*, **128**, 559 (1966).

138. K. Wulff, D. Mecke, and H. Holzer, *Biochem. Biophys. Res. Commun.*, **28**, 740 (1967).

139. H. S. Kingdon, B. M. Shapiro, and E. R. Stadtman, *Proc. Natl. Acad. Sci. U.S.*, **58**, 1703 (1967).

140. E. R. Stadtman, B. M. Shapiro, A. Ginsberg, H. S. Kingdon, and M. D. Denton, *Brookhaven Symp. Biol.*, **21**, 328 (1968).

141. H. S. Kingdon and E. R. Stadtman, *J. Bacteriol.*, **94**, 949 (1967).

142. H. Holzer, H. Schutt, Z. Masek, and D. Mecke, *Proc. Natl. Acad. Sci. U.S.*, **60**, 721 (1968).

143. L. Heilmeyer, D. Mecke, and H. Holzer, *European J. Biochem.*, **2**, 399 (1967).

144. B. M. Shapiro and E. R. Stadtman, *Biochem. Biophys. Res. Commun.*, **30**, 32 (1968).

145. L. Heilmeyer, F. Battig, and H. Holzer, *European J. Biochem.*, **9**, 259 (1969).

146. B. M. Shapiro, *Biochemistry*, **8**, 659 (1969).

147. W. B. Anderson, S. B. Hennig, A. Ginsburg, and E. R. Stadtman, *Proc. Natl. Acad. Sci. U.S.*, **67**, 1417 (1970).

148. S. B. Hennig, W. B. Anderson, and A. Ginsberg, *Proc. Natl. Acad. Sci. U.S.*, **67**, 1761 (1970).

149. M. Kapoor and D. Bray, *Biochemistry*, **7**, 3583 (1968).

150. A. Becker, G. Lyn, M. Gefter, and J. Hurwitz, *Proc. Natl. Acad. Sci. U.S.*, **58**, 1996 (1967).

151. B. Weiss and C. C. Richardson, *Proc. Natl. Acad. Sci. U.S.*, **57**, 1021 (1967).

152. J. W. Little, S. B. Zimmerman, C. K. Oshinsky, and M. Gellert, *Proc. Natl. Acad. Sci. U.S.*, **58**, 2004 (1967).

153. Y. Nishizuka, K. Ueda, T. Honjo, and O. Hayaishi, *J. Biol. Chem.*, **245**, 3765 (1968).

154. C. A. Chelala, L. Hirschbein, and H. N. Torres, *Proc. Natl. Acad. Sci. U.S.*, **68,** 152 (1971).

155. M. Colvin and V. H. Bono, *J. Cancer Res.*, **30,** 1516 (1970).

156. C. Voit, *Z. Biol.*, **2,** 307 (1866).

157. O. Folin, *Am. J. Physiol.*, **13,** 117 (1905).

158. R. Schoenheimer, *The Dynamic State of Body Constituents*, Harvard University Press, Cambridge, Mass., 1942.

159. M. V. Simpson, *J. Biol. Chem.*, **202,** 143 (1953).

160. C. O. Brostom and H. Jeffay, *J. Biol. Chem.*, **245,** 4001 (1970).

161. M. Rechcigl, Jr., in *Handbook of Biochemistry* (Ed.: H. A. Sober), Chemical Rubber Co., Cleveland, 1970.

162. J. D. Kemp and D. W. Sutton, *Biochemistry*, **10,** 81 (1971).

163. R. R. Klevecz, *Biochem. Biophys. Res. Commun.*, **43,** 76 (1971).

164. R. D. Feigin, W. R. Beisel, and R. W. Wannemacher, *Am. J. Clin. Nutr.*, **24,** 329 (1971).

165. D. J. Millward, *Clin. Sci.*, **39,** 577 (1970).

166. A. L. Goldberg, *J. Biol. Chem.*, **244,** 3223 (1969).

167. A. L. Goldberg, *Proc. Natl. Acad. Sci. U.S.*, **68,** 362 (1971).

168. M. Rechcigl, Jr., and W. E. Heston, *Biochem. Biophys. Res. Commun.*, **27,** 119 (1967).

169. J. S. Bond, *Biochem. Biophys. Res. Commun.*, **43,** 333 (1971).

170. H. L. Segal, T. Matsuzawa, M. Haider, and G. J. Abraham, *Biochem. Biophys. Res. Commun.*, **36,** 764 (1969).

171. R. W. Gray, C. Arsenis, and H. Jeffay, *Biochim. Biophys. Acta*, **222,** 627 (1970).

172. D. Richter, in *Molecular Basis of Some Aspects of Mental Activity*, Vol. 1 (Ed.: O. Walaas), Academic Press, London–New York, 1966, p. 115.

173. L. Kuehl and E. N. Sumsion, *J. Biol. Chem.*, **245,** 6616 (1970).

174. F. T. Kenney, *Science*, **156,** 525 (1967).

175. R. R. Crichton, *Biochem. J.*, **119,** 40P (1970).

176. D. K. McClintock and G. Markus, *J. Biol. Chem.*, **243,** 2855 (1968).

177. L. Polgar, *Biochim. Biophys. Acta*, **118,** 276 (1966).

178. R. T. Schimke, E. W. Sweeney, and C. M. Berlin, *J. Biol. Chem.*, **240,** 322 (1965).

179. D. G. Smyth, *Biochim. Biophys. Acta*, **200,** 395 (1970).

180. J. J. T. Gerding, A. Koppers, P. Hagel, and H. Bloemendal, *Biochim. Biophys. Acta*, **243,** 374 (1971).

181. J. Kennedy and S. Grisolia, *Biochim. Biophys. Acta*, **96,** 102 (1965).

182. D. H. Russell and S. H. Snyder, *Mol. Pharm.*, **5,** 253 (1969).

183. R. M. Nalbandian, G. Schultz, J. M. Lusher, J. W. Anderson, and R. L. Henry, *Am. J. Med. Sci.* **6,** 309 (1971).

184. P. A. Bromberg and W. N. Jensen, *J. Clin. Invest.*, **44,** 1031 (1965).

185. A. Cerami and J. M. Manning, *Proc. Natl. Acad. Sci., U.S.*, **68,** 1180 (1971).

186. L. M. Kraus and A. P. Kraus, *Biochem. Biophys. Res. Commun.*, **44,** 1381 (1971).

187. C. M. Allen, Jr. and M. E. Jones, *Biochemistry*, **3,** 1238 (1964).

CHAPTER 7

Regulation of "Active Isoprene" Biosynthesis

DONALD J. McNAMARA AND
VICTOR W. RODWELL

Department of Biochemistry, Purdue University, Lafayette, Indiana

1. INTRODUCTION

This chapter will consider the regulation of the initial stages of polyiso-prenoid biosynthesis—specifically the reactions which lead to synthesis of "active isoprene" (Δ^3- and Δ^2-isopentenylpyrophosphate). Control points beyond "active isoprene" exist but are not considered here. Figure 1 outlines the scope of our review. Topics considered here include HMG-CoA* synthesis and catabolism and the conversion of HMG-CoA to mevalonate and thence to "active isoprene."

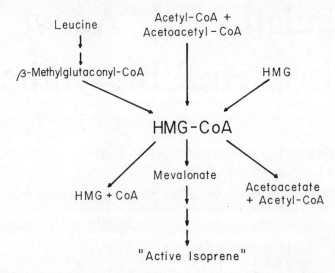

Fig. 1.   HMG-CoA as a branch point compound. Principal routes for synthesis and further catabolism of HMG-CoA are shown.

2. BIOGENESIS AND METABOLISM OF HMG-CoA

We shall consider HMG-CoA the starting point for "active isoprene" bio-synthesis. As shown in Fig. 1, the pathway leading from HMG-CoA to "active isoprene" is unbranched, although HMG-CoA itself is situated at the intersection of multiple metabolic pathways and thus is a "branch-point metabolite." We shall begin, therefore, by discussing the reactions which

* Abbreviations used are HMG (3-hydroxy-3-methylglutaric acid) and ACP (acyl-carrier protein).

determine the intracellular steady-state concentration of HMG-CoA. Two enzymes specifically concerned with "active isoprene" synthesis (HMG-CoA synthase and HMG-CoA reductase) are present primarily in the particulate fraction of cellular homogenates, while all the subsequent reactions leading from mevalonate to "active isoprene" occur in the soluble fraction or cytosol (Fig. 2).

HMG-CoA can be formed in three distinct ways:

1. From two- and four-carbon precursors.
2. By hydration of β-methyglutaconyl-CoA.
3. By conversion of HMG to its thioester.

Although all three are discussed below, only route 1 appears to be general for the synthesis of "active isoprene" in all life forms studied.

A. HMG-CoA Synthase

Acetyl-CoA + acetoacetyl-CoA → HMG-CoA

The reaction catalyzed by HMG-CoA synthase [EC 4.1.3.5: 3-hydroxy-3-methylglutaryl-CoA acetoacetyl-CoA lyase (CoA-acetylating)] is analogous to the condensations of acetyl-CoA with oxaloacetate, pyruvate, glyoxylate, or α-ketobutyrate except that acetoacetyl-CoA lacks the favorable polarizing effect of the carboxylate ion. In common with similar reactions, it may be visualized as proceeding via nucleophilic attack of the enzyme-bound enolate of acetyl-CoA or of acetyl-enzyme on the carbonyl carbon of a ketone, acetoacetyl-CoA.

Conversion of the methyl and methylene carbon atoms of acetoacetate to cholesterol by rat liver slices (1, 2), together with the demonstrated conversion of HMG to cholesterol (3), suggested to Rabinowitz and Gurin (4) and to Rudney (5) that HMG might arise via condensation of an acetyl with an acetoacetyl moiety. Rat liver homogenates fortified with ATP were shown to convert the methyl carbon of acetate to the methyl and methylene carbons of HMG (5). The 4-carbon moiety which condenses with acetyl-CoA was later shown to be acetoacetyl-CoA in extracts of mammalian liver (6) or of baker's yeast (6–9). In yeast, the thioester bond of HMG-CoA is that of acetoacetyl-CoA, as shown by the incorporation of isotopically labeled CoA from acetoacetyl-CoA, but not from acetyl-CoA, into HMG-CoA (10).

Sources

YEAST.   Baker's yeast contains 200 iu* per gram wet weight of HMG-CoA synthase. The synthase, purified 400-fold to a specific activity of 5780

---

* Enzyme activities are expressed throughout as international units (iu) or as micromoles of substrate turned over per minute, generally at 30°.

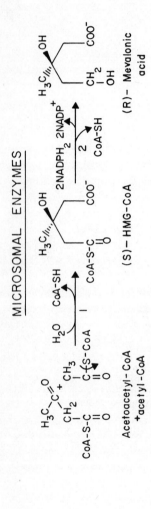

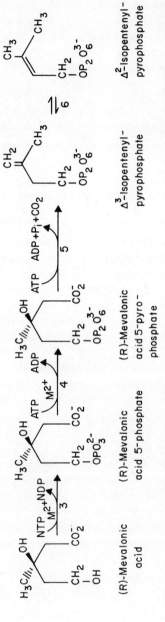

Fig. 2. Reactions in the synthesis of "active isoprene." Numbered reactions are catalyzed by (1) HMG-CoA synthase, (2) HMG-CoA reductase, (3) mevalonate kinase, (4) phosphomevalonate kinase, (5) pyrophosphomevalonate decarboxylase, (6) isopentenylpyrophosphate isomerase.

iu/mg, catalyzes the above reaction essentially irreversibly. There are no known cofactors, and activity is optimal at pH 8–9 (7–9). The synthase is accompanied by acetoacetyl-CoA thiolase activity throughout purification, and these activities are not readily resolved. Since the ratios of the activities vary during purification, it appears unlikely that both are activities of a single protein, although their close association, possibly as a complex, has been documented. The thiolase activity may be destroyed by treatment with iodoacetamide or by papain digestion in the presence of acetoacetyl-CoA (9).

Acetoacetyl-CoA appears to act as an inhibitor competitive with acetyl-CoA (9). Acetoacetyl-CoA stabilizes the enzyme toward digestion by trypsin or chymotrypsin, and acetyl-CoA protects against heat denaturation. These observations have been interpreted by Rudney and his coworkers to indicate a substrate-induced conformation change. These workers have reported that HMG-CoA synthase is subject both to substrate inhibition by acetoacetyl-CoA and to product inhibition by CoA (9) and also by HMG-CoA (8, 11).

Acetyl glutathione or acetyl pantetheine substitute for acetyl-CoA with 30 and 20% the activity of acetyl-CoA, respectively. Although the corresponding thioesters of acetoacetate are not substrates (9), acetoacetyl-ACP prepared from *Escherichia coli* ACP is about one-sixth as active as acetoacetyl-CoA for synthesis of a protein-bound form of HMG thought to be HMG-ACP. Protein-bound HMG is not formed when acetoacetyl-CoA is the substrate (12). The ability of acetoacetyl-ACP to substitute for acetoacetyl-CoA cannot be attributed either to transacylation of acetoacetyl-ACP to acetoacetyl-CoA or of HMG-CoA to HMG-ACP (12). These observations led Rudney et al. (12) to propose a pathway for HMG-CoA synthesis which incorporates features of the malonyl-CoA pathway. This is discussed in Section 3.C.

LIVER. The distribution of HMG-CoA synthase activity in rat tissues is shown in Table 1. Liver tissue is by far the richest source of the enzyme. Study of mammalian liver HMG-CoA synthase has long been hampered by difficulties inherent in its assay. The enzyme appears to be present in mitochondria, microsomes, and the soluble fraction of liver homogenates (6, 14–16), each of which presents problems for assay. Mitochondrial assays, performed by addition of excess HMG-CoA lyase with subsequent determination of acetoacetate (14), may give high results because of the presence of acetoacetyl-CoA deacylase or low results because of the presence of HMG-CoA deacylase. Microsomes, which have low levels of deacylases, contain levels of HMG-CoA synthase too low to detect by this procedure, and the soluble fraction converts the bulk of the added substrate to products other than HMG-CoA (6).

Using different methods of assay, Rudney (6) and Bucher et al. (14) reach quite different conclusions regarding both the total activity present and its distribution within the cell. Thus conclusions as to the site and activity of HMG-CoA synthase cannot be stated with confidence at this time. It seems likely, however, that although the major site of HMG-CoA synthesis is in the mitochondria (14) it may be the microsomal synthase that is concerned with "active isoprene" synthesis. Hepatic microsomal HMG-CoA synthase, like that of yeast, is inhibited by acetoacetyl-CoA (D. J. McNamara, unpublished observations).

Table 1.   Distribution of HMG-CoA Synthase and HMG-CoA Lyase in Various Tissues of the Rat[a]

From McGarry and Foster (13).

Tissue	Activity, iu/g	
	HMG-CoA Synthase	HMG-CoA Lyase
Liver	2.4	11.0
Kidney cortex	0.26	6.5
Heart	0.11	1.6
Intestine	0.097	0.40
Brain	0.073	0.00
Muscle	0.000	0.00

[a] The data are for crude homogenates assayed by measuring acetoacetate production from acetyl-CoA. Activities are expressed per gram of liver.

MYCOPLASMA.   On the basis of nutritional requirements, Smith (17) distinguished between the strains of *Mycoplasma* which require a hydroxylated steroid such as cholesterol (e.g., *M. hominis* and *M. gallisepticum*) and those which do not (e.g., *M. laidlawii*). The latter group can synthesize nonsaponifiable sterols from acetate. Although HMG-CoA synthase activity has been detected directly in *M. laidlawii* and indirectly in *M. gallisepticum*, it is absent from *M. hominis*. The metabolic block in *M. gallisepticum* thus lies beyond HMG-CoA (18).

PLANTS.   Plant tissues have long been known to synthesize HMG derivatives (19), and HMG-CoA synthase activity has been detected in orange peel (20) and sweet potato root tissue (21, 22).

Regulation

The position of HMG-CoA at the intersection of several metabolic pathways (Fig. 1) suggests that HMG-CoA synthase may play a role in the regulation both of "active isoprene" and of ketone body production. The available evidence, although not conclusive, is consistent with this view. Thus HMG-CoA synthase appears to catalyze the rate-limiting reaction in hepatic ketone body synthesis (13, 23), and hepatic HMG-CoA synthase activity is increased by fasting, high-fat diets, and alloxan diabetes—all of which promote ketone body production. Possible modes of regulation suggested by Rudney include

Table 2. Physiological Manipulations Reported to Affect Rat Liver HMG-CoA Synthase Activity

Manipulation	Fold Increase (+) or Decrease (−) in HMG-CoA Synthase Activity		
	Mitochondria	Microsomes	Soluble Fraction
Fasting (14)	+2.1	+1.7	+8.4
Fasting (23)	+1.6		+2.9
Fasting (16)		+1.8	+1.0
Triton injection (14)	+1.8	+2.0	+11.5
Triton injection (16)		+1.2	+1.1
High-fat diet (23)	+2.6		+1.7
High-cholesterol diet (16)		−0.4	−0.5
Alloxan diabetes (23)	+1.9		+2.3
Injection of 17-β-estradiol (24)		−2.5	

substrate and product inhibition and repression or inhibition by end products such as cholesterol.

In tissues or organisms which do not produce ketone bodies, HMG-CoA synthase catalyzes the first reaction unique to polyisoprenoid synthesis. Thus it might be expected to be subject to regulation by factors known to affect polyisoprenoid biosynthesis. Little or no information is available, however, other than for mammalian liver (Table 2). Fasting or Triton injection, which stimulates overall cholesterol synthesis, appears to increase hepatic HMG-CoA synthase activity. Although the effect on HMG-CoA reductase is more profound (see Section 3.A), rat liver HMG-CoA synthase activity is significantly decreased by dietary cholesterol (25). The decreased levels of hepatic microsomal HMG-CoA synthase reported after administration of estrogen are interesting, although of uncertain significance. At the dose levels employed, male rats are known to exhibit profound physiological changes that include cessation of growth and pituitary and adrenal atrophy (26).

The highly malignant, undifferentiated Morris rat hepatoma 3924A, in contrast to the well-differentiated minimal-deviation hepatoma 7787, cannot synthesize acetoacetate and synthesizes cholesterol only at very low rates (13). This biosynthetic lesion appears to be due to the absence from 3924A of detectable HMG-CoA synthase activity. Two other tumors, 9921 and 7787, synthesize cholesterol at rates considerably greater than liver tissue. This is not attributable, however, to increased levels of synthase (13) but suggests, rather, that HMG-CoA synthase may not perform a regulatory function for cholesterol synthesis by these tumors.

Although HMG-CoA synthase may play a minor role in the regulation of "active isoprene" synthesis, the regulation of this enzyme may prove significant under conditions where intracellular levels of acetoacetyl-CoA are elevated. Since both the yeast and the mammalian microsomal synthases are exquisitely sensitive to inhibition by acetoacetyl-CoA, cells would appear to have the capacity to regulate "active isoprene" synthesis at this locus.

Sweet potato root tissue infected with *Ceratocystis fimbriata*, the fungus causing black rot, forms and accumulates large quantities of terpenoids such as ipomearone in the infected region (21). The ability of normal and infected tissue to catalyze the HMG-CoA synthase reaction has been studied in cell-free extracts. Enzyme levels rise about six-fold 2 days after infection. Although the synthase of sweet potato is inhibited by sulfhydryl alkylating reagents, ipomearone has no inhibitory effect (22).

The activity of HMG-CoA synthase has been studied in postnatal rats in conjunction with an investigation of ketone body synthesis during development (26a). Mitochondrial and cytosol, but not microsomal synthase levels, were studied. At all stages of development, synthase levels were higher in mitochondria. Levels of the mitochondrial enzyme rise immediately after birth, peak at 15 days, then decline to adult levels. These changes appear to correlate well with ketone body synthesis during development, and suggest that mitochondrial HMG-CoA synthase may be the rate-limiting enzyme in rat postnatal ketogenesis. The synthase activity of the cytosol increased steadily after birth, reaching a maximum at 40 days. Since the HMG-CoA synthase activity of microsomes was not studied, these observations appear, however, to have relevance solely to ketone body synthesis rather than to the synthesis of "active isoprene."

B. β-Methylglutaconyl-CoA Hydratase

$$
\underset{\substack{|\\ \mathrm{CH_3}}}{\mathrm{HOOC-CH_2-C=CH-\overset{\overset{\displaystyle O}{\|}}{C}-S-CoA}} + H_2O \longrightarrow
$$

β-Methylglutaconyl-CoA

$$
\underset{\substack{|\\ \mathrm{CH_3}}}{\mathrm{HOOC-CH_2-\overset{\overset{\displaystyle OH}{|}}{C}-CH_2-\overset{\overset{\displaystyle O}{\|}}{C}-S-CoA}}
$$

HMG-CoA

β-Methylglutaconyl-CoA, an intermediate in leucine catabolism by mammalian liver tissue (27), is reversibly hydrated to HMG-CoA by β-methylglutaconyl-CoA hydratase (EC 4.2.1.18: 3-hydroxy-3-methylglutaryl-CoA hydro-lyase). The hydratase, present in sheep (28) and chicken liver (29) and in a species of *Mycobacterium* (30), has been purified 50- (28) and 100-fold (29) and is distinct from crotonase (EC 4.2.1.17: L-3-hydroxyacyl-CoA hydro-lyase).

C. Acylation of HMG

Although free HMG is not thought to lie on the direct pathway from acetyl-CoA to "active isoprene," HMG present in the diet or arising via the HMG-CoA deacylase reaction (see Section 2.D) can be converted to HMG-CoA and thence to "active isoprene." It has been shown that HMG is metabolized by mammals, for intact rats form $^{14}CO_2$ from [3-^{14}C]HMG (31). This activity resides primarily in kidney mitochondria. Activation of HMG appears to involve transesterification from succinyl-CoA and to be catalyzed by a β-ketoacid CoA transferase (EC 2.8.3.5: succinyl-CoA:3-oxoacid CoA-transferase) (31). Conversion of HMG to HMG-CoA by an HMG-specific, ATP-dependent CoA ligase reaction has been reported in extracts of a *Pseudomonas* species isolated from soil by elective culture on HMG as the sole source of carbon (M. A. Siddiqi and N. Ahmad, unpublished observations). In both of these instances the reactions forming HMG-CoA from HMG appear to serve catabolic rather than anabolic functions, and hence would not be expected to be subject to regulation by intermediates or end products of polyisoprenoid biosynthesis.

D. HMG-CoA Deacylase

$$HMG\text{-}CoA + H_2O \rightarrow HMG + CoA$$

This reaction is catalyzed by HMG-CoA deacylase (EC 3.1.2.5: 3-hydroxy-3-methylglutaryl-CoA hydrolase). Although HMG-CoA deacylase and HMG-CoA lyase do not catalyze reactions on the direct pathway leading to "active isoprene," some consideration of their properties is indicated by their potential for determining the availability of substrate for the synthesis of mevalonate and hence of "active isoprene."

Sources

LIVER. The HMG-CoA deacylase of chicken liver has been studied but not extensively purified because of its instability. Activity is optimal at pH 7.8, and only one stereoisomer of HMG-CoA is hydrolyzed. Although no data are available concerning the distribution of HMG-CoA deacylase within the cell, it would not appear to be a microsomal enzyme, judging from the methods employed (32). Thus it may play no role in the regulation of "active isoprene" synthesis by the microsomal enzymes forming and utilizing HMG-CoA.

YEAST. Crude yeast extracts catalyze the deacylation of both isomers of DL-HMG-CoA, apparently because of the presence of deacylases specific for one or the other diastereoisomer (33). These activities have been partially resolved, and the one acting on (S)-HMG-CoA is specifically inactivated by p-hydroxymercuribenzoate. Whether this indicates a mechanistic difference for the activity hydrolyzing (R)-HMG-CoA is not known. It is possible that yeast HMG-CoA deacylase is a particulate enzyme, as it copurifies with HMG-CoA reductase (33). The (S)-HMG-CoA deacylase thus could affect the overall rate of "active isoprene" synthesis in yeast.

OTHER. Activity of HMG-CoA deacylase has been detected in crude extracts of pig liver, heart, and brain, *Neurospora crassa*, *Escherichia coli*, and *Tetrahymena pyriformis* and hence appears to be widely distributed in nature (32).

Regulation

Although HMG-CoA deacylase may perhaps regulate "active isoprene" synthesis by lowering the intracellular levels of HMG-CoA, no evidence implicating this process in physiological regulation is presently available.

E. HMG-CoA Lyase

$$HMG\text{-}CoA \rightarrow acetyl\text{-}CoA + acetoacetate$$

The (S) isomer of HMG-CoA formed by HMG-CoA synthase is cleaved to acetyl-CoA plus acetoacetate by a specific HMG-CoA lyase (EC 4.1.3.4: 3-hydroxy-3-methylglutaryl-CoA acetoacetate-lyase).

Sources

MAMMALIAN TISSUES.   As first suggested by Lynen et al. (34), HMG-CoA lyase of mammalian liver appears to function in conjunction with HMG-CoA synthase for the synthesis of ketone bodies from acetoacetyl-CoA (13, 34–37). Unlike HMG-CoA synthase, which performs a dual role in "active isoprene" and ketone body synthesis, HMG-CoA lyase functions solely in the latter process.

It is possible to assay HMG-CoA lyase activity by observing the formation of acetoacetate (14, 38) or by coupling the release of acetyl-CoA to condensation with oxaloacetate formed from malate and observing the rate of NADH formation in the presence of malate dehydrogenase and citrate-condensing enzyme (39). As shown in Table 1, HMG-CoA lyase is found in greatest abundance in the liver, an observation consistent with its function in ketogenesis in the primary ketogenic organ. Bucher et al. (14) found that HMG-CoA lyase of rat liver is associated primarily with the mitochondrial fraction and is present in considerable excess over HMG-CoA synthase. When purified over 2000-fold to an apparently homogeneous state, as judged by cellulose acetate electrophoresis (39), HGM-CoA lyase from beef liver is unstable, has a molecular weight of 48,000 ± 3000 (gel filtration), and has a turnover number at the optimum pH of 9.5 of 3600 moles/mole/min. A divalent metal and a thiol are required for activity, and preincubation with thiol is needed for the reaction to proceed without a lag. Thiol-binding agents and metal-chelating compounds are inhibitory. The K_m values are as follows: $Mg^{2+} = 1 \times 10^{-3}\ M$, $Co^{2+} = 4 \times 10^{-4}\ M$, and (S)-HMG-CoA = $8 \times 10^{-6}\ M$ (with Mg^{2+}) or $1.2 \times 10^{-5}\ M$ (with Co^{2+}) (39).

BACTERIA.   Activity of HMG-CoA lyase has been detected in *Mycobacteric* (40) and *Pseudomonas* (41). In these organisms the enzyme serves a catabolia function apparently unrelated to "active isoprene" biosynthesis.

Regulation

Despite the role of hepatic HMG-CoA lyase in ketone body synthesis, the levels of this enzyme appear to respond little (14) or not at all (37) to fasting or to experimental diabetes. This observation is relevant also to the regulation of "active isoprene" biosynthesis, for fasting decreases both cholesterol synthesis and HMG-CoA reductase activity (see Section 3.A). Presumably it is the microsomal lyase activity rather than that associated with mitochondria that can affect mevalonate synthesis, but this activity is similarly insensitive to change under conditions known to influence sterol synthesis and HMG-CoA reductase activity (14). Indeed, compartmentation appears to play a key role in the regulation of HMG-CoA lyase activity. In this regard Bucher et al. (14) have stated:

The pattern of distribution of HMG-CoA lyase in the cellular fractions was closely similar to that of HMG-CoA synthase, indicating that it too functions mainly in the mitochondria. Possibly acetoacetate production is primarily a mitochondrial function, carried out by HMG-CoA synthase and HMG-CoA lyase that are abundant in these particles, whereas the small fraction of HMG-CoA synthase that is in the microsomes produces the HMG-CoA that is converted to mevalonic acid by microsomal HMG-CoA reductase. This latter HMG-CoA would be more readily available to the reductase because of the low level of HMG-CoA lyase in the microsomes. Such compartmentation could explain how cholesterol synthesis from acetate can occur in a tissue with a many times greater potential capacity to split HMG-CoA than to synthesize it or to convert it to mevalonic acid.

3. CONVERSION OF HMG-CoA TO "ACTIVE ISOPRENE"

Although HMG-CoA may have multiple fates, the only known fate of mevalonate is conversion to "active isoprene." The reactions discussed below are therefore concerned solely with "active isoprene" synthesis.

A. HMG-CoA Reductase

$$(S)\text{-HMG-CoA} + 2\text{NADPH} + 2\text{H}^+ \rightarrow 2\text{NADP}^+ +$$

$$\text{CoA} + (R)\text{-mevalonate}$$

The reduction of (S)-HMG-CoA to (R)-mevalonic acid catalyzed by HMG-CoA reductase [EC 1.1.1.34: mevalonate:NADP oxidoreductase (acylating CoA)] of liver tissue represents the first biosynthetic reaction unique to polyisoprenoid biosynthesis and is the rate-limiting and major regulated enzyme of hepatic "active isoprene" biosynthesis. Whether HMG-CoA reductase rather than HMG-CoA synthase of yeast and of other tissues can properly be considered to catalyze the first reaction leading to "active isoprene" and whether the reductase fulfills a regulatory role in these systems are less certain.

Sources

YEAST. The initial impetus for the study of nonmammalian HMG-CoA reductases was derived in no small measure from the difficulties inherent in studying the relatively inactive, particle-bound reductase of hepatic microsomes (Table 3). HMG-CoA reductase is localized in the mitochondria of

Table 3. Comparison of HMG-CoA Reductases from Various Sources

Source of Enzyme	Reductive Deacylation of HMG-CoA	
	Specific Activity, iu/mg	K_m Calculated for (S)-HMG-CoA, $M \times 10^5$
Pseudomonas[a]	28.5	2.5
Baker's yeast[b]	0.70	1.2
Neurospora crassa[c]	0.01	2.5
Rat liver microsomes[d]	8.0×10^{-5}	3.3
Rat liver microsomes[e]	4.0×10^{-4}	7.5
Purified rat liver[f]	3.0×10^{-3}	0.60
Mouse liver microsomes[g]	1.3×10^{-5}	0.42

[a] 30°, pH 7.1, mol. wt. = $2.6–2.8 \times 10^5$ (41).
[b] 25°, pH 6.5, enzyme purified 117-fold, mol. wt. = $1.5–2.0 \times 10^5$ (44).
[c] 30°, pH 6.8 (R. L. Imblum, unpublished data).
[d] 37°, pH 7.2 (81).
[e] 37°, pH 6.5 (48, 49).
[f] 37°, pH 7.4, mol. wt. = 2.2×10^5 (50).
[g] 37°, pH 6.0 (51).

yeast and is adaptively induced when cells are transferred from an anaerobic to an aerobic environment (41a). The HMG-CoA reductase of baker's yeast, which is sufficiently active to be assayed by spectral means (42–45), has been purified 200–350-fold (42–44) to a specific activity of 0.7 iu/mg (44). NADPH is specifically used as a reductant, and the reaction proceeds essentially irreversibly in the direction of mevalonate synthesis (43). The reverse reaction is, however, demonstrable (42). The purified reductase has a molecular weight of $1.5–2.0 \times 10^5$ and is inactivated by preincubation with free coenzyme A or with CoA thioesters of acetate, acetoacetate, or HMG. Thiol-binding reagents also are inhibitory, suggesting involvement of an essential thiol group (44). Although free mevaldate is not formed during reduction of HMG-CoA (42, 43), the CoA hemimercaptal of mevaldate may be an intermediate (45). Mevaldate-coenzyme A hemimercaptal is reduced to mevalonate, but its formation from HMG-CoA remains to be demonstrated.

Preliminary mechanistic studies (44) suggest a ping-pong mechanism with no ternary complexes. NADP+ inhibits competitively with NADPH and uncompetitively with HMG-CoA. Mevalonate inhibits uncompetitively with NADPH. Although an acyl-enzyme intermediate has been proposed (46, 47), the available evidence (44, 45) does not support the view that HMG-CoA exchanges with enzyme-SH, forming HMG-S-enzyme.

PSEUDOMONAS. An inducible HMG-CoA reductase has been purified 21 fold from soluble extracts of *Pseudomonas* grown on mevalonate as the sole source of carbon to a state judged homogeneous by disc electrophoresis. The estimated molecular weight is 2.8×10^5. In common with many bacterial nicotinamide-nucleotide oxidoreductases, but unlike other HMG-CoA reductases, the *Pseudomonas* enzyme specifically uses NADH as reductant. The purified enzyme also catalyzes the mevaldate reductase [EC 1.1.1.X: mevalonate:NAD oxidoreductase] and mevaldate:NAD oxidoreductase (acylating CoA) [EC 1.1.1.X] reactions. Parallel enrichment of the three activities suggests that all are activities of a single protein.

Although the HMG-CoA reductase reaction is catalyzed reversibly, mevalonate synthesis is strongly favored (equilibrium constant $= 10^{13}$–10^{15} M^{-1}). Mevaldate appears not to be a free intermediate either in HMG-CoA reduction or in mevalonate oxidation. Optimum activity for mevalonate oxidation occurs at pH 9.2–9.6. The K_m values are 0.025 mM [calculated for (S)-HMG-CoA], 0.032 mM (NADH), 0.30 mM (mevalonate), 0.27 mM (NAD$^+$), and 0.039 mM (CoA). The HMG is a competitive inhibitor both of mevalonate oxidation ($K_i = 0.22$ mM) and of HMG-CoA reduction ($K_i = 3.9$ mM), although the former reaction is more effectively inhibited (K_i: $K_m = 0.73$) than is the latter (K_i:$K_m = 79$). It appears that HMG-CoA reduction proceeds by a mechanism other than that proposed for yeast HMG-CoA reductase (41).

NEUROSPORA CRASSA. Activity of HMG-CoA reductase of *Neurospora crassa* is associated with the cellular fraction sedimenting at $10^5 \times g$ and specifically uses NADPH as reductant (R. L. Imblum, unpublished observations).

RAT LIVER. Until comparatively recently, substantial difficulties were inherent in the assay of hepatic HMG-CoA reductase. Although the reductases of yeast or *Pseudomonas* may be assayed by spectral methods, this is not possible with crude hepatic microsomal HMG-CoA reductase, on account both of its low catalytic efficiency (Table 3) and of the presence of interfering reactions utilizing NADPH. Assay requires measuring the incorporation of isotope from ^{14}C-HMG-CoA into mevalonate, and this necessitates methods for separating mevalonate from HMG-CoA and/or HMG. Techniques employed include column chromatography (15, 16, 43), gas-liquid chromatography (51–54), and thin-layer chromatography (55, 56).

It has been found that HMG-CoA reductase is associated with the microsomal fraction of rat liver tissue (14, 16, 46, 49, 50, 55, 57–59). Although Linn (55) reported solubilization of HMG-CoA reductase by preparation of an acetone powder, a partially purified preparation has only recently been obtained using solubilization with bile salts (50). The purified enzyme is optimally active at pH 7.0, is specific for NADPH as reductant (K_m = 8.7 × 10^{-5} M), and has a K_m for HMG–CoA [calculated for (S)-HMG-CoA] of 6 × 10^{-6} M. Inhibitory are Fe^{2+}, Fe^{3+}, p-mercuribenzoate, and deoxy-cholate. Acetyl-CoA, acetoacetyl-CoA, and CoA are inhibitory if pre-incubated with the enzyme but are otherwise without effect (50).

Sterol synthesis from acetate is known to be low in the early morning and to rise significantly by late afternoon (51, 58). Hamprecht et al. (59) and Shapiro and Rodwell (49, 49a) reported that HMG-CoA reductase activity undergoes a circadian rhythm with minimal activity at noon and maximal activity at around midnight.

RAT INTESTINE. Louw and Mosbach (59a) have demonstrated the presence of HMG-CoA reductase in rat intestines and report that about one-third of the total activity resides in crypt cells and two-thirds in villi. In both cell types, reductase activity appears to be about equally distributed between the microsomal and the mitochondrial fractions. Dietary cholestyramine enhances both microsomal and mitochondrial reductase activity of both crypts and villi.

MOUSE LIVER. Mouse liver microsomal HMG-CoA reductase activity is optimal at pH 6.5–7.5. The K_m for NADPH, which is specifically required, is 1.9 mM and that for HMG-CoA [calculated for the (S)-isomer] is 4.2 × 10^{-6} M. Reductase levels undergo a diurnal rhythm and are greatly stimulated by injection of Triton WR 1339 (51).

MOUSE BRAIN. Within the first 40 days of postnatal life, sterol synthesis in the brains of rats and mice increases during the period of myelination and then decreases to low, adult levels (60–62). Mice carrying the sex-linked recessive gene "jimpy" suffer from a defect in myelination that is fatal early in life (63). In brain tissue from normal mice HMG-CoA reductase is highest shortly after birth (0.01 iu per brain), declines to about 0.001 iu per brain at 40 days, and appears to limit the overall rate of sterol synthesis from acetate after about 11 days of age. Mice bearing the "jimpy" gene exhibit decreased HMG-CoA reductase levels at all times (about 50% that of normal mice at 11 days) and a corresponding decrease in sterol synthesis from acetate, but not from mevalonate.

MYCOPLASMA. Smith and Henrikson (18) have demonstrated the presence of reductase activity in *Mycoplasma laidlawii* and *M. gallisepticum*. *Mycoplasma hominis*, which lacks HMG-CoA synthase, also lacks HMG-CoA reductase.

Regulation

The reduction of HMG-CoA to mevalonic acid is considered to be the rate-limiting step in hepatic polyisoprenoid biosynthesis. Regulation at the level of HMG-CoA reductase, implicated by several investigators (64–66) using indirect techniques, has been substantiated by direct assay of HMG-CoA reductase activity. Although the majority of these studies have utilized mammalian hepatic tissue, generally that of the rat, end-product regulation of steroid synthesis appears to occur in many other species (Table 4).

DIURNAL RHYTHM. Kandutsch and Saucier (51) observed a rhythmic change both in mouse liver sterol production from acetate and in HMG-CoA reductase activity. Activity was significantly higher at 8:30 P.M. than at 8:30 A.M., and this rise in activity was blocked by injection of puromycin. Similar observations concerning rat liver sterol synthesis (58) were followed by the report of Hamprecht et al. (59) that HMG-CoA reductase activity was minimal at noon and rose steadily to a maximum at 2 A.M. (8 P.M.– 8 A.M. dark cycle). Activity then declined steadily to minimum values at noon–2 P.M. Similar diurnal variations were observed in fasted rats, although absolute activities were greatly reduced. Shapiro and Rodwell (49, 49a) observed a diurnal variation in HMG-CoA reductase activity with a peak at

Table 4. Sensitivity of Cholesterol Synthesis to Dietary Cholesterol[a]

From Siperstein (67), except where otherwise noted.

Species	Catalysis of Cholesterol Synthesis	Cholesterol Synthesis Depressed by Dietary Cholesterol
Yeast	+	−
Yeast (68–70)[b]	+	+
Planaria, land snail, earthworm, lobster, starfish	−	
Trout (71)	+	+
Frog	+	+
Iguana	+	+
Chicken	+	+
Rat	+	+
Man	+	+

[a] Based on measurement of acetate incorporation into 3β-hydroxy steroids by slices of liver, hepatopancreas, or intact organism.

[b] Depressed by ergosterol or an ergosterol metabolite.

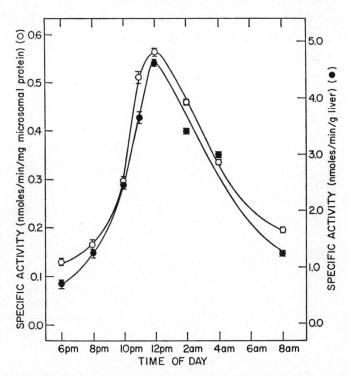

Fig. 3. Cyclic rhythm of rat liver microsomal HMG-CoA reductase activity. Rats were kept in a windowless room with the lights off from 6 P.M. to 6 A.M. (EST). Microsomes isolated from rats killed at the indicated times were frozen overnight and then assayed for HMG-CoA reductase activity (56). Activities are expressed both as activity per milligram of microsomal protein (○) and as activity per gram of liver (●) (49a).

midnight (6 P.M.–6 A.M. dark cycle) (Fig. 3). Cycloheximide prevented the rise in activity and also blocked the fall from the midnight maximum. These data suggest that both the increase and the decrease in enzymatic activity are dependent on protein synthesis. Direct measurements of the incorporation of ³H-leucine into highly purified HMG-CoA reductase have shown that the diurnal variation in activity observed in intact microsomes is due to synthesis of new enzyme protein for approximately 6 hr, followed by complete cessation of synthesis for approximately 15 hr. The rate of HMG-CoA reductase degradation does not appear to be significantly altered (71b).

The diurnal rhythm of HMG-CoA reductase has recently been shown to consist of two distinct peaks in activity (Fig. 4), which occur at about 12 P.M. and 1:45 A.M. (71a). Both the first and the second rise are blocked by injection of cycloheximide and hence appear to require protein synthesis.

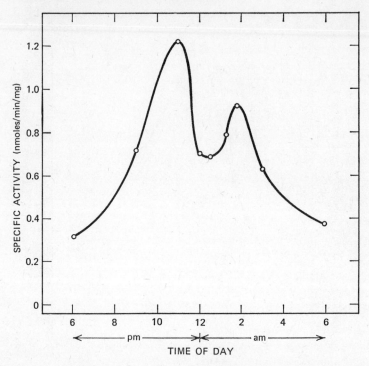

Fig. 4. Double peak in the cyclic rhythm of HMG-CoA reductase. The data are for the reductase activity of microsomes from livers of rats killed at the indicated times (71a).

Rhythmic changes have not been observed for other enzymes of "active isoprene" biosynthesis. Dugan et al. (71c) studied the activity of rat liver for each enzyme between HMG-CoA and "active isoprene" throughout a 24-hr day. Only HMG-CoA reductase exhibited a significant diurnal rhythm.

DIETARY CHOLESTEROL. Although dietary cholesterol has been implicated as a regulator of hepatic sterol synthesis and subsequently of HMG-CoA reductase (64–66, 72–76), the first direct evidence for reduced reductase activity in microsomes from livers of cholesterol-fed rats was that of Linn (77). Since reduced activity persisted after acetone and ether treatment, Linn inferred that the reduced activity was not due to the presence of an acetone- or ether-soluble inhibitor. Additions of various solubilized preparations of cholesterol to the reductase were without inhibitory effect. Depressed reductase activity in the livers of cholesterol-fed rats was also observed by others (16, 49). Shapiro and Rodwell (49, 49a) reported that the K_m's for HMG-CoA reductase from livers of cholesterol-fed and normal-fed rats did

Table 5. Failure of Cholesterol-Rich Liver Lipoproteins to Inhibit Hepatic Microsomal HMG-CoA Reductase Activity[a]

From Shapiro and Rodwell (49a).

Cholesterol Added as Lipoprotein, mM	Specific Activity of HMG-CoA Reductase, nmoles/min/mg	Fraction of Control, %
0	1.16	(100)
0.01	1.11	96
0.1	1.11	96
1.0	1.25	108
10	1.30	112

[a] Cholesterol-rich lipoproteins isolated from livers of rats fed a 5% cholesterol diet were added to HMG-CoA reductase incubations at the indicated final concentrations of cholesterol.

not differ significantly, while V_{max} decreased from 0.33 (normal-fed) to 0.04 nmoles/min/mg microsomal protein (cholesterol-fed). Mixing experiments failed to suggest the presence of an inhibitor in the livers of cholesterol-fed rats, and cholesterol-rich lipoproteins do not inhibit HMG-CoA reductase activity (49a) (Table 5). A variety of other steroids depress HMG-CoA reductase activity when present in the diet or when injected intraperitoneally (Table 6).

In part because all attempts to demonstrate inhibition of HMG-CoA reductase activity by cholesterol have failed, the view that cholesterol acts as a repressor of HMG-CoA reductase synthesis has attracted attention. The approximate half-life of HMG-CoA reductase in normal rats appears to be sufficiently short to permit the rapid effects of dietary cholesterol to be mediated via processes involving protein synthesis. The K_m values for both the midnight and the noon reductase are similar, and cholesterol-fed rats exhibit a greatly reduced diurnal rise in reductase activity. Thus cholesterol may act either by repressing synthesis of new reductase and/or by degrading existing reductase.

More recently Shapiro and Rodwell (49a) have shown that the fall in HMG-CoA reductase activity after cholesterol feeding is accompanied by a concomitant drop in the overall synthesis of cholesterol from acetate (Fig. 5). During the same time interval, cholesterol synthesis from mevalonate and HMG-CoA synthesis from acetyl-CoA are unaffected. These authors therefore propose that the short-term effects of dietary cholesterol can be

Table 6. Effect of Dietary and Injected Steroids on Mouse Liver HMG-CoA Reductase

Data of Kandutsch and Packie (78).

Steroid	HMG-CoA Reductase Activity as Fraction of Control, %	
	In Diet[a]	Injected[b]
Cholestanol	18	
Cholestanone	39	
Cholesterol	10	
Cholest-4-en-3-one	6	42
Testosterone	86	29
Androsta-4,6-diene-17β-ol-3-one	39	2
Cholesta-4,6-diene-3-one	9	
Progesterone		49

[a] Steroids were present at a 1.0% level in the diet for 20 hr.
[b] Three milligrams of steroids was injected intraperitoneally 8.5 hr before sacrifice.

accounted for solely as a result of effects on HMG-CoA reductase, and that it is not necessary to postulate other points of control in order to explain the effects of cholesterol feeding on hepatic cholesterol synthesis.

BILE ACIDS.   Despite an early report (78a) that bile acids inhibit conversion of acetate to mevalonate in crude rat liver homogenates, at present bile acids do not appear to be specific inhibitors of HMG-CoA reductase. Hamprecht et al. (78b, 78c) have shown that the concentrations of bile acids required to significantly inhibit HMG-CoA reductase greatly exceed known physiological levels. These investigators conclude that the *in vitro* effects of bile acids reflect a nonspecific detergent effect on the membrane-bound enzyme. The same claim cannot be made, however, for *in vivo* effects of bile acids. Feeding a 1% cholic acid diet for 20 hr raised the liver bile acid concentration to 0.46 mM and significantly decreased reductase activity. The effect on the diurnal rhythm was even more pronounced. Noon activities were 40% of control values, while midnight reductase activity was only 13% that of controls. Since similar effects were noted in animals with diversion of the lymphatic flow, it appears that bile acids do not exert their effect solely by increasing the intestinal absorption of cholesterol. The exact mechanism by which bile acids act in this case has not, however, been determined.

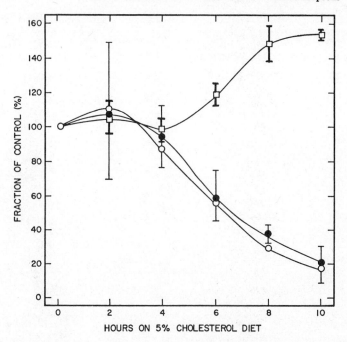

Fig. 5. Effect of dietary cholesterol on HMG-CoA reductase activity and on cholesterol synthesis from acetate. Paired female Sprague-Dawley rats (90–115 g) were fed a 5% cholesterol diet from zero hour (9 A.M.) and sacrificed every 2 hr. Livers were removed and passed through a Harvard tissue press; the mince was used to evaluate cholesterol synthesis from acetate-2-^{14}C (●). Washed microsomes were isolated from a second portion of minced tissue and assayed for HMG-CoA reductase activity (○). Cholesterol content of the liver (□) extracts was determined on a third portion of minced tissue (49a).

CHOLESTEROL REGULATION IN TUMOR TISSUE. Rat and mouse liver hepatic tumor tissues readily catalyze cholesterol synthesis from acetate (71, 79, 79a). Although 5% dietary cholesterol decreases cholesterol synthesis from acetate in normal and regenerating liver tissue from normal or tumor-bearing rats, the tumor tissue itself appears to be insensitive to dietary cholesterol (Table 7). Since dietary cholesterol exerts its principal effect on HMG-CoA reductase, these data suggest that tumor tissue reductase lacks this regulation (79, 80). The absence of regulation by cholesterol is common to both poorly differentiated and minimal-deviation tumors (79). These observations have recently been confirmed by Goldfarb and Pitot (80a) by direct assay of tumor HMG-CoA reductase activity. Dietary cholesterol depressed HMG-CoA reductase activity in all host livers but in none of the hepatomas tested. Similarly, dietary cholestyramine markedly increased reductase activity in host livers but had little or no effect on the reductase of hepatoma tissue.

Table 7. Effect of Dietary Cholesterol on Mevalonate and Cholesterol Synthesis in Mouse Hepatic and Tumor Tissue

Data from Siperstein and Fagan (79).

Animal	Tissue	Cholesterol in Diet, %	Acetate-2-^{14}C, %, Converted to	
			Mevalonate	Cholesterol
Normal	Liver	0	0.55	0.16
Normal	Liver	5	0.03	0.007
Tumor-bearing	Liver	0	0.54	0.16
Tumor-bearing	Liver	5	0.03	0.003
Tumor-bearing	Hepatoma	0	1.08	0.42
Tumor-bearing	Hepatoma	5	1.24	0.28

The apparent absence of regulation of HMG-CoA reductase in transplantable mouse hepatoma tissue has also been confirmed by Kandutsch and Hancock (79a). Hepatic HMG-CoA reductase levels of mice with a high degree of spontaneous hepatomas were within the control range, and reductase levels responded in a normal manner to dietary cholesterol. In tumor-bearing animals, while the host liver responded normally, the HMG-CoA reductase of tumor tissue was insensitive not only to dietary cholesterol, but also to starvation or to Triton induction (Table 8). No significant differences were detected in the K_m-values or heat stabilities of reductases from tumor or host liver tissue (79a).

The defect in regulation of HMG-CoA reductase of hepatic tumor tissue may, however, be more apparent than real. The view has been advanced by several investigators that the insensitivity of tumor HMG-CoA reductase to regulation by dietary cholesterol may result from the failure of cholesterol or its metabolites to reach their site of action in the tumor cell. The most convincing evidence in support of this view is that of Harry et al. (80b), who demonstrated that transplantable hepatomas exhibit impaired uptake and storage of free cholesterol and of cholesterol esters. Additional evidence reported by Sabine (80c) suggests that, while hepatic tumor may be impermeable to bile acids, cholesterol synthesis from acetate is sensitive to inhibition by 0.5 mM deoxycholate both in normal and in tumor tissue homogenates. Thus, although the apparent defect in regulation of HMG-CoA reductase of hepatic tumor tissue remains an intriguing observation, the possibility that it may reflect the inability of a metabolic regulator to reach its intracellular site of action merits serious consideration.

Table 8.   Regulation of HMG-CoA Reductase Activity in Two Transplanted Mouse Hepatomas

From Kandutsch and Hancock (79a).

Hepatoma	Treatment	HMG-CoA Reductase Activity, %, Relative to Untreated Control	
		Hepatoma	Host Liver
BW-7756	None	100	100
	Starved 48 hr	110	8
	1% Cholesterol, 7 days	57	<3
	Triton, 0.8 days	145	270
H4	None	100	100
	Starved 48 hr	110	10
	1% Cholesterol, 7 days	100	<3
	Triton, 0.8 days	110	860

FASTING.   Fasting decreases cholesterol synthesis from acetate (64), and fasted rats have greatly reduced HMG-CoA reductase levels (14, 16, 59, 81). Regen et al. (81) observed that reductase activity decreased to less than 50% of that of fed controls after only 5 hr. Even lower levels were found in animals fasted for longer times. Restoration of activity accompanied refeeding, with a significant rise within 5 hr. The diurnal variation of reductase activity is still seen in fasted rats (59), although the absolute activities are greatly reduced.

TRITON.   Injection of the detergent Triton increases rat hepatic steroidogenesis from acetate but not from mevalonate (64). The inference that Triton gives rise to increased HMG-CoA reductase levels was confirmed by direct assay of reductase activity by Bucher et al. (14) and later by others (16). Kandutsch and Saucier (51) have shown that puromycin blocks the effect of Triton in raising mouse liver HMG-CoA reductase levels, suggesting that Triton may act by increasing the synthesis of HMG-CoA reductase.

THYROID HORMONE.   Thyroidectomy lowers rat hepatic HMG-CoA reductase activity to less than 50% of control values. Injection of 3,3',5'-triiodothyronine into hypothroid rats restores reductase activity to normal levels after a latent period of 30 hr (82).

INHIBITORY PROTEIN PRESENT IN BILE.   Ogilvie and Kaplan (83) reported the presence in rat bile of a low-molecular-weight protein which appeared to inhibit HMG-CoA reductase activity. Sterol synthesis from acetate was profoundly decreased, whereas that from mevalonate was relatively unaffected.

Since neither fatty acid synthesis nor CO_2 production from acetate was inhibited, the protein does not appear to be a general inhibitor of acetate metabolism. An approximate molecular weight of 19,000 was reported for a substantially purified but inhomogeneous preparation. The inhibitory activity of bile was increased by fasting.

We have recently shown that the inhibitory activity of bile acts directly on HMG-CoA reductase of rat liver microsomes, as proposed by Ogilvie and Kaplan (83). An apparently similar protein present in bovine bile also inhibits (D. J. McNamara, unpublished observations). Conceivably this protein may play a physiological role in the regulation of reductase activity. Shapiro and Rodwell (49) have provided evidence implicating a short-lived protein in the normal diurnal fall in activity occurring during the early morning hours. It is possible that the short life of this protein results from its excretion into the bile.

X-IRRADIATION. X-irradiation has long been known to stimulate hepatic cholesterol synthesis from acetate, but not from mevalonate (64). The site of action has recently been definitively established as HMG-CoA reductase. Seventy-two hours after total body irradiation the HMG-CoA reductase activity of hepatic microsomes of fasted mice exceeds that of fasted controls by six- to sevenfold. An even greater effect is observed on cholesterol synthesis from acetate. Both increases are largely blocked by administration of puromycin simultaneously with or after irradiation (83a).

These effects, although re-emphasizing the central role of HMG-CoA reductase in the regulation of hepatic cholesterol synthesis, are not observed until 72 hours after irradiation. We suggest that they may well represent a response to acute stress rather than an effect of radiation per se.

DEVELOPMENTAL PATTERN OF RAT HEPATIC HMG-CoA REDUCTASE. The postnatal developmental pattern of HMG-CoA reductase parallels that reported for acetate incorporation into sterols (83b). Reductase levels at birth approximate adult levels, fall at 8 to 10 days, and remain depressed until weaning. At weaning, reductase activity overshoots adult levels and returns to normal adult values within 4 days (McNamara, unpublished observations).

OTHER FACTORS REPORTED TO AFFECT HMG-CoA REDUCTASE ACTIVITY. Before the development of methods for its quantitative assay, indirect evidence was obtained for a variety of factors affecting microsomal HMG-CoA reductase activity. In most cases cholesterol synthesis was measured simultaneously from acetate and from mevalonate in Bucher homogenates (64) of liver tissue. Factors affecting cholesterol synthesis from acetate but

Table 9. Hormones Implicated as Affecting HMG-CoA Reductase Activity

| Tissue | Hormone | Effect on Cholesterol Synthesis | |
		From Acetate	From Mevalonate
Rat liver	Alloxan diabetes (84)	↓	→
Rat liver	Deoxycorticosterone (85)	↑	→
Rat liver	Growth hormone (86)	↑	→
Rat liver	Hypothyroidism (87)	↓	→
Rat skin	Hypothyroidism (87)	→	→
Rat liver	Thyroid hormone (87)	↑	→
Rat skin	Thyroid hormone (87)	→	→
Rat liver	Noradrenalin (88)	↑	→

not from mevalonate were assumed to affect HMG-CoA reductase. Apparent increases and decreases in reductase activity reported on the basis of the above evidence are given in Tables 9–11. In the case of fasting, dietary cholesterol, and Triton injection the indirect evidence has been confirmed by direct assays of reductase activity (49, 77, 81), but similar evidence for other factors is lacking. That caution is indicated in the interpretation of indirect data is suggested by the failure of direct assays to confirm the suggestion that bile acids inhibit HMG-CoA reductase of rat liver microsomes (103). The reductase is relatively insensitive to bile salts present at concentrations which significantly inhibit the conversion of acetate to mevalonate (50, 57; D. J. McNamara, unpublished observations). In the absence of direct evidence, therefore, the tabulated observations should be regarded as suggestive rather than conclusive evidence for effects on the reductase.

Table 10. Vitamins Implicated as Affecting HMG-CoA Reductase Activity

| Tissue | Vitamin | Effect on Cholesterol Synthesis | |
		From Acetate	From Mevalonate
Rat liver	B_6 deficiency (89, 90)	↑	→
Chicken liver	B_6 deficiency (91)	↓	→
Rat liver	Addition of vitamin E (92)	↓	→

Table 11.   Miscellaneous Factors Implicated as Affecting HMG-CoA Reductase Activity

		Effect on Cholesterol Synthesis	
Tissue	Factor	From Acetate	From Mevalonate
Rat liver	L-Aspartate (93)	↓	→
Rat liver	Glycerol (94)	↓	→
Rat liver	Mitochondrial extract (95)	↓	→
Rat liver	Nicotinic acid (96–98)	↓	→
Rat liver	Ubiquinone-9 (99–100)	↓	→
Rat intestine	Nicotinic acid (95)	↑	→
Rat liver	Medium-chain triglycercides (101)	↓	→
Rat liver	Biliary drainage (102)	↑	→
Rat liver	X-irradiation (64)	↑	→

B.   Mevaldate Reductase

$$\text{Mevaldate} + \text{NADPH} + \text{H}^+ \rightarrow \text{mevalonate} + \text{NADP}^+$$

The presence of mevaldate reductase activity (EC 1.1.1.32: mevalonate: NADP oxidoreductase) in mammalian tissues has been reported by many groups (104–108). Although mevaldate appears not to be a free intermediate in the reduction of HMG-CoA to mevalonate by either yeast (43) or *Pseudomonas* HMG-CoA reductase (41), these reductases can catalyze the conversion of mevaldate to mevalonate. *Pseudomonas* HMG-CoA reductase catalyzes this reduction of mevaldate to mevalonate (in the presence of NADH) and the oxidation of mevaldate to HMG-CoA (in the presence of CoA and NAD$^+$) (41). The yeast enzyme catalyzes the reduction both of mevaldate and (more rapidly) of synthetically prepared mevaldyl-CoA hemimercaptal to mevalonate (45).

Although the activity of rat liver tissue originally reported as mevaldate reductase (106) may be due, at least in part, to alcohol dehydrogenase activity (107), rat liver contains an NADPH-specific mevaldate reductase (107). No metals are required, and activity is optimal over the pH range 6.5–7.5. Since neither HMG-CoA reductase not the malonyl-CoA pathway for mevalonate synthesis (see Section 3.C) appears to involve mevaldate as a free intermediate, the physiological significance of this activity is not clear at present.

C. The Malonyl-CoA Pathway for Mevalonate Synthesis

An alternative pathway for mevalonate biosynthesis was established in the early 1960's by Brodie et al. (108–113). The reactions occur in soluble extracts of pigeon liver extensively purified as a fatty acid-synthesizing system (112) and in the cytosol of rat liver homogenates (114).

Sources

PIGEON LIVER. Malonyl-CoA is an obligatory intermediate, and enzyme-bound acetoacetate and enzyme-bound HMG are formed en route to mevalonate. A ^{14}C-malonyl-S-enzyme complex was isolated and shown to incorporate isotope both into HMG (in the absence of NADPH) and into fatty acids (in the presence of NADPH) (113). The malonyl-CoA pathway appears to account for the greater part of the mevalonate synthesized by pigeon liver (109, 113).

RAT LIVER. Crude rat liver homogenates also convert malonyl-CoA to HMG and to mevalonate without prior cleavage to two-carbon compounds. As in pigeon liver, the activity is associated with the cytosol (114). Using conditions designed to prevent malonyl-CoA formation, Fimognari and Rodwell (114) concluded that conversion of acetate or acetyl-CoA to mevalonate by the malonyl-CoA pathway is of minor quantitative importance for rat liver. This contrasts with the situation in pigeon liver. However, as discussed under "Regulation" below, this should not be taken as implying that the malonyl-CoA pathway is of minor physiological significance in mammalian liver.

FUNGI. A possible role for the malonyl-CoA pathway in fungi is suggested by the observation of Neujahr and Björk (114a) that malonyl-CoA is efficiently utilized for the synthesis of ergosterol and of β-carotene by cell-free extracts of *Blakeslea trispora*.

Comparison of the Two Pathways for Mevalonate Synthesis

Figure 6 depicts the reactions, intermediates, and intracellular location of the classical and of the malonyl-CoA or bound intermediate pathways for mevalonate synthesis. In addition to differences of intracellular compartmentation, of intermediates, and of participating enzymes, these pathways differ with respect to both their requirement for ATP (the malonyl-CoA pathway is energetically favored over the classical pathway by an amount equivalent to the free energy of hydrolysis of one ATP per mole of HMG-CoA synthesized) and their regulation (see "Regulation," below).

Rudney et al. (12) have proposed that the malonyl-CoA pathway may

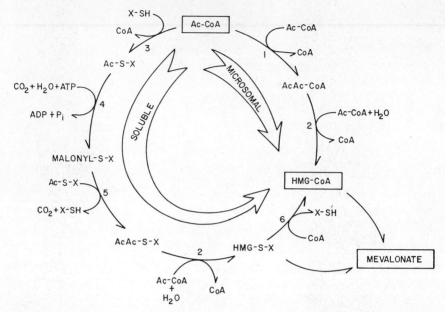

Fig. 6. A unified view of the classical and malonyl-CoA pathways for mevalonate synthesis. Adapted from Rudney et al. (12). As originally proposed for yeast, X—SH and acyl-S—X represent acyl-carrier protein and acylated acyl-carrier protein. ACP derivatives have not been shown to function in the malonyl-CoA pathway of hepatic tissue, although thiol esters of enzymes which perform analogous functions are involved. For the sequence of reactions proposed by Rudney et al. (12) for yeast, the reaction catalysts are (1) acetoacetyl-CoA thiolase (EC 2.3.1.9: Acetyl-CoA:acetyl-CoA C-acetyltransferase), (2) HMG-CoA synthase [EC 4.1.3.5: 3-hydroxy-3-methylglutaryl-CoA acetoacetyl-CoA-lyase (CoA-acetylating)], (3) acetyltransacylase, (4) acetyl-CoA carboxylase [EC 6.4.1.2: acetyl-CoA: CO_2 ligase (ADP)], (5) malonyltransacylase, (6) nonenzymic exchange reaction. Whether HMG-ACP may be converted to mevalonate directly or only after prior conversion to HMG-CoA is not known. For the malonyl-CoA pathway, however, HMG-enzyme appears to be directly convertible to mevalonate.

actually share many common features with the synthesis of HMG-CoA via acyl-carrier protein derivatives by yeast. These investigators propose that X-SH and acyl-S-X in Fig. 6 may actually be free or acylated acyl-carrier proteins or acyl-carrier-protein-like portions of an enzyme.

Regulation

Since the classical pathway for mevalonate synthesis is confined to the microsomes, the small but significant capacity of rat liver cytosol to form mevalonate (80, 114) probably reflects synthesis via the malonyl-CoA pathway. Siperstein and Fagan (80) showed that synthesis by cytosol, unlike that by microsomes, is insensitive to dietary cholesterol. Fimognari and Rodwell (114) therefore inferred that the malonyl-CoA pathway is concerned with the synthesis of mevalonate destined for fates other than sterols. Coenzyme Q and N^6-(Δ^2-isopentenyl)adenosine of transfer RNA (115) represent two key products of mevalonate in mammals. Compartmentation of two pathways for mevalonate synthesis could permit unrestricted synthesis of nonsteroid isoprenoids under conditions (such as fasting or cholesterol feeding) where steroid synthesis is curtailed. Hence the physiological significance of the malonyl-CoA pathway may ultimately be shown to relate to the synthesis of isoprenoids other than sterols (114).

D. Mevalonate Kinase

$$\text{(R)-Mevalonate} + \text{XTP} \xrightarrow{\text{Mg}^{2+}} \text{(R)-mevalonate-5-phosphate} + \text{XDP}$$

The observation of Amdur et al. (116) that incorporation of isotope from ^{14}C-mevalonate into squalene catalyzed by dialyzed yeast extracts required ATP was followed by the demonstration (117, 118) of mevalonate kinase (EC 2.7.1.36: ATP:mevalonate 5-phosphotransferase) activity in yeast. Kinase activity subsequently was detected in mammalian liver (119), plants (120–123), insects (123a, 123b) *Neurospora crassa* (R. L. Imblum, unpublished observations), *Mycoplasma* (124), and *Euglena* (125). Mevalonate kinase requires a divalent metal, as well as a nucleoside triphosphate as phosphate donor, and is specific for (R)-mevalonate.

Sources

YEAST.   Mevalonate kinase purified 48-fold from yeast to a specific activity of 5.4 × 10^{-2} iu/mg is optimally active at pH 6.4–6.7 and can utilize ATP, GTP, CTP, or UTP as cosubstrate. The use of Mn^{2+} is preferred over that of other divalent cations. *p*-Chloromercuribenzoate inhibits, and this inhibition is reversed by thiol-containing reagents.

MAMMALIAN LIVER.   Mevalonate kinase has been extensively purified from pig (126–129) and rabbit (119) liver. The pig liver enzyme, purified about 1000-fold to a specific activity of 0.6–1.0 iu/mg, requires Mn^{2+} or Mg^{2+} and a thiol for activity and uses ATP or ITP equally well as cosubstrate (126, 127). The only active cosubstrate for the rabbit liver enzyme is ATP (119). The substrate is (R)-mevalonate, which is converted to (R)-mevalonate 5-phosphate. (S)-Mevalonate, mevalonic acid lactone, *cis*-5-hydroxy-3-methylpent-2-enoate, 5-hydroxy-3-methylpent-2-enoate, 3-oxobutan-1-ol, and farnesol are not substrates, and (S)-mevalonate is not an inhibitor. Activity is inhibited by farnesoate and by its saturated analog, 3:7:11-trimethyl-dodecanoate (126, 127).

The reaction catalyzed by purified pig liver enzyme (specific activity 17 iu/mg) proceeds by a sequential, apparently ordered mechanism with (R)-mevalonate adding before ATP and with release of mevalonate-5-phosphate before ADP. Mevaldate is an inhibitor competitive with mevalonate (128, 129).

RAT OVARY.   Mevalonate kinase activity has been detected in rat ovarian tissue. The properties appear to resemble those of the liver kinase (K_m for DL-mevalonate = 3.6 μM; K_m for MgATP = 120 μM). Administration of lutenizing hormone does not affect activity assayed *in vitro* (129a).

PLANTS AND EUGLENA.   Mevalonate kinase has been studied in pumpkin (120) and bran seedlings (122) and in *Hevea brasiliensis* latex (121). The latex enzyme is optimally active at pH 7.5 and has a K_m for mevalonate [calculated for the (R) isomer] of $7.5 \times 10^{-5} M$. French bran seedlings are reported to contain isozymes of mevalonate kinase differing with respect to their location within the cell and in their kinetic properties (122). The cytoplasmic enzyme (pH optimum 5.5) resembles that of yeast and pumpkin seedlings, and the chloroplasmic enzyme (pH optimum 7.5) that of *Hevea* latex and of rabbit liver.

The properties of *Euglena gracilis* mevalonate kinase generally resemble those of the yeast enzyme. The K_m values are $3 \times 10^{-5} M$ [(R)-mevalonate] and $6 \times 10^{-3} M$ (ATP), and the optimum pH is 7.5 (125).

NEUROSPORA.   Mevalonate kinase of *Neurospora crassa* has been purified 200-fold from soluble extracts. Preliminary data indicate that the enzyme resembles that of mammalian liver (R. L. Imblum, unpublished observations).

MYCOPLASMA.   Mevalonate kinase activity is present in *Mycoplasma laidlawii* strain B. It is absent, however, from *Mycoplasma* sp. avian strain J, which contains enzymes catalyzing the synthesis of mevalonate, and from *M. hominis*, which lacks the capability for mevalonate synthesis (124).

INSECTS.   Mevalonate kinase purified 110-fold from extracts of aseptic cultures of *Sarcophaga bullata* has an apparent molecular weight of 113,000. The K_m for mevalonate is 0.62 mM and for ATP is 4.7 mM (123a).

Regulation

PLANTS.   The demonstration of mevalonate kinase activities in different compartments of plant cells (122) suggests that enzyme compartmentation of mevalonate kinase may play a role in regulating the synthesis of various polyisoprenoids. The authors couple this observation to other evidence for the existence in plant tissues of two pathways leading to polyisoprenoid biosynthesis.

MAMMALIAN LIVER. The work of Gould and Swyryd (130) established the presence of a regulatory site between mevalonate and farnesyl pyrophosphate. This site appears now to have been localized as mevalonate kinase (128). Highly purified pig liver mevalonate kinase is specifically inhibited by the polyisoprenyl pyrophosphates geranyl and farnesyl pyrophosphate. Inhibition is uncompetitive with respect to mevalonate and competitive with respect to MgATP^{2+}. It is known that ATP adds to an enzyme-mevalonate complex, and, consistent with its competition with MgATP^{2+}, geranyl pyrophosphate combines with enzyme-mevalonate, giving a catalytically inactive complex. The authors suggest that geranyl and farnesyl pyrophosphate may be acting as isosteric inhibitors and note that comparison of space-filling models of ATP and farnesyl pyrophosphate reveals that the two can attain quite similar configurations (128). The relation of these observations to cellular regulation *in vivo* remains to be shown, although Dorsey and Porter (128) note:

A mechanism for regulating farnesyl-pyrophosphate synthesis would offer an obvious advantage to the cell since 9 moles of ATP are required for the biosynthesis of one mole of farnesyl pyrophosphate from mevalonate . . . farnesyl-pyrophosphate can control its own synthesis by acting as a feedback inhibitor of mevalonate kinase. As a consequence it can also control the utilization of ATP for this and subsequent processes.

RAT OVARY. Like the hepatic enzyme, the enzyme of ovarian tissue is inhibited by geranyl and farnesyl pyrophosphate ($K_i = 1.3$ and $1.0 \ \mu M$, respectively). The inhibition is competitive with MgATP^{2+} (129a).

INSECTS. Barnes and Goodfellow (123b) have reported that the mevalonate kinase of *Sarcophaga bullata* undergoes changes in activity during development with significant peaks at specific times in the larval, pupal, and adult stages.

E. Phosphomevalonate Kinase

Mevalonate-5-phosphate + ATP → mevalonate-5-pyrophosphate + ADP

Mevalonate phosphate kinase activity (EC 2.7.4.2: ATP:5-phosphomevalonate phosphotransferase) has been detected in the soluble fraction of yeast (118, 131, 132), mammalian liver (126, 127, 133), higher plants (123, 134, 135), and *Mycoplasma* (124).

Sources

YEAST. Phosphomevalonate kinase has been separated from mevalonate kinase and obtained in a partially purified state (118, 131, 132) from baker's yeast at a specific activity of 0.06 iu/mg (118, 132). A divalent metal and

ATP are required, and activity is essentially independent of pH over the range 5.5–10.0. High concentrations of ATP are inhibitory.

MAMMALIAN LIVER. The phosphomevalonate kinase of pig liver, purified fifteen-fold to a specific activity of 0.078 iu/mg, is closely associated with but distinct from mevalonate kinase. High concentrations of ATP are not inhibitory, and ITP cannot replace ATP. The presence of Zn^{2+}, an activator for the yeast enzyme, is inhibitory. 5-Phosphomevalonate ($K_m = 3 \times 10^{-4}$ M) is not inhibitory at high concentrations, and activity shows a sharp optimum at about pH 7 (133).

PLANTS. Phosphomevalonate kinase from *Hevea brasiliensis* latex is optimally active from pH 7.0 to 7.5 and is irreversibly inactivated beyond this pH range. A thiol and either Mg^{2+} or Mn^{2+} are required for activity (134). K_m for 5-phosphomevalonate is 0.042 mM and for ATP is 0.19 mM. The activation energy is 14.8 Kcal/mole (134a). Potty and Bruemmer (123, 135) have demonstrated phosphomevalonate kinase activity in the orange.

Regulation

Although no evidence is available implicating phosphomevalonate kinase as a regulated enzyme, Hellig and Popják (133) estimate that this catalyzes the rate-limiting step between mevalonate and squalene in pig liver.

F. Pyrophosphomevalonate Decarboxylase

$$ATP + 5\text{-pyrophosphomevalonate} \rightarrow ADP + P_i + CO_2 +$$
$$\Delta^3\text{-isopentenylpyrophosphate}$$

Conversion of 5-pyrophosphomevalonate to Δ^3-isopentenylpyrophosphate, catalyzed by pyrophosphomevalonate decarboxylase [EC 4.1.1.33: ATP:5-pyrophosphomevalonate carboxy-lyase (dehydrating)], has been

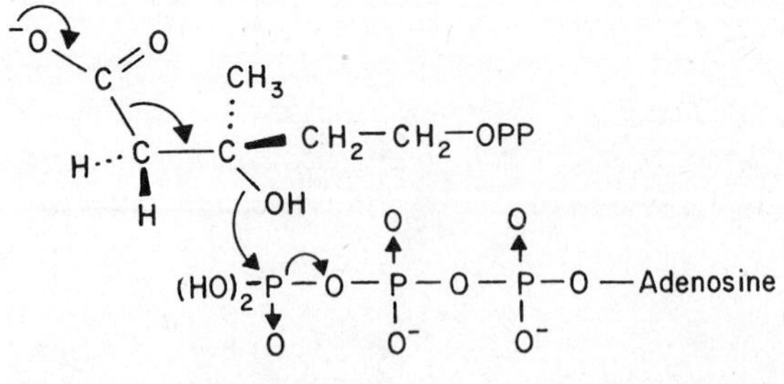

Fig. 7. Suggested concerted mechanism for the pyrophosphomevalonate decarboxylase reaction (137).

demonstrated in yeast (118, 136), liver tissue (137), higher plants (134a, 135, 138, 139), and *Mycoplasma* (124). Since a 3-phosphorylated mevalonate-5-pyrophosphate has not been isolated as an intermediate, the reaction is visualized as proceeding via a concerted mechanism (Fig. 7). The decarboxylase from pig liver has been purified fourteenfold to a specific activity of 0.025 iu/mg but is highly unstable at all stages of purity. No evidence suggesting a regulatory role for this enzyme is available.

G. Δ^3-Δ^2-Isopentenylpyrophosphate Isomerase

Δ^3-Isopentenylpyrophosphate $\rightleftharpoons \Delta^2$-isopentenylpyrophosphate

Lynen et al. (140) reported that part of the Δ^3-isopentenylpyrophosphate must isomerize to an allylic pyrophosphate, dimethylallyl pyrosphosphate (Δ^2-isopentenylpyrophosphate), before polymerization to geranyl pyrophosphate and higher isoprenoids. This conversion is catalyzed by isopentenylpyrophosphate isomerase (141) (EC 5.3.3.2: isopentenylpyrophosphate Δ^3-Δ^2-isomerase). Isomerase activity has been detected in yeast (140, 141) mammalian liver (141, 142), kidney and brain (141), and *Mycoplasma* (124). The reaction is freely reversible and is readily inhibited by thiol-alkylating reagents.

Sources

YEAST. The yeast enzyme requires Mg^{2+} ($K_m = 3.6 \times 10^{-5}\ M$), has a broad pH optimum from pH 5.5 to 9.3, and is inhibited by iodoacetamide or *p*-hydroxymercuribenzoate.

HOG LIVER. The properties and mechanism of action of partially purified hog liver isomerase have been studied by Shah et al. (142). Formation of dimethylallyl pyrophosphate is favored (ratio of product to reactant at equilibrium = 6.7), Mg^{2+} ($K_m = 1 \times 10^{-3}\ M$) is specifically required and cannot be replaced by Mn^{2+}, and optimal activity is observed over the pH range 4.0–8.3. Isomerase activity is strongly inhibited by sulfhydryl-alkylating reagents. Attempts to isolate an enzyme-bound intermediate showed that a covalent bond, possibly a thioester linkage, was formed between Δ^3-isopentenylpyrophosphate and protein. Studies using isotopically labeled substrate indicated that a stereospecific addition and removal of protein occurs at carbon 2 during isomerization. A mechanism has been proposed (142).

Regulation

The isomerase of *Mycoplasma* (124) is inhibited by cholesterol (143), which raises the K_m for Δ^3-isopentenylpyrophosphate from 0.58 to $1.5 \times 10^{-4}\ M$.

Growth of *Mycoplasma* in the presence of cholesterol and of mevalonate-2-^{14}C was accompanied by decreased isotope incorporation into lipid but normal incorporation into the crude nucleic acid fraction, specifically the transfer RNA fraction. Further examination revealed that the isotope was associated with a component of transfer RNA chromatographically similar to N^6-(Δ^2-isopentenyl)adenosine (144). High concentrations of cholesterol totally suppress the growth of *Mycoplasma*, presumably in part because of inhibition of the isomerase. Since the isomerase is needed for synthesis of N^6-(Δ^2-isopentenyl)adenosine, inhibition of the isomerase by cholesterol is probably incomplete. The isomerase may thus be viewed as a secondary point of control beyond HMG-CoA reductase.

4. CONCLUDING REMARKS

The regulation of "active isoprene" synthesis has only recently advanced from studies of general synthetic capabilities to the gathering of data on changes in individual enzymic activities. As yet, the latter studies are in their infancy. Of various systems, the most intensively studied has been rat hepatic steroidogenesis. Although cholesterol was recognized early as an end-product regulator of its own biosynthesis, preliminary speculations as to its mechanism of action centered about the concept of feedback inhibition. Although this may be valid for regulation of isopentenylpyrophosphate isomerase in *Mycoplasma*, currently available data suggest that HMG-CoA reductase, the major point of control in hepatic steroidogenesis, is regulated by mechanisms involving protein synthesis. Rapid induction and repression of HMG-CoA reductase synthesis and possibly altered rates of enzyme inactivation or destruction appear to be capable of accounting for the greatly reduced rates of steroidogenesis in cholesterol-fed animals. Changes in rates of synthesis and/or degradation as control mechanisms in mammalian systems are well documented (145), and HMG-CoA reductase appears to be subject to regulation by these mechanisms. Microsomal HMG-CoA reductase is therefore of interest not only from the viewpoint of isoprenoid biosynthesis but also as an example of a tightly regulated, mammalian, biosynthetic enzyme.

REFERENCES

1. R. O. Brady and S. Gurin, *J. Biol. Chem.*, **189,** 371 (1951).
2. G. L. Curran, *J. Biol. Chem.*, **191,** 775 (1951).
3. K. Bloch, L. C. Clark, and I. Harary, *J. Biol. Chem.*, **211,** 687 (1954).

4. J. L. Rabinowitz and S. Gurin, *J. Biol. Chem.*, **208**, 307 (1954).

5. H. Rudney, *J. Am. Chem. Soc.*, **76**, 2595 (1954).

6. H. Rudney, *J. Biol. Chem.*, **227**, 363 (1957).

7. H. Rudney and J. J. Ferguson, Jr., *J. Biol. Chem.*, **234**, 1076 (1959).

8. J. J. Ferguson, Jr. and H. Rudney, *J. Biol. Chem.*, **234**, 1072 (1959).

9. P. R. Stewart and H. Rudney, *J. Biol. Chem.*, **241**, 1212 (1966).

10. P. R. Stewart, and H. Rudney, *J. Biol. Chem.*, **241**, 1222 (1966).

11. F. Sauer and J. D. Erfle, *J. Biol. Chem.*, **241**, 30 (1966).

12. H. Rudney, P. R. Stewart, P. W. Majerus, and P. R. Vagelos, *J. Biol. Chem.*, **241**, 1226 (1966).

13. J. D. McGarry and D. W. Foster, *J. Biol. Chem.*, **244**, 4251 (1969).

14. N. L. R. Bucher, P. Overath, and F. Lynen, *Biochim. Biophys. Acta*, **40**, 491 (1960).

15. L. W. White and H. Rudney, *Biochemistry*, **9**, 2713 (1970).

16. L. W. White and H. Rudney, *Biochemistry*, **9**, 2725 (1970).

17. P. F. Smith, *Bacteriol. Rev.*, **28**, 97 (1964).

18. P. F. Smith and C. V. Henrikson, *J. Bacteriol.*, **89**, 146 (1965).

19. J. A. Johnson, D. W. Racusen, and J. Bonner, *Proc. Natl. Acad. Sci. U.S.*, **40**, 1031 (1954).

20. V. H. Potty, *J. Food Sci.*, **34**, 231 (1969).

21. K. Oshima and I. Uritani, *Agr. Biol. Chem.* (*Tokyo*), **31**, 1105 (1967).

22. K. Oshima and I. Uritani, *J. Biochem.* (*Tokyo*), **63**, 617 (1968).

23. D. H. Williamson, M. W. Bates, and H. A. Krebs, *Biochem. J.*, **108**, 353 (1968).

24. S. Mukherjee and A. Bhose, *Biochim. Biophys. Acta*, **164**, 357 (1968).

25. M. D. Siperstein and V. M. Fagan, *Advan. Enzyme Regulation*, **2**, 249 (1964).

26. G. S. Boyd, *Federation Proc.*, **21**, Suppl. 11, 86 (1962).

26a. E. A. Lockwood and E. Bailey, *Biochem. J.*, **124**, 249 (1971).

27. V. W. Rodwell, in *Metabolic Pathways*, 3rd ed., Vol. 3 (Ed.: D. M. Greenberg), Academic Press, New York, 1969, p. 191.

28. H. Hilz, J. Knappe, E. Ringelman, and F. Lynen, *Biochem. Z.*, **229**, 476 (1958).

29. A. Del Campillo-Campbell, E. E. Dekker, and M. J. Coon, *Biochim. Biophys. Acta*, **31**, 290 (1959).

30. J. Knappe and F. Lynen, *Intern. Union Biochem.*, *Abstr. 4th. Intern. Congr.*, *Vienna*, 1958, p. 49.

31. R. E. Burch, H. Rudney, and J. J. Irias, *J. Biol. Chem.*, **239**, 4111 (1964).

32. E. E. Dekker, M. J. Schlesinger, and M. J. Coon, *J. Biol. Chem.*, **233**, 434 (1958).

33. M. E. Kirtley, H. Rudney, and I. F. Durr, *J. Biol. Chem.*, **237**, 1781 (1962).

34. F. Lynen, U. Henning, C. Bublitz, B. Sörbo, and L. Kröplin-Rueff, *Biochem. Z.*, **330**, 269 (1958).

35. F. J. R. Hird and R. H. Symonds, *Biochim. Biophys. Acta*, **46**, 457 (1961).

36. I. C. Caldwell and G. I. Drummond, *J. Biol. Chem.*, **238**, 64 (1963).

37. J. D. McGarry and D. W. Foster, *Biochim. Biophys. Acta*, **177**, 35 (1969).

38. B. K. Bachhawat, W. G. Robinson, and M. J. Coon, *J. Biol. Chem.*, **216**, 727 (1955).

39. L. D. Stegink and M. J. Coon, *J. Biol. Chem.*, **243**, 5272 (1968).

40. M. A. Siddiqi and V. W. Rodwell, *J. Bacteriol.*, **93**, 207 (1967).

41. W. R. Bensch and V. W. Rodwell, *J. Biol. Chem.*, **245**, 3755 (1970).

41a. I. Shimizu, J. Nagai, H. Hatanaka, E. Saito, and H. Katsuki, *J. Biochem.* (*Tokyo*), **70**, 175 (1971).

42. J. Knappe, E. Ringelmann, and F. Lynen, *Biochem. Z.*, **332**, 195 (1959).

43. I. F. Durr and H. Rudney, *J. Biol. Chem.*, **235**, 2572 (1960).

44. M. E. Kirtley and H. Rudney, *Biochemistry*, **6**, 230 (1967).

45. J. Retey, E. Von Stetten, U. Coy, and F. Lynen, *European J. Biochem.*, **15**, 72 (1970).

46. H. J. Knauss, J. W. Porter, and G. Wasson, *J. Biol. Chem.*, **234**, 2835 (1959).

47. G. Popjak and J. W. Cornforth, *Advan. Enzymol.*, **22**, 281 (1960).

48. T. C. Linn, Ph.D. thesis, University of Minnesota, 1965.

49. D. J. Shapiro and V. W. Rodwell, *Biochem. Biophys. Res. Commun.*, **37**, 867 (1969).

49a. D. J. Shapiro and V. W. Rodwell, *J. Biol. Chem.*, **246**, 3210 (1971).

50. T. Kawachi and H. Rudney, *Biochemistry*, **9**, 1700 (1970).

51. A. A. Kandutsch and S. E. Saucier, *J. Biol. Chem.*, **244**, 2299 (1969).

52. M. D. Siperstein, V. M. Fagan, and J. M. Dietschy, *J. Biol. Chem.*, **241**, 597 (1966).

53. R. B. Guchhait and J. W. Porter, *Anal. Biochem.*, **15**, 509 (1966).

54. C. M. Hinse and P. J. Lupien, *Anal. Biochem.*, **19**, 392 (1967).

55. T. C. Linn, *J. Biol. Chem.*, **242**, 984 (1967).

56. D. J. Shapiro, R. L. Imblum, and V. W. Rodwell, *Anal. Biochem.*, **31**, 383 (1969).

57. B. Hamprecht, *Naturwissenschaften*, **56**, 398 (1969).

58. P. Back, B. Hamprecht, and F. Lynen, *Arch. Biochem. Biophys.*, **133**, 11 (1969).

59. B. Hamprecht, C. Nussler, and F. Lynen, *FEBS Letters*, **4**, 117 (1969).

59a. A. I. Louw and E. H. Mosbach, *Federation Proc.*, **30**, 347 Abs. (1971).

60. D. Kritchevsky, S. Tepper, N. W. Ditullio, and W. Holmes, *J. Am. Oil Chemists' Soc.*, **42**, 1024 (1965).

61. H. S. Maker and G. Hauser, *J. Neurochem.*, **14**, 457 (1967).

62. M. T. Kelley, R. T. Aexel, B. L. Herndon, and H. J. Nicholas, *J. Lipid Res.*, **10**, 166 (1969).

63. A. A. Kandutsch and S. E. Saucier, *Arch. Biochem. Biophys.*, **135**, 201 (1969).

64. N. L. R. Bucher, K. McGarrahan, E. Gould, and A. V. Loud, *J. Biol. Chem.*, **234**, 262 (1959).

65. M. D. Siperstein and M. J. Guest, *J. Clin. Invest.*, **38**, 1043 (1959).

66. M. D. Siperstein and V. M. Fagan, *J. Biol. Chem.*, **241**, 602 (1966).

67. M. D. Siperstein, in *Developmental and Metabolic Control Mechanisms and Neoplasia*, Williams and Wilkins, Baltimore, 1965, p. 427.

68. A. Kawaguchi, H. Hatanaka, and H. Katsuki, *Biochem. Biophys. Res. Commun.*, **33**, 463 (1968).

69. A. Kawaguchi, *J. Biochem.* (*Tokyo*), **67**, 219 (1970).

70. H. Hatanaka, A. Kawaguchi, and H. Katsuki, *Biochem. Biophys. Res. Commun.*, **40**, 786 (1970).

71. M. D. Siperstein, *Proceedings of the 7th Canadian Cancer Conference, Honey Harbour, Ontario*, Vol. 7, Pergamon Press, Toronto, 1966, p. 152.

71a. D. J. Shapiro and V. W. Rodwell, *Federation Proc.*, **30**, 1229 Abs. (1971).

71b. M. Higgins, T. Kawachi, and H. Rudney, *Biochem. Biophys. Res. Commun.*, **45**, 138 (1971).

71c. R. E. Dugan, L. L. Slakey, and J. W. Porter, *Federation Proc.*, **30**, 1229 Abs. (1971).

72. R. G. Gould, *Am. J. Med.*, **11**, 209 (1951).

73. W. T. Beher, G. D. Baker, W. L. Anthony, and M. E. Beher, *Henry Ford Hosp. Med. Bull.*, **9**, 201 (1961).

74. R. G. Gould, C. B. Taylor, J. S. Hagerman, I. Warner, and D. J. Campbell, *J. Biol. Chem.*, **201**, 519 (1953).

75. H. Sakakida, C. C. Shediac, and M. D. Siperstein, *J. Clin. Invest.*, **42**, 1521 (1963).

76. E. P. M. Bhattathiry and M. D. Siperstein, *J. Clin. Invest.*, **42**, 1613 (1963).

77. T. C. Linn, *J. Biol. Chem.*, **242**, 990 (1967).

78. A. A. Kandutsch and R. M. Packie, *Arch. Biochem. Biophys.*, **140**, 122 (1970).

78a. G. M. Fimognari and V. W. Rodwell, *Science*, **147**, 1038 (1965).

78b. B. Hamprecht, C. Nüssler, G. Waltinger, and F. Lynen, *European J. Biochem.*, **18**, 10 (1971).

78c. B. Hamprecht, R. Roscher, G. Waltinger, and C. Nüssler, *European J. Biochem.*, **18**, 15 (1971).

79. M. D. Siperstein and V. M. Fagan, *Cancer Res.*, **24**, 1108 (1964).

79a. A. A. Kandutsch and R. L. Hancock, *Cancer Res.*, **31**, 1396 (1971).

80. M. D. Siperstein and V. M. Fagan, *J. Biol. Chem.*, **291**, 602 (1966).

80a. S. Goldfarb and H. Pitot, *Cancer Res.* (in press).

80b. D. S. Harry, H. P. Morris, and N. McIntyre, *J. Lipid Res.*, **68**, 315 (1971).

80c. J. R. Sabine, *Biochim. Biophys. Acta*, **176**, 600 (1969).

81. D. Regen, C. Riepertinger, B. Hamprecht, and F. Lynen, *Biochem. Z.*, **346**, 78 (1966).

82. W. Guder, I. Nolte, and O. Wieland, *European J. Biochem.*, **4**, 273 (1968).

83. J. W. Ogilvie and B. H. Kaplan, *J. Biol. Chem.*, **241**, 4722 (1966).

83a. J. Berndt and R. Gaumont, *FEBS Letters*, **13**, 49 (1971).

83b. F. J. Ballard and R. W. Hansen, *Biochem. J.*, **102**, 952 (1967).

84. R. Carenburg and I. L. Chaikoff, *Am. J. Physiol.*, **210**, 37 (1966).

85. J. S. Wilmer and T. S. Foster, *Can. J. Biochem. Physiol.*, **38**, 1387 (1960).

86. R. Leal, *Rev. Port. Quim.*, **4**, 3 (1962).

87. K. Fletcher and N. B. Myant, *J. Physiol.*, **144**, 361 (1958).

88. W. M. Bortz, *Biochim. Biophys. Acta*, **152**, 619 (1968).

89. S. N. Shah, P. V. Johnston, and F. A. Kummerow, *J. Nutr.*, **72**, 81 (1960).

90. P. J. Lupien, C. M. Hinse, and M. Avery, *Can. J. Biochem.*, **47**, 631 (1969).

91. P. J. Lupien and B. B. Migicovsky, *Can. J. Biochem.*, **42**, 1161 (1964).

92. C. D. Eskelson and H. P. Jacobi, *Physiol. Chem. Phys.*, **1**, 487 (1969).

93. H. Nakamura, *Nippon Seirigaku Zasshi*, **26**, 445 (1964).

94. B. B. Migicovsky, *Can. J. Biochem.*, **46**, 859 (1968).

95. B. B. Migicovsky, *Can. J. Biochem. Physiol.*, **38**, 339 (1960).

96. K. Nakamura, T. Masuda, and H. Nakamura, *J. Atherosclerosis Res.*, **7**, 253 (1967).

97. W. F. Perry, *Metab. Clin. Exptl.*, **9**, 686 (1960).

98. W. Gamble and L. D. Wright, *Proc. Soc. Exptl. Biol. Med.*, **107**, 160 (1961).

99. K. V. Krishnaiah and T. Ramasarma, *Biochim. Biophys. Acta*, **202**, 332 (1970).

100. K. V. Krishnaiah, A. R. Inamdar, and T. Ramasarma, *Biochem. Biophys. Res. Commun.*, **27**, 474 (1967).

101. D. Kritchevsky and S. A. Tepper, *J. Nutr.*, **86**, 67 (1965).

102. N. B. Myant and H. A. Eder, *J. Lipid Res.*, **2**, 363 (1961).

103. G. M. Fimognari and V. W. Rodwell, *Biochemistry*, **4**, 2086 (1965).

104. L. D. Wright, M. Cleland, B. N. Dutta, and J. S. Norton, *J. Am. Chem. Soc.*, **79**, 6572 (1957).

105. M. J. Schlesinger and M. J. Coon, *J. Biol. Chem.*, **236**, 2421 (1961).

106. H. Nakamura and D. M. Greenberg, *Arch. Biochem. Biophys.*, **93**, 153 (1961).

107. H. J. Knauss, J. D. Brodie, and J. W. Porter, *J. Lipid Res.*, **3**, 197 (1962).

108. J. D. Brodie, Ph.D. thesis, University of Wisconsin (1962).

109. J. D. Brodie and J. W. Porter, *Biochem. Biophys. Res. Commun.*, **3**, 173 (1960).

110. J. D. Brodie, G. W. Wasson, and J. W. Porter, *Biochem. Biophys. Res. Commun.*, **8**, 76 (1962).

111. J. D. Brodie, G. W. Wasson, and J. W. Porter, *Biochem. Biophys. Res. Commun.*, **12**, 27 (1963).

112. J. D. Brodie, G. Wasson, and J. W. Porter, *J. Biol. Chem.*, **238**, 1294 (1963).

113. J. D. Brodie, G. Wasson, and J. W. Porter, *J. Biol. Chem.*, **239**, 1346 (1964).

114. G. M. Fimognari and V. W. Rodwell, *Lipids*, **5**, 104 (1970).

114a. H. Y. Neujahr and L. Björk, *Acta Chem. Scand.*, **24**, 2361 (1970).

115. M. J. Robins, R. H. Hall, and R. Thedford, *Biochemistry*, **6**, 1837 (1967).

116. B. H. Amdur, H. Rilling, and K. Bloch, *J. Am. Chem. Soc.*, **79**, 2646 (1957).

117. T. T. Tchen, *J. Biol. Chem.*, **233**, 1100 (1958).

118. K. Bloch, S. Chaykin, A. H. Phillips, and A. Dewaard, *J. Biol. Chem.*, **234**, 2595 (1959).

119. K. Markley and S. Smallman, *Biochim. Biophys. Acta*, **47**, 327 (1961).

120. W. D. Loomis and J. Battaile, *Biochim. Biophys. Acta*, **67**, 54 (1963).

121. I. P. Williamson and R. G. O. Kekwick, *Biochem. J.*, **88**, 18P (1963).

122. L. J. Rogers, S. P. J. Shah, and T. W. Goodwin, *Biochem. J.*, **100**, 14C (1966).

123. V. H. Potty and J. H. Bruemmer, *Phytochemistry*, **9**, 1229 (1970).

123a. R. D. Goodfellow and F. J. Barnes, *Insect Biochem.*, **1**, 271 (1971).

123b. F. J. Barnes and R. D. Goodfellow, *J. Insect Physiol.*, **17**, 1415 (1971).

124. C. V. Henrickson and P. F. Smith, *J. Bacteriol.*, **92**, 701 (1966).

125. C. Z. Cooper and C. R. Benedict, *Plant Physiol.*, **42**, 515 (1967).

126. H. R. Levy and G. Popják, *Biochem. J.*, **75**, 417 (1960).

127. G. Popják, in *Methods in Enzymology*, Vol. XV (Ed.: R. B. Clayton), Academic Press, New York, 1969, p. 393.

128. J. K. Dorsey and J. W. Porter, *J. Biol. Chem.*, **243**, 4667 (1968).

129. E. Beytia, J. K. Dorsey, J. Marr, W. W. Cleland, and J. W. Porter, *J. Biol. Chem.*, **245**, 5450 (1970).

129a. A. P. E. Flint, *Biochem. J.*, **120**, 145 (1970).

130. R. G. Gould and E. A. Swyryd, *J. Lipid Res.*, **7**, 698 (1966).

131. U. Henning, E. M. Moslein, and F. Lynen, *Arch. Biochem. Biophys.*, **83**, 259 (1959).

132. T. T. Tchen, in *Methods in Enzymology*, Vol. 5 (Eds.: S. P. Colowick and N. O. Kaplan), Academic Press, New York, 1962, p. 489.

133. H. Hellig and G. Popják, *J. Lipid Res.*, **2**, 235 (1961).

134. D. N. Skilleter, I. P. Williamson, and R. G. O. Kekwick, *Biochem. J.*, **98**, 27P (1966).

134a. D. N. Skilleter and R. G. O. Kekwick, *Biochem. J.*, **124**, 407 (1971).

135. V. H. Potty and J. H. Bruemmer, *Phytochemistry*, **9**, 99 (1970).

136. M. Lindberg, C. Yuan, A. DeWaard, and K. Bloch, *Biochemistry*, **1**, 182 (1962).

137. H. Hellig and G. Popják, *Biochem. J.*, **80**, 42P (1961).

138. C. J. Chesterton and R. G. O. Kekwick, *Arch. Biochem. Biophys.*, **125**, 76 (1968).

139. P. Valenzuela, E. Beytia, O. Cori, and A. Yudelevich, *Arch. Biochem. Biophys.*, **113**, 536 (1966).

140. F. Lynen, H. Eggerer, U. Henning, and I. Kessel, *Angew. Chem.*, **70**, 738 (1958).

141. B. W. Agranoff, H. Eggerer, U. Henning, and F. Lynen, *J. Biol. Chem.*, **235**, 326 (1960).

142. D. H. Shah, W. W. Cleland, and J. W. Porter, *J. Biol. Chem.*, **240**, 1946 (1965).

143. P. F. Smith and M. R. Smith, *J. Bacteriol.*, **103**, 27 (1970).

144. R. H. Hall, *Biochemistry*, **3**, 769 (1964).

145. R. T. Schimke and D. Doyle, *Ann. Rev. Biochem.*, **39**, 929 (1970).

CHAPTER 8

Control Mechanisms in the Biosynthesis of Aliphatic Amines in Eukaryotic Cells

H. G. WILLIAMS-ASHMAN

Ben May Laboratory for Cancer Research and Department of Biochemistry, University of Chicago, Chicago, Illinois

1. PREAMBLE

Spermidine and spermine, together with their constituent diamine putrescine, are three related aliphatic amines (Fig. 1) that appear to be present in all nucleated eukaryotic cells (1–4). Although spermine was originally discovered in human seminal fluid and is found in high concentrations in the prostatic secretion of man and certain other mammalian species (1, 3, 5, 6), neither spermidine nor spermine normally leach out of most mammalian cells, and these polyamines are not detectable in blood plasma, cerebrospinal fluid, or pancreatic juice (1, 3) or in human saliva (7). Today it remains questionable whether putrescine, spermidine, or spermine serves any truly *essential* functions in living eukaryotic cells. There are, nevertheless, many indications that these bases, and especially spermidine, may act as regulators of the biosynthesis of polynucleotides and proteins (1–4, 8–11), and also can stabilize nucleic acids, ribosomes, and certain enzymes, as well as mitochondrial and other biological membranes (1–4, 8–12, 12a). The available evidence in favor of such regulatory or stabilizing functions of aliphatic polyamines is, however, based largely on (*a*) correlations between the levels of spermidine and spermine during the growth of tissues vis-à-vis the formation of ribonucleic acids and other macromolecules, or (*b*) experiments concerned with the influence of polyamines on isolated nucleic acids, polynucleotide polymerizing enzymes, polyribosomal protein synthesizing systems, mitochondria, and so on (1–12). Extrapolation of findings from the latter types of cell-free systems to the functions of polyamines in living eukaryotic cells cannot yet be made with any great surety.

One experimental approach to the physiological significance of aliphatic polyamines in animal tissues is to study the enzymic pathways responsible for their biosynthesis. If the nature and properties of the various enzymes

$$H_2N(CH_2)_4NH_2$$
Putrescine

$$H_2N(CH_2)_3NH(CH_2)_4NH_2$$
Spermidine

$$H_2N(CH_2)_3NH(CH_2)_4NH(CH_2)_3NH_2$$
Spermine

Fig. 1.

and cofactors involved in the manufacture of these substances can be established, then it might be possible to develop specific inhibitors of one or more of the intermediary reactions. Application of such potential inhibitors might in turn enable the production of polyamines in eukaryotic organs to be prevented in unique and predictable ways. Any physiological consequences of such induced changes in the steady-state concentrations and/or intracellular distributions of putrescine, spermidine, and spermine could then perhaps point to at least some of their normal intracellular functions.

The following overview of the mechanisms of enzymic synthesis of aliphatic polyamines in eukaryotic organisms centers around recent investigations made in the author's laboratory. When these researches were first begun late in 1967, it was already known that the amino acids L-ornithine and L-arginine could act as precursors for the formation of putrescine in liver, and also that L-methionine could provide the propylamine moiety of spermidine in this tissue (1, 2, 13, 14). But no information was available concerning the nature of the intermediary enzymic processes, at least in eukaryotic cells. Thanks to the pioneering studies of Tabor, Rosenthal, and Tabor (15), published in 1958, however, a series of reactions that resulted in spermidine synthesis in *Escherichia coli* had already been uncovered.

2. SPERMIDINE BIOSYNTHESIS IN BACTERIA

Most bacteria neither contain nor synthesize spermine, but many gram-negative bacteria elaborate considerable amounts of putrescine and spermidine (1). In *E. coli*, spermidine can be formed by the following four linked enzyme reactions: (*a*) L-ornithine decarboxylase EC 4.1.1.17: (L-ornithine carboxy-lyase), which catalyzes the synthesis of putrescine and carbon dioxide from L-ornithine; (*b*) methionine adenosyltransferase (EC 2.5.1.6: ATP:L-methionine S-adenosyltransferase), which forms S-adenosyl-L-methionine (S-Ado-met)* from ATP and L-methionine, with elimination of P_i and PP_i; (*c*) S-Ado-met decarboxylase, catalyzing the release of carbon dioxide from S-Ado-met with the formation of 5'-deoxy-5'-S-(3-methylthiopropylamine)sulfonium adenosine (so-called "decarboxylated S-Ado-met"); and (*d*) spermidine synthase (a propylamine transferase), which promotes transfer of a propylamine group from decarboxylated S-Ado-met to putrescine to yield spermidine, 5'-methylthioadenosine (MTA), and a proton. The overall synthesis of spermidine by the sequential actions of

---

* The following abbreviations are employed: S-Ado-met: S-adenosyl-L-methionine; S-Ado-et: S-adenosylethionine; P_i: inorganic orthophosphate; PP_i: inorganic pyrophosphate; MTA: 5'-methylthioadenosine.

these four enzymes can be represented by the following equations:

$$\text{L-Ornithine} \rightarrow \text{putrescine} + CO_2, \tag{1}$$

$$\text{L-Methionine} + \text{ATP} \rightarrow S\text{-Ado-met} + P_i + PP_i, \tag{2}$$

$$S\text{-Adomet} \rightarrow \text{decarboxylated } S\text{-Ado-met} + CO_2, \tag{3}$$

$$\text{Decarboxylated } S\text{-Ado-met} + \text{putrescine} \rightarrow \text{spermidine} + \text{MTA} + H^+. \tag{4}$$

$$Sum: \text{L-Ornithine} + \text{L-methionine} + \text{ATP} \rightarrow \text{spermidine} + 2CO_2 + \text{MTA} + P_i + PP_i \tag{5}$$

The enzymic decarboxylation of L-ornithine (reaction 1) in *E. coli* was discovered by Gale (16) in 1940. The enzyme requires pyridoxal phosphate. Morris and Pardee (17, 18) showed that *E. coli* contains distinct "biosynthetic" and "inducible" L-ornithine decarboxylases that differ in their heat sensitivities, pH optima, and conditions for formation. An alternative pathway for putrescine biosynthesis in *E. coli* involves the base agmatine, the product of the enzymic decarboxylation of L-arginine (Fig. 2). L-Arginine decarboxylases are present in certain bacteria (18) and higher plants (19). Agmatine can, in turn, be converted to putrescine, either directly by the action of an agmatine ureohydrolase found in certain strains of *E. coli* (18), or by a two-step process in which *N*-carbamylputrescine is an intermediate, as in some higher plants (19) (see Fig. 2). There is, however, no evidence that this "agmatine pathway" for putrescine biosynthesis operates in higher animal cells, from which L-arginine decarboxylases appear to be absent (3).

The *S*-Ado-met decarboxylase (reaction 3) of *E. coli* was purified to a state of homogeneity by Wickner et al. (20). The enzyme, which has a molecular weight of 113,000, is inactivated by incubation in the absence of *S*-Ado-met with certain carbonyl reagents or with sodium borohydride, and

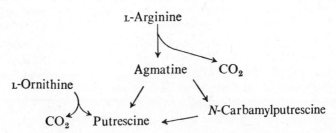

Fig. 2. The L-ornithine decarboxylase and agmatine pathways for putrescine biosynthesis. The role of the direct decarboxylation of L-ornithine in animal tissues, and additionally of the agmatine pathways in bacteria and higher plants, in the biosynthesis of putrescine is reviewed in the text.

was shown to contain pyruvate bound to the protein in covalent linkage. The enzyme-bound pyruvate almost certainly functions as a prosthetic group; the purified decarboxylase is devoid of pyridoxal phosphate. The S-Ado-met decarboxylase of E. coli exhibits a complete requirement for Mg^{2+} ions and is virtually unaffected by the addition of putrescine or spermidine in the absence or the presence of Mg^{2+} (3, 20, 21). A similar Mg^{2+}-dependent and putrescine-insensitive S-Ado-met decarboxylase can be extracted from Azotobacter vinelandii (22). Celia Tabor (23), working with E. coli extracts, was able to separate from the S-Ado-met decarboxylase another enzyme (spermidine synthase) that catalyzes the formation of spermidine from putrescine and decarboxylated S-Ado-met (reaction 4), but not from S-Ado-met added as such.

Feedback inhibitions of E. coli L-ornithine and L-arginine decarboxylases by putrescine and spermidine, and the influence of putrescine deprivation, as well as other factors, on the levels of these enzymes in E. coli, are described by Morris et al. (24). Tabor and Tabor (25, 26) made the interesting observation that, when the putrescine content of cells was diminished by growth of E. coli in an arginine-limited chemostat, the levels of L-ornithine and L-arginine decarboxylases could be depressed by the addition of putrescine; even under conditions of severe ornithine-arginine restriction, a substantial proportion of the ornithine and arginine entering cells was still used for polyamine biosynthesis, despite the demands on L-arginine for the biosynthesis of proteins. And there have been many other studies (see refs. 3, 4, 8, 10, 12a, 24) on factors affecting the concentrations of putrescine and spermidine in intact bacterial cells.

3. FORMATION OF PUTRESCINE IN ANIMAL TISSUES

In 1968, Pegg and Williams-Ashman (27) reported that the ventral lobe of the rat prostate gland, an organ that synthesizes and secretes large amounts of spermidine and spermine, contains an active L-ornithine decarboxylase (reaction 1). This was the first demonstration of an enzyme of this class in higher animal tissues. The presence of L-ornithine decarboxylase in normal and regenerating rat liver was described independently by Jänne and Raina (28) and by Russell and Snyder (29). Many studies on this enzyme in various animal tissues have since been published from a number of laboratories (30–38). Mammalian L-ornithine decarboxylases require pyridoxal phosphate, which very readily dissociates from the enzyme surfaces. Unlike some of the corresponding bacterial enzymes, mammalian L-ornithine decarboxylases do not seem to be dependent on any metal cofactors. A remarkable property of the enzyme in the partially purified form from regenerating livers or ventral prostates of rats is that it is both markedly activated and stabilized by certain dithiols, notably 1,4-dithiothreitol (Cleland's reagent) (37, 38). Inactivation of purified prostatic L-ornithine decarboxylase in the absence of appropriate dithiols seems to be due to polymerization of the enzyme to forms (mainly dimers) that are catalytically inactive unless dithiothreitol is added to the assay media. Whether there exist multiple forms (isoenzymes) of L-ornithine decarboxylase in various regions of animal cells remains an open question, although no evidence supporting this eventuality has yet been obtained in studies on the rat ventral prostate enzyme (27, 38).

Prostatic L-ornithine decarboxylase is inhibited competitively by putrescine (K_i ca. 1 mM), and to a lesser extent by spermidine (K_i ca. 3 mM) and spermine (K_i ca. 9 mM) (27, 38). The inhibition constants for these aliphatic amines are sufficiently high that it is doubtful whether feedback inhibition of this decarboxylase reaction by its product (putrescine) or by the end products of the polyamine biosynthetic pathway (spermidine and spermine) is of profound significance *in vivo*. Inhibition of rat liver L-ornithine decarboxylase by putrescine has also been observed (31). The K_m for L-ornithine for the soluble mammalian enzymes is close to 0.1 mM, with obeyance of classical Michaelis-Menten kinetics. Mammalian L-ornithine decarboxylases seem to be peculiarly unresponsive to fine controls by small molecules that participate in neighboring metabolic pathways, such as the urea cycle in liver, and more ubiquitous transmethylation reactions for which S-Ado-met serves as a methyl group donor (38).

The soluble L-ornithine decarboxylase of animal tissues is, in contrast, subject to extensive regulation at the level of synthesis (and possibly degradation) of the enzyme protein. Experiments involving the administration of inhibitors of protein biosynthesis hint that the half-life of this enzyme in regenerating rat liver (34, 39) and in Morris rat hepatoma 7800 (40) is less than 20 min, which is considerably less than the half-life of any other animal enzyme which has up to now been examined in this regard. But it should be emphasized that conclusions as to the half-lives of mammalian L-ornithine decarboxylases have been based solely on measurements of enzyme *activities*; more definitive experiments (e.g., involving assay of actual enzyme protein levels by suitable radioimmunological procedures) to prove that the coarse control of L-ornithine decarboxylase activities in mammalian tissues genuinely operates via alternations in the rates of synthesis of the enzyme protein have yet to be carried out.

Enormous increases, sometimes as much as fiftyfold or even greater, in the activity of L-ornithine decarboxylases occur in a variety of animal tissues as concomitants of early states of induced growth or as a result of hormonal stimulations. This is the case, for example, with respect to androgen-induced growth of the ventral prostate in orchiectomized or hypophysecto-mized rats (3, 41), regenerating rodent liver (28–31), cultures of embryonic cells after simulation by epidermal growth factor (32), rodent ovary after treatment with luteinizing hormone (35), chick oviduct or the uterus of castrated or hypophysectomized rats after injection of estrogens (36), rat liver after thioacetamide administration (42), and hypophysectomized rat liver as a result of treatment with growth hormone (33). Such elevations in tissue L-ornithine decarboxylase activities not only are impressive in magnitude, but in addition usually occur swiftly after application of the growth stimuli. They can be prevented by administration of inhibitors of protein biosynthesis. Nevertheless, as is underscored below, heightened L-ornithine decarboxylase neither is an invariable correlate of rapid growth nor is it always associated with large elevations in the steady-state concentration of putrescine in tissues, the values of which, in mammalian cells, are nearly always much lower than those of spermidine and spermine (see below). Cohen et al. (36) made the important observation that, after exposure of immature chick oviducts grown in organ culture to the estrogen diethylstilbestrol for 3.5 hr at 37°, there was a fifteenfold increase in the L-ornithine decarboxylase activity of extracts of the explants.

4. ENZYMIC DECARBOXYLATION OF *S*-ADENOSYL-L-METHIONINE IN EUKARYOTIC ORGANISMS

Pegg and Williams-Ashman (21, 43) were the first to show that particle-free supernatant fluids obtained by ultracentrifugation of homogenates of various rat tissues catalyzed decarboxylation of *S*-Ado-met-[14]COOH (reaction 3). The release of carbon dioxide from *S*-Ado-met was almost abolished by dialysis of the soluble tissue extracts, but could be restored completely by addition of saturating concentrations (1 mM) of putrescine. This stimulatory effect of putrescine on mammalian *S*-Ado-met decarboxylase was highly specific. Of a large number of related aliphatic amines tested (including the next lower and higher homologs of putrescine, namely, 1,3-diaminopropane and cadaverine), only spermidine appeared to have any activity, and at saturating concentrations (3 mM) spermidine induced a considerably lower rate of decarboxylation of *S*-Ado-met than did putrescine at pH values near to neutrality (21, 44).* Unlike the corresponding *E. coli* enzyme, the *S*-Ado-met decarboxylase of animal tissues does not require Mg^{2+} or any other metal-ion cofactor. The activity of this enzyme in the ventral prostate gland was much higher than that of a large number of other embryonic and

---

* A very recent reexamination of the specificity of activation of rat prostate S-Ado-met decarboxylase by aliphatic diamines revealed that high concentrations of cadaverine and 1,3-diaminopropane enhance the reaction to some extent (92).

adult rat tissues that were examined (21). *S*-Ado-met decarboxylase activities in the ventral prostates of rats decrease greatly in adult animals after castration or hypophysectomy, and are rapidly elevated after injection of testosterone into the androgen-deficient creatures (3, 41).

The *S*-Ado-met decarboxylase of rat ventral prostate was recently purified many hundredfold by Jänne and Williams-Ashman (44). The purification procedure permitted complete separation of this *S*-Ado-met decarboxylase from other soluble prostatic enzymes capable of synthesizing spermidine (in the presence of putrescine) or spermine (with spermidine added) from either *S*-Ado-met or from added decarboxylated *S*-Ado-met (44, 45). Even after elimination of all of the latter propylamino group transferase activities, the purified prostatic *S*-Ado-met decarboxylase is still activated manyfold by putrescine, and to a smaller degree by spermidine.

Activation of mammalian *S*-Ado-met decarboxylases by putrescine depends critically on a number of factors (3, 21, 22, 44). It is increased if the proton concentration of the assay medium is raised over the pH range of 5.8–8.7, or if the temperature at which the reaction is conducted is elevated over the range of 10–45° (44, 46). In the presence of saturating quantities of putrescine the *S*-Ado-met decarboxylase of rat prostate obeys classical Michaelis-Menten kinetics near neutral pH with respect to *S*-Ado-met concentrations, the K_m for *S*-Ado-met under these circumstances being about 0.05 mM at 37° (21, 44). In the absence of putrescine, however, the K_m for *S*-Ado-met is considerably higher than the latter value, and the kinetics of decarboxylation as a function of initial *S*-Ado-met concentrations presents a more complex picture (44, 47). A possible explanation for these findings is that prostatic *S*-Ado-met decarboxylase is an allosteric enzyme with at least two distinct binding sites, one of which interacts with the *S*-Ado-met substrate, and the other of which, when occupied by the activator putrescine, possibly allows the protein to fold into configuration(s) that permit more vigorous catalytic activity than would otherwise take place in the absence of activating aliphatic amines. Alternatively, mammalian *S*-Ado-met decarboxylase could conceivably be composed of separate and dissimilar catalytic and regulatory subunits, and activators like putrescine might possibly overcome putative inhibitory actions of the regulatory subunit.

Sucrose gradient ultracentrifugation and gel filtration experiments indicate that rat ventral prostate *S*-Ado-met decarboxylase behaves as if its molecular weight, in the presence of putrescine, is in the same range as that of the L-ornithine decarboxylase of this tissue (47).

The nature of the prosthetic group of mammalian *S*-Ado-met decarboxylases remains uncertain. The prostatic enzyme is strongly inhibited by certain pyridoxal phosphate antagonists, notably 4-bromo-3-hydroxy-benzyloxyamine (NSD-1055). Depression of the decarboxylase reaction in the presence of putrescine by NSD-1055 (which does not inhibit the *E. coli S*-Ado-met decarboxylase) can be reversed by addition of pyridoxal phosphate, but not by free pyridoxal (21, 43). These facts are consistent with the view that mammalian *S*-Ado-met decarboxylases are pyridoxal phosphate-requiring enzymes. It has not been possible, however, to resolve the prostatic enzyme from any bound functional pyridoxal phosphate in the absence of inhibitors such as NSD-1055, and even the most highly purified preparations do not have their activity enhanced by exogenous pyridoxal phosphate, provided that due corrections are applied for the considerable nonenzymic decarboxylation of *S*-Ado-met by pyridoxal phosphate, a reaction that is enhanced by Mn^{2+} and certain other metal ions in phosphate buffers at pH 7 (48). Rat ventral prostate *S*-Ado-met decarboxylase is inhibited by incubation with sodium borohydride in the absence of *S*-Ado-met (47), and the possibility that this enzyme, like its *E. coli* counterpart (20), contains protein-bound pyruvate (or some other keto acid) as a prosthetic group, rather than pyridoxal phosphate, should still be borne in mind.

Activation of carbon dioxide release from *S*-Ado-met by putrescine has been observed with extracts of amphibian, reptilian, avian, and crustacean tissues, as well as of mammalian organs (22). Similarly, baker's yeast (*Saccharomyces cerevisiae*) contains an *S*-Ado-met decarboxylase which, like the mammalian but in contrast to the bacterial enzyme, is enhanced by putrescine and does not require any metal cofactor (22, 49). The *S*-Ado-met decarboxylase of baker's yeast was recently purified to a high degree (49) and thereby largely separated from the spermidine synthase of the same organism. Table 1 compares and contrasts the properties of *S*-Ado-met decarboxylase from *E. coli*, rat ventral prostate, and baker's yeast. These characteristics may suggest, at first sight, that activation by putrescine and lack of stimulation by metal ions are features of the enzyme from eukaryotic cells, since the enzyme from *E. coli* and also *A. vinelandii* displays the converse properties. Recent studies on the decarboxylation of *S*-Ado-met by extracts of mung bean shoots, however, make this hypothesis untenable, because this higher plant enzyme is not activated at all by putrescine but does require Mg^{2+} (22). During the latter studies, it was noticed (48) that the horse radish peroxidase-catalyzed decarboxylation of methionine in the presence of Mn^{2+} and pyridoxal phosphate, which was originally discovered by Mazelis (50), provides a model for an analogous reaction with *S*-Ado-met as substrate. This underscores the need for caution in the interpretation of experiments on the release of carbon dioxide from *S*-Ado-met by extracts of plant or other tissues that are rich in peroxidases.

Table 1.   Comparison of the Properties of S-adenosyl-L-methionine Decarboxylases from E. Coli, Rat Ventral Prostate, and Baker's Yeast

Characteristic	Enzyme Source		
	E. coli	Prostate	Yeast
Activation by Mg^{2+}	+	−	−
Activation by putrescine	−	+	+
Inhibition by 4-bromo-3-hydroxybenzyloxyamine	−	+	Not determined
Inhibition by $NaBH_4$	+	+	+
Prosthetic group	Enzyme-bound pyruvate	Unknown	Unknown
Requirement for sulfhydryl groups	+	+	+
K_m for S-Ado-met	0.09 mM	0.05 mM (putrescine present)	0.06 mM (putrescine present)
Inhibition by decarboxylated S-Ado-met	Weak	Strong	Strong
Molecular weight	113,000	ca. 70,000	Not determined

Citations to the relevant publications are provided in the text.

5.   ENZYMIC SYNTHESIS OF SPERMIDINE IN EUKARYOTIC CELLS

Enzymes catalyzing the formation of sperimidine from decarboxylated S-Ado-met and putrescine (reaction 4) have been described in rat ventral prostate (21, 45), liver (39, 39a), and brain (39b), as well as in baker's yeast (49) and E. coli (23). Initial attempts (21) to separate this spermidine synthase from S-Ado-met decarboxylase in rat ventral prostate were unfruitful. More recently, however, prostatic spermidine synthase was purified about 600-fold and completely separated from the decarboxylase (47). By adding highly refined preparations of prostatic spermidine synthase to S-Ado-met decarboxylase purified from the same tissue, stochiometric coupling between carbon dioxide release and spermidine synthesis, which is readily demontrable with crude prostatic extracts (21) but not with each purified enzyme alone, can be re-established (45). The facile coupling between the prostatic S-Ado-met decarboxylase and spermidine synthase reactions, resulting in the equimolar production of spermidine, CO_2, and MTA without accumulation of significant quantities of decarboxylated S-Ado-met, is made possible by the extremely low K_m for decarboxylated S-Ado-met (*ca.* 10 μM) for the prostatic spermidine synthase (47). Both the prostatic and yeast spermidine synthases are inhibited by high concentrations of their substrate (decarboxylated S-Ado-met), the extent of this inhibition of the prostatic enzyme being critically dependent on the ambient levels of the other substrate, putrescine (47). As yet, no dissociable or bound cofactors have been shown to be necessary for the action of spermidine synthases from any biological source.

6. SPERMINE SYNTHASE

Animal tissues (39, 39b, 51) and baker's yeast (47) contain enzymes (spermine synthases) that catalyze the formation of spermine from decarboxylated S-Ado-met and spermidine:

$$\text{Decarboxylated } S\text{-Ado-met} + \text{spermidine} \rightarrow \text{spermine} + \text{MTA} + \text{H}^+. \quad (6)$$

Such a reaction has been said not to occur in *E. coli* (23). Mammalian spermidine and spermine synthase activities can be partially separated from one another (39, 39a, 39b, 45), contrary to statements in the first account of the discovery of the spermine synthase reactions (21, 51). A unique property of the spermine synthase reaction in rat prostate (51) and liver (39) is that it is powerfully inhibited by putrescine; this inhibition is competitive with respect to the spermidine substrate, and the K_i for putrescine near pH 7 is about 0.1 mM for the prostatic enzyme (51). The K_m for spermidine in the rat ventral prostate spermine synthase reaction is about 1 mM, and the approximate K_m for decarboxylated S-Ado-met is roughly 0.025 mM (51). No prosthetic group or dissociable cofactor has been found to be necessary for the spermine synthase reaction.

7. METABOLISM OF 5′-METHYLTHIOADENOSINE IN ANIMAL TISSUES

One molecule of MTA is formed in both the spermidine synthase (reaction 4) and the spermine synthase (reaction 6) equations. Although MTA may accumulate in certain microorganisms under some circumstances (52), this nucleoside does not appear to be present in more than trace amounts in higher animal cells (53–55). The ventral prostate gland of adult rats, for example, which contains 5–7 μmoles each of spermidine and spermine and around 0.2 μmole of putrescine per gram fresh weight of tissue (41), holds less than 0.2 μmole of MTA/g (55). From the standpoint of metabolic economy, any piling up of MTA in cells as a result of polyamine formation would probably be deleterious, as it would deplete the pool of adenine nucleotides, since S-Ado-met and its decarboxylated derivative are ultimately produced from ATP.

A number of MTA-degrading enzymes are found in microorganisms, (52, 54). Pegg and Williams-Ashman (54) found that homogenates of rat liver and ventral prostate readily destroy added MTA. One of the principal enzymes involved in this process is localized in the cytosol and catalyzes the release of free adenine from MTA. With preparations of the prostate enzyme that were purified about thirtyfold, it was observed (54) that the reaction was almost completely dependent on P_i. The requirement for inorganic orthophosphate could be satisfied to some extent by inorganic arsenate, but not by other anions such as citrate, maleate, or sulfate. This suggested that the enzyme promoted a phosphorolytic cleavage of MTA at the adenine-ribose linkage according to the following equation:

$$5'\text{-Methylthioadenosine} + P_i \rightarrow \text{adenine} + 5\text{-methylthioribose-1-P.} \quad (7)$$

The optimum pH of this reaction is close to 7.5. It does not require any metal ions, and the K_m for MTA is around 0.6 mM. Evidence obtained with partially purified prostatic enzymes indicated that, in addition to adenine (which was identified rigorously) the other product of the reaction was a sugar with the chromatographic properties of 5-methylribose. These enzyme preparations contained phosphatases, and it seems reasonable to assume that the latter enzymes cleaved the expected immediate product of the MTA-degrading enzyme, that is, 5-methylribose-1-P. The prostatic enzyme that phosphorolytically cleaves MTA is clearly not identical with a number of known purine nucleoside phosphorylases of animal origin (54), and neither adenosine deaminase nor 5'-AMP deaminase attacks MTA (54).

The MTA-cleaving enzyme of animal tissues provides a mechanism for the rescue of adenine from MTA produced as a result of polyamine biosynthesis, so that the adenine can be reutilized in cellular metabolism. When assayed under optimal conditions, the rate of enzymic decomposition of MTA (reaction 7) by crude extracts of rat ventral prostate is about 25 μmoles MTA destroyed per gram fresh weight of tissue per hour at 37°; this is orders of magnitude faster than the activity of enzymes catalyzing the synthesis of putrescine, spermidine, and spermine by this organ (41, 54). Worthy of note here is a recent report (56) that MTA is a potent inhibitor of certain enzymic transmethylations that utilize S-Ado-met as a methyl group donor; such inhibitions could be overcome in living cells by the action of the MTA-cleaving enzyme. The biological fate of the 5-methylthioribose-1-P product of the latter reaction remains completely unknown and is certainly deserving of experimental scrutiny. Of course, the possibility that other enzymic pathways for the degradation of MTA may operate in animal cells (e.g., pyrophosphorolytic rather than phosphorolytic breakdown of MTA at the adenine-ribose bond) should not be ignored.

8. METHIONINE ANALOGS AND THE BIOSYNTHESIS OF SPERMIDINE AND SPERMINE

Certain congeners of L-methionine are well known to be substrates for methionine adenosyltransferases, thus giving rise to the corresponding analogs of S-Ado-met (57–60). The chemical structures of some of these analogs in comparison with those of S-Ado-met and decarboxylated S-Ado-met are depicted in Fig. 3. One of the most thoroughly studied of the analogs

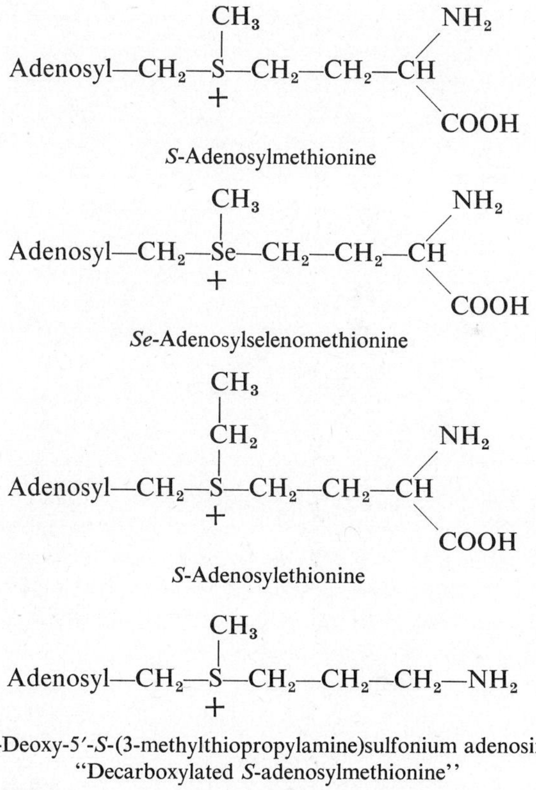

$$\underset{\substack{|\\+}}{\text{Adenosyl—CH}_2\text{—S—CH}_2\text{—CH}_2\text{—CH}} \overset{\text{CH}_3}{\underset{\text{COOH}}{\overset{\text{NH}_2}{<}}}$$

S-Adenosylmethionine

$$\text{Adenosyl—CH}_2\text{—Se—CH}_2\text{—CH}_2\text{—CH}$$

Se-Adenosylselenomethionine

$$\text{Adenosyl—CH}_2\text{—S—CH}_2\text{—CH}_2\text{—CH}$$

S-Adenosylethionine

$$\text{Adenosyl—CH}_2\text{—S—CH}_2\text{—CH}_2\text{—CH}_2\text{—NH}_2$$

5'-Deoxy-5'-S-(3-methylthiopropylamine)sulfonium adenosine
"Decarboxylated S-adenosylmethionine"

Fig. 3. Chemical structure of enzymically synthesized analogs of S-adenosyl-methionine in comparison with the formula of decarboxylated S-adenosyl-methionine.

of L-methionine that react with methionine adenosyltransferases is the substance ethionine, in which a methyl group attached to the sulfur atom of methionine is replaced by an ethyl moiety. Administration of ethionine can induce marked alterations in various hepatic metabolic processes and may also lead to the development of cancers in some species after prolonged feeding (57, 61). Many of the toxic actions of ethionine are believed to be due to the enzymic synthesis and accumulation of S-adenosylethionine (S-Ado-et) in tissues. Although S-Ado-et can ethylate certain cellular constituents as a result of its utilization by some transmethylating enzymes, the rates of these reactions are often much slower with the ethyl analog than with the natural S-Ado-met substrate, so that large amounts of S-Ado-et accumulate in tissues, leading, among other things, to a severe depletion of total tissue ATP levels (57, 61).

The central precursor functions of S-Ado-met in polyamine biosynthesis made it of interest to examine possible interactions of S-Ado-et with enzymes responsible for the formation of spermidine and spermine in mammalian tissues. Pegg (62, 63) showed with prostate enzyme preparations that S-ado-et is only slowly decarboxylated, although carbon dioxide release was enhanced by putrescine as a function of pH in a manner similar to that observed with S-Ado-met. Moreover, the rates of spermidine production from S-Ado-et and putrescine, and of spermine formation from S-Ado-et and spermidine, were very much lower than those observed with S-Ado-met as the ultimate propylamino group donor (51, 62). Not only is S-Ado-et but a feeble precursor for the biosynthesis of spermidine and spermine; in addition, it inhibits the enzymic production of spermidine from S-Ado-met and putrescine in the prostate (62). These findings suggest that administration of ethionine may cause a decline in tissue polyamine levels, and this has indeed been observed in liver (2, 64). On continued treatment with ethionine, however, hepatic spermidine levels gradually return to normal. It seems improbable that the minuscule rates of spermidine synthesis from S-Ado-et could account for the latter phenomenon, and alternative pathways for polyamine biosynthesis (see below) may conceivably come into play under these circumstances.

Another analog of methionine that exerts striking biological actions is selenomethionine, in which the sulfur atom is replaced by selenium (Fig. 3). Selenomethionine is known to accumulate in some plants grown on selenium-rich soils, and it is extremely toxic to certain higher animals (59). Selenomethionine and the corresponding selenoethionine are converted into analogs of S-Ado-met by mammalian methionine adenosyltransferases (Fig. 3). As a methyl group donor, Se-Adenosylselenomethionine is almost as active as S-Ado-met itself in a number of biological methyltransferase reactions (58, 59). Pegg (62, 63) found that Se-adenosylselenomethionine was almost as active a precursor for spermidine synthesis by prostatic enzyme preparations as was S-Ado-met. In contrast, Se-adenosylseleneoethionine was only very feebly active as a propylamino group donor for the formation of spermidine in rat ventral prostate. This is comparable to the very poor precursor function of S-Ado-et in relation to S-Ado-met for spermidine synthesis by the same enzyme preparations; however, the actual decarboxylation of the two selenium-containing compounds has not yet been examined. (It would seem unlikely that substitution of an ethyl group for the methyl attached to the sulforium pole of S-Ado-met would alter greatly the "bond energy" of the link between the sulfur atom and the C-4 methylene group derived from methionine or ethionine in the decarboxylated derivatives of S-Ado-met or S-Ado-et.)

Pegg's experiments (62, 63) suggest that, although Se-adenosylselenomethionine produced from selenomethionine in living cells may not directly disturb the synthesis of polyamines, 5'-methylselenoadenosine might be expected to be formed in tissues as a result of Se-adenosylselenomethionine serving as a precursor for spermidine. Thus it can be imagined that 5'-methylselenoadenosine and/or its further enzymic degradation products might be responsible for at least some of the toxic actions of selenomethionine (59). Other biochemical reactions undergone by methionine analogs such as ethionine and selenomethionine (e.g., their possible activation by tRNA-aminoacyl synthetases and their subsequent incorporation into proteins by polyribosomes) may, of course, also contribute to the toxicity of these methionine congeners to eukaryotic cells (57, 59, 61).

9. THE PARAMOUNT ROLES OF PUTRESCINE IN POLYAMINE BIOSYNTHESIS

The foregoing considerations point to a key role of putrescine as a regulator as well as a substrate for the biosynthesis of spermidine and spermine in eukaryotic cells. Starting with L-ornithine and S-Ado-met as the ultimate sources of the carbon, hydrogen, and nitrogen atoms of spermine, it is evident that putrescine (a) is both the product and a weak competitive inhibitor of reaction 1 (L-ornithine decarboxylase); (b) functions as an (allosteric?) activator of S-Ado-met decarboxylase (reaction 3); (c) is the natural substrate for the spermidine synthase (reaction 4); and (d) acts as a powerful inhibitor of the enzymic conversion of spermidine into spermine (reaction 6). These multifarious functions of putrescine are in accord with the observation of Raina et al. (39) that very low levels (0.02 mM) of putrescine evoke small increases in the formation of spermine from S-Ado-met but not from added decarboxylated S-Ado-met as catalyzed by crude soluble extracts of regenerating liver, whereas, if the concentration of putrescine is progressively raised above a value of 0.1 mM, the rate of synthesis of spermidine continues to increase while the formation of spermine becomes more and more inhibited.

It seems safe to predict that the putrescine/spermidine ratio may be crucial, in living animal cells, in determining the extent of spermine formation, because putrescine enhances the decarboxylation of S-Ado-met but inhibits the spermine synthase reaction. In many resting adult mammalian tissues, the steady-state concentration of putrescine is considerably less than 0.1 μmole/g wet weight, either in tissues that exhibit the usual range of levels of spermidine and spermine (0.2–1.5 μmoles/g), or in organs where the polyamine concentrations may be much higher, such as the rat ventral prostate [in the latter case, a considerable proportion of the total spermidine and spermine may be present in prostatic secretions stored extracellularly in the lumina of the gland (41)]). However, in the early phases of the induced growth of some tissues, such as regenerating liver prompted by partial hepatectomy (64–69), or androgen-induced ventral prostate growth in castrated rats (41), there occur transient but significant rises in the overall concentrations of putrescine that more or less parallel the heightened L-ornithine decarboxylase activities. In both of the latter situations, tissue spermidine levels also increase fairly quickly, in concert with elevations in the total (largely ribosomal) RNA. Nevertheless, under the same conditions, in regenerating liver the tissue spermine concentrations either remain constant or even decline slightly (39, 67, 68), and in the prostate only start to become elevated many days after the first signs of enhancement of spermidine levels (41). In the latter organ, the pronounced delay in any rise in the spermine/spermidine ratio after daily treatment of castrated rats with testosterone correlates with both the earlier increase and the subsequent return toward the norm of putrescine concentrations (41) and the depression of the enzymic transformation of spermidine into spermine by putrescine (51).

10. POSSIBLE ALTERNATIVE PATHWAYS FOR POLYAMINE BIOSYNTHESIS IN EUKARYOTIC CELLS

Obviously the four linked enzyme reactions now shown to result in the formation of spermidine and spermine in animal tissues may not represent the only pathway for polyamine biosynthesis in eukaryotic organisms. It has been pointed out that the "agmatine pathways" for the production of putrescine in certain bacteria and higher plants do not seem to occur in animal tissures. It may be profitable, nevertheless, to keep a lookout for alternative routes for the biosynthesis of spermidine and/or spermine in eukaryotes.

Experiments suggesting that certain mammalian species can convert labeled spermine into spermidine (69) prompted a guess (3) that animal tissues could conceivably contain an enzyme that catalyzes the reversible dismutation:

$$2 \text{ Spermidine} \rightleftharpoons \text{spermine} + \text{putrescine}. \tag{8}$$

No evidence has yet been forthcoming that this reaction takes place in eukaryotic cells, either as written or with the participation of some propylamino group carrier of low molecular weight (3). This hypothetical reaction has sufficient intellectual attraction, however, that it may be wise to continue to search for its occurrence in animals. Apart from providing an alternative pathway for spermine formation and, additionally, a mechanism for the conversion (via the back reaction) of spermine to spermidine, this putative dismutation could conceivably function, at the border of secretory cells, as a device for active extrusion of spermine with retention of putrescine intracellularly, with the utilization of spermidine. Noteworthy in this context is the secretion of large amounts of spermine, but very little spermidine or putrescine, by the human (1, 5, 6) and the rhesus monkey (48) prostate gland.

Another hypothesis that stems from considerations of known enzymic mechanisms for polyamine biosynthesis may be raised here. It is well established that S-Ado-met is a methyl group donor for many enzymic transmethylations involving as acceptors not only countless small molecules but also various macromolecules such as tRNAs, ribosomal RNAs, DNA, and certain proteins (70). On analogical grounds, it is possible that decarboxylated S-Ado-met could donate its "activated" propylamine group not only to putrescine and spermidine as in the biosynthesis of polyamines, but also to various acceptor residues in nucleic acids or proteins?

11. TURNOVER AND DEGRADATION OF POLYAMINES IN ANIMAL TISSUES

Putrescine is degraded at the expense of molecular oxygen by diamine oxidases found in mammalian kidney and in the placenta and blood serum of pregnant females (1, 3, 71, 72). And spermidine and spermine are oxidized by other, specific oxidases present in the blood plasma of ruminants, as well as a few closely related species like the hippopotamus and the tree hyrax (1, 3, 73, 74). These ruminant plasma oxidases convert spermidine and spermine into primary aldehyde products as follows:

$$H_2N(CH_2)_3NH(CH_2)_4NH_2 + H_2O + O_2 \rightarrow$$
$$OHC(CH_2)_2NH(CH_2)_4NH_2 + NH_3 + H_2O_2, \quad (9)$$

$$H_2N(CH_2)_3NH(CH_2)_4NH(CH_2)_3NH_2 + 2H_2O + 2O_2 \rightarrow$$
$$OHC(CH_2)_2NH(CH_2)_4NH(CH_2)_2CHO + 2NH_3 + 2H_2O_2. \quad (10)$$

Tabor et al. (75) and Bachrach (76) consider the chemistry of the immediate aldehyde products of these reactions, which can, in turn, undergo all sorts of secondary nonenzymic transformations, including the formation of cyclized derivatives, of acrolein ($CH_2{=}CHCHO$), and other compounds. The aldehydes produced by the action of ruminant plasma oxidases on spermidine and spermine and/or derivatives thereof such as acrolein can interact with nucleic acids, proteins, viruses, and so forth, and thereby elicit some profound biological effects (76). Human seminal plasma is also said to oxidize the spermine in this fluid (77, 78). However, whether comparable oxidizing enzymes that attack spermidine and spermine are of widespread occurrence inside various differentiated animal cells is still a moot point. Caldarera's group (79) has documented evidence for the presence of an aliphatic polyamine-oxidizing enzyme system in the brain of developing chick embryos. Bachrach (76) has advanced the interesting hypothesis that the aldehyde product of the oxidation of spermidine (reaction 9) by such a brain enzyme might undergo further oxidation to form the substance putreanine:

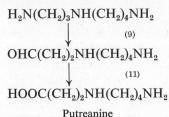

Putreanine

Putreanine [i.e., N-(4-aminobutyl)-3-aminopropionic acid] was shown by Kakimoto et al. (80) to be present in the central nervous systems, but not in many other organs, of a variety of avian and mammalian species. The neurophysiological functions, if any, of putreanine are obscure.

Whether spermidine and permine within animal cells can undergo any of the various conjugation reactions that occur with spermidine in certain bacteria (81, 82) is also unknown. Russell et al. (83) have interpreted their experiments to mean that spermidine turns over in regenerating rat liver with a half-life of about 4–5 days. Biochemical pathways for the disposal of spermidine and spermine in higher animal cells remain shrouded in mystery; recently the enzymic oxidation of polyamines has been demonstrated in some higher plants (19). In the prostate glands of some mammals, spermine is removed via its secretion by the epithelial cells (3, 41).

12. POLYAMINE BIOSYNTHESIS IN MALIGNANT TISSUES

Apart from a few sporadic observations on the spermidine and spermine content (1, 2, 84–87), and on the activity of L-ornithine decarboxylase (29, 85, 86) and S-Ado-met decarboxylase (86) of certain cancerous tissues, hardly any studies have been carried out in a systematic fashion on the synthesis, turnover, and possible functions of polyamines in malignant cells.

In collaboration with Professor George Weber there was recently conducted a comprehensive survey (40) of the concentrations of aliphatic amines and of the activities of certain polyamine biosynthetic enzymes in a spectrum of Morris rat hepatomes of widely different growth rates. The following conclusions were reached:

1. In comparison with levels in livers from animals of the same age, sex, and dietary regimen, the L-ornithine decarboxylase of fast-growing hepatomas (e.g., lines 3924A and 7777) was extremely active (as much as fifteenfold greater than in the control liver specimens), whereas a number of slow-growing and more differentiated hepatomas exhibited L-ornithine decarboxylase activities that were only slightly higher than that of normal rat liver.

2. None of the hepatomas examined, regardless of its speed of growth, had significantly elevated S-Ado-met decarboxylase activities (release of carbon dioxide from S-Ado-met catalyzed by all of the Morris hepatoma extracts was enhanced manyfold by putrescine, as in normal liver).

3. The steady-state concentrations of putrescine in some but not all of the fast-growing neoplasms (e.g., 0.25–0.45 μmole of putrescine per gram fresh weight) were much greater than the level in the normal control livers (less than 0.05 μmole/g of putrescine). The putrescine content of all slow-growing hepatomas was only slightly elevated.

4. There was no simple relationship between the L-ornithine decarboxylase activities of the various Morris hepatomas and their putrescine contents.

5. The spermidine/spermine ratio tended to be higher in all of the hepatomas studied than in the corresponding normal livers; the overall concentrations of these two polyamines in all tumors were, however, within the range seen in the normal livers (0.6–1.5 μmoles/g).

6. Experiments involving administration of the protein synthesis inhibitor cycloheximide indicated that the half-lives of both L-ornithine and S-Ado-met decarboxylases in Morris hepatoma 7800 were less than 30 min.

These results indicate that enhanced growth rates, coupled with progressive morphological and biochemical dedifferentiations in a series of Morris rat hepatomas, are not accompanied by any decline in the formation of spermidine or spermine, or in the levels of two key decarboxylating enzymes required for their biosynthesis. In fact, in the more anaplastic neoplasms, which multiply very quickly, L-ornithine decarboxylase activities and the spermidine/spermine ratios are elevated. Two considerations are germane to these findings. First, recent studies in a number of laboratories (reviewed in ref. 70) have shown that the enzymic methylation of tRNAs and possibly of other types of nucleic acid molecules may be abnormally high in many malignant animal cells; these are reactions which, like the synthesis of spermine and spermidine, utilize S-Ado-met as a substrate. Second, the activity of L-ornithine carbamyltransferase—a key enzyme in the urea cycle, one of the specialized functions of adult mammalian liver—falls dramatically and as a function of increasing rates of growth and dedifferentiation in the same spectrum of Morris hepatomas that was employed in the aforementioned studies on polyamine biosynthesis (88).

13. PHARMACOLOGICAL PROSPECTIVES

Now that something has been learned about the properties of each of the four enzymes necessary for what is clearly a major if not the sole pathway for spermidine and spermine synthesis in eukaryotes, the way is open for the development, on at least semirational grounds, of specific competitive or active-site-directed (89) inhibitors of one or more of the intermediary reactions. Hopefully such inhibitors could perturb fairly specifically the patterns of polyamine production in various organs of living animals, and thus might be of value in deciding among the many current hypotheses regarding the intracellular functions of polyamines in various tissues (12a).* Perhaps it may not be out of place to close with a brief speculative consideration of which step or steps in the biosynthesis of spermidine and spermine in eukaryotic organisms are the most likely to be vulnerable to this sort of potential pharmacological interference.

S-Adenosyl-L-methionine is of central importance in the formation of polyamines. If the enzymic synthesis of S-Ado-met was diminished or prevented, the production of spermidine and spermine would obviously be

---

* The physiological value, if any, of the high concentrations of spermine present extracellularly in prostatic secretion, and hence in seminal plasma, of man and some other mammalian species remains problematical. Possible functions of seminal spermine in regard to the well-being and the transport of spermatozoa in the male and female genital tracts have been discussed by Williams-Ashman and Lockwood (91).

halted. A variety of inhibitors of methionine adenosyltransferase (reaction 2)—albeit none of them exceptionally potent—has been described (58, 60). Nevertheless, it seems most unlikely that selection of methione adenosyltransferase as a target for specific pharmacological inhibition is going to be fruitful if the aim is to obtain drugs that selectively depress polyamine biosynthesis or, for that matter, any other particular avenues of utilization of S-Ado-met, rather than substances of high toxicity that simply kill cells. The reason why this approach lacks promise is that S-Ado-met is an obligatory reactant in many vital biological transmethylations and also is essential for some other types of enzymic reactions.

Putrescine is a substrate (reaction 4), activator (reaction 3), and inhibitor (reaction 6) of enzymes in the polyamine biosynthetic sequence, so that any stopping of putrescine formation *in vivo* clearly would disturb the manufacture of spermidine and spermine. The only known enzyme that synthesizes putrescine in higher animal tissues is L-ornithine decarboxylase, whose substrate can be derived by the arginase-catalyzed splitting of L-arginine [it has been pointed out elsewhere (3) that the function of arginase known to be present in many tissues that do not carry out the urea cycle may be to provide L-ornithine for polyamine biosynthesis]. But there are reasons to believe that neither L-ornithine decarboxylase nor arginase will be a suitable target for specific attack by drugs if the aim is to selectively depress polyamine biosynthesis. In tissues of central importance to the organism, such as liver and kidney, arginase is an essential component of the urea cycle, the interruption of which would be catastrophic. To be sure, L-ornithine, unlike the twenty amino acids commonly found in proteins, is not cognate to any code word in the universal genetic dictionary, and tRNAs or tRNA-aminoacyl synthetases that react with L-ornithine have never been detected; it would thus appear that potential inhibitors of the utilization of L-ornithine for polyamine production would not suffer from the disadvantage that they would specifically interfere with key reactions in protein biosynthesis. However, L-ornithine as a substrate for L-ornithine carbamyltransferase also plays a central role in the urea cycle, and additionally is a substrate for another mitochondrial enzyme, L-ornithine-keto acid aminotransferase. Cogeners of L-ornithine may therefore have relatively little chance of turning out to be specific inhibitors of L-ornithine decarboxylase, and thus of interfering in a truly selective fashion with the biosynthesis of spermidine or spermine.

At the moment of writing, interference with the functions of putrescine as an activator of S-Ado-met decarboxylase (reaction 3), as a substrate for spermidine synthase (reaction 4), and as an inhibitor of spermine synthase (reaction 6) appears to offer more promise for the development of drugs that might specifically interfere with polyamine formation in animal tissues. The procedures developed by Israel et al. (90) for the chemical synthesis of aliphatic amines may prove most useful in producing analogs of putrescine that may be active in the aforementioned respects. Putrescine cogeners that could interact with the sites on mammalian S-Ado-met decarboxylases responsible for activation of the enzyme by putrescine might prove to be especially interesting from a pharmacological standpoint, as specific enhancement of this reaction by substances that might not serve as substrates for the spermidine or spermine synthase reactions, as well as selective inhibition of S-Ado-met decarboxylase, would perhaps in living cells provide novel insight into the normal operation of polyamine biosynthetic enzymes. It is of interest in this respect that methyl glyoxal bis(guanylhydrazone) was recently shown (92) to be an exceptionally potent inhibitor of putrescine-activated S-Ado-met decarboxylases from yeast and mammalian tissues.

ACKNOWLEDGMENTS

Studies from the author's laboratory were supported in part by a Research Grant HD-04592 from the U.S. Public Health Service. These investigations were made possible by the excellent collaborations of Drs. James B. Rhodes, Anthony E. Pegg, Dean H. Lockwood, Juhani Jänne, Gordon L. Coppoc, and Mrs. Mary E. Geroch.

REFERENCES

1. H. Tabor and C. W. Tabor, *Pharmacol. Rev.*, **16**, 245 (1964).

2. J. Jänne, *Acta Physiol. Scand.*, Suppl. 300, 1 (1967).

3. H. G. Williams-Ashman, A. E. Pegg, and D. H. Lockwood, *Advan. Enzyme Regulation*, **7**, 291 (1969).

4. L. Stevens, *Biol. Rev.*, **45**, 1 (1970).

5. T. Mann, *The Biochemistry of Semen and of the Male Reproductive Tract*, 2nd ed., John Wiley, New York, 1964.

6. H. G. Williams-Ashman, *Invest. Urol.*, **2**, 605 (1965).

7. G. L. Coppoc and P. Kallio, personal communication.

8. S. S. Cohen and A. Raina, in *Organizational Biosynthesis* (Eds.: H. J. Vogel, J. O. Lampen, and V. Bryson), Academic Press, New York, pp. 157–182.

9. S. S. Cohen, *Ann. N.Y. Acad. Sci.*, **171**, 869 (1970).

10. M. Inouye and A. B. Pardee, *Ann. N.Y. Acad. Sci.*, **171**, 901 (1970).

11. P. S. Leboy, *Ann. N.Y. Acad. Sci.*, **171**, 895 (1970).

12. S. Silver, L. Wendt, P. Battacharyya, and R. S. Beauchamp, *Ann. N.Y. Acad. Sci.*, **171**, 838 (1970).

12a. S. S. Cohen, *Introduction to the Polyamines*, Prentice-Hall, New Jersey, 1971.

13. A. Raina, J. Jänne, and M. Siimes, *Biochim. Biophys. Acta*, **123**, 197 (1966).

14. W. G. Dykstra, Jr. and E. J. Herbst, *Science*, **149,** 428 (1965).

15. H. Tabor, S. M. Rosenthal, and C. W. Tabor, *J. Biol. Chem.*, **233,** 907 (1958).

16. E. F. Gale, *Biochem. J.*, **34,** 392 (1940).

17. D. R. Morris and A. B. Pardee, *Biochem. Biophys. Res. Commun.*, **20,** 697 (1965).

18. D. R. Morris and A. B. Pardee, *J. Biol. Chem.*, **241,** 3129 (1966).

19. T. A. Smith, *Biol. Rev.*, **46,** 201 (1971).

20. R. B. Wickner, C. W. Tabor, and H. Tabor, *J. Biol. Chem.*, **245,** 2132 (1970).

21. A. E. Pegg and H. G. Williams-Ashman, *J. Biol. Chem.*, **244,** 682 (1969).

22. G. L. Coppoc, P. Kallio, and H. G. Williams-Ashman, *Int. J. Biochem.* **2,** 673 (1971).

23. C. W. Tabor, *Methods Enzymol.*, **5,** 761 (1962).

24. D. R. Morris, W. H. Wu, D. Applebaum, and K. L. Koffron, *Ann. N.Y. Acad. Sci.*, **171,** 968 (1970).

25. H. Tabor and C. W. Tabor, *J. Biol. Chem.*, **244,** 2286 (1969).

26. H. Tabor and C. W. Tabor, *J. Biol. Chem.*, **244,** 6383 (1969).

27. A. E. Pegg and H. G. Williams-Ashman, *Biochem. J.*, **108,** 533 (1968).

28. J. Jänne and A. Raina, *Acta Chem. Scand.*, **22,** 1349 (1968).

29. D. H. Russell and S. H. Snyder, *Proc. Natl. Acad. Sci. U.S.*, **60,** 1420 (1968).

30. N. Fausto, *Biochim. Biophys. Acta*, **190,** 1420 (1969).

31. T. R. Schrock, N. J. Oakman, and N. L. R. Bucher, *Biochim. Biophys. Acta*, **204,** 504 (1970).

32. M. Stastny and S. Cohen, *Biochim. Biophys. Acta*, **204,** 578 (1970).

33. J. Jänne and A. Raina, *Biochim. Biophys. Acta*, **174,** 769 (1969).

34. D. H. Russell and S. H. Snyder, *Mol. Pharmacol.*, **5,** 253 (1969).

35. Y. Kobayashi, J. Kupelian, and D. V. Maudsley, *Science*, **172,** 379 (1971).

36. S. Cohen, B. W. O'Malley, and M. Stastny, *Science*, **170,** 336 (1970).

37. J. Jänne and H. G. Williams-Ashman, *Biochem. J.*, **119,** 595 (1970).

38. J. Jänne and H. G. Williams-Ashman, *J. Biol. Chem.*, **246,** 1725 (1971).

39. A. Raina, J. Jänne, P. Hannonen, and E. Hölttä, *Ann. N.Y. Acad. Sci.*, **171,** 697 (1970).

39a. A. Raina and P. Hannonen, *Acta Chem. Scand.*, **34,** 3061 (1970).

39b. A. Raina and P. Hannonen, *FEBS Letters*, **16,** 1 (1971).

40. G. L. Coppoc, H. G. Williams-Ashman, and G. Weber, unpublished observations.

41. A. E. Pegg, D. H. Lockwood, and H. G. Williams-Ashman, *Biochem. J.*, **117,** 17 (1970).

42. N. Fausto, *Cancer Res.*, **30,** 1947 (1970).

43. A. E. Pegg and H. G. Williams-Ashman, *Biochem. Biophys. Res. Commun.*, **30,** 76 (1968).

44. J. Jänne and H. G. Williams-Ashman, *Biochem. Biophys. Res. Commun.*, **42,** 222 (1971).

45. J. Jänne, A. Schenone, and H. G. Williams-Ashman, *Biochem. Biophys. Res. Commun.* **42,** 758 (1971).

46. H. G. Williams-Ashman, J. Jänne, and G. L. Coppoc, *Federation Proc.*, **30,** 1114 Abs. (Abstr. 358) (1971).

47. J. Jänne, A. Schenone, and H. G. Williams-Ashman, unpublished observations.

48. G. L. Coppoc, P. Kallio, and H. G. Williams-Ashman, unpublished observations.

49. J. Jänne, H. G. Williams-Ashman, and A. Schenone, *Biochem. Biophys. Res. Commun.* **43**, 1362 (1971).

50. M. Mazelis, *J. Biol. Chem.*, **237**, 104 (1962).

51. A. E. Pegg and H. G. Williams-Ashman, *Arch. Biochem. Biophys.*, **137**, 156 (1970).

52. F. Schlenk, in *Transmethylation and Methionine Biosynthesis* (Eds.: S. K. Shapiro and F. Schlenk), University of Chicago Press, Chicago, 1965, pp. 48–65.

53. R. L. Smith, E. E. Anderson, R. N. Overland, and F. Schlenk, *Arch. Biochem. Biophys.*, **42**, 72 (1953).

54. A. E. Pegg and H. G. Williams-Ashman, *Biochem. J.*, **115**, 241 (1969).

55. J. B. Rhodes and H. G. Williams-Ashman, *Med. Exptl.*, **10**, 281 (1964).

56. V. Zappia, C. R. Zydek-Cwick, and F. Schlenk, *J. Biol. Chem.*, **244**, 4499 (1969).

57. E. Farber, *Advan. Cancer Res.*, **7**, 382 (1964).

58. J. A. Stekol, in *Transmethylation and Methionine Biosynthesis* (Eds.: S. K. Shapiro and F. Schlenk), University of Chicago Press, Chicago, 1965, pp. 231–248.

59. L. D. Fowden, D. Lewis, and H. Tristram, *Advan. Enzymol.*, **29**, 89 (1967).

60. J. B. Lombardini, A. W. Coulter, and P. Talalay, *Mol. Pharmacol.*, **6**, 481 (1970).

61. E. Farber, *Advan. Lipid Res.*, **5**, 119 (1967).

62. A. E. Pegg, *Biochim. Biophys. Acta*, **177**, 361 (1969).

63. A. E. Pegg, *Ann. N.Y. Acad. Sci.*, **171**, 977 (1970).

64. A. Raina, *Acta Physiol. Scand.*, 60, *Suppl.* **218**, 1 (1963).

65. A. Raina, J. Jänne, and M. Siimes, *Acta Chem. Scand.*, **18**, 1804 (1964).

66. A. Raina and J. Jänne, *Federation Proc.*, **29**, 1568 (1970).

67. O. Heby and L. Lewan, *Virchows Arch. Abt. B Zellpath.*, **8**, 58 (1971).

68. A. Raina and J. Jänne, *Acta Chem. Scand.*, **22**, 2375 (1968).

69. M. Siimes, *Acta Physiol. Scand.*, *Suppl.* **298**, 1 (1967).

70. P. N. Magee, *Cancer Res.*, **31**, 599 (1971).

71. E. A. Zeller, in *The Enzymes* (Eds.: P. D. Boyer, H. Lardy, and K. Myrbäck), Academic Press, New York, 1963, pp. 313–335.

72. H. Tabor, *J. Biol. Chem.*, **188**, 125 (1951).

73. H. Yamada and K. T. Kasunobu, *J. Biol. Chem.*, **237**, 1511 (1962).

74. H. Blaschko and R. Bonney, *Proc. Roy. Soc.*, Ser. B, **156**, 268 (1962).

75. C. W. Tabor, H. Tabor, and U. Bachrach, *J. Biol. Chem.*, **239**, 2194 (1964).

76. U. Bachrach, *Ann. N.Y. Acad. Sci.*, **171**, 939 (1970).

77. E. A. Zeller and C. A. Joël, *Helv. Chim. Acta*, **24**, 117 (1941).

78. E. A. Zeller and C. A. Joël, *Helv. Chim. Acta*, **24**, 968 (1941).

79. C. M. Caldarera, M. S. Moruzzi, C. Rossoni, and B. Barbiroli, *J. Neurochem.*, **16**, 309 (1969).

80. Y. Kakimoto, T. Nakajima, A. Kumon, Y. Matsuoka, N. Imaoka, I. Sano, and A. Kanazawa, *J. Biol. Chem.*, **244**, 6003 (1969).

81. C. W. Tabor, *Biochem. Biophys. Res. Commun.*, **30**, 339 (1968).

82. C. W. Tabor and H. Tabor, *Federation Proc.*, **30**, 1068 Abs. (Abst. 89) (1971).

83. D. H. Russell, V. J. Medina, and S. H. Snyder, *J. Biol. Chem.*, **245,** 6732 (1970).

84. L. T. Kremzner, R. E. Barrett, and M. J. Terrano, *Ann. N.Y. Acad. Sci.*, **171,** 735 (1970).

85. N. J. P. Neish and L. Key, *Comp. Biochem. Physiol.*, **27,** 1709 (1968).

86. D. H. Russell, *Cancer Res.*, **31,** 248 (1971).

87. M. Siimes and J. Jänne, *Acta Chem. Scand.*, **21,** 815 (1967).

88. G. Weber and S. F. Queener, personal communication.

89. B. R. Baker, *Design of Active-Site Directed Irreversible Enzyme Inhibitors*, John Wiley, New York, 1967.

90. M. Israel, J. S. Rosenfield, and E. J. Modest, *J. Med. Chem.*, **7,** 710 (1964).

91. H. G. Williams-Ashman and D. H. Lockwood, *Ann. N.Y. Acad. Sci.*, **171,** 882 (1970).

92. H. G. Williams-Ashman and A. Schlenone, *Biochem. Biophys. Res. Commun.*, **46,** 288 (1972).

CHAPTER 9

Regulation of Succinate Dehydrogenase in Mitochondria

THOMAS P. SINGER, EDNA B. KEARNEY, AND M. GUTMAN*

Molecular Biology Division, Veterans Administration Hospital, San Francisco, California, and Department of Biochemistry and Biophysics, University of California, San Francisco, California

* On leave of absence from Tel-Aviv University, Tel-Aviv, Israel.

1. INTRODUCTION

Succinate dehydrogenase is an ubiquitous enzyme, the properties of which reflect the metabolic needs of the cell (1). In aerobic cells the enzyme is designed to favor the oxidation of succinate to fumarate with concomitant energy conservation. The aerobic enzyme is regulable; it is always membrane-bound and, in all cases examined, contains FAD covalently linked to histidine. In anaerobic cells the properties of the enzyme also match the physiological needs of the organism: in such cases the reduction of fumarate to succinate is the favored process, providing a mechanism for the reoxidation of DPNH generated in glycolysis or fermentation (1). In the anaerobic enzyme FAD is noncovalently linked to the protein, and regulation appears to be absent. In facultative anerobes both types of enzyme may be present (2–4), and in such instances they are under separate genetic control.

Although the fact that the aerobic enzyme is activable (regulable) but the anaerobic type is not in any of the instances studied had been known for many years, its significance did not become clear until recently, when the existence of multiple control mechanisms for the aerobic enzyme and the rapid regulation of the activity of the enzyme by the metabolic state of the mitochondria were discovered (5–7). Until now, in fact, with few exceptions, articles dealing with metabolic regulation in mitochondria did not dwell on the control of succinate dehydrogenase activity, perhaps because this enzyme is not a pacemaker in the Krebs cycle. It is very likely, however, that rapid and extensive modulation of succinate dehydrogenase activity is a requisite for efficient energy conservation.

This chapter maps the tortuous path that research has taken to the point where the multiple types of control of succinate dehydrogenase were recognized, and ends with a hypothesis to account for the existence of the complex regulatory mechanisms that operate on this enzyme.

2. MODULATION OF SUCCINATE DEHYDROGENASE BY SUBSTRATES AND COMPETITIVE INHIBITORS

In 1955 Kearney et al. (8) reported that inorganic phosphate potentiates the activity of mammalian succinate dehydrogenase. A number of other anions tested failed to produce activation, and arsenate competitively inhibited activation by phosphate. Activation by P_i is most readily observed if the assay is carried out at low temperature (e.g., 15°; Fig. 1). Under these conditions the measured rate in the phenazine methosulfate (PMS) assay

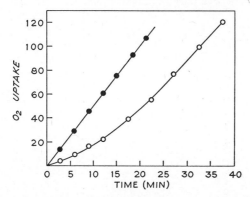

Fig. 1. Effect of preincubation with phosphate on succinic dehydrogenase activity. Lower curve: standard manometric assay at 15°, with 0.61 mg of enzyme preparation per vessel. Upper curve: 0.86 mg of the same preparation was incubated in 2.8 ml of 0.075 M PO$_4$, pH 7.6, for 20 min at 25°, then chilled to 0°, and an aliquot equivalent to 0.61 mg of enzyme was assayed at 15° as above. From Kearney (9), reproduced by permission.

increases slowly with time until it reaches linearity. However, samples preincubated with P$_i$ at a higher temperature and then assayed at 15° show linear kinetics from the start.

It was soon recognized (9, 10) that the activation by P$_i$ was a consequence of its structural similarity to the substrate [it is known to bind at the substrate site (11)] and that succinate, fumarate, and particularly malonate were far better activators. Although the concentrations required for activation varied with the nature of the activator, the final extent of activation was the same with each agent (Fig. 2)

By modification of the Michaelis-Menten equation, K_A, the activation constant could be derived; this is formally similar but not identical with K_m. The values of K_A were significantly lower for succinate and malonate than the corresponding K_I or K_m values, and the K_A for fumarate was higher than the corresponding K_I. In retrospect this might be taken to suggest that the combining site of succinate, malonate, and fumarate, when acting as modulators of the enzyme, may not be the same as that involved in catalysis.

Many of the features of this activation suggested that a conformational change in the enzyme is involved. Thus the conversion from inactive to active enzyme followed first-order kinetics with a very high activation energy (35–36 kcal/mole) and appeared to be accompanied by small but significant changes in the absorption spectrum (9, 10).

Conversion of the enzyme from inactive to active form under the influence of substrates and substrate analogs was observed in all soluble and particulate preparations in mitochondria from mammalian cells and aerobic yeast, and

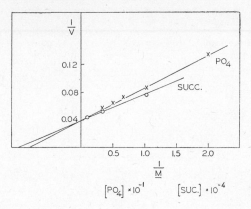

Fig. 2. Comparison of succinate and phosphate as activators. Aliquots of a purified enzyme sample were incubated in 1.5 ml of 0.075 M imidazole buffer, pH 7.6, with the amounts of succinate or phosphate indicated on the abscissa for 15 min at 25°. A control sample was similarly treated but without activator present to ascertain the basal activity of the nonactivated enzyme. After cooling to 0°, aliquots were placed in Warburg vessels held in ice and assayed at 15° by the phenazine methosulfate method. The ordinate represents the reciprocal of the O_2 uptake per 9 min after subtraction of the basal activity. From Kearney (9), reproduced by permission.

in a variety of assays, including succinate-cytochrome c reductase and succinoxidase measurements (9, 12–16). Periodic reports in the literature that butanol-extracted preparations were not activable were later satisfactorily resolved by the demonstration that such preparations also require activation if care is taken to remove the succinate used during extraction (17, 18). It should be noted, however, that the fumarate reductase activity of the dehydrogenase (fumarate-FMNH$_2$ assay) seemed not to require activation (Table 1) (17). Since in this assay only the covalently bound FAD moiety of the enzyme, not the iron-sulfur components, is thought to participate, this was taken as supporting Kearney's suggestion (9) that the structural change involved affects the iron-sulfur moieties of the enzyme.

In early studies of the phenomenon the activation seemed irreversible (9). This conclusion was based on the finding that soluble preparations activated by malonate, after dialysis to remove the malonate, remained fully activated, although most of the malonate must have been removed since no inhibition remained. Failure to reverse activation by malonate in dialysis experiments was evidently due to the retention of at least an equimolar amount of malonate tightly bound to the enzyme and maintaining it in the activated state; this residual malonate was probably displaced by succinate in activity measurements, thus escaping detection. With the advent of more rapid and efficient methods for separating proteins from low-molecular-weight metabolites

Table 1. Effect of Activation on the Fumarate-FMNH$_2$ Reaction

From Kimura et al. (17), reproduced by permission.

Sample	Temperature of Assay	Specific Activity Succinate-PMS Reaction	Specific Activity Fumarate-FMNH$_2$ Reaction
ETP, washed 4 times	25°	0.100	0.032
Same, activated by 5 mM malonate	25°	0.942	0.036
Keilin-Hartree, washed 4 times	25°	0.124	0.023
Same, activated by 2 mM malonate	25°	0.404	0.018
Soluble enzyme, Wang et al. preparation	15°	0.123	0.022
Same, activated by 5 mM succinate	15°	1.31	0.022

An ETP preparation was washed 4 times by centrifugation with 0.05 M imidazole buffer, pH 7.4, and was resuspended in the same buffer at a protein concentration of 19.8 mg/ml. For batch activation, a 5-ml aliquot was incubated with 25 μl of 1 M malonate under N$_2$ for 30 min at 38°. After cooling to 0°, aliquots of both the malonate-activated and the unactivated samples were assayed without further activation. The Keilin-Hartree preparation was treated similarly except that the protein concentration during malonate activation was 24.9 mg/ml. The preparation of Wang et al. was the gel eluate, precipitated with 0.55-saturated (NH$_4$)SO$_4$ and desalted on Sephadex G-25 (equilibrated with 0.02 M imidazole buffer, pH 7.5 at 0°). A 1.56-ml aliquot (8.55 mg of protein/ml) was incubated with 0.04 ml of 0.2 M succinate, pH 7.6, for 15 min at 25° for batch activation. In the fumarate reductase assay imidazole buffer was substituted for phosphate. Activities are corrected for inhibition by residual malonate, where present.

(such as gel exclusion), the activation proved to be fully reversible (17). Moreover, deactivation, like activation, was shown to proceed at finite rates, with an appreciable temperature coefficient. The concept which eventually emerged (13, 17) visualizes a mobile equilibrium between the inactive (E_U) and the active (E_A) enzyme, the latter being stabilized by the presence of bound activator (C):

$$E_U \rightleftharpoons E_A \underset{-C}{\overset{+C}{\rightleftharpoons}} E_A C.$$

An interesting corollary of this scheme is the suggestion of Thorn (13) that the equilibrium between inactive and active succinate dehydrogenase is governed primarily by temperature and that the role of the activator is primarily one of stabilizing the active conformation. Thorn has, in fact, reported experiments with Keilin-Hartree preparations in which some activation was noted in the absence of added activator. In view of the complex nature of the preparation, the possibility that bound activator was present cannot be excluded; in fact, suggestive evidence has been presented (17) that this may be the explanation of the "activation without added activator."

Before leaving this subject an as-yet-unsettled aspect of the process deserves mention. Kearney (9), in her first detailed paper on the regulation of succinate dehydrogenase, documented the spectral changes accompanying the activation of the dehydrogenase by malonate. Later Dervartanian and Veeger (19) confirmed and extended this observation to other competitive inhibitors but concluded that the spectral changes were due to the formation of enzyme-inhibitor complexes, perhaps of a charge-transfer type, and not to activation. Indeed, on reversal of the activation, the changed spectrum appears to remain (17). On the other hand, Kimura et al (17) have shown that the inactive (unactivated) enzyme does not form a colored complex with malonate, although it combines with the latter, and that there is substantial agreement between the rate of formation of the colored complex and the rate of activation on adding malonate to the unactivated enzyme, as well as in the apparent dissociation constants for malonate derived from the two processes. Thus it appears that the activation is by no means unrelated to the spectral changes observed but is, in fact, a prerequisite thereof. It remains uncertain, however, whether E_AC, the activated conformation stabilized by the analog, represents the same structure as the colored inhibitor-enzyme complex.

3. MODULATION BY REDUCED CoQ_{10}

The possibility that the reversible activation of succinate dehydrogenase may represent a physiological control mechanism was pointed out some years ago (20). The relatively rapid activation-deactivation at 37° was compatible with the requirements of fine regulation, but it was not easy to envisage that succinate itself would serve as the positive modifier in physiological conditions, for the fluctuation in succinate concentration in different metabolic states is not thought to be sufficiently great to induce major changes in succinate dehydrogenase activity.

The probable nature of one of the physiologically important regulators of succinate dehydrogenase was established in recent times in two laboratories which arrived at the same conclusion from entirely different types of evidence.

Rossi et al (21) compared the properties of succinate dehydrogenase in inner membrane preparations before and after extraction of the endogenous CoQ$_{10}$ and after reincorporation of the quinone. They found that on extraction of CoQ$_{10}$ the activity (V_{max}) was approximately halved and the K_m for succinate fell to about one-fourth of the value seen in normal particles; the original kinetic constants were restored on the reincorporation of CoQ$_{10}$. Furthermore, whereas in the original and reconstituted particles about 50% of the succinate-PMS activity was inhibited by thenoyltrifluoro-acetone (TTFA) or by incubation with cyanide, in accord with earlier data (22), in the CoQ-depleted preparations neither TTFA nor KCN affected dehydrogenase activity. The interpretation of these data was aided by earlier work on the mechanism of the inactivation of succinate dehydrogenase by cyanide (22). In particulate preparations incubation with cyanide gradually leads to complete loss of the ability of the enzyme to transfer electrons to the respiratory chain, as may be ascertained with any oxidant that does not react directly with the dehydrogenase, such as methylene blue (Fig. 3). On the other hand, only about 50% of the activity with PMS* is lost and the fumarate-FMNH$_2$ activity (which involves only the flavin at the active center) remains constant. In soluble preparations, which have about half the activity of particulate ones, cyanide does not lower the activity in the PMS assay. Giuditta and Singer (22) interpreted their observations to suggest that cyanide attacks some, but not all, of the nonheme irons of the dehydro-genase, leading to a conformational change evident from the altered K_m values for PMS (Fig. 4) and FMNH$_2$. The conclusion that the nonheme iron, but not the flavin region, was involved was based on the fact that only activities thought to proceed via the nonheme iron were affected. This interpretation was first debated (23), then accepted (24), by King. It was further postulated (22) that the reason why the soluble enzyme has a lower activity than the membrane-bound form and is resistant to cyanide and TTFA might be due to structural changes during extraction, which render non-functional the nonheme irons that are attacked by cyanide.

Rossi et al. (21) interpreted their recent studies along the lines of this hypothesis. Since removal of CoQ$_{10}$ from membrane preparations leads to properties resembling those of the soluble enzyme or the particulate form after cyanide treatment, they reasoned that the presence of CoQ$_{10}$ is essential for the normal functioning of some of the iron-sulfur moieties of the de-hydrogenase. They concluded, therefore, that CoQ exerts a regulatory influence on the dehydrogenase, but did not suggest how CoQ might function in regulating the enzyme under physiological conditions. A totally different

* These results were based on manometric PMS assays, which do not measure the full activity. In spectrophotometric assays the inactivation is somewhat greater (60–70%).

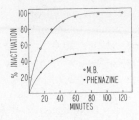

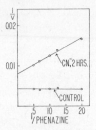

Fig. 3. Time course of the inactivation by 0.02 M cyanide at pH 7.85, 30°. Activity determinations were based on V_{max} values with both dyes at 38°. M.B.-methylene blue. From Giuditta and Singer (22), reproduced by permission.

Fig. 4. Inactivation of succinic dehydrogenase by cyanide in Keilin-Hartree preparation. Two samples of the enzyme preparation were incubated in 0.1 M phosphate buffer, pH 7.6, at 30°, one with 0.03 M cyanide, and one without cyanide (control). The curves show the activity of aliquots removed after 2 hr. Abscissa: reciprocal concentration of phenazine methosulfate, expressed as milliliters of 1% dye per 3 ml of final volume; ordinate: reciprocal activity, expressed as microliters of O_2 uptake per 5 min at 38° per 0.1 ml of enzyme preparation. From Giuditta and Singer (22), reproduced by permission.

effect of the quinone on succinate dehydrogenase, which is more readily related to physiological regulation, came to light as a result of a different experimental approach.

Gutman et al. (25, 26) noted that on adding DPNH to an inner membrane preparation (ETP$_H$), in which succinate dehydrogenase is about 80–90% in the unactivated state, the enzyme became gradually activated. Figure 5 illustrates this behavior; the experiment was conducted in aerobic conditions and sufficient antimycin A was present to inhibit DPNH oxidation by 98%, so as to permit the maintenance of DPNH for a sufficient period to observe the activation of succinate dehydrogenase. If oxidative removal of DPNH was completely prevented, the same maximal activation of succinate dehydrogenase was reached as with succinate or malonate as activator. On depletion of the DPNH (dashed line in Fig. 5) succinate dehydrogenase activity rapidly returned to the ground (unactivated) state. A second addition of DPNH caused activation once again. At 37° both activation and deactivation are much more rapid, so that severalfold changes occur in less than a minute.

Several lines of evidence show that DPNH itself is not the actual activator but serves only to reduce a component further along the respiratory chain.

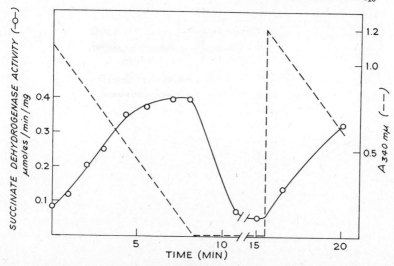

Fig. 5.   Activation of succinate dehydrogenase by DPNH. An ETP$_H$ preparation (succin-oxidase activity = 1.18 μmoles succinate/min/mg at 30°) was washed by centrifugation in STM buffer and resuspended in the same buffer to 1 mg of protein/ml. Antimycin 4 (1 nM/mg of protein) was added to slow the rate of aerobic DPNH oxidation, followed by 0.25 mM DPNH. Oxidation of the latter at 23° was monitored spectrophotometrically (dashed line). Samples were removed periodically and assayed immediately for succinate dehydrogenase activity in the presence of 0.33 mg phenazine methosulfate/ml (solid line). At 16 min a second aliquot of 0.25 mM DPNH was added. From Gutman et al. (25), reproduced by permission.

Thus, in purified preparations devoid of DPNH dehydrogenase, DPNH does not activate succinate dehydrogenase. Furthermore, as shown in Fig. 6, rhein, a competitive inhibitor of DPNH dehydrogenase (27), prevents activation by DPNH, as does piericidin A. On the other hand, neither antimycin A nor 0.23 mM TTFA (sufficient to block by 99% succinoxidase activity and the reduction of fumarate by DPNH) interferes with the activation by DPNH (Fig. 6). The experiment with TTFA shows that normal electron flux between succinate dehydrogenase and the CoQ pool does not seem essential for the activation, while the results with the other inhibitors implicate a component on the O$_2$ side of DPNH dehydrogenase and on the substrate side of the antimycin block as the actual activator. This leaves CoQ$_{10}$ and cytochrome b as possible candidates for the role of activator.

Evidence that the reduced form of CoQ$_{10}$ may be the activator is presented in Figs. 7 and 8. Figure 7 shows that in anaerobic conditions nearly the same level of activation is reached with 175 μM CoQ$_{10}$H$_2$ as with succinate, while the oxidized quinone does not activate. Figure 8 illustrates the effect of

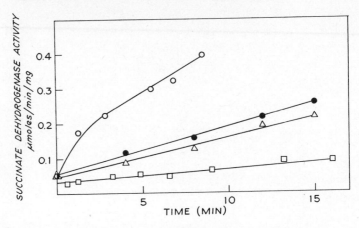

Fig. 6. Effect of various inhibitors on the DPNH-induced activation of succinate dehydrogenase. Experimental conditions were as in Fig. 5. Open circles: 1 nM of antimycin/mg plus 0.233 mM TTF; solid circles: 630 pM of piericidin A/mg of protein preincubated with the particles for 30 min at 0° before adding DPNH; triangles: piericidin as above plus 0.233 mM TTF; squares: the ETP$_H$ was diluted to 0.33 mg of protein/ml and 0.66 mM rhein was added before DPNH. Note that piericidin inhibition was entirely due to unspecifically bound inhibitor. From Gutman et al. (25), reproduced by permission.

depletion of the endogenous CoQ$_{10}$ content and of its restoration on the activation of succinate dehydrogenase by DPNH. In unextracted ETPH preparations the same level of activation is reached with DPNH as with succinate (level *A*). In pentane-extracted particles (squares) DPNH no longer activates significantly, but succinate does. Maximal activation by succinate (level *B*) is lower than in unextracted particles, not because CoQ is essential for activation by succinate, but because after removal of the CoQ the turnover number of the enzyme declines by 50%, as shown by Rossi et al. (21) and discussed above. On replacing CoQ$_{10}$, activation by DPNH is restored, the level reached (*C*) being the same as with succinate. (Incomplete restoration of the CoQ$_{10}$ content may be responsible for the difference between unextracted and reconstituted samples.)

The results of these experiments indicate strongly that CoQ$_{10}$H$_2$ serves as the regulator of succinate dehydrogenase in activation initiated by DPNH. In line with this conclusion, it was observed in the authors' laboratory that all metabolites which can reduce CoQ$_{10}$ in mitochondria (DPN-linked substrates in heart and liver mitochondria, α-glycerophosphate in brain mitochondria) activate succinate dehydrogenase in cyanide- or antimycin-blocked preparations.

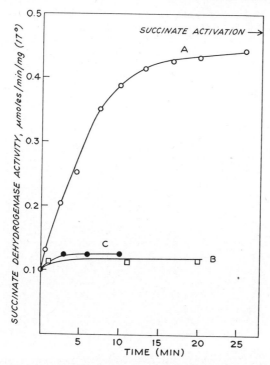

Fig. 7. Activation of succinate dehydrogenase by reduced CoQ$_{10}$·ETP$_H$, washed with 0.25 M STM buffer, pH 7.4, was resuspended in 0.18 M STM buffer, pH 7.4, at 4 mg of protein/ml. Antimycin A (1 nM/mg) and cyanide (1 mM) were added, and the sample was placed under an atmosphere of N$_2$ to prevent autooxidation of CoQ$_{10}$H$_2$. CoQ$_{10}$ was reduced with borohydride, neutralized with dilute acetic acid, and shaken till the first appearance of the yellow color of oxidized CoQ$_{10}$ to ensure removal of unreacted borohydride, all at 0°. Activation of succinate dehydrogenase was started by adding 50 μl of either CoQ$_{10}$H$_2$ (curve A) or of CoQ$_{10}$ (curve B) in absolute ethanol to 3 ml of enzyme, giving 175 mM final concentration of the quinone (curve C: no addition). Samples were withdrawn at intervals and assayed immediately at 17°. The horizontal arrow indicates the maximal activation reached with succinate as activator. From Gutman et al. (26), reproduced by permission.

Succinate dehydrogenase, whether activated by substrate or by DPNH via CoQ$_{10}$H$_2$, has the same activity (V_{max}) and K_m for PMS. Moreover, the energy of activation is the same for activation by either type of activator (26). Activation initiated by DPNH has the distinct advantage that the activator can be rapidly removed at any time (by nonenzymatic oxidation with PMS), thus permitting study of the kinetics of the deactivation. The energy of activation for the deactivation process has been found to be 16.3 kcal/mole

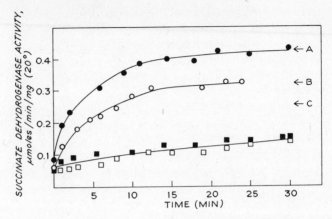

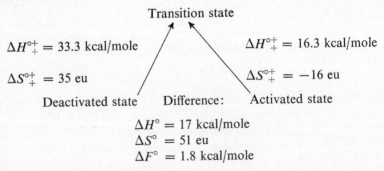

Fig. 8. Effect of pentane extraction and of reincorporation of CoQ_{10} on activation of succinate dehydrogenase by DPNH. ETP_H particles were washed, as described in Methods, lyophilized, and 5 times extracted with cold anhydrous pentane as in Szarkowska's procedure. Pentane was removed in high vacuum. Samples were then activated at $30°$ with succinate or DPNH, in the presence of 1 mM KCN and 1 nM of antimycin A/mg. Solid circles: lyophilized ETP_H activated by DPNH. The arrow at A represents maximal activation of the same sample by succinate. Solid squares and open squares: activation of pentane-extracted particles by DPNH in the presence and the absence of cyanide and antimycin A, respectively. The arrow at B denotes maximal activation of these samples by succinate. Open circles: pentane-extracted ETP_H, following readdition of 50 μM CoQ_{10} in ethanol activated by DPNH. The arrow at C denotes maximal activation of this sample by succinate. Activity was determined at $20°$. From Gutman et al. (26), reproduced by permission.

(26). The thermodynamic parameters for the activation and its reversal and the differences in $\Delta H°$, $\Delta S°$, and $\Delta F°$ between the two states of the enzyme are as follows:

Transition state

$\Delta H°{}^{+}_{+} = 33.3$ kcal/mole

$\Delta S°{}^{+}_{+} = 35$ eu

Deactivated state

$\Delta H°{}^{+}_{+} = 16.3$ kcal/mole

$\Delta S°{}^{+}_{+} = -16$ eu

Activated state

Difference:

$\Delta H° = 17$ kcal/mole
$\Delta S° = 51$ eu
$\Delta F° = 1.8$ kcal/mole

It may be seen that the activation is characterized by a relatively small free-energy change but a large change in entropy, compatible with the conformational change previously postulated (9). Of this, conversion of the unactivated enzyme to the excited state appears to account for the major part of the entropy change ($\Delta S^{\circ\ddagger}_{\ddagger} = 35$ eu), and the rest is generated during conversion of the latter to the activated form. Although the unactivated form is clearly favored thermodynamically, the fact that in the presence of activator all the enzyme is in the active state which suggests that the activator reacts with and stabilizes the active form of the enzyme.

Before discussing to what extent these effects noted in membrane preparations and with the purified enzyme are applicable to intact mitochondria and *in vivo* situations, it should be emphasized that not only overall succinoxidase activity or succinate dehydrogenase activity assayed with artifical electron acceptors, but also the rate of partial reactions of the respiratory chain linked to the dehydrogenase is affected by the state of activation of the enzyme. Figure 9 shows the dependence of the ATP-driven reduction of DPN$^+$ by succinate on activation of the dehydrogenase, and Fig. 10 demonstrates that over a wide range (from 0 to about 50% activation) succinate dehydrogenase activity and the rate of reduction of endogenous cytochrome *b* by succinate go hand in hand. The dependence of the rate of reduction of endogenous CoQ$_{10}$ on activation has also been documented (26).

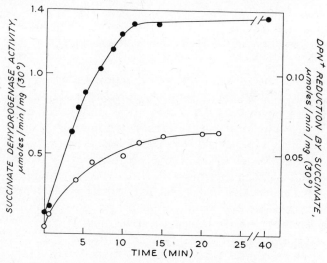

Fig. 9. Effect of activation of succinate dehydrogenase on the energy-linked reduction of DPN$^+$ by succinate. ETP$_H$ was washed and resuspended in 0.18 M STM buffer, pH 7.4, as in previous experiments at 3.3 mg protein concentration/ml. Activation by 16 mM succinate was carried out at 30° in the presence of 2 nM of antimycin A/mg and 1 mM KCN. Samples withdrawn periodically were assayed at 30° for succinate dehydrogenase activity (solid circles) and for the ATP-driven reduction of DPN$^+$ by succinate (open circles). From Gutman et al. (26), reproduced by permission.

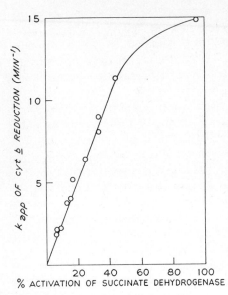

Fig. 10. Relation of the extent of activation of succinate dehydrogenase to the rate of the succinate-cytochrome b reaction of ETP_H. The particles were washed as in the experiment of Fig. 5, resuspended to 2 mg/ml, and activated for varying periods with 0.33 mM malonate at 24° to obtain different degrees of activation of succinate dehydrogenase. Samples were periodically removed and rapidly cooled to 0° to halt the activation, antimycin A (1 nM/mg) and KCN (1 mM) were added to prevent oxidase activity, and succinate dehydrogenase activity was determined as in Fig. 5. The results are plotted on the abscissa as per cent of the maximal activity reached. The ordinate represents the rate constant for the reduction of cytochrome b by 20 mM succinate at 11°, measured at 563–575 mμ. From Gutman et al. (25), reproduced by permission.

4. REGULATION OF SUCCINATE DEHYDROGENASE IN MITOCHONDRIA BY THE REDOX STATE OF CoQ_{10}

It is known (28) that the $CoQ_{10}/CoQ_{10}H_2$ ratios may change as much as tenfold in the transition from the controlled to the active state and that almost complete oxidation of $CoQ_{10}H_2$ occurs in the presence of uncouplers. If the state of activation of the dehydrogenase is a function of the fraction of CoQ that at any given moment is in the reduced state, a major decline in succinate dehydrogenase activity would be expected in the state $4 \rightarrow 3$ transition and on adding uncouplers.

Figure 11 demonstrates that in rat liver mitochondria respiring on α-ketoglutarate extensive deactivation of the dehydrogenase occurs on adding ADP, concurrently with the increase in respiration. When the ADP is exhausted, succinate dehydrogenase becomes activated again, as expected from the fact that under these conditions (state $3 \rightarrow 4$) CoQ_{10} becomes reduced, increasing the concentration of the activator. Figure 12 (left side) illustrates the effect of ADP on the dehydrogenase in rat liver mitochondria oxidizing endogenous substrate. Deactivation is more marked in this case since (*a*) the oxidation of reduced CoQ_{10} is more nearly complete when no external substrate is added, and (*b*) less ATP is generated under these conditions: as will be discussed later, ATP also activates the enzyme and thus

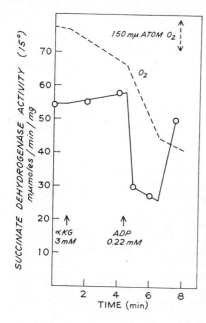

Fig. 11. Variation of succinate dehydrogenase activity in rat liver mitochondria in states 3 and 4. Conditions: 2.55 mg protein in 1.5 ml of 230 mM mannitol-70 mM sucrose-20 mM Tris-5 mM KP$_i$-5 μM EDTA, pH 7.4, at 30°. Dashed line: respiration on 3 mM α-keto-glutarate; solid line: succinate dehydrogenase activity (at 15°) of aliquots removed at intervals and assayed immediately. From Gutman et al. (5), reproduced by permission.

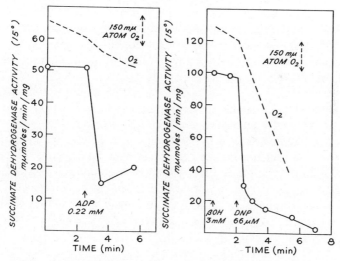

Fig. 12. Left side: variation of succinate dehydrogenase activity of rat liver mitochondria on endogenous substrate. Conditions were as in Fig. 11, except that the protein concentration was 3.5 mg/ml and no substrate was added. Right side: effect of 2,4-dinitrophenol on succinate dehydrogenase activity of rat liver mitochondria respiring on β-hydroxybutyrate. Conditions were as in Fig. 11 except that the protein concentration was 2.3 mg/ml. From Gutman et al. (5), reproduced by permission.

285

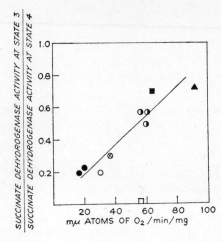

Fig. 13. Correlation between the rate of respiration in state 3 and the extent of deactivation in state $4 \rightarrow 3$ transition. Rat liver mitochondria in the presence of 3 mM substrates. After addition of 0.3 mM ADP samples were removed for immediate succinate dehydrogenase assay. Symbols: ●, endogenous substrate; ○, choline; ⊗, α-glycerophosphate; ◑, α-ketoglutarate; ■, β-OH butyrate; ▲, succinate; □, β-OH butyrate in presence of 66 μM dinitrophenol and no ADP.

its accumulation leads to reversal of the deactivation process. Figure 13 summarizes the results of many experiments on the state $4 \rightarrow 3$ transition in rat liver mitochondria in the presence of various substrates (each denoted by a different symbol). An inverse relation is apparent between the increased O_2 uptake and the loss of succinate dehydrogenase activity elicited by ADP.

The rapid deactivation of the enzyme in the transition from the controlled to the active state provides an explanation for the findings that the steady-state concentration of succinate is lower (29) and the labeling of malate by ^{14}C-succinate higher (30, 30a, 30b) in state 4 than in state 3.

The right side of Fig. 12 shows that the release of respiratory control by dinitrophenol is accompanied by a sharp decline in succinate dehydrogenase activity, as expected from the fact that $CoQ_{10}H_2$ becomes extensively oxidized on adding uncouplers (28). This result may provide an explanation for the reports (31, 32) that uncouplers cause succinate accumulation and, in the presence of antimycin, prevent the reduction of quinones by succinate.

It should be noted that the decline in succinate dehydrogenase activity induced by dinitrophenol or ADP is not an inactivation but a deactivation, since the addition of succinate (in the presence of cyanide) caused a rapid return of the original activity.

5. MODULATION BY ATP AND ITP

The experiments described in Section 4 suggest that in tightly coupled mitochondria the activity of succinate dehydrogenase is controlled by the $CoQ_{10}H_2/CoQ_{10}$ ratio, which is in turn a function of the DPNH/DPN$^+$ ratio and of the phosphate potential. In many experiments designed to follow the state of activation of the enzyme during the state $4 \to 3 \to 4$ transitions, it was noted that the higher the respiration in state 3, the less extensive was the deactivation of succinate dehydrogenase. This suggested that, in addition to $CoQ_{10}H_2$ and succinate, the ATP generated may also play a role in determining what fraction of the dehydrogenase is in the activated state at any given moment. This possibility was reinforced by a report in the literature (33) that ATP increases the rate of succinate oxidation in yeast mitochondria, although this effect was considered by others (34) to be an indirect one, involving the removal of inhibitory oxalacetate. The problem, then, was to demonstrate an ATP effect in mammalian mitochondria under conditions which preclude both oxalacetate removal and reduction of CoQ_{10} and, conversely, to demonstrate that CoQ_{10} activates the dehydrogenase in the absence of ATP synthesis. The latter requirement is readily met by the inclusion of an inhibitor of the respiratory chain during activation by a DPN-linked substrate, which would reduce CoQ_{10} without generating ATP.

Figure 14 is an experiment of this general design. Rat heart mitochondria were chosen, since, in these, succinate dehydrogenase is largely in the unactivated state upon isolation, whereas in isolated liver mitochondria the enzyme is largely or entirely in the activated form. Pyruvate and malate were used to reduce endogenous CoQ_{10}, and cyanide was included to prevent cycling and consequent oxidative phosphorylation. Normal activation, to the same level as observed on incubation with succinate, was evident. The inclusion of glutamate to transaminate with any oxalacetate present or formed did not alter the results.

Figure 15 demonstrates activation by ATP in mitochondria. No exogenous substrate was added, and the oxidation of endogenous substrates was inhibited with piericidin A. It may also be seen that oligomycin did not interfere with the activation of succinate dehydrogenase; hence the phosphorylation system is not involved in this effect. Under these conditions activation is not likely to be mediated by $CoQ_{10}H_2$, since no significant reduction of the quinone is expected.

The activation initiated by ATP is not readily interpreted by oxalacetate removal in this experiment for two reasons. First, the kinetics of the activation is compatible with the rate of activation at 30° but is far too slow for enzymatic removal of any endogenous oxalacetate. Second, as shown in Fig. 16, in the presence of glutamate included to transaminate with any oxalacetate present, activation by ATP occurs normally. Arsenite and piericidin were included in order to prevent the oxidation of any α-keto-glutarate that might have been formed, since this could lead to the reduction of DPN^+ and, hence, of CoQ_{10} and to the formation of succinate. Despite these precautions, activation triggered by ATP was clearly demonstrable even at 6 μM concentration.

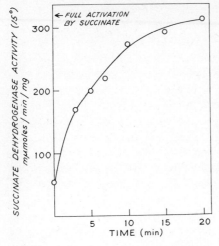

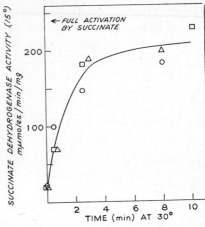

Fig. 14. Activation of succinate dehydrogenase in rat heart mitochondria initiated by pyruvate + malate in the presence of KCN. The mitochondria were suspended in the buffer given in Fig. 11 at 0.48 mg protein/ml; 1 mM KCN was added, and activation was initiated by 3 mM each of pyruvate and malate at 15°. At intervals aliquots were removed and immediately assayed for succinate dehydrogenase activity (shown on ordinate). The O_2 uptake of the preparation on pyruvate + malate at 30° was 500 natoms O_2/min/mg, and the respiratory control ratio 6. From Gutman et al. (5), reproduced by permission.

Fig. 15. Activation of succinate dehydrogenase in rat heart mitochondria initiated by ATP. The mitochondria were suspended in the buffer of Fig. 11 at 1 mg protein/ml at 30°, and 1 mM ATP was added at 0 time. Aliquots were assayed for succinate dehydrogenase activity as in Fig. 1. Without ATP no change in succinate dehydrogenase activity occurred. Symbols: $\triangle$ = ATP alone; $\square$ = ATP + oligomycin (2 μg/mg of protein); $\bigcirc$ = ATP + piericidin A (1.25 nM/mg of protein). From Gutman et al. (5), reproduced by permission.

An approach to estimating the relative efficiencies of activation by endogenous $CoQ_{10}H_2$ and ATP, respectively, is represented in Fig. 17. In this experiment rat heart mitochondria were preincubated with dinitrophenol to deplete endogenous substrate and ATP as much as possible; then oligomycin was added to block the ATPase. Since respiration was uncoupled and no respiratory chain inhibitor was present, malate and glutamate (in the presence of arsenite) caused no activation (curve 1). The addition of ATP along with substrate, however, resulted in considerable activation (curve 2). When respiration was inhibited by cyanide, activation initiated by substrate was evident (curve 3); under these conditions extensive reduction of CoQ_{10} occurs. Curve 4 shows that, when ATP was added along with substrate and cyanide, activation was maximal. Curve 4 is essentially a summation of

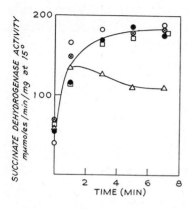

Fig. 16. Activation of succinate dehydrogenase in rat heart mitochondria by various concentrations of ATP. The mitochondria suspended in the medium of Fig. 11 (1 mg/ml) were incubated for 1 min with 1.2 nM of piericidin A/mg of protein, 1 mM arsenite, and 2.5 μg of oligomycin, after which 10 mM glutamate was added and activation was started by addition of ATP. ATP concentrations: ○, 3.15 mM; ●, 0.315 mM; ⊗, 0.189 mM; ⊠, 0.019 mM; △, 0.006 mM.

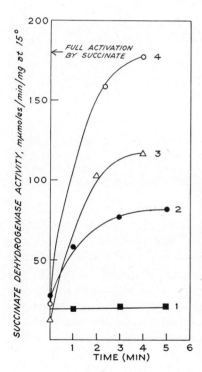

Fig. 17. Activation of succinate dehydrogenase by DPN-linked substrates and ATP. Rat heart mitochondria, 1 mg/ml, suspended in the medium of Fig. 11, were treated for 1 min with 20 μM dinitrophenol and 0.3 mM arsenite; then 5 μg of oligomycin/mg of protein was added. After 5 min malate (5 mM) plus glutamate (10 mM) and/or ATP (3 mM) was added as follows. Curve 1: malate + glutamate; curve 2: malate + glutamate + ATP; curve 3: malate + glutamate + KCN (1 mM); curve 4: malate + glutamate + KCN + ATP. Incubations were at 30°, succinate dehydrogenase assays on aliquots at 15°.

curves 2 (ATP effect) and 3 ($CoQ_{10}H_2$ effect). It may be noted that because of the presence of glutamate plus arsenite complications due to oxalacetate or succinate formation are reasonably well eliminated.

The insensitivity of the activation by ATP to oligomycin implies that this mechanism does not operate through the phosphorylating system, leading to reversed electron transport and thus to partial reduction of the quinones. This conclusion is reinforced by the experiment illustrated in Fig. 18. In

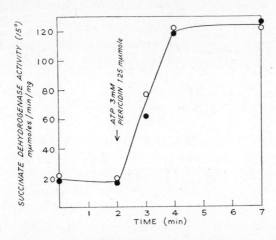

Fig. 18. Activation of succinate dehydrogenase by ATP in the presence and the absence of ADP. Rat heart mitochondria (1 mg/ml) in the buffer of Fig. 11 were treated with 1.23 nM of piericidin A/mg of protein and incubated for 2 min at 30°. Activation at 30° was initiated by adding 3 mM ATP (closed circles) or 3 mM ATP + 0.34 mM ADP (open circles).

this experiment oxidation of endogenous substrate was blocked by piericidin and no exogenous substrate was added. The mitochondria were then activated by ATP (3 mM) either with or without 2 min preincubation with ADP (0.34 mM). Since the phosphate potential as defined by Chance and Williams (35), [ATP]/[ADP][Pi], is less than 2×10^3 under these conditions, while state 3 is assumed to prevail up to values of 10^4 (35), clearly the sample which included ADP was in state 2, and activation of the succinate dehydrogenase took place under conditions where the phosphate potential would have prevented reduction of the CoQ_{10} pool.

The present authors, in their previous reports on the ATP effect (5, 6), emphasized that ATP might not be the actual agent responsible for the activation, but that a compound in equilibrium with it might be the activator.

In submitochondrial particles from heart mitochondria (ETP and ETP_H) little or no activation of succinate dehydrogenase was noted with ATP or

GTP, but both ITP and IDP are good activators; IDP activates at somewhat lower concentrations (Fig. 19). Activation by moderately high concentrations of IDP and ITP reaches nearly the same level as activation by succinate. At 5–10 mM concentrations, 5'-AMP, 5'-IMP, cyclic 3',5'-AMP, and cyclic 3',5'-IMP gave virtually no activation. Since ITP and IDP do not penetrate, they could not be tested in intact mitochondria.

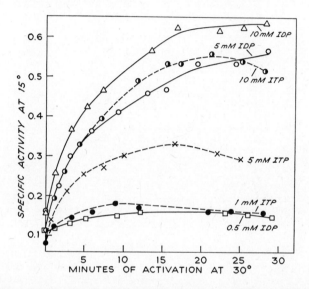

Fig. 19. Activation of succinate dehydrogenase in ETP$_H$ by ITP and IDP. ETP$_H$, deactivated by washing and suspended in STM buffer (Fig. 5), was incubated at 30° with the indicated concentrations of nucleotides. Samples were removed and immediately assayed for succinate dehydrogenase activity at 15°. Note that no correction was made for the possible inhibitory effects of ITP and IDP in succinate-PMS assays. Maximal activation reached was close to that obtained with succinate as activator.

Since ETP$_H$ preparations contain considerable ATPase activity, one could not be sure that the effectiveness of ITP was due to conversion to IDP. For this reason the effect of ITP on the activation of succinate dehydrogenase in Complex II was also examined. This particle (36) is a highly purified form of succinate dehydrogenase which may reasonably be expected to be free from enzymes that hydrolyze ITP. Figure 20 shows that the succinate dehydrogenase in a Complex II sample, which had been 17% activated at the start of the experiment, was fully activated by ITP, reaching the same level as achieved by activation with succinate. (The slow decline in the activity of the succinate-activated control in this experiment is due to *in*activation at 30°.)

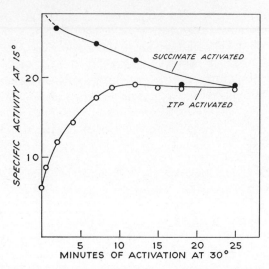

Fig. 20. Activation of succinate dehydrogenase in Complex II by ITP. The Complex II preparation was 83% deactivated by passage through Sephadex G-25 in the absence of succinate, suspended in 0.18 M STM buffer at 0.2 mg/ml concentration, and warmed for 5 min at 30° ensure deactivation before the addition of 10 mM ITP where shown. The upper curve shows the decay of a succinate-activated sample at 30° (no DTT present). Note that no correction was made for possible interference of ITP in succinate-PMS assays and that values are not V_{max}.

Preliminary results with soluble, highly purified succinate dehydrogenase from beef heart suggest that the enzyme may be activated by ITP also in this form. Because of the relatively high concentrations of inosine nucleotides required for activation in membrane preparations and Complex II, as compared with the very low ATP concentration that produces activation in intact mitochondria, the conclusion that ATP acts only after conversion to inosine nucleotides seems premature at this time. It appears clear, nevertheless, that certain purine di- and trinucleotides can activate succinate dehydrogenase without mediation by reduced CoQ_{10} or succinate.

6. CONTROL OF SUCCINATE DEHYDROGENASE IN MITOCHONDRIA

The data presented in the foregoing sections indicate that at least three types of positive modulators of succinate dehydrogenase exist in mitochondria:

substrate, reduced CoQ (the concentration of which reflects the phosphate potential in the particular metabolic state), and certain purine di- or tri-nucleotides, activation by which does not hinge on the removal of oxalacetate inhibition, is not mediated by the respiratory chain, and does not involve the energy conservation mechanism. The list of activators may not be complete, nor can the possibility be excluded that negative modulators of the enzyme [other than oxalacetate, which is compartmented separately from succinate dehydrogenase (37)] may also exist. It is of interest in this connection that, although both ATP and metabolites which reduce CoQ_{10} can fully activate the dehydrogenase in intact mitochondria, in many experiments activation by DPN-linked substrates and ATP did not reach completion and even tended to reverse itself after a time.

Activation of succinate dehydrogenase by $CoQ_{10}H_2$ and especially by ATP appears to be much faster in mitochondria than in membranal or soluble preparations. Thus activation in intact heart mitochondria by ATP was usually complete in 2 min at 30°, and in much less time at 38°, and occurred at appreciable rates even at 15°, where soluble preparations undergo activation only exceedingly slowly (9). Although an accurate temperature coefficient for ATP activation could not be measured in mitochondria because of complications imposed by the temperature sensitivity of the adenine nucleotide translocase system (38), rough estimates of the activation energy for activation initiated by pyruvate plus malate in intact mito-chondria indicate a value of 10 kcal/mole, much less than in membranal or soluble preparations. Thus the absolute rate of activation appears to be higher and its energy of activation lower than in submitochondrial systems. Conceivably, the conformational change, which is regarded as the rate-limiting step in the activation, is facilitated in the environment of the intact mitochondrion.

Many attempts were made to see whether deactivation of succinate dehydrogenase in transitions from state 4 to 3 is ever extensive enough for this enzyme to become rate limiting in the Krebs cycle. No instance has been found in which this is the case. Thus ADP-triggered deactivation appears to halt at a residual level of succinate dehydrogenase activity, which is still sufficient to keep up with full operation of the cycle, possibly because synthesis of ATP tends to reverse the deactivation caused by the ADP-initiated oxidation of $CoQ_{10}H_2$. Thus, for all intents and purposes, succinate dehydrogenase does not regulate the cycle, but rather is regulated by the metabolic state of the mitochondria, fluctuating in a five- to tenfold range of activity between extremes. One may well inquire what the physiological purpose is of this intricate, highly refined kinetic regulation of the enzyme. Section 8 considers this question.

7. MOLECULAR MECHANISM OF REGULATION

Until recently only indirect indications were available that the reversible activation of succinate dehydrogenase is a conformational alteration (9). As to the modulators, ATP (or ITP) would have to be considered an allosteric modifier in the usual sense. On the other hand, in the case of succinate and $CoQ_{10}H_2$, which are also substrates of the enzyme, the applicability of the term is open to question, since, as already mentioned, it is not certain whether they react at the same sites on the enzyme when acting as electron donors and as modulators. It is certain, however, that none of these regulators acts by simply changing the affinity for the substrate; hence the analogy with asparatate transcarbamylase or hemoglobin is not close.

Until a short time ago all known preparations of the enzyme could be activated and deactivated. Recently, however, Davis and Hatefi (39) described a novel method for the extraction of succinate dehydrogenase, involving incubation of Complex II with 0.8 M $NaClO_4$ in the presence of succinate and dithiothreitol (DTT). Examination of the properties of this preparation in the authors' laboratory (40) revealed that it was fully activated, as expected from the presence of succinate during isolation, and that it could not be readily *deactivated* either by dialysis or by repeated passage through Sephadex G-25. The possibility that the enzyme remained fully activated because of the presence of a small amount of succinate tightly bound at a hypothetical regulatory site was excluded by the experiment of Fig. 21. This purified preparation was isolated in the presence of succinate-2,3-^{14}C, and the labeled enzyme was slowly passed through a long column of Sephadex G-50. All the radioactivity was recovered in a fraction well separated from the enzyme, which was excluded on the column; the protein component contained no radioactive succinate, although much less than 1 mole/mole of enzyme would have been detectable. Figure 21 also shows that the activity of the enzyme recovered from the column could not be increased by activation with succinate.

The first question that arose in connection with this unexpected difference between the Davis-Hatefi preparation and all others discussed in the literature concerned the nature of the agent or treatment that causes this profound change in the properties of the enzyme. Figure 22 illustrates an experiment designed to answer this question. A sample of Complex II was 92% deactivated by passage through a Sephadex G-25 column, suspended in Tris buffer, pH 8.0, containing DTT, and divided into two parts. To one $NaClO_4$ was added, whereas the other received no $NaClO_4$ (control); both were incubated at 0°. As seen in Fig. 22, the control sample decayed somewhat during the prolonged incubation (inactivation). The sample containing

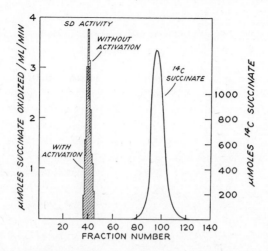

Fig. 21. Absence of succinate from nonregulable preparations of succinate dehydrogenase. Four-tenths milliliters of the Davis-Hatefi SDB preparation (39) (6.2 mg protein/ml) in Tris-DTT buffer, pH 8.0, was equilibrated for 20 min at 0° with 0.5 ml Na succinate-1,3-^{14}C (33.7 mM, 0.247 μC/μmole), applied to a Sephadex G-50 column (58 × 1.5 cm, coarse, equilibrated with 50 mM Tris-5 mM DTT, pH 8.0, at 0° under N$_2$), and eluted with the same buffer under N$_2$. The fractions collected were assayed for succinate dehydrogenase activity (fixed PMS concentration) and counted for radioactivity. Note that (a) no radio activity was detected in the enzyme, and (b) the dehydrogenase was not deactivated on the column, as judged by assays with and without activation by succinate.

NaClO$_4$, however, gradually increased in activity. Hence perchlorate appears to activate the enzyme during the process of extraction; this explains why NaClO$_4$-extracted preparations are fully activated. An aliquot removed from the NaClO$_4$-treated sample at 85 min was then dialyzed against Tris-DTT buffer at 0°, and samples were removed periodically for assay with and without activation with succinate. It is seen that on removal of the NaClO$_4$ by dialysis the enzyme gradually returns to the unactivated form, but it can be fully activated at any time by brief treatment with succinate (cf. vertical arrows).

It appears from this experiment that high concentrations of NaClO$_4$ fully activate the enzyme, even at 0°, and that this type of activation is also reversible. When both succinate and 0.8 M NaClO$_4$ were present during extraction, however, as in the procedure of Davis and Hatefi (39), the enzyme became activated, but deactivation following removal of succinate and perchlorate was exceedingly slow (40, 40a). Similar anomalous properties were conferred (40) by exposure to NaClO$_4$ and succinate on preparations ordinarily readily activated and deactivated (41). These studies clarify why

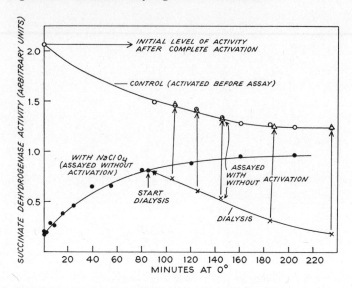

Fig. 22. Activation of succinate dehydrogenase in Complex II and conversion to non-regulable form under the influence of NaClO$_4$-DTT. A sample of Complex II was passed through Sephadex G-25, suspended at 12 mg/ml in the Tris-DTT buffer used in Fig. 21, and heated for 6 min at 38° to obtain extensive deactivation. The specific activity in the PMS assay at 38° and V_{max} was 43.6 with activation and 3.6 without activation with succinate. The enzyme (2.4 ml) was incubated with 0.25 ml of saturated NaClO$_4$ at 0°, aliquots being removed for PMS assay at 38°. Each sample was assayed with and without activation with succinate at fixed PMS concentration. At the point indicated by the arrow an aliquot of the perchlorate-treated enzyme was removed and dialyzed at 0° against the same Tris-DTT buffer in order to remove NaClO$_4$. Samples were again taken for succinate-PMS assay with and without activation.

the Davis-Hatefi preparation of the enzyme differs from all others in not requiring activation. At high concentrations Br$^-$ acts like ClO$_4^-$ in activating the enzyme, even at 0°, and in altering its regulatory properties (40a).

Two recent reports appear to pave the way to chemical and physical studies of the conformation changes involved in the reversible activation of the enzyme. Sanborn et al. (40b) found that in the deactivated form the dehydrogenase does not combine with bromopyruvate and *N*-ethylmaleimide to form an inhibited enzyme. This implies that the —SH groups, long believed to be present at the substrate-binding site, become reactive toward these —SH reagents only after activation. The second observation (40c) is that in membrane-bound and soluble preparations succinate dehydrogenase is gradually activated in the temperature range of 20–38° by merely lowering the pH. The extent of the resulting activation is a function of the pH, and the activation is readily reversed by raising the pH. The pH range in which these changes occur covers approximately two pH units, with the point of half-maximal activation at pH 6.3–6.6 at 30°. This may indicate that the ionization of a histidyl residue governs activation by pH. Certain anions displace the equilibrium in favor of the active form (18a). Their order of effectiveness appears to be ClO_4^-, formate, NO_3^-, $I^- > Br^- > Cl^-$. Acetate, CN^-, and F^- do not show the effect; in fact the latter two anions inhibit pH activation (40a).

These interesting effects of pH and ions not only provide a convenient means of studying the structural basis of the activation process, but also raise important questions about their possible relation to the regulatory effects that have been observed in mitochondria and to the "temperature activation" (i.e., activation without added activators) of Thorn (13).

8. WHY SUCCINATE DEHYDROGENASE ACTIVITY IS REGULATED IN MITOCHONDRIA

Several years ago a series of papers appeared (42–46) showing that substrates of the various respiratory chain-linked flavoproteins inhibit the oxidation of each other. Thus succinate inhibits DPNH, choline, and α-glycerophosphate oxidase activity, and vice versa. An example of this effect is seen in Table 2 from an early paper of Ringler and Singer (42), showing that in brain mitochondria the oxidation of α-glycerophosphate and that of succinate mutually interfere with each other. This is not merely a consequence of exceeding the capacity of the respiratory chain by the combined electron flux from two dehydrogenases, for the activity on two substrates that are oxidized at different rates may be well below the rate observed with the faster substrate alone (44, 45). Moreover, when PMS is used in the experimental conditions of Table 2, the rates of oxidation become strictly additive: thus the competition involves electron transport from the dehydrogenase to the respiratory chain, presumably the CoQ pool. This curious behavior was interpreted at the time (46) to suggest that the electron transport system can orient itself toward whichever flavoprotein is in the reduced form at a given moment, thereby being unavailable to accept electrons from other flavo-proteins. In light of the subsequent demonstration of rapid conformational changes in mitochondrial membranes this interpretation seems to merit serious consideration.

Extending this hypothesis and the experimental facts on which it rests to the question of why succinate dehydrogenase requires close regulation in mitochondria, and considering that the unconstrained activity of succinate

Table 2. Mutual Interference of Succinic and α-Glycerophosphate Oxidases in Pig Brain Mitochondria

From Ringler and Singer (42).

Substrate	Oxygen Uptake, μatoms	Products Formed	
		DHAP, μmoles	Fumarate, μmoles
α-Glycero-P	4.0	4.8	...
Succinate	15.0	...	16.5
Succinate + α-glycero-P	13.8	3.3	8.4

Reaction carried out in conventional Warburg apparatus in the presence of 0.05 M phosphate, pH 7.6, 2.0 mg of cytochrome c, and 0.15 ml of mitochondria (5.8 mg protein) in a total volume of 3 ml at 38°. Succinate ($3.3 \times 10^{-2} M$) present where indicated. The reactions were stopped after 21 min with perchloric acid, and the products were determined.

dehydrogenase in mitochondria [turnover number = 18,000/min at 38° (20)] is much higher than the activity of the respiratory chain it feeds, it is clear that uncontrolled succinate oxidation would restrict the oxidation of DPNH. Considering further that in each turn of the Krebs cycle 10 moles of ATP are generated as a result of DPNH oxidation but only 2 moles result from the succinate → fumarate step, it is also apparent that when the maximal rate of ATP synthesis is called for a degree of constraint on succinate dehydrogenase is required, so as to permit unimpeded DPNH oxidation (7). The constraint on succinate dehydrogenase, however, must not be great enough to depress its activity below the rate required for unhampered operation of the Krebs cycle. As we have seen, this is indeed the situation in the state 4 → 3 transition in mitochondria: on addition of ADP there is an immediate sharp drop in succinate dehydrogenase activity, but the deactivation does not proceed to a point where succinate dehydrogenase becomes rate limiting; as ATP accumulation begins, the flavoprotein again becomes activated. This may be the reason for the multiple types of control of the enzyme, for while the ATP/ADP ratio determines the state of activation by regulating the redox state of CoQ_{10}, ATP can operate also as a more direct activator, preventing excessive deactivation in state 3 conditions. Accumulated succinate and succinyl CoA act synergistically with ATP in this respect.

The proposed regulation would then operate as follows. When the ATP/ADP ratio is low, succinate dehydrogenase would be deactivated, partly because of the lack of ATP and partly because in state 3 $CoQ_{10}H_2$ becomes oxidized (28); thus the two main activators are removed and the enzyme returns to the deactivated form. This permits the maximal rate of DPNH oxidation and hence ATP synthesis. During this phase succinate accumulates (29) in mitochondria. As the ATP/ADP ratio rises, succinate dehydrogenase is turned on, partly by ATP, partly by accumulated succinate, and partly by the resulting high $CoQ_{10}H_2/CoQ_{10}$ ratio characteristic of state 4. The increased succinate dehydrogenase activity would then result in oxidative removal of accumulated succinate.

In order for this hypothesis to be plausible, sources of succinate other than the one resulting from α-ketoglutarate oxidation would have to exist; otherwise the supply of succinate generated in the α-ketoglutarate step would automatically regulate succinate dehydrogenase activity. It is known, however, that considerable amounts of succinyl-CoA and, hence, succinate arise from the oxidation of odd-numbered fatty acids (47), of branched amino acids (48), and of methionine via methylmalonyl-CoA.

It may be argued that neither pure state 3 nor state 4 occurs under physiological conditions. Indeed, except for conditions of extreme stress, the average mammalian tissue may, at any given moment, be somewhere between states 3 and 4.

The situation in individual cells in the tissue and in individual mitochondria in a given cell may be more extreme: it is likely that rapid oscillations between conditions approaching the extremes represented by states 3 and 4 occur, resulting in rapid fluctuations in requirements for ATP synthesis. The intricate regulation of succinate dehydrogenase is visualized as serving the ever-varying needs of the mitochondrion for ATP synthesis, acting in conjunction with the respiratory control machinery.

This hypothesis is not the only one which can be formulated for the multiple regulation of the dehydrogenase; other alternatives have been considered elsewhere (7). It provides a framework that seems to fit known facts, however, and may hopefully stimulate further experimental work.

ACKNOWLEDGMENTS

This work was supported by Grant BC 46A from the American Cancer Society, Grant GB 08214 from the National Science Foundation, and Grant HE 10027 from the National Institutes of Health.

REFERENCES

1. T. P. Singer, in *Biochemical Evolution and the Origin of Life* (Ed.: E. Schoffeniels), North-Holland, Amsterdam, 1971, p. 203.

2. C. A. Hirsch, M. Raminsky, B. D. Davis, and E. C. C. Lin, *J. Biol. Chem.*, **238**, 3370 (1963).

3. J. Hauber and T. P. Singer, *European J. Biochem.*, **3**, 107 (1967).

4. H. Tisdale, J. Hauber, G. Prager, P. Turini, and T. P. Singer, *European J. Biochem.*, **4**, 472 (1968).

5. M. Gutman, E. B. Kearney, and T. P. Singer, *Biochem. Biophys. Res. Commun.*, **44**, 526 (1971).

6. T. P. Singer, E. B. Kearney, and M. Gutman, in *Biochemistry and Biophysics of Mitochondrial Membranes* (Eds.: G. F. Azzone, E. Carafoli, A. L. Lehninger, E. Quagliariello, and N. Siliprandi), Academic Press, New York, 1972, in press.

7. T. P. Singer, M. Gutman, and E. B. Kearney, *FEBS Letters*, **17**, 11 (1971).

8. E. B. Kearney, T. P. Singer, and N. Zastrow, *Arch. Biochem. Biophys.*, **55**, 579 (1955).

9. E. B. Kearney, *J. Biol. Chem.*, **229**, 363 (1957).

10. E. B. Kearney, in *Proceedings of the International Symposium of Enzyme Chemistry, Tokyo and Kyoto, 1957*, Maruzen, Tokyo, 1958, p. 340.

11. E. C. Slater and W. D. Bonner, Jr., *Biochem. J.*, **52**, 185 (1952).

12. T. Y. Wang, C. L. Tsou, and Y. L. Wang, *Sci. Sinica*, **5**, 73 (1956).

13. M. B. Thorn, *Biochem. J.*, **85**, 116 (1962).

14. T. Kimura, J. Hauber, and T. P. Singer, *Biochem. Biophys. Res. Commun.*, **11**, 83 (1963).

15. T. P. Singer, in *Comprehensive Biochemistry*, Vol. 14 (Eds.: M. Florkin and E. H. Stotz), Elsevier, Amsterdam, 1966, p. 127.

16. T. P. Singer, E. Rocca, and E. B. Kearney, in *Flavins and Flavoproteins* (Ed.: E. C. Slater), Elsevier, Amsterdam, 1966, p. 391.

17. T. Kimura, J. Hauber, and T. P. Singer, *J. Biol. Chem.*, **242**, 4987 (1967).

18. W. P. Zeijlemaker, D. V. Dervartanian, C. Veeger, and E. C. Slater, *Biochim. Biophys. Acta*, **178**, 213 (1969).

19. D. V. Dervartanian and C. Veeger, *Biochim. Biophys. Acta*, **92**, 233 (1964).

20. T. P. Singer, in *Biological Oxidations* (Ed.: T. P. Singer), John Wiley-Interscience, New York, 1968, p. 339.

21. E. Rossi, B. Norling, B. Persson, and L. Ernster, *European J. Biochem.*, **16**, 508 (1970).

22. A. Giuditta and T. P. Singer, *J. Biol. Chem.*, **234**, 666 (1959).

23. D. Keilin and T. E. King, *Biochem. J.*, **69**, 32P (1958).

24. T. E. King, *Advan. Enzymol.*, **28**, 155 (1966).

25. M. Gutman, E. B. Kearney, and T. P. Singer, *Biochem. Biophys. Res. Commun.*, **42**, 1016 (1971).

26. M. Gutman, E. B. Kearney, and T. P. Singer, *Biochemistry*, **10**, 4763 (1971).

27. E. A. Kean, M. Gutman, and T. P. Singer, *J. Biol. Chem.*, **246**, 2346 (1971).

28. M. Klingenberg, in *Biological Oxidations* (Ed.: T. P. Singer), John Wiley-Interscience, New York, 1968, p. 3.

29. K. LaNoue, W. J. Nicklas, and J. R. Williamson, *J. Biol. Chem.*, **245**, 102 (1970).

30. A. F. McElory, and G. R. Williams, *Arch. Biochem. Biophys.*, **126**, 492 (1968).

30a. G. Schäfer, P. Balde, and W. Lamprecht, *Nature*, **214**, 20 (1967).

30b.   R. W. VonKorff, *Nature*, **214,** 23 (1967).

31.   S. Tsuiki, T. Sukeno, and H. Takeda, *Arch. Biochem. Biophys.*, **126,** 436 (1968).

32.   A. Kröger and M. Klingenberg, *Biochem. Z.*, **344,** 317 (1966).

33.   C. Gregolin and P. Scallella, *Biochim. Biophys. Acta*, **99,** 185 (1965).

34.   S. Papa, J. M. Tager, and E. Quagliariello, in *Regulatory Functions of Biological Membranes* (Ed.: J. Järnefelt), Elsevier, Amsterdam, 1968, p. 264.

35.   B. Chance and G. R. Williams, *Advan. Enzymol.*, **16,** 65 (1956).

36.   D. Ziegler and J. S. Rieske, in *Methods in Enzymology*, Vol. 10 (Eds.: R. W. Estabrook and M. E. Pullman), Academic Press, New York, 1967, p. 231.

37.   A. E. Jones and H. Gutfreund, *Biochem. J.*, **87,** 639 (1963).

38.   M. Klingenberg and E. Pfaff, in *Metabolic Roles of Citrate* (Ed.: T. W. Goodwin), Academic Press, New York, 1968, p. 105.

39.   K. A. Davis and Y. Hatefi, *Biochemistry*, **10,** 2509 (1971).

40.   C. J. Coles, H. Tisdale, and T. P. Singer, article to be published.

40a.   T. P. Singer, E. B. Kearney, and W. C. Kenney, *Advan. Enzymol.*, **36,** (in press).

40b.   B. M. Sanborn, N. T. Feldberg, and T. C. Hollocher, *Biochim. Biophys. Acta*, **227,** 210 (1971).

40c.   E. B. Kearney, M. Mayr, and T. P. Singer, *Biochem. Biophys. Res. Commun.* (in press).

41.   T. P. Singer, E. B. Kearney, and P. Bernath, *J. Biol. Chem.*, **223,** 599 (1956).

42.   R. L. Ringler and T. P. Singer, *J. Biol. Chem.*, **234,** 2211 (1959).

43.   C. Y. Wu and C. L. Tsou, *Sci. Sinica*, **4,** 137 (1955).

44.   T. Kimura and T. P. Singer, *Nature*, **184,** 791 (1959).

45.   T. Kimura, T. P. Singer, and C. J. Lusty, *Biochim. Biophys. Acta*, **44,** 284 (1960).

46.   R. L. Ringler and T. P. Singer, *Arch. Biochem. Biophys.*, **77,** 229 (1958).

47.   H. R. Mahler, in *Fatty Acids* (Ed.: K. S. Markley), Part 3, 2nd ed., John Wiley, New York, 1968, p. 1487.

48.   A. L. Lehninger, *Biochemistry*, Worth Publishers, New York, 1970, p. 447.

CHAPTER 10

Cytoplasmic Regulation of Mitochondrial Bioenergetics by a Specific Cellular Factor

ERNEST KUN*

Department of Pharmacology and Biochemistry and Biophysics and the Cardiovascular Research Institute, University of California, San Francisco, California

* Recipient of the Research Career Award of the U.S. Public Health Service.

1. INTRODUCTION

Il n'y a rien de plus convaincant qu'une grande conviction.

Alexander Dumas

The purpose of this chapter is to describe a rather unusual series of experiments which led to the recognition of a novel cytoplasmic factor. This factor, in trace amounts, acts on isolated mitochondria under unorthodox experimental circumstances and seems to regulate a variety of energy-dependent processes. The specific cytoplasmic regulation of mitochondria by a hitherto unknown natural product is a new observation, although the *problem* of intracellular control of mitochondrial processes is not. The purpose here is to correlate the new information with more general mechanisms of metabolic regulation in order to project further research.

Differentiated cells are recognizable by their specific cellular structure; therefore it seems reasonable to expect that specificity of subcellular architecture should be contributory to cell-specific metabolic regulation. As pointed out by Schmitt (1) in 1959, "This (structural) organization or ordering of enzymes permits cycles of reactions to occur that would be highly improbable or impossible if the individual steps in the reaction were catalyzed by individual enzymes randomly distributed in the cell." This important statement predicts that different cell types, which have apparently similar enzymatic composition—as determined by conventional enzyme isolation procedures—may contain specific multienzymatic functional organizations, depending on the spatial arrangement of enzymes within the cell.

Control mechanisms known to regulate the activity of water-soluble enzymes are more readily applicable to the metabolic control of cellular systems if the enzymes in question are known to be operative within the cell in an environment which resembles an aqueous solution (e.g., in the cytosol). Many examples dealt with in the present book confirm this conclusion. An understanding of the super-macromolecular organization of enzymes built into hydrophobic cellular structures depends in part on the resolution (isolation) of these water-insoluble macromolecular components. If it is assumed that cellular structure determines the catalytic function, especially the coupling of multienzyme systems, then it would be expected that *changes induced in the structural components only* can significantly alter apparent metabolic functions without direct interference with the catalytic activity of enzyme proteins themselves. It seems pertinent to consider this indirect interference with metabolic systems by way of *structural regulators* in connection with the short-term action of drugs and hormones, agents which seldom act directly on enzyme molecules. Since macromolecular constituents of subcellular structures largely resist conventional protein isolation procedures, their molecular biology (biosynthesis, degradation) is especially difficult to explore. Consequently, effects that are mediated through biosynthesis of structural macromolecules and are noticed by way of indirectly induced metabolic alterations in cellular systems are equally difficult to analyze.

It has become increasingly evident that regulation of cellular metabolism is critically dependent on the catalytic function of membrane systems, which not only control anisotropic ion flux (cf. refs. 2, 3, 4, and 5), but also act as energy transducers (6, 7). Mitochondria (8, 9) have been clearly identified as the most versatile subcellular structures. The catalytic function of structurally bound enzymes and poorly understood membrane-associated reactions of mitochondria exert powerful effects on cellular metabolism. The experimentally demonstrable existence of manifold energy-coupled reactions in mitochondria has been amply documented and tends to create the impression that the study of *isolated* mitochondria necessarily yields information which has direct cell physiological significance. However, it was recognized that metabolic connections between extra- and intramitochondrial systems may have to be discovered before cellular metabolism can be understood (8). The potentially misleading (from the viewpoint of cell physiology) nature of studies with isolated mitochondria has been clearly stated by Racker (cf. p. 4 of ref. 10): ". . . studies with 'intact mitochondria' separated from the rest of the cell may also lead us astray, particularly if conducted with the illusion of physiological reality." It is obvious, as Racker points out, that the analytically oriented experimental approach, that is, isolation of mitochondrial components, must *not* be abandoned and remains an essential part of any mechanism-oriented biochemical research. On the other hand, the apparent deadlock of arguments surrounding the unexplained coupling mechanism between electron and cation flux and ATP synthesis (11) in mitochondria and chloroplasts may eventually gain refreshing impetus if some attention is paid to the more real function of mitochondria in a cellular environment.

During the past five years, a study of intracellular regulation of mitochondrial functions by cellular substances was undertaken in the author's laboratory. Initially the purpose of experiments was to find an explanation for the effects of artificial uncouplers when mitochondrial metabolism was studied in the presence of cytosol. This work led to the recognition of the existence of a specific cytoplasmic metabolic factor (CMF) which was eventually shown to influence mitochondrial functions in the presence of other physiological cellular constituents (e.g., Ca^{2+}). Many new problems emerged in the course of this work, and many of them are, at present, far from being solved. For this reason, this survey is intended to set the background for new investigations rather than to conclude an already established area.

There is extensive, but unfortunately inconsistent, scientific literature dealing with the bioenergetic functions of mitochondria. This topic cannot be discussed in this chapter, although such review would aid the nonspecialist reader. Since no attempt will be made to compile a complete bibliography of this diverse subject, only selected references will be listed. Essential information related to oxidative phosphorylation can be found in the papers of Lardy (12), Chance and Williams (13), Lehninger (14), Racker (15), Boyer (16), and Green (17).

2. APPLICATION OF ENZYME INHIBITORS AND UNCOUPLERS AS METABOLIC PROBES

> · · · *one should be as artificial as possible—to be natural is a pose*
>
> Oscar Wilde

A study of cellular metabolic control should be conducted under conditions which permit reasonable assurance that the order characteristic of the chosen intracellular system will remain as little disturbed as possible. At the same time, it should be possible to abstract specific answers to questions related to cellular regulation in terms of enzymatic or other biochemical mechanisms. With this aim in mind, some time ago the synthesis of *F*-substituted enzyme substrate homologs was undertaken (18) and the mode of action of fluoro analogs of citric acid cycle intermediates investigated (19). Inhibition mechanisms of several key enzymes that are known components of important metabolic pathways were studied first under conventional *in vitro* conditions and later in increasingly complex multienzyme systems. It was assumed that the rate-limiting role of individual enzymes present in structured multienzymes (e.g., mitochondria or isolated cells) can be explored by inhibition of individual enzymatic components with a specific inhibitor. If the inhibition occurs in accordance with kinetics previously determined with isolated enzymes, inhibition of metabolism should be predictable from enzyme kinetics. Gross deviations from predicted inhibition patterns presumably indicate the operation of systems which may be recognizable by the difference between calculated (from enzyme kinetics) and observed experimental results.

There are several anticipated modes of responses to substrate-site-reactive competitive inhibitors. If enzymatic components of biochemically feasible metabolic pathways are connected in a linear multienzyme system (cf. ref. 20), inhibition of any enzymatic component should disrupt metabolic flux in accordance with the K_i of the inhibited enzyme. Ineffectivity of a site-specific inhibitor on metabolism may indicate that the pathway contains branching points and that the operation of the multienzyme system can be channeled in more than one direction. Simultaneous analyses of various predictable intermediary products, or tracer experiments, may settle this question. Permeability barriers (e.g., energy-coupled translocations) and unexpected reversal of metabolic flow (e.g., energy-coupled reverse electron transfer) are among the more complex reasons which may explain the ineffectivity of a specific inhibitor on the metabolic activity of structured multienzyme systems (e.g., mitochondria). At first approximation, it appears possible to distinguish between energy-coupled processes and enzyme mechanisms that operate at substrate levels by the combined use of uncouplers and substrate analogs (18, 19); however, certain calculated risks must be considered. The mode of action of uncouplers is as yet unknown (21), and it seems difficult to distinguish between inhibition of energy transfer on a macromolecular level and inhibition of "energy-coupled" cation and anion translocations.

The purpose of our experiments was to explore the metabolic consequences of uncoupling as well as to understand the nature of uncoupling itself; therefore, for experimental reasons, simplified assumptions were made. Applications of uncouplers were presumed to reduce the functional complexity of mitochondrial systems to a level where substrate-analog-type inhibitors should produce a kinetically predictable pattern. Specific inhibitors of enzymes of the metabolic pathways of glutamate were first tested. It was shown earlier (19) that monofluorooxalacetate is a powerful inhibitor of malate dehydrogenase, and that at low concentrations difluorooxalacetate inhibits only glutamate-aspartate transaminase. Combination of these fluorooxalacetates inhibits the transaminative pathway of glutamate, as shown in experiments with kidney mitochondria (22). The oxidative deamination of glutamate by glutamate dehydrogenase is specifically inhibited by monofluoroglutarate (19). Application of inhibitors of the oxidative or transaminative pathway in the absence and the presence of an uncoupler (2,4-dinitrophenol) should indicate how energy coupling may interfere with substrate level regulation of enzymatic pathways. The influence of extra-mitochondrial cellular components on mitochondrial metabolism was also measured by comparing glutamate metabolism by isolated liver mitochondria and by liver homogenates.

The following complex phenomena were observed. As shown in Table 1, the simultaneous inhibition of malate dehydrogenase and glutamate-aspartate transaminase by mono- and difluorooxalacetates, or the inhibition of glutamate dehydrogenase by fluoroglutarate alone, had no significant apparent effect on glutamate metabolism by isolated rat liver mitochondria, as measured by O_2 uptake and NH_3 evolution. Variation in experimental conditions, such as preincubation of mitochondria with the inhibitors for 15 min before the addition of glutamate (Expts. 4–6 and 10–12) had no significant effect on the rates of O_2 uptake, except that 2,4-dinitrophenol (DNP) increased respiration and NH_3 evolution. A variety of experiments, performed independently from the present study, showed that the fluorine-containing inhibitors actually exerted their effect on specific target enzymes in mitochondria, as measured by a large decrease in aspartate accumulation, when glutamate was the added substrate. Complete inhibition of glutamate metabolism could be achieved when both inhibitors of the transaminative pathway (i.e., fluorooxalacetates) and the inhibitor of glutamate dehydrogenase (i.e., fluoroglutarate) were present simultaneously. It became apparent that liver mitochondria were still capable of metabolizing glutamate when only one of the two enzymatically feasible pathways was inhibited. Since the alternating behavior of the two modes of glutamate metabolism by liver mitochondria was observed with the aid of specific enzyme inhibitors, it is probable that this multienzymatic mechanism is actually operative under cellular conditions.

The dithiol reagent arsenite, as well as oligomycin, a well-known inhibitor of energy transfer (50–53), were often applied for the study of enzymatic mechanisms of glutamate pathways (23). It seems probable that metabolic mechanisms based on the effect of arsenite and oligomycin are more difficult to interpret than results obtained with site-specific enzyme inhibitors. The distributive glutamate metabolism of liver mitochondria is in contrast to the multienzymatic behavior of kidney mitochondria, where inhibition of

Table 1. The Effects of Fluorine-Containing Inhibitors on Glutamate Metabolism by Liver Mitochondria under Varying Conditions[a]

Expt. No.	Experimental Conditions	ΔO_2	ΔNH_3
No preincubation			
1	Mitochondria + ADP	346	32
2	Mitochondria + ADP + F_1 + F_2	324	49
3	Mitochondria + ADP + F-glutarate	324	37
15-min preincubation			
4	Mitochondria + ADP	374	34
5	Mitochondria + ADP + F_1 + F_2	328	40
6	Mitochondria + ADP + F-glutarate	430	39
No preincubation			
7	Mitochondria + ADP + DNP	584	46
8	Mitochondria + ADP + DNP + F_1 + F_2	593	53
9	Mitochondria + ADP + DNP + F-glutarate	575	67
15-min preincubation			
10	Mitochondria + ADP + DNP	570	48
11	Mitochondria + ADP + DNP + F_1 + F_2	436	70
12	Mitochondria + ADP + DNP + F-glutarate	565	61

[a] Unpublished experiments.

NOTE TO TABLE 1

Each respirometer flask contains 20–30 mg (protein) of rat liver mitochondria, 0.15 M KCl, 3×10^{-2} M phosphate (pH 7.3), 10^{-3} M Mg^{2+}, 2.3×10^{-3} M ADP in a final volume of 3 ml; CO_2 is absorbed by KOH. Results are expressed as micromoles (O_2 or NH_3) per 1 g protein/30 min. Incubation is carried out at 30° for $\frac{1}{2}$ hr. When fluoro inhibitors are present, their concentration is 5×10^{-4} M; F_1 = monofluorooxalacetate, F_2 = difluorooxalacetate, F-glutarate = monofluoroglutarate. When indicated, mitochondria were preincubated for 15 min at 30° with inhibitors and/or 5×10^{-5} M 2,4-dinitrophenol before addition of substrate-glutamate (final concentration = 10^{-2} M). Rates of endogenous metabolism were subtracted.

any one of the key enzymes of either the oxidative or the transaminative pathway blocks glutamate utilization (22). The reasons for these apparently organ-specific differences between kidney and liver mitochondria are as yet unknown but may be related to the higher capacity of liver mitochondria to produce endogenous keto acid, thus supplying the second substrate required for transaminative glutamate utilization when the initiating reaction of oxidative deamination is inhibited.

A significantly different inhibition pattern of glutamate metabolism was observed when, instead of isolated mitochondria, liver homogenates were used. As shown in Table 2 (Expt. 5), inhibition of the transaminative pathway by fluorooxalacetates resulted in significant depression of O_2 uptake and a

Table 2. The Effect of Fluorine-Containing Inhibitors on Glutamate Metabolism by Liver Homogenates under Varying Conditions[a]

Expt. No.	Experimental Conditions	ΔO_2	ΔNH_3
No preincubation			
1	Homogenate + ADP	231	2
2	Homogenate + ADP + F_1 + F_2	190	9
3	Homogenate + ADP + F-glutarate	240	6
No preincubation			
4	Homogenate + ADP + DNP	308	53
5	Homogenate + ADP + DNP + F_1 + F_2	164	144
6	Homogenate + ADP + DNP + F-glutarate	325	60
15-min preincubation			
7	Homogenate + ADP	234	5
8	Homogenate + ADP + F_1 + F_2	170	38
9	Homogenate + ADP + F-glutarate	224	25
15-min preincubation			
10	Homogenate + ADP + DNP	300	118
11	Homogenate + ADP + DNP + F_1 + F_2	50	335
12	Homogenate + ADP + DNP + F-glutarate	40	240

[a] Unpublished experiments.

NOTE TO TABLE 2

Experimental conditions are the same as described in the Note to Table 1 except that liver homogenate, prepared in 0.15 M KCl (0.05 M Tris, pH 7.4), equivalent to 30 mg protein, replaced mitochondria.

large increase in NH_3 formation when the substrate analogs (fluorooxalacetates) were applied in the presence of 2,4-dinitrophenol (Expt. 5). Preincubation of homogenates for 15 min with fluoro inhibitors without an uncoupler also inhibited O_2 uptake (Expt. 8, Table 2), but there was no large increase in NH_3 production. The augmenting effect of 2,4-dinitrophenol on NH_3 production was reinvestigated separately and was traced to an increased rate of deamination of AMP in homogenates (24). Under these conditions, NH_3 is derived from both oxidative deamination of glutamate and deamination of AMP. When homogenates were preincubated with fluorine-containing inhibitors in the presence of 2,4-dinitrophenol (Expts. 11 and 12, Table 2), selective inhibition of either the oxidative or the transaminative pathways of glutamate resulted in almost complete inhibition of O_2 uptake. In other words, *an uncoupler modified the regulation of glutamate pathways as operative in homogenates only, and apparently converted the distributive liver system to the linear multienzymatic path as found in kidney mitochondria.*

At this stage of experimentation, it became evident that modification of mitochondrial glutamate metabolism by an uncoupler required extramitochondrial cellular components. Furthermore, a relatively simple experimental variable—preincubation with the uncoupler—reinforced the apparent enzyme inhibition by substrate-type inhibitors. These results were hard to explain on the basis of known metabolic mechanisms. A systematic search led to recognition of the existence of a cytoplasmic component which, under specific circumstances, acts on mitochondrial energy-coupled reactions.

3. ISOLATION OF THE CYTOPLASMIC METABOLIC FACTOR (CMF)

> *Comme c'est curieux, comme c'est bizarre, et quelle coincidence!*
>
> E. Ionesco

When isolated liver mitochondria were preincubated for 15 min in phosphate, 0.15 M KCl-Tris medium (25), in the presence of 2.3 mM ATP, 1 mM $MgCl_2$, and varying concentrations of 2,4-dinitrophenol and the rate of metabolism of subsequently added glutamate was followed, a 2,4-dinitrophenol concentration-dependent biphasic effect of the uncoupler was observed. The biphasic effect of 2,4-dinitrophenol depended critically on intactness of the liver mitochondria and was modified markedly by small amounts of rat liver cytosol, as shown in Fig. 1.

In the presence of cytosol (equivalent to 3.5 mg protein) or a cytoplasmic factor equivalent to the crude cytosol, the stimulatory effect of 2,4-dinitrophenol was sustained at higher concentrations of the uncoupler than in the

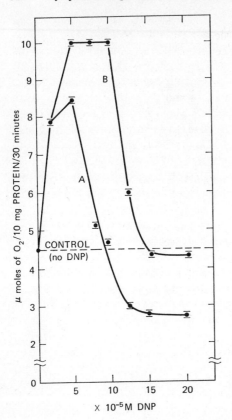

Fig. 1. The effect of CMF or cytosol on the oxidation of glutamate by liver mitochondria in the presence of $2.3 \times 10^{-3}\,M$ ATP, $10^{-3}\,M$ $MgCl_2$, $3 \times 10^{-2}\,M$ phosphate, and various concentrations of DNP. Curve A: no CMF or cytosol present; curve B: CMF (1.2 micromoles of reducing equivalents) or cytosol (3.5 mg of protein) was present in all reaction vessels, containing $0–20 \times 10^{-5}\,M$ DNP. The experimental procedure was the same as described in the text.

absence of the cytosol. This remarkable effect of liver cytosol was critically dependent on preincubation of mitochondria with specific concentrations of 2,4-dinitrophenol. It was soon realized that crude cytosol could be replaced by a relatively heat-stable cytoplasmic fraction obtained after heat treatment and chromatography of the cytosol (25). The increments of the rates of glutamate metabolism in the presence of varying amounts of cytosol were, within a certain range of added cytosol, proportional to its amount, provided glutamate oxidation was measured at one concentration of 2,4-dinitrophenol. The optimal concentration of 2,4-dinitrophenol had to be determined for each preparation of mitochondria and was characteristic for mitochondria prepared from different organs. This highly complex bioassay served as a guide for initial attempts to isolate the cytoplasmic component responsible for the stimulation of glutamate metabolism under the specifically defined experimental conditions (cf. ref. 25). For convenience, the active factor is now called cytoplasmic metabolic factor (CMF). Despite the variability due to the different uncoupler sensitivities of various batches of mitochondria and to many other experimental difficulties, a definite kinetic relationship could be established between the concentration of 2,4-dinitrophenol and the amount of cytoplasmic extract (CMF), provided all other experimental conditions were kept constant. This is shown in Figs. 2 and 3.

As the result of laborious experiments, involving many thousands of manometric tests, the existence of CMF in liver, kidney, and heart cytosol could be established with reasonable certainty (25). On the basis of gel filtration, the active cytoplasmic fraction was first thought to be macromolecular (4000–6000 daltons), but it was subsequently realized that Sephadex-type gels fail to indicate molecular size in this range. The active material passes through ordinary dialysis membranes and membrane filters, which retain particles having a molecular weight above 1000 daltons.

During the following period of work, the nature of the effect of CMF on mitochondria was elucidated more extensively than its chemical composition. The difficulty in chemical identification was due primarily to manufacturing problems, as is illustrated by the following description of a routinely followed method of isolation of CMF. Active preparations of CMF were obtained

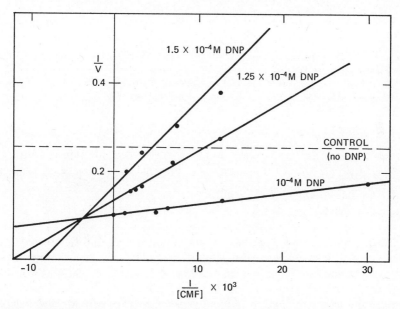

Fig. 2. Kinetic relationship between DNP and CMF. CMF concentration was varied at three levels of DNP. Ordinate: O_2 uptake (micromoles of O_2/10 mg of protein per 30 min); abscissa: CMF (micromoles of reducing equivalents). Substrate was glutamate. Experimental conditions were the same as described in the legend to Fig. 1.

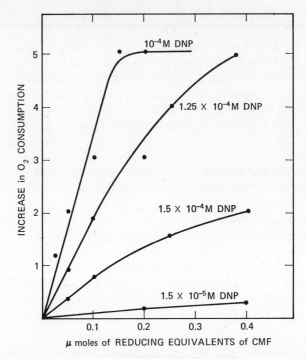

Fig. 3. Kinetic relationship between DNP and CMF. CMF concentration was varied at three levels of DNP. Ordinate: O_2 uptake (micromoles of O_2/10 mg of protein/30 min); abscissa: CMF (micromoles of reducing equivalents). Substrate was glutamate. Experimental conditions were the same as described in the legend to Fig. 1.

from a variety of animal tissues (rat liver, pig liver, kidney, heart, Ehrlich ascites tumor cells*) and yeast.† Since most experiments were carried out with pig liver preparations, the method of purification will be described from this source (cf. ref. 26).

ISOLATION PROCEDURE

Freshly obtained pig liver was chilled in cracked ice and then sliced and homogenized with an equal amount of ice cold (4°) distilled water in a Sorvall Omnimixer for 45 sec, in batches of 300 g, while chilled in an ice

---

* We are indebted to Dr. Julian L. Ambrus (Roswell Park Memorial Institute, Buffalo, N.Y.) for several hundred grams of freeze-dried Ehrlich ascites tumor cells used in the isolation of CMF.
† We are indebted to Dr. Jack Siegel of P-L Biochemicals for the supply of aqueous yeast extract used in these studies.

bath. The crude sediment was removed by centrifugation at 16,300 × g (at 4°), for 1 hr. The supernatant was decanted, and in 70-ml batches (in Erlenmeyer flasks with agitation) submerged in a boiling water bath for 2.75 min, in order to coagulate the proteins. The combined batches of heat-treated material were centrifuged at 34,800 × g for 40 min, and the volume of the clear yellow-red supernatant was reduced to about one-fifth of its original by freeze drying.

This extract was fractionated first by passing it through a large Sephadex G-25 column equilibrated with distilled water (at 4°). The size of the column suitable for separation of this amount of crude extract was 9.5 cm (diameter) × 50 cm (height), containing Sephadex G-25 of a bed volume of 3.5 liters (void volume = 1.2 liters). The volume of the heat-treated liver extract was usually 250 ml. Fractions of 15 ml were collected by an automated fraction collector. The flow rate was 10 ml/min.

Three distinct fractions were obtained, as illustrated in Fig. 4. The first

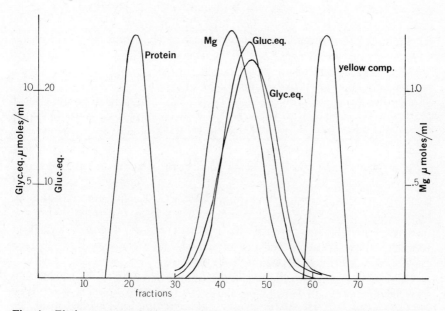

Fig. 4. Elution pattern of Sephadex G-25 column, developed with distilled H_2O. Left ordinate: glycine equivalents, calculated from colorimetric ninhydrin analysis, and glucose equivalents, calculated from colorimetric determination of reducing material (Cf. ref. 25). Right ordinate: micromoles Mg^{2+}/ml effluent, determined by atomic absorption (with the Perkin–Elmer Model 403 instrument). The volume of each fraction is 15 ml. Unpublished results.

fraction was the residual protein remaining in the extract after heat treatment. The second fraction contained almost quantitatively the CMF present in the original homogenate, as estimated by the metabolic assay (cf. ref. 27), and almost all of the Mg^{2+} present in the liver extract, as well as ninhydrin-positive and reducing material, which coincided with CMF activity. It became apparent that the organic constituents of this fraction were a mixture of Mg^{2+} chelates. This mixture behaved as one peak in smaller-size G-25 columns and simulated macromolecular behavior in earlier experiments (25). The total volume of the second fraction, as shown in Fig. 4, was 800 ml. Its volume was reduced to 100 ml by freeze drying. At this stage of purity, the liver extract contained less than enzymatically detectable ATP or ADP (less than 10^{-5} M), as measured by specific spectrophotometric enzyme assays. Enzymatic assays revealed the presence of glucose-6-PO$_4$, but this sugar phosphate had no CMF-like effect. The third fraction obtained by Sephadex G-25 chromatography was bright yellow, and since it had no CMF activity it was not analyzed further.

The second, active fraction obtained by the Sephadex G-25 column, after concentration to 100 ml, was passed through a Chelex (Bio-Rad) column in order to remove metal ions. The Mg^{2+}-binding capacity of commercial Chelex resins had to be determined for each batch. The average size of the Chelex column, which contained on the average 35 g resin as the Na$^+$ salt, at pH 7.4 (Tris), was 3.5 cm × 4.5 cm. Besides Mg^{2+} and Ca^{2+}, the Chelex column removed about 20% inert organic material.

Each 100 ml of effluent from the Chelex column was subsequently passed through a Dowex-1-bicarbonate column in order to remove material which absorbs at 260 nm. A column of 3 cm × 25 cm removed about 90% of material which absorbed light at 260 nm without appreciable loss of CMF activity. The eluate was evaporated to dryness, leaving 4–6 g of organic material* (from 1.8 kg fresh liver) which contained about 74% or more of the CMF activity present in the original homogenate. This material could be kept at −15° in the dry form for months without loss of activity and, for most experiments dealing with the mode of action of CMF, was used routinely. Maximal *in vitro* effects were elicited at a concentration range of 0.5–0.9 mg CMF/ml test system (27–30, 34, 48).

The CMF preparation obtained in the manner described contains reducing material and ninhydrin-positive contaminants, which can be removed by chromatography on DOWEX-50-Na$^+$ (pH 3.6), thin layer chromotography (cf. ref. 29), or by ethanol fractionation as the Ba^{2+} salt.

---

* Total organic material was determined by a micro modification of the dichromate method of Johnson (cf. ref. 26).

4. THE EFFECTS OF CMF IN THE PRESENCE OF ARTIFICIAL UNCOUPLERS

A study of experimental conditions necessary for demonstration of the effect of CMF in the presence of artificial uncouplers led to some degree of understanding of the mode of action of both the uncouplers and CMF. It was evident that CMF did not act directly on an enzymatic component of mitochondria. Since preincubation of mitochondria with uncouplers was a necessary experimental requirement for the effectivity of CMF,* it seemed important to study the criteria of mitochondrial stability, which was eventually related to Mg^{2+} binding of the inner membrane.

A. Uncoupler-Induced Mg^{2+} Ejection and Its Relationship to Mitochondrial Respiration

The following experiment provided an approach to the study of mitochondrial stability. As shown earlier (25), incubation of mitochondria with ATP or ADP and varying concentrations of an artificial uncoupler had a biphasic effect on the rates of glutamate oxidation (see Fig. 1). These experiments were repeated in a simplified test system, composed of 0.15 M KCl-Tris (pH 7.4), 2.3 mM ADP, 0.1 mM $MgCl_2$, and varying concentrations of 2,4-dinitrophenol. When the efflux of inner-membrane-bound Mg^{2+} was monitored simultaneously with O_2 uptake, it was found that the ability of mitochondria to respond with an increase in O_2 consumption to a certain concentration of 2,4-dinitrophenol was inversely proportional to Mg^{2+} efflux. Retention of mitochondrial Mg^{2+} by CMF was essential for the stimulatory effect of 2,4-dinitrophenol on glutamate metabolism. This is shown in Fig. 5.

As found later (28), ejection of mitochondrial Mg^{2+} by 2,4-dinitrophenol + ADP and its prevention by CMF occur without externally added Mg^{2+}. On the other hand, the prevention of Mg^{2+} ejection by CMF in *respiring mitochondria* requires 0.1 mM Mg^{2+} in addition to CMF. In the respiring mitochondria, the Mg^{2+}-stabilizing effect of CMF consists of an apparent increase of the affinity of mitochondria toward externally added Mg^{2+}. This is illustrated in Figs. 6 and 7. As shown in Table 3 (30), the same range

* More recent experiments showed that CMF has an immediate effect as an antagonist of DNP on endogenous oxidative phosphorylation of liver mitochondria (cf. ref. 29, and D. C. Lin and E. Kun, *Fed. Amer. Soc. Exp. Biol.*, 1972, Atlantic City, N.J., Abstract).

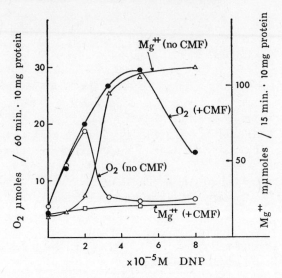

Fig. 5. Inverse relationship between Mg^{2+} retention and the stimulating effect of 2,4-dinitrophenol. The efflux of Mg^{2+} was induced by preincubation of mitochondria with varying concentrations of 2,4 dinitrophenol + 2.3 mM ADP in the absence and the presence of CMF. When the oxidation of added glutamate was followed after preincubation, 0.1 mM $MgCl_2$ was also required for stimulation of O_2 uptake by 2,4 dinitrophenol + CMF. The presence of 0.1 mM Mg^{2+} had no effect on Mg^{2+} ejection by 2,4 dinitrophenol + ADP and did not influence Mg^{2+} retention in the presence of CMF. Unpublished experiments.

of uncouplers is required to eject bound Mg^{2+} from mitochondria as to uncouple oxidative phosphorylation. Furthermore, uncoupler-induced ATPase requires bound mitochondrial Mg^{2+}, as shown in Fig. 8.

It may be concluded from these results that some correlation exists between bound Mg^{2+}, energy transfer, electron flux, and ATP synthesis. A direct connection between uncoupling and labilization of bound Mg^{2+} is complicated by the uncertain criteria of the process of uncoupling. As Fig. 5 shows, maximal stimulation of respiration by 2,4-dinitrophenol occurs at a concentration of the uncoupler which is lower than required for Mg^{2+} ejection. It may be concluded from Fig. 5 that bound Mg^{2+} is *necessary* for increased respiration after exposure of mitochondria to a certain concentration of an uncoupler. The general belief is that increased respiration caused by an uncoupler is synonymous with uncoupling, but there is no assurance that the process of "uncoupling" does not consist of a series of successive reactions. The ability to bind Mg^{2+} may be a step in the sequence of reactions leading to energy coupling. Conversely, release of Mg^{2+} by an uncoupler may be an isolated step of the uncoupling process.

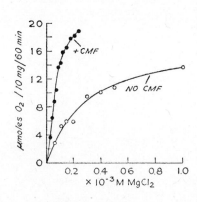

Fig. 6. Relationship between rates of glutamate metabolism, concentration of $MgCl_2$, and concentration of Mg^{2+} present in CMF. Experimental conditions were the same as described for Fig. 1, curve *B*. Reaction rates (ordinate) are expressed as micromoles of O_2/10 mg of mitochondrial protein per 60 min. Abscissa: concentration of added Mg^{2+}. The lower curve shows the $MgCl_2$ concentration–reaction rate relationship, while the upper curve indicates the effect of Mg^{2+} present in increasing amounts of CMF. The stock liver CMF contained 9.7 mM Mg^{2+}. The reaction system contained 5.9 mg of mitochondrial protein and components of the standard assay (2.3 mM ADP and $5 \times 10^{-5}\,M$ DNP). Rates of O_2 uptake were corrected for controls containing only DNP, no $MgCl_2$ or CMF.

An accurate appraisal of sequential steps leading to the inhibition of ATP synthesis by uncouplers is necessary before the term "uncoupling" can be considered meaningful. At the present time, it seems more appropriate to conclude that the process of uncoupling (and therefore energy coupling itself) is related to bound mitochondrial Mg^{2+}. If this conclusion is correct, there should be a correlation between the Mg^{2+} content of submitochondrial inner membrane preparations and their ability to perform energy-coupled reactions. As shown in Table 4, inner membrane preparations isolated in

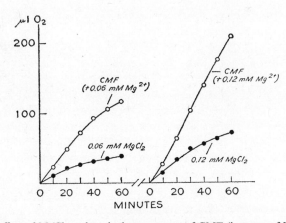

Fig. 7. The effect of $MgCl_2$ and equivalent amounts of CMF (in terms of Mg^{2+}) on glutamate metabolism as a function of time. Experimental conditions were the same as described in the legend of Fig. 6, except that two fixed concentrations of Mg^{2+} (as $MgCl_2$ or Mg^{2+} present in CMF) were employed.

Table 3. Concentrations of Uncouplers Required for ADP-Dependent Mg^{2+} Ejection and for Disruption of Oxidative Phosphorylation

	Concentration of Uncoupler (in M) Required for:			
Uncoupler	$0<$	Mg^{2+} Ejection	$\sim 100\%$	Uncoupling
DNP	2	to	5×10^{-5}	2.7×10^{-5}
PCP	2	to	8×10^{-5}	2.5×10^{-5}
TTFB	2	to	10×10^{-7}	3.0×10^{-7}
DCIP	2	to	5×10^{-5}	4.0×10^{-5}

NOTE TO TABLE 3

Ejection of Mg^{2+} was measured at 30° in the presence of 2.3 mM ADP and varying amounts of uncouplers and oxidative phosphorylation (at 30°) polarographically with glutamate as substrate.

Abbreviations: DNP = 2,4-dinitrophenol, PCP = pentachlorophenol, TTFB = 4,5,6,7-tetrachlorotrifluoromethylbenzimidazole, DCIP = 2,6-dichlorophenolindophenol.

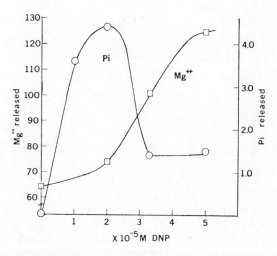

Fig. 8. DNP concentration dependence of ATPase of mitochondria and induced Mg^{2+} ejection. Conditions are the same as described in Table 3.

Table 4. Distribution of Mg^{2+} in Rat Liver Mitochondria

Name of Fraction	Amount of Fraction, mg Protein	Per Cent of Mitochondrial Protein/Fraction	Moles × 10⁻⁶ Mg^{2+}/Fraction	Moles × 10⁻⁶ Mitochondrial Mg^{2+} Recovered in Fractions
Expt. 1				
Whole mitochondria	315	...	10.5	...
LSP	210	67	8.2	78
Expt. 2				
Whole mitochondria	332	...	5.25	...
Inner membrane H₂O ghosts	82	24.5	1.00	19
Inner membrane succinate ghosts	132	40.0	2.52	48
Inner membrane EDTA ghosts	110	33.0	3.18	61

NOTE TO TABLE 4

In Expt. 1, the inner membrane + matrix (LSP) fraction was isolated by the digitonin technique; in Expt. 2, inner membrane ghosts were prepared by distilled water treatment (cf. ref. 31), in 10 mM succinate and 10 mM EDTA.

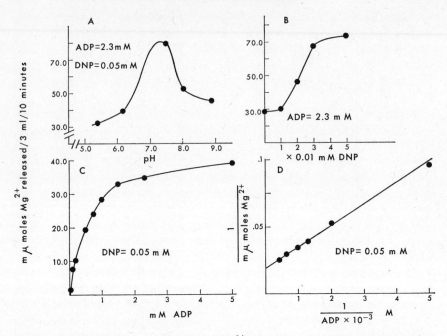

Fig. 9. The pH dependence and kinetics of Mg^{2+} release from liver mitochondria induced by DNP + ADP. The rate of Mg^{2+} ejection during 10 min of aerobic incubation at 30° of mitochondria (5 mg protein) in 3 ml KCl-Tris was determined as described in Ref. 1. In A and B, endogenous Mg^{2+} release (30 nM/10 min) was not subtracted, while C and D were corrected for this background value.

the presence of succinate or EDTA (31) retain significantly higher concentrations of Mg^{2+} than mitochondrial fragments isolated in H_2O. Inner membrane fragments prepared in succinate or EDTA performed oxidative phosphorylation, whereas the H_2O preparations did not; therefore positive correlation exists between Mg^{2+} retention and the ability to carry out oxidative phosphorylation. Inner membrane fragments prepared in EDTA have been extensively employed for the study of energy-coupled reactions (32), but the correlation between Mg^{2+} content and energy coupling was not recognized.

If bound Mg^{2+} is essential for energy transfer, it is of importance to study the membrane-associated ligand system for Mg^{2+}. As an approach to this question, the pH dependence and the kinetics of uncoupler-induced Mg^{2+} efflux were determined, as shown for 2,4-dinitrophenol in Fig. 9.

In the range of 2,4-dinitrophenol concentration which is effective in the metabolic assay of CMF (27), a second component, ADP, is specifically required for Mg^{2+} ejection. Both hyperbolic and sigmoidal correlations can be observed between rates of Mg^{2+} efflux and uncoupler or ADP concentration, as illustrated in Figs. 9 and 10. When valinomycin was the Mg^{2+}-labilizing agent, between 10^{-7} and 10^{-8} M, not only ADP but also, characteristically, K^+ was necessary to induce Mg^{2+} efflux (Fig. 11) by this antibiotic. Sigmoidal kinetics of membrane-associated reactions, presumably characteristic for the cooperativity between macromolecular membrane components, has been proposed (33). It seems possible that the results shown in Figs. 9–11 can be explained by interactions between membrane components.

The phosphorylation site specificity of the uncoupler-CMF antagonism has not yet been studied extensively. Preliminary results suggest, however, that bound Mg^{2+} plays the same role at all three sites of phosphorylation. Energy transfer at any phosphorylating site is probably mediated by the same cation-binding macromolecular system.

If the stability of Mg^{2+} binding in mitochondria can influence metabolic rates, it is logical to expect that inhibition of electron transfer or addition of oxidizable substrates should influence Mg^{2+} binding in mitochondria. This assumption was verified experimentally (Table 5), and the results show that specific inhibitors of electron flux, although not affecting Mg^{2+} binding directly, can inhibit the Mg^{2+}-stabilizing effect of CMF (see Expts. 14–17, Table 5). Labilization of Mg^{2+} by 2,4-dinitrophenol + ADP is temporarily inhibited by externally added NAD^+-requiring substrates. These added substrates are ineffective, however, in the presence of valinomycin + K^+ as

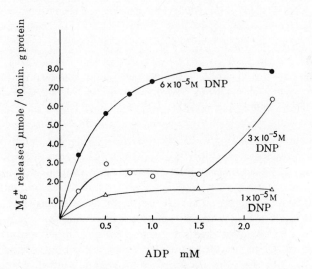

Fig. 10. Hyperbolic and sigmoidal relationship between rates of Mg^{2+} efflux and ADP concentration as a function of the concentration of 2,4-dinitrophenol (DNP). Conditions are the same as those described in Fig. 9.

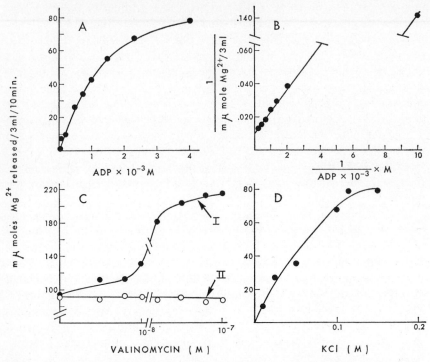

Fig. 11. Kinetics of Mg^{2+} release induced by valinomycin + K^+. In *A* and *B*, valinomycin concentration was 1.5×10^{-8} *M*, and the concentration of ADP was varied. In *C* the effect of variation in valinomycin concentration is shown in 0.15 *M* KCl (curve I) and in Tris-sucrose (curve II). In *C* the amount of mitochondria was increased to 8.3 mg/3 ml in order to improve the assay of endogenous Mg^{2+} release. This value is not subtracted in *C* and corresponds to the valinomycin- and K^+-independent rate found in Tris-sucrose-mannitol medium (curve II). In *D*, the valinomycin level was 1.5×10^{-8} *M*, and the suspending medium Tris-sucrose. In both *C* and *D*, ADP was 2.3 m*M*.

Mg^{2+} labilizers (Expts. 18 and 19). In this case, only CMF is effective (Expt. 20). In the presence of uncouplers, the Mg^{2+} *stabilizing* effect of oxidizable substrates is abolished by inhibitors of respiration (Expts. 13–16).

Despite the presence of uncouplers, CMF seems to orient energy derived from respiration to support Mg^{2+} retention by sustaining maximal rates of electron transfer to O_2. Whereas a causal relationship is apparent between cation retention and preservation of energy-requiring mitochondrial functions, it is not known to what extent cation retention contributes to the molecular process of energy transfer. A practical application of this hypothesis is routinely used in this laboratory to predict the energetic "intactness" of isolated rat liver mitochondria. It has been found that high concentrations

Table 5. Influence of Substrates, Inhibitors, and CMF on the Initial Rate of Mg^{2+} Ejection from Mitochondria

Expt. No.	Inducing Agent	Variable Components	Rate of Mg^{2+} Loss[a]
1	ADP + DNP	...	103
2	ADP + DNP	10^{-2} M glutamate	30
3	ADP + DNP	10^{-2} M glutamate + 2.4×10^{-7} M rotenone	106
4	ADP + DNP	10^{-2} M glutamate + 5×10^{-4} M arsenite	85
5	ADP + DNP	10^{-2} M glutamate + antimycin[b]	100
6	ADP + DNP	10^{-2} M α-ketoglutarate + 2.4×10^{-7} M rotenone	56
7	ADP + DNP	10^{-2} M α-ketoglutarate	106
8	ADP + DNP	10^{-2} M α-ketoglutarate + 5×10^{-4} M arsenite	94
9	ADP + DNP	10^{-2} M α-ketoglutarate + antimycin[b]	90
10	ADP + DNP	10^{-2} M succinate	42
11	ADP + DNP	10^{-2} M succinate + 2.4×10^{-7} M rotenone	57
12	ADP + DNP	10^{-2} M succinate + 5×10^{-4} M arsenite	42
13	ADP + DNP	10^{-2} M succinate + antimycin[b]	90
14	ADP + DNP	CMF	3
15	ADP + DNP	CMF + 2.4×10^{-7} M rotenone	119
16	ADP + DNP	CMF + antimycin[b]	154
17	ADP + valinomycin + K^+ (0.1 M)		98
18	ADP + valinomycin + K^+ (0.1 M)	10^{-2} M glutamate	98
19	ADP + valinomycin + K^+ (0.1 M)	10^{-2} M succinate	90
20	ADP + valinomycin + K^+ (0.1 M)	CMF	44

[a] Nanomoles/10 min/10 mg protein at 30°. [b] 0.1 μg antimycin/mg protein.

of K+ and Mg²⁺ in mitochondria invariably coincide with the capability of the mitochondria to perform energy-dependent catalytic processes.

B.   Decrease of Mitochondrial Affinity for ADP by 2,4-Dinitrophenol and Its Prevention by CMF

The experiments so far described were concerned with the apparent relationship between metabolic rates and Mg²⁺ movements in the presence of artificial uncouplers (+ADP) and CMF. The cytoplasmic factor CMF consistently preserved Mg²⁺ retention by acting as an antagonist to uncouplers. In metabolic experiments, CMF increased the affinity or effectivity of low concentrations (0.1 mM) of externally added Mg²⁺. When the uncoupler concentration was kept constant (50 μM 2,4-dinitrophenol) and the concentration of ADP varied during preincubation, the rate of glutamate oxidation by liver mitochondria was unaffected by variations in ADP concentration. However, in the presence of CMF, respiration became dependent on the concentration of externally added ADP and was maximally stimulated by 1 mM ADP. The apparent affinity of mitochondria for ADP was re-established by CMF (top curve of Fig. 12). This effect of CMF can

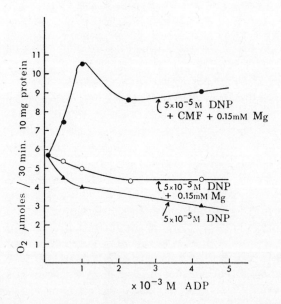

Fig. 12. The effect of CMF on the apparent affinity of mitochondria toward ADP. Experimental conditions were the same as in Fig. 5, except that DNP concentration was kept constant at 5×10^{-5} M and ADP concentration was varied, as shown in the abscissa.

be demonstrated only by comparing the effect of preincubation with an uncoupler in the absence and the presence of CMF.

According to a simplified interpretation, 2,4-dinitrophenol reduces the metabolic efficiency of ADP, and CMF restores or preserves the apparent high affinity of mitochondria for ADP and, as shown previously, also for externally added Mg^{2+}. The ADP and Mg^{2+} sites of mitochondria which are affected by uncouplers and CMF are presumably in the inner membrane. This system responds to uncouplers with a progressive decay of energy transduction. Loss of bound Mg^{2+} or decrease of affinity for externally added Mg^{2+} and ADP is a sign of decay. Since CMF can counteract these symptoms of disintegration of the mitochondrial energy transfer apparatus, it seems reasonable to define CMF as a "stability cofactor" of energy-coupled mitochondrial functions. From results so far discussed, the protective or stabilizing effect of CMF has only pharmacological or toxicological significance. Controlled degradation of the inner-membrane-associated energy-transducing system by artificial uncouplers offers a new experimental approach to the study of the functional components of this system.

C. Modification of Glutamate Metabolism by Mitochondria in the Presence of 2,4-Dinitrophenol and CMF

The discovery of CMF originated in experiments concerned with the effectivity of fluorine-containing enzyme inhibitors on metabolic pathways of glutamate (see Section 4). So far, little attention has been paid to the metabolic fate of glutamate during the conditions which permitted the recognition and estimation of the effect of CMF. Direct metabolic studies may shed some light on the significance of the previously discussed stabilizing influence of CMF on the mitochondrial energy transfer system (34).

The steady-state concentrations of extramitochondrial α-ketoglutarate, NH_3, and aspartate were determined at various concentrations of 2,4-dinitrophenol. Under the usual experimental conditions employed for the assay of CMF, the results illustrated in Fig. 13 were obtained (34). The typical increase in O_2 consumption that occurred at $2-5 \times 10^{-5}$ M 2,4-dinitrophenol was followed by a decline in respiration to the "control" (containing no uncoupler) level. The rate of aspartate formation ran parallel with the biphasic rate of O_2 uptake. As soon as both O_2 consumption and aspartate accumulation declined above 3×10^{-5} M 2,4-dinitrophenol, NH_3 and α-ketoglutarate began to accumulate in the extramitochondrial suspending medium. Diminution of aspartate formation, O_2 uptake, and accumulation of α-ketoglutarate and NH_3 *coincided with loss of mitochondrial* Mg^{2+} (see Section 4.B).

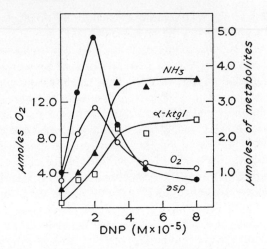

Fig. 13. The influence of varying concentrations of DNP on the rate of O_2 consumption (left ordinate) and product formation (right ordinate) in the presence of glutamate (30 μM/3 ml). Standard manometric assays were carried out as described in ref. 27. Mitochondria (5 mg protein) were incubated in the Tris-KCl-ADP medium with varying concentrations of DNP for 15 min before addition of glutamate. Results are expressed as micromoles/60 min/3 ml reaction system.

When CMF was present during preincubation, increased O_2 uptake and aspartate accumulation were sustained. This effect of CMF is illustrated in Fig. 14. Since the consumption of glutamate was difficult to determine in the presence of relatively high concentrations of this substrate, the efficiency with which glutamate was converted to aspartate was expressed as the ratio of α-ketoglutarate and the sum of NH_3 and aspartate (Fig. 14, 2). At 5 × 10^{-5} M 2,4-dinitrophenol concentration, about 60% of glutamate was oxidatively deaminated and appeared as extramitochondrial NH_3 + α-ketoglutarate. The accumulation of α-ketoglutarate coincided with diminished O_2 uptake. The presence of CMF prevented the accumulation of α-ketoglutarate and simultaneously raised the O_2 uptake to maximal rates, but did not diminish NH_3 accumulation (Fig. 14, 4). A critical concentration of 2,4-dinitrophenol inhibited the ability of mitochondria to retain bound Mg^{2+}, and this effect coincided with the failure to retain α-ketoglutarate in the metabolic compartment. Extrusion of α-ketoglutarate from mitochondria would be expected to diminish oxalacetate formation, and thus the concentration of the second substrate required for transaminative aspartate accumulation should also decline. This mechanism explained the diminished rate of aspartate formation. The present results illustrate the profound metabolic consequences of the labilization of energy transfer systems by uncouplers and the opposing effect of CMF to uncouplers.

It was now possible to reinvestigate the combined effects of fluorooxalace-tates, 2,4-dinitrophenol, and CMF on the mitochondrial metabolism of glutamate. As shown in Table 6, after preincubation of mitochondria with 2.3 mM ADP and 5×10^{-5} M 2,4-dinitrophenol, CMF stimulated O_2 consumption (5-fold), but aspartate accumulation was augmented even more (17-fold). This effect suggested that transamination was critically rate-limiting under these conditions. It follows that inhibitors of the transam-inative pathway should inhibit the increment in glutamate metabolism induced by CMF, and this prediction was verified as shown in Expt. 6, Table 6. On the other hand, inhibitors of the transaminative pathway (i.e., fluorooxalacetates) had no effect on glutamate metabolism in the *absence* of CMF. These results confirm those of earlier experiments (Tables 1 and 2), and show that liver mitochondria posses an alternating mechanism of glutamate utilization.

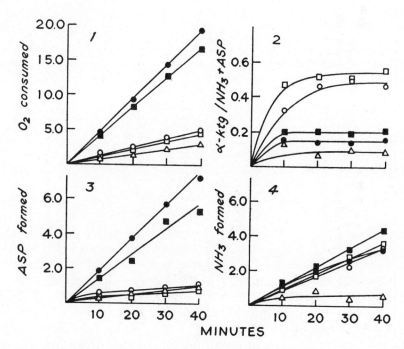

Fig. 14. The effect of CMF on rates of O_2 consumption and product formation from glutamate. Experimental conditions were the same as described in legend of Fig. 13 except analyses for reaction products in the extramitochondrial compartment were carried out in 10 min intervals. Symbols: —△— = control, containing no DNP; —□— = $+5 \times 10^{-5}$ M DNP; —○— = 8×10^{-5} M DNP; —■— and —●— = 5 and 8×10^{-5} M DNP + CMF (partially purified = 3.4 μequiv. reducing sybstance and 0.12 mM Mg^{2+}). All values are expressed as micromoles/3 ml reaction system.

Table 6. The Effects of Fluorooxalacetates and CMF on Glutamate Metabolism under Varying Conditions

Expt. No.	Variable Components	O₂ Consumption	α-Keto-glutarate	NH₃	Aspartate	Malate
			Rate of Formation of:			
1	Tris-ADP + Glu	2.4	0.18	1.3	1.6	0.4
2	Tris-ADP + Glu + CMF	2.1	...	1.3	1.8	...
3	Tris-ADP + Glu + 5 × 10⁻⁵ M DNP	3.7	2.1	3.2	0.5	0.7
4	Tris-ADP + Glu + 5 × 10⁻⁵ M DNP + CMF	15.1	1.3	3.2	8.7	0.5
5	Tris-ADP + Glu + 5 × 10⁻⁵ M DNP + F_1 + F_2	3.5	2.0	3.5	0.6	0.5
6	Tris-ADP + Glu + 5 × 10⁻⁵ M DNP + F_1 + F_2 + CMF	3.5	1.1	3.2	1.0	1.1
7	Tris-ADP + Glu + OAA	2.8	7.5	1.2	8.0	2.9
8	Tris-ADP + Glu + OAA + 5 × 10⁻⁵ M DNP	2.0	11.0	3.1	10.0	2.1
9	Tris-ADP + Glu + OAA + 5 × 10⁻⁵ M DNP + CMF	14.6	6.0	0.3	15.0	1.6
10	Tris-ADP + Glu + OAA + 5 × 10⁻⁵ M DNP + F_1 + F_2	1.9	10.5	3.2	10.0	2.5
11	Tris-ADP + Glu + OAA + 5 × 10⁻⁵ M DNP + F_1 + F_2 + CMF	3.2	10.0	2.0	11.0	2.5

NOTE TO TABLE 6

Abbreviations: F_1 = monofluorooxalacetate (final concentration 5×10^{-4} M); F_2 = difluorooxalacetate (final concentration 5×10^{-4} M); DNP = 2,4-dinitrophenol; CMF = same as in legend of Fig. 14; Glu = 10^{-2} M glutamate; OAA = 10^{-2} M oxalacetate. All numerical values are in micromoles/60 mins/3 ml reaction volume. For details of the manometric system see ref. 27.

330

When glutamate metabolism was studied in the presence of *added oxalace-tate*, a condition which would be expected to result in a maximal rate of transamination, CMF still augmented O_2 uptake but had no influence on the already high rate of aspartate formation. It follows that stimulation of O_2 uptake by CMF must be due to the oxidation of intramitochondrially formed α-ketoglutarate, which is retained in mitochondria in the presence of CMF and thus is made available for oxidative metabolism (Fig. 14). The apparent stimulation of transamination-dependent accumulation of aspartate by CMF is measurable only when the rate of intramitochondrial oxalacetate formation limits the rate of transamination. This condition is created by preincubation of mitochondria with 2,4-dinitrophenol + ADP.

5. THE EFFECT OF CMF IN THE PRESENCE OF POTENTIALLY TOXIC METABOLITES

A. Oleic Acid-CMF Antagonism

Certain natural products present in cellular systems have been recognized as exerting effects that are similar, in many ways, to the uncoupling of oxidative phosphorylation by artificial uncouplers. It was shown in Lardy's laboratory (35) that oleic acid at low concentrations uncouples oxidative phosphoryla-tion. Recently these results were confirmed and extended (36). We observed that oleic acid below 2 μM concentration increases glutamate oxidation but that higher concentrations inhibit respiration. Like 2,4-dinitrophenol, oleate, at a concentration which depressed O_2 uptake, induced Mg^{2+} and Ca^{2+} ejection. Also, CMF antagonized oleate-induced cation efflux. These effects are illustrated in Fig. 15, where O_2 uptake and Mg^{2+} ejection are correlated in the upper section and Ca^{2+} efflux is shown in the lower.

It has been suggested by Lehninger (8) that fatty acids (U factor) may play a cellular regulatory function by way of uncoupling and mitochondrial swelling. Antagonism by CMF to these destructive effects of oleate may be important in maintaining intact mitochondrial function in a cellular environ-ment under abnormal circumstances.

B. Bilirubin-CMF Antagonism

This degradation product of hemoproteins has been assumed to play a role in chemical pathology. If detoxification of bilirubin by conjugation with uridine diphosphate-glucuronate is impaired, or its oxidative metabolism is inhibited, free bilirubin may exert a toxic effect which is not well understood

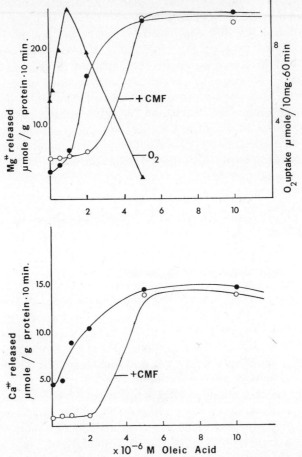

Fig. 15. The effects of increasing concentrations of oleate on Mg^{2+} efflux, rates of O_2 uptake (upper graph), and Ca^{2+} efflux (lower graph). Rat liver mitochondria (2.1 mg protein/ml) were suspended in 30 mM Tris-HCl (pH 7.4) + 150 mM KCl, and cation efflux was determined during 10 min incubation at 30° in air. Rates of cation efflux are shown in ordinate when O_2 uptake was measured. After 10 min preincubation (with oleate) 10 mM glutamate was tipped in from the side arm, and O_2 uptake (state 4 respiration) was measured for 60 min in Warburg respirometers. Unpublished results.

(37, 38). It was reported that bilirubin uncouples oxidative phosphorylation (see ref. 39), but it has also been suggested that bilirubin acts on mitochondrial membranes as a detergent (40). According to a report from King's laboratory (39), 20 μM bilirubin increases and 50 μM bilirubin inhibits the respiration of liver and heart mitochondria. On the basis of a variety of experiments, King and his colleagues also conclude that bilirubin may act as an ion-transport-inducing agent (cf. p. 6412 ref. 39). In our experience, bilirubin is a very potent Mg^{2+}-labilizing agent, highly effective at less than 10 μM concentrations (Fig. 16).

The pH maximum for the effectivity of bilirubin as a Mg^{2+}-efflux-inducing agent is about 7.5, close to the pH maximum for Mg^{2+} ejection by 2,4-dinitrophenol + ADP (see Fig. 9). This pH profile is shown in Fig. 17. It is probable that this pH maximum reflects some property of the ligand system of Mg^{2+} in mitochondria. Bilirubin is antagonized not only by CMF but also by orthophosphate and ATP, as shown in Fig. 18: ADP is ineffective. In the presence of 8.3 μM bilirubin, half-maximal protection against Mg^{2+} ejection was obtained with 360 μM orthophosphate or 80 μM ATP and CMF, equivalent to 9 μM reducing substance. The active component of CMF was estimated to be 1–5% of the total organic substance present in the preparation used. On a weight basis, the estimated protective efficiency

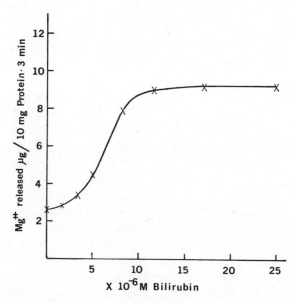

Fig. 16. Relationships between rates of Mg^{2+} release in 3 min and the concentrations of bilirubin. The reaction mixture contained Tris-HCl, 30 mM, pH 7.5; KCl, 150 mM, mitochondria, 5.6 mg protein; bilirubin, as indicated in the graph. The reaction mixtures were incubated in Dubnoff shaker, and ejection of Mg^{2+} measured as reported previously (cf. ref. 27). In all experiments where the effect of bilirubin on Mg^{2+} ejection was determined, the spontaneously released Mg^{2+} (in the absence of bilirubin) was subtracted. This was in the range of 30–46 $\times$ 10^{-9} g Mg^{2+}/10 mg protein/3 min at 30°. Unpublished experiments.

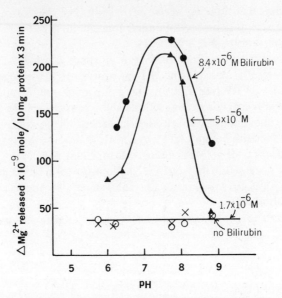

Fig. 17. The pH dependence of Mg^{2+} release from rat liver mitochondria induced by bilirubin during incubation at $30°$ in 3 min. Experimental conditions were the same as described in the legend of Fig. 16. The pH of each incubation medium was adjusted with the aid of a glass electrode with dilute KOH or HCl at $30°$. Unpublished experiments.

of CMF is much greater than that of orthophosphate or ATP against bilirubin-induced Mg^{2+} efflux, sug-gesting a specific role of CMF the in maintenance of the functional integrity of mitochondria within the cell.

6. CYTOPLASMIC REGULATION OF MITOCHONDRIAL BIOENERGETICS BY Ca^{2+}-CMF ANTAGONISM

It is generally recognized that Ca^{2+} plays the role of a messenger for the initiation of various processes essential for intracellular communication. As pointed out by Rasmussen (41), many important processes, such as hormonal mechanisms, neural transmission, cyclic AMP-linked reactions, and cation pumps in cellular membranes, utilize Ca^{2+} as a chemical mediator. These Ca^{2+}-initiated reactions generally involve cell surfaces and cytoplasmic processes. Although the uptake of Ca^{2+} by mitochondria has been observed by many investigators (42, 43), and its effect on mitochondrial respiration extensively studied (44), the cell physiological role of energy-dependent

Ca^{2+} uptake has not been explained (45). The rapid process of Ca^{2+} uptake by mitochondria was considered in a teleological sense to "maintain cytosol-Ca^{2+} at a low level" (41), but it remained unclear whether or not Ca^{2+} uptake by mitochondria is part of a more involved cellular control process. The carrier-mediated mechanism of Ca^{2+} transport was clarified to a considerable extent by identification of a Ca^{2+}-binding protein in Lehninger's laboratory (46, 47). The process of rapid Ca^{2+} uptake itself appears to compete with the phosphorylation of ADP in mitochondria (47); however, this effect was not considered to be equivalent with the uncoupling of oxidative phosphorylation as seen with artificial uncouplers (44).

A. Ca^{2+}-CMF Antagonism

Our interest in the role of Ca^{2+} in CMF-controlled mitochondrial processes was initiated by the experimental results illustrated in Fig. 19. Preincubation

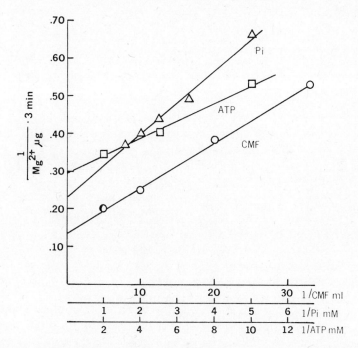

Fig. 18. Antagonistic effects of CMF, orthophosphate, and ATP on the ejection of Mg^{2+} from mitochondria in the presence of $8.3 \times 10^{-6} M$ bilirubin. The amount of CMF is expressed as reducing equivalent/1 ml incubation mixture, while P_i and ATP are given in molar concentrations. Unpublished experiment.

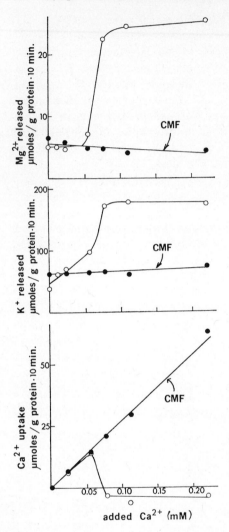

Fig. 19. The effect of increasing concentrations of externally added Ca^{2+} and CMF on Mg^{2+} efflux (upper) K^+ efflux (middle), and Ca^{2+} uptake (lower). CMF = 0.6 mg/ml. Ordinate: rate of cation movements; abscissa: concentrations of added Ca^{2+}. For experimental details, see ref. 48.

of mitochondria for 10 min at 30° with low concentrations of Ca^{2+}, actually found to be present in the cytosol of liver homogenates (0.18 micromole of Ca^{2+} per cytosol corresponding to 1 g of liver, or 10–60 nanomoles of Ca^{2+} per 1 mg of motochondrial protein), induced an efflux of almost all bound Mg^{2+} and of K^+. There is Ca^{2+} uptake at very low concentrations of externally added Ca^{2+}. An increase in the concentration of externally added Ca^{2+} resulted in the efflux of all accumulated Ca^{2+}, including that originally present in mitochondria. The presence of CMF counteracted Ca^{2+}-induced Mg^{2+} and K^+ efflux, and all externally added Ca^{2+} was taken up by mitochondria when CMF was present (48). There is a connection between Ca^{2+}-induced cation movements and metabolism. The effect of CMF was inhibited by 2,4-dinitrophenol, indicating that the CMF-Ca^{2+} antagonism *depends* on energy coupling. At 50 μM concentration, 2,4-dinitrophenol itself induced Ca^{2+} efflux from mitochondria, which was not prevented by CMF. Furthermore, oligomycin but not aurovertin simulated the effect of CMF (see Tables 7 and 8).

The effect of rotenone on Ca^{2+}-induced cation movements was tested in order to determine the role of endogenous respiration as a supportive mechanism to cation flux. Inhibition of respiration by rotenone counteracts the Mg^{2+} and K^+ efflux induced by Ca^{2+} (Expts. 3 and 4, Table 8). The protective effect of CMF against Ca^{2+}-induced K^+ efflux is diminished, and the augmentation of Ca^{2+} uptake by CMF is also depressed by rotenone. Endogenous metabolism of mitochondria contributes as an energy source to Ca^{2+}-initiated processes, and the effect of CMF is also dependent on endogenous metabolism.

The time course of Ca^{2+}-induced cation movements is illustrated in Fig. 20 (cf. ref. 26). When the time course of Ca^{2+} movements was followed, after an initial rapid Ca^{2+} uptake a latency of 5–6 min was observed before cation ejection started. During this latent period, only small K^+ and even less Mg^{2+} efflux took place. After this sharply defined latency had ended, rapid ejection of all cations occurred within 2–3 min, including some of the originally intramitochondrial Ca^{2+}. It was observed that CMF prevented cation ejection induced by Ca^{2+} and promoted the uptake of all externally added Ca^{2+}; CMF was an effective antagonist of Ca^{2+} when it was added either simultaneously with Ca^{2+} or within 3 min after exposure of mitochondria to Ca^{2+}. It may be of cell physiological importance that reversibility of Ca^{2+}-induced events by CMF is limited to this relatively short period.

B. Ca^{2+}-Induced Permeability to NAD^+

The Ca^{2+}-CMF antagonism is not restricted to cation movements. Addition of cytoplasmic levels of Ca^{2+} to mitochondria renders the inner membrane permeable to NAD^+, and this effect is completely prevented, or counteracted, by CMF. Intermittent Ca^{2+}-induced permeability changes of mitochondria to NAD^+ may be of physiological importance, since this control mechanism could account for the accumulation of extramitochondrially synthesized NAD^+ in mitochondria. The effects of Ca^{2+} and CMF on mitochondrial NAD^+ content are shown in Table 9.

Table 7. The Effects of Ca^{2+}, CMF, and Reagents Influencing Energy Transfer on Ion Movements

Batch	Expt. No.	Variable Conditions	Mg^{2+} Ejection	Ca^{2+} Uptake	Ca^{2+} Ejection	K^+ Ejection
A	1	Ca^{2+} (0.1 mM)	19.4	7.5	⋯	210
	2	Ca^{2+} (0.1 mM) + CMF	1.6	56.3	⋯	80
	3	Ca^{2+} (0.1 mM) + DNP (50 μM)	18.0	0.8	⋯	180
	4	Ca^{2+} (0.1 mM) + DNP (50 μM) + CMF	24.5	1.6	⋯	190
B	5	Ca^{2+} (0.15 mM)	31.2	⋯	4.2	230
	6	Ca^{2+} (0.15 mM) + CMF	3.6	61.5	⋯	70
	7	Ca^{2+} (0.15 mM) + oligomycin	10.4	49.0	⋯	163
	8	Ca^{2+} (0.15 mM) + oligomylcin + CMF	3.9	59.9	⋯	71
	9	Ca^{2+} (0.15 mM) + aurovertin	29.1	1.4	⋯	218
	10	Ca^{2+} (0.15 mM) + aurovertin + CMF	3.8	59.8	⋯	70
C	11	No Ca^{2+}	3.0	⋯	1.2	70
	12	No Ca^{2+} + CMF	4.6	⋯	1.2	70
	13	No Ca^{2+} + DNP (50 μM)	3.0	⋯	10.6	100
	14	No Ca^{2+} + DNP (50 μM) + CMF	5.3	⋯	9.0	110

NOTE TO TABLE 7

For experimental details, see ref. 48.

Results are expressed as micromoles of ions/1 g mitochondrial protein, translocated in 10 min at 30°.

Table 8. The Influence of Rotenone, CMF, and Oligomycin on Ca²⁺-Induced Ion Movements

Expt. No.	Experimental Conditions	Mg²⁺ Ejection, μmoles/1 g/10 min	K⁺ Ejection, μmoles/1 g/10 min	Ca²⁺ Movements, μmoles/1 g/10 min	
				Uptake	Discharge
1	No additions	3.0	75.0	...	3.8
2	Rotenone	3.1	70.0	...	6.0
3	0.11 mM Ca²⁺	19.8	178.0	6.1	...
4	0.11 mM Ca²⁺ + rotenone	3.8	85.0	10.8	...
5	0.11 mM Ca²⁺ + CMF	2.9	11.0	47.2	...
6	0.11 mM Ca²⁺ + CMF + rotenone	4.0	26.0	28.7	...
7	0.066 mM Ca²⁺	19.5	149.0	...	5.9
8	0.066 mM Ca²⁺ + oligomycin	4.0	71.0	23.3	...

NOTE TO TABLE 8

The rates of cation movements (in 10 min at 30°) were determined under the conditions described in ref. 48. The final concentration of rotenone was 3.3×10^{-7} M and of oligomycin 2.4 μg/mg mitochondrial protein. Mitochondrial protein was 2.3 mg/ml of incubation medium (see ref. 48).

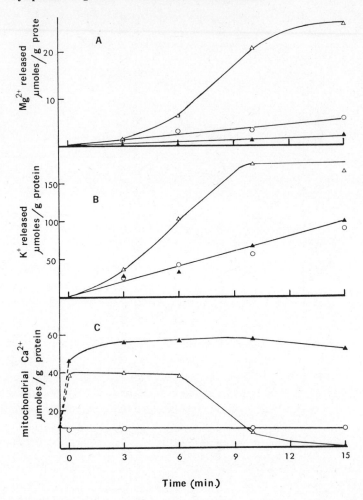

Fig. 20. Time course of Ca²⁺-induced cation movements. Each incubation mixture contained 2.8 mg mitochondrial protein/ml suspended in 0.25 M sucrose $+$ 30 mM Tris-HCl (pH 7.4) in a final volume of 6 ml. At 0 time Ca²⁺ was added (final concentration of 0.17 mM, or 60 nM/mg protein). When added, CMF was equivalent to 0.6 mg organic matter/ml. Incubation was carried out in beakers at 30° in a Dubnoff shaker. The incubation was stopped at 0, 3, 6, 10, 12, and 15 min, and the mitochondria separated by rapid centrifugation at 0–4°, followed by cation analyses in both mitochondria and supernatant (see ref. 48). Cation movements were calculated as micromoles/g mitochondrial protein (ordinate) and plotted against time (abscissa). Symbols: $\bigcirc$ = no Ca²⁺; $\triangle$ = +Ca²⁺; $\blacktriangle$ = Ca²⁺ + CMF. In A and B, Mg²⁺ and K⁺ ejection (i.e., cations appearing in the supernatant) is illustrated; in C, the initial Ca²⁺ uptake is shown by Ca²⁺ values above controls (which is the endogenous Ca²⁺ content) (—$\bigcirc$—), while the loss of mitochondrial Ca²⁺ is shown by points which fall below these control values (after 9 min).

Table 9. The Effect of Preincubation of Liver Mitochondria for 10 min with Physiological Levels of Ca^{2+} on NAD$^+$ Ejection

Experimental Conditions	NAD$^+$ Ejection	NAD$^+$ Content of Mitochondria
Incubation without additions	0	2.46
Incubation with 0.11 mM Ca^{2+}	1.56	0.90
Incubation with 0.11 mM Ca^{2+} + CMF	0	2.46

NOTE TO TABLE 9

The NAD$^+$ content of mitochondria, or the NAD$^+$ ejection, is calculated as micromoles NAD$^+$/g mitochondrial protein. CMF = 0.6 mg organic matter/ml; mitochondrial protein-2.5 mg/ml. Unpublished experiment.

C. Regulation of the Oxidation of Glutamate and Succinate by Ca^{2+}-CMF

It follows from these observations that the metabolism of NAD$^+$-linked mitochondrial substrates should be regulated by extramitochondrial Ca^{2+} + CMF. A comparison of glutamate and succinate oxidation by liver mitochondria preincubated with Ca^{2+} in the presence and the absence of CMF is shown in Fig. 21. Preincubation with Ca^{2+} inhibits glutamate oxidation almost completely, and this inhibition is prevented by CMF. Inhibition of glutamate oxidation is reversed by externally added NAD$^+$ in the *absence* of CMF, in agreement with predictions made from results shown in Table 9. There is a large stimulation of succinate oxidation when liver mitochondria are preincubated with Ca^{2+}; this effect is prevented by CMF. It is uncertain as yet whether or not the regulation of succinate oxidation by the Ca^{2+}-CMF system is related to NAD$^+$ or to cation movements induced by Ca^{2+}. Other aspects of mitochondrial control of succinate oxidation are discussed by Singer et al. in Chapter 9 of this book.

It may be presumed that the effect of CMF on metabolism could be due to prevention of uncoupler-induced loss of mitochondrial NAD$^+$. This is not the case, however, since *2,4-dinitrophenol, in the presence of 2.3 mM ADP*, does not cause a loss of mitochondrial NAD$^+$ (cf. ref. 27); hence the metabolic effects of CMF in the ADP-2,4-dinitrophenol system (27) cannot be explained on the basis of NAD$^+$ retention in the presence of CMF. The effect of CMF in this system is related to Mg^{2+} retention (see Fig. 5).

D. Ca^{2+}-Oligomycin Antagonism

As shown in Tables 7 and 8, CMF, as an antagonist of Ca^{2+}, can be replaced by oligomycin. The assumption that CMF acts on mitochondrial energy-coupled functions is supported by the apparent similarity between CMF and

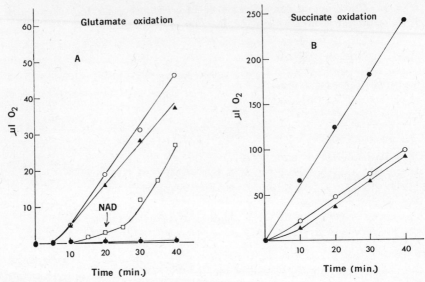

Fig. 21. The effects of preincubation of mitochondria with 0.11 mM Ca^{2+} and CMF (0.6 mg/ml) on the rates of glutamate (A) and succinate (B) oxidation, measured in Gilson respirometers in the absence of added ADP and P_1. The suspending medium was 3 ml of 0.25 M sucrose + 30 mM Tris-HCl (pH 7.4). Glutamate or succinate as Tris salts (final concentration 10 mM) was tipped in from side arm after 10 min preincubation of mitochondria with Ca^{2+} or Ca^{2+} + CMF at 30°, gas phase = air, Co_2 was trapped in 0.15 ml 15% KOH in center well. When added after 20 min (see A), NAD^+ was present in a concentration of 2 mM. Symbols: O = no Ca^{2+}; ● = +Ca^{2+}; ▲ = Ca^{2+} + CMF; □ = +NAD^+. Mitochondrial protein was 2.7 mg/ml suspending medium. Unpublished experiment.

the toxic antibiotic oligomycin (48, 26, 49). It is well known from the work of Lardy (50–53) that oligomycin specifically inhibits oxidative phosphorylation and electron transfer which is *coupled* to ATP synthesis. Another group of investigators (54–57) have demonstrated that oligomycin, under conditions in which ATP synthesis cannot occur, appears to stabilize an energized intermediate or conformation of a macromolecular system required for energy transfer. Under these experimental conditions, energy-coupled reactions require the toxic antibiotic as a stabilizing cofactor. This effect of oligomycin may indicate that inhibition of phosphorylative respiration by the toxic antibiotic is due to the binding of oligomycin to an essential component of the energy transfer system, which, in the absence of ADP and orthophosphate, can still catalyze partial reactions of energy transduction.

As illustrated in Fig. 22, oligomycin, like CMF, counteracts the cation-labilizing influence of increasing concentrations of externally added Ca^{2+}, and oligomycin also supports Ca^{2+} uptake (26). It is obvious that oligomycin, although imitating one effect of CMF, cannot act in every respect in the same manner. It has been shown that CMF stabilizes oxidative phosphorylation, whereas oligomycin inhibits it (see Section 6.E). However, the partial similarity of action between these agents suggests that the oligomycin site of mitochondria probably reacts under physiological conditions with CMF.

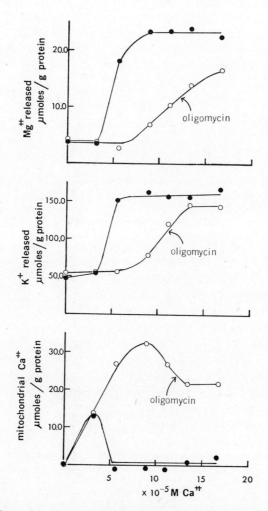

Fig. 22. The effects of varying concentrations of Ca^{2+} (abscissa) on cation movements, expressed per 10 min (ordinate), in the presence and the absence of a constant amount of oligomycin (2.7 μg/1 mg protein). Standard conditions are described in ref. 26. Mitochondrial protein was 2.4 mg/1 ml incubation medium.

E.   Stabilization of Acceptor Control by CMF

During the period of preincubation (10 min at 30°) with cytoplasmic concentrations of Ca^{2+}, after rapid Ca^{2+} uptake a progressive decline in the efficiency of oxidative phosphorylation of mitochondria takes place. As shown earlier, this decay of oxidative phosphorylation coincides with the loss of mitochondrial cations. The loss of respiratory control is prevented by CMF if it is present during preincubation with Ca^{2+}, or if it is added within a few minutes after exposure of mitochondria to Ca^{2+}. This effect of CMF is illustrated in Fig. 23. Stabilization of the mitochondrial energy transfer system by CMF results in restored oxidative phosphorylation.

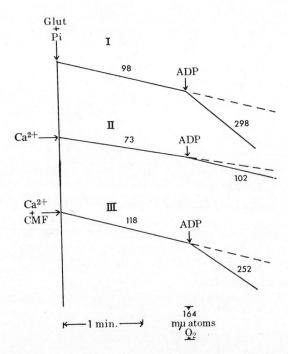

Fig. 23.   Preservation of respiratory control by CMF in the presence of 0.1 mM Ca^{2+}. The O_2 uptake of liver mitochondria (2.6 mg/ml) was measured polarographically. The suspending medium was 0.25 mM sucrose + 30 mM HCl, pH 7.4, 2 mM orthophosphate. The substrate = 10 mM glutamate, ADP = 0.2 mM; temperature = 30°. Preincubation was carried out in the absence of substrate for 10 min at 30°. Curve I = control; curve II = after preincubation with 0.11 mM Ca^{2+}; curve III = after preincubation with 0.11 mM Ca^{2+} + CMF (0.6 mg/ml). Rates of O_2 uptake are given in natoms O_2/12 mg mitochondrial protein/min. Unpublished results.

F. The Mode of Action of EDTA

Raaflaub (58) considered EDTA as a "cofactor" of mitochondria, although it was apparent that the *in vitro* effect of EDTA merely suggests the "deleterious" influence of Ca^{2+} on mitochondrial stability. As illustrated in Table 10, incubation of mitochondria with increasing concentrations of EDTA for 10 min at 30° results in significant loss of mitochondrial Ca^{2+}

Table 10. The Effect of Incubating Mitochondria with EDTA On Mg^{2+}, K^+, and Ca^{2+} Release[a]

Expt. No.	Experimental Conditions	Cation Release, moles/1 g protein/10 min		
		Mg^{2+}	K^+	Ca^{2+}
1	No additions	3.7	44	2.4
2	0.1 m*M* EDTA	4.6	48	9.3
3	0.01 m*M* EDTA	4.1	53	3.9
4	0.001 m*M* EDTA	4.4	53	2.1

[a] Unpublished results.

NOTE TO TABLE 10

Liver mitochondria (2.2 mg/ml) were incubated for 10 min at 30° in 0.25 *M* sucrose + 0.03 *M* Tris (pH 7.4) in the presence of increasing concentrations of EDTA. Cation release into the suspending medium during 10 min incubation was measured by atomic absorption analysis.

only. Removal of mitochondrial Ca^{2+} by EDTA or complexation of externally added Ca^{2+} by EDTA eliminates Ca^{2+} as a cellular metabolic signal. It is understandable that "stabilization" of mitochondria by EDTA precluded the possibility of discovering a real cellular stability cofactor of mitochondria.

G. pH Profile of Ca^{2+}-Induced Cation Efflux

The effect of extramitochondrial pH on Ca^{2+}-induced cation movements was also determined. As illustrated in Fig. 24, Ca^{2+} induced a significantly higher rate of Mg^{2+} ejection at pH 6.0 than at 8.0; the pH profile of Ca^{2+} movements was sigmoidal in this range. At pH 6.0 there was some Ca^{2+} release by added Ca^{2+}, while toward alkalinity the rate of Ca^{2+} uptake markedly increased. Added Ca^{2+} increased K^+ efflux, but this reaction did not vary between pH 6.0 and 8.0.

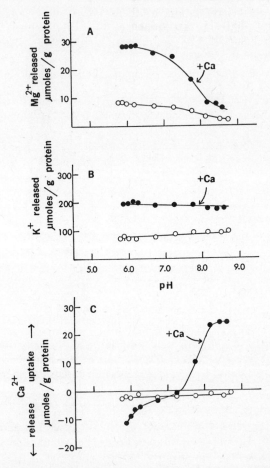

Fig. 24. In *A* and *B*, the effect of variation in pH between 6 and 8 on mitochondrial Mg^{2+} and K^+ ejection with (—●—) and without (—○—) added Ca^{2+} (0.11 mM) is given. In *C*, Ca^{2+} ejection and uptake after addition of Ca^{2+} (0.11 mM) (—●—) and without added Ca^{2+} (—○—) are shown. Mitochondrial protein was 2.7 mg/ml; total incubation volume-6 ml; suspending medium was 250 mM sucrose + 30 mM Tris-HCl; pH was adjusted with a glass electrode (at 30°) for each incubation system; and rates of cation movements were determined in beakers during 10 min aerobic incubation at 30° in a shaking incubator. Rates (ordinate) are plotted against pH (abscissa). Unpublished results.

A comparison of artificial uncouplers (Fig. 9) and of bilirubin (Fig. 17) with Ca^{2+} reveals significant differences in the pH profiles of cation movements. Although interpretation of these phenomena is not possible at present, the dissimilarity of pH profiles supports the view that the modes of action of artificial uncouplers, of bilirubin, and of Ca^{2+} are different.

7. ANALYSIS OF THE MODE OF ACTION OF CMF

> *The girl who sells melons beside the stream*
> *Gathers her melons in the fields on the hillsides*
> *She does not need to spin hemp.*
> *She has handsfull of bronze money.*
>
> Mei Yao Ch'en

A general survey of experimental results indicates that CMF becomes effective under conditions in which an extramitochondrial stress is applied on the energy transfer system of the inner mitochondrial membrane. The stress signal can be cytoplasmic Ca^{2+}, potentially toxic cellular metabolites (bilirubin, oleic acid), or artificial uncoupling agents. Under conditions which are characteristic for each agent, mobilization of bound Mg^{2+}, failure of cation retention, and uncoupling of respiration from phosphorylation can be experimentally induced. Although it is highly probable that the molecular mode of action of each of these energy-dissipation-inducing agents is different, CMF is capable of preventing their effects on certain energy-coupled reactions. The degree of effectivity of CMF as an antagonist to various energy transfer labilizing agents varies, however, both in magnitude and in mode, depending on experimental conditions.

It would appear that CMF acts on some common intermediate, or on a system shared by all energy-coupled reactions. From a variety of experiments it was deduced that the site common to all energy transfer reactions may be an inner-membrane-bound form of Mg^{2+} (26–30, 48, 49). The customary notation $\sim X$, indicating an energized intermediate, could then be replaced by $\sim X\text{-}Mg^{2+}$, and CMF could be presumed to act on the energizable ligand system of Mg^{2+}. This interpretation appears to imply that CMF may not participate in the molecular mechanism of energy transfer processes itself, but rather serves as an external stabilizing cofactor of this membrane system. This conclusion, however, may not be fully justified. Although CMF could not be extracted from isolated mitochondria by distilled water, the procedure routinely used for its isolation from cytosol, the possibility cannot be excluded at present that a water-insoluble (hydrophobic) metabolite of CMF exists in the inner membrane system. This bound form of CMF could serve as a component of the Mg^{2+}-binding ligand system assumed to be an essential part of the mitochondrial energy transfer apparatus. At present, only indirect evidence, deduced mainly from kinetics, tends to support the existence of a hypothetical membrane-bound form of CMF.

Except for oxidative phosphorylation (cf. ref. 29), all mitochondrial reactions that can be preserved by CMF are made CMF-sensitive during a characteristic time lag, which occurs after the exposure of mitochondria to an extramitochondrial energy-dissipating agent. The time lag, before CMF sensitivity appears, depends on the concentration of the energy-dissipating agent. During this time lag, various mitochondrial functions are stimulated; for example, a large stimulation of glutamate metabolism is observed when mitochondria are preincubated for 15 min with less than 50 μM 2,4-dinitrophenol in the presence of 2.3 mM ADP. At this concentration of 2,4-dinitrophenol, Mg^{2+} is still present in mitochondria in the bound form. This is in agreement with the conclusion that Mg^{2+} binding is essential for the ability of mitochondria to respond with hypermetabolism to an uncoupler. The requirement for CMF appears only at higher concentrations of the uncoupler, sufficient to induce Mg^{2+} efflux, which is then antagonized by CMF (Fig. 5). The apparent affinity of mitochondria, not only for Mg^{2+}, but also for ADP, is restored by CMF (Fig. 12) under these conditions.

The time required for the induction of apparent CMF deficiency by an uncoupler depends on its concentration. It follows that almost any concentration of an uncoupler should be able to dislodge bound Mg^{2+} and to produce apparent "CMF deficiency" if given sufficient time to act on the energy transfer system. The fixed time of 15 min for preincubation was chosen for experimental convenience only. It is easier to measure the effects of varying concentrations of uncouplers than to follow the decay of mitochondrial energy coupling at one concentration of an uncoupler. The time-dependent effect of 2,4-dinitrophenol could be due to the decay of a membrane-bound form of CMF, followed by loss of binding of cations and of ADP. Early kinetic studies (25) indicated that a double reciprocal plot of metabolic rates and of CMF concentration at various concentrations of 2,4-dinitrophenol produced an intercept:

$$= - \frac{(k_{DNP})}{(k_{DNP \times CMF})},$$

suggesting some form of interaction between the uncoupler and a CMF site.

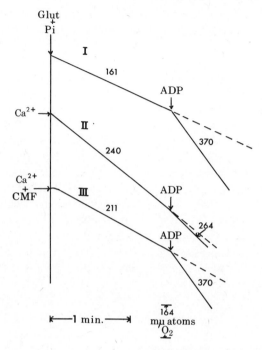

Fig. 25. Loss of respiratory control by preincubation of liver mitochondria with 5 μM 2,4-dinitrophenol for 10 min at 30°, and its prevention by CMF. All experimental conditions were the same as given in the legend of Fig. 23. Substrate was glutamate. Curve I = control; curve II = after preincubation with 5 μM 2,4-dinitrophenol; curve III = after preincubation with 5 μM 2,4-dinitrophenol + CMF (0.6 mg/ml). Unpublished results.

More recently, direct competition between DNP and CMF was found when endogenous oxidative phosphorylation was studied (cf. ref. 29). If uncouplers act by displacing CMF (or by hastening its decomposition), then very low concentrations of an uncoupler should have an effect on energy transfer, if given enough time. This assumption was verified experimentally. 2,4-Dinitrophenol, at a concentration as low as 5 μM, if incubated for 10 min with mitochondria, produced uncoupling, and this effect was prevented by added CMF. These results are illustrated in Fig. 25.

Interpretation of the mode of action of CMF depends critically on its specificity. Under properly controlled assay conditions (cf. ref. 29) the effects of CMF are highly specific. CMF cannot be replaced by inositol or its known phosphorylated derivatives (including the cyclic phosphate). The similarity of action between CMF and oligomycin on Ca^{2+}-induced cation flux is a notable exception. Analysis of the problem of specificity leads to the necessity for critically evaluating some commonly used experimental models. It is generally known that many cellular substances tend to stabilize mitochondria. Millimolar concentrations of externally applied mitochondrial substrates, Mg^{2+}, and adenine nucleotides are effective stabilizing agents of isolated mitochondria. Connelly and Hallstrom (59) have emphasized that ADP can act as an *in vitro* allosteric stabilizer of mitochondria. It is also known that externally added orthophosphate, carboxylic acid substrates, and adenine di- or triphosphate cause increased Ca^{2+} uptake in mitochondria, similarly to CMF; thus any one of these agents can, under certain *in vitro* conditions, simulate some effects of CMF. By careful consideration of experimental conditions, however, the effects of these nonspecific mitochondrial stabilizers can be readily distinguished from the effect of CMF, which, in its purest form, is half maximally active in 8 to 10 $\mu g/ml$ concentration.

ACKNOWLEDGMENTS

The research discussed in this review was supported by Grants HD-01239 and CA-07955 of the National Institute of Health and by Grant E-493A of the American Cancer Society.

The valuable help of Miss Carol Fegté in the editing of this text is acknowledged with thanks.

We are grateful also for permission from *Biochemistry*, *Biochemical and Biophysical Research Communications*, and the *Journal of Physiological Chemistry and Physics* to reproduce certain graphs and tables.

REFERENCES

1. F. O. Schmitt, *Rev. Mod. Phys.*, **21**, 5 (1959).
2. D. C. Tosteson (Ed.), *The Molecular Basis of Membrane Function*, Prentice-Hall, Englewood Cliffs, N.J., 1968.
3. R. M. Dowben (Ed.), *Biological Membranes*, Little Brown, Boston, 1969.
4. J. Järnfelt (Ed.), *Regulatory Functions of Biological Membranes*, BBA Library, Vol. 11, Elsevier, Amsterdam, 1968.
5. W. D. Stein (Ed.), *The Movement of Molecules Across Cell Membranes*, Academic Press, New York, 1967.

6. E. Racker (Ed.), *Membranes of Mitochondria and Chloroplasts*, ACS Monograph 165, Van Nostrand Reinhold, New York, 1970.

7. A. Katchalsky and R. Spangler, *Quart. Rev. Biophys.* **1**, 127 (1968).

8. A. L. Lehninger, *The Mitochondrion*, W. A. Benjamin Press, New York, 1964.

9. W. W. Wainio, *The Mammalian Mitochondrial Respiratory Chain*, Academic Press, New York, 1970.

10. E. Racker, "The Two Faces of the Inner Mitochondrial Membrane," in *Essays in Biochemistry*, Vol. 6 (Eds.: P. N. Dampbell and F. Dickens), The Biochemical Society, Academic Press, London, 1970, p. 1.

11. G. D. Greville, "A Scrutiny of Mitchell's Chemiosmotic Hypothesis of Respiratory Chain and Photosynthetic Phosphorylation," in *Current Topics in Bioenergetics*, Vol. 3 (Ed.: D. Rao Sanadi), Academic Press, New York, 1969, p. 1.

12. H. A. Lardy, Address to the 3rd International Congress of Biochemistry, Brussels, 1955.

13. B. Chance and G. R. Williams, *Advan. Enzymol.*, **17**, 65 (1956).

14. A. L. Lehninger, *Rev. Mod. Phys.*, **31**, 136 (1959).

15. E. Racker, *Mechanisms of Bioenergetics*, Academic Press, 1965.

16. P. D. Boyer, "Oxidative Phosphrylation," in *Biological Oxidations* (Ed.: T. P. Singer), John Wiley-Interscience, New York, 1968, p. 193.

17. D. E. Green, "The Mitochondrial Electron Transfer System," in *Comprehensive Biochemistry*, Vol. 14 (Eds.: M. Florkin and E. H. Stotz), Elsevier, Amsterdam, 1966.

18. E. Kun and R. J. Dummel, "Chemical Properties and Synthesis of Fluoro Analogues of Compounds Related to Substrates of the Citric Acid Cycle," in *Methods in Enzymology*, Vol. XIII (Eds.: S. P. Colowick, N. O. Kaplan, and J. M. Lowenstein), Academic Press, New York, 1969.

19. E. Kun, "Mechanism of Action of Fluoro Analogues of Citric Acid Cycle Compounds: An Essay in Biochemical Tissue Specificity," in *The Citric Acid Cycle* (Ed.: J. M. Lowenstein), Dekker, New York, 1969, chapter 6.

20. J. L. Webb, *Enzyme and Metabolic Inhibitors*, Vol. 1, Academic Press, New York, 1963, Chapter 7.

21. H. A. Lardy and S. M. Ferguson, *Ann. Rev. Biochem.*, **38**, 713 (1969).

22. E. Kun, J. E. Ayling, and B. G. Baltimore, *J. Biol. Chem.*, **239**, 2896 (1964).

23. J. M. Tager, "Nicotinamide Nucleotide-Linked Oxidoreductions in Rat Liver," in *Regulation of Metabolic Processes in Mitochondria* (Eds.: J. M. Tager, S. Papa, E. Quagliariello, and E. C. Slater), BBA Library, Vol. 7, Elsevier, Amsterdam, 1966, p. 203.

24. E. Kun, H. H. Loh, and S. B. I. El-Fiky, *Mol. Pharmacol.*, **2**, 481 (1966).

25. H. H. Loh, P. Volfin, and E. Kun, *Biochemistry*, **7**, 726 (1968).

26. N. M. Lee, I. Wiedemann, and E. Kun, *FEBS Lett.*, **18**, 81 (1971).

27. E. Kun, E. B. Kearney, I. Wiedemann, and N. M. Lee, *Biochemistry*, **8**, 4443 (1969).

28. E. Kun, E. B. Kearney, N. M. Lee, and I. Wiedemann, *Biochem. Biophys. Res. Commun.*, **38**, 1002 (1970).

29. E. Kun, N. M. Lee, D. C. Lin, I. Wiedemann, K. L. Johnson, and R. J. Dummel, in *Molecular Basis of Electron Transport* (Ed.: Schultz), Academic Press, New York, 1972 (in press)

30. N. M. Lee, I. Wiedemann, K. L. Johnson, D. N. Skilleter, and E. Kun, *Biochem. Biophys. Res. Commun.*, **40**, 1058 (1970).

31. A. I. Caplan and J. W. Greenawalt, *J. Cell Biol.*, **31**, 455 (1966); *ibid.*, **36**, 15 (1968).

32. C. P. Lee, L. Ernster, and B. Chance, *European J. Biochem.*, **8**, 153 (1969).

33. J. P. Changeux, J. Thiery, Y. Tung, and C. Kittel, *Proc. Natl. Acad. Sci. U.S.*, **57**, 335 (1967).

34. E. B. Kearney, N. M. Lee, I. Wiedemann, and E. Kun, *J. Physiol. Chem. Phys.*, **1**, 575 (1969).

35. B. Pressman and H. A. Lardy, *Biochim. Biophys. Acta*, **18**, 482 (1955).

36. S. G. Van den Bergh, C. P. Modder, J. H. M. Souverijn, and H. C. J. M. Pierrot, "Mitochondria: Struct. Funct.," in *Fed. Eur. Biochem. Soc. 5th Meeting 137* (Ed.: L. Ernster), Academic Press, London, 1968.

37. I. A. D. Bouchier and B. H. Billing (Eds.:) *Bilirubin Metabolism*, Blackwell, Oxford, 1967.

38. T. K. With, *Bile Pigments: Chemical, Biological and Clinical Aspects*, Academic Press, New York, 1968.

39. M. G. Mustafa, M. L. Cowger, and T. E. King, *J. Biol. Chem.*, **244**, 6403 (1969).

40. L. Ernster, in *Kernicterus* (Ed.: A. Sass-Kortsak), University of Toronto Press, Toronto, 1961, p. 174.

41. H. Rasmussen, *Science*, **170**, 404 (1970).

42. W. Hasselbach, *Biochim. Biophys. Acta*, **25**, 562 (1957); *Progr. Biophys. Mol. Biol.*, **1**, 167 (1964).

43. A. L. Lehninger, E. Carafoli, and C. S. Rossi, *Advan. Enzymol.*, **29**, 259 (1967).

44. B. Chance, *J. Biol. Chem.*, **240**, 2729 (1965).

45. H. Rasmussen, N. Hangaard, N. H. Hangaard, N. H. Lee, and S. Horn, *Federation Proc.*, **28**, 1657 (1969).

46. A. L. Lehninger and E. Carafoli, in *Biochemistry of the Phagocytic Process* (Ed.: J. Schultz), North-Holland, Amsterdam, 1970, p. 9.

47. A. L. Lenhinger, *Biochem. J.*, **119**, 129 (1970); *Biochem. Biophys. Res. Commun.*, **42**, 312 (1971).

48. N. M. Lee, I. Wiedemann, and E. Kun, *Biochem. Biophys. Res. Commun.*, **42**, 1030 (1971).

49. E. Kun, N. M. Lee, and I. Wiedemann, *Biophys. Soc. Abstr.*, **121a** (1971), New Orleans.

50. H. A. Lardy, D. Johnson, and W. C. McMurray, *Arch. Biochem.*, **78**, 587 (1958).

51. H. A. Lardy, J. L. Connelly, and D. Johnson, *Biochemistry*, **3**, 1961 (1964).

52. J. L. Connelly and H. A. Lardy, *Biochemistry*, **3**, 1969 (1964).

53. H. A. Lardy, P. Witonsky, and D. Johnson, *Biochemistry*, **4**, 552 (1965).

54. C. P. Lee, and L. Ernster, *Biochem. Biophys. Res. Commun.*, **18**, 523 (1965); *European J. Biochem.*, **3**, 391 (1968).

55. C. P. Lee, L. Ernster, and B. Chance, *European J. Biochem.*, **8**, 153 (1969).

56. J. M. Fessenden and E. Racker, *J. Biol. Chem.*, **241**, 2483 (1966).

57. K. Nordenbrand and L. Ernster, *European J. Biochem.*, **18,** 258 (1971).

58. J. Raaflaub, "Complexing Agents as Cofactors in Isolated Mitochondria," in *Biological Phosphorylations* (Ed.: H. A. Kalckar), Prentice-Hall, Englewood Cliffs, N.J., 1969, p. 262.

59. J. L. Connelly and C. H. Hallstrom, *Biochemistry*, **6,** 1567 (1967).

CHAPTER 11

Kinetic and Thermodynamic Aspects of Biological and Biochemical Control Mechanisms*

CHARLES F. WALTER

Biomathematics Department and the Biochemistry Department, M. D. Anderson Hospital and Tumor Institute, University of Texas at Houston, Houston, Texas

* Received May 15, 1969 and in revised form March 26, 1970.

1. INTRODUCTION

Molecular biology and experimental biochemistry have provided us with some fairly precise concepts about the control of macromolecular synthesis and the metabolism of smaller molecules. Models for the control of gene activity, RNA synthesis, and protein synthesis have been proposed and are constantly being refined; models for the control of enzyme activity and of material fluxes through the various metabolic and diffusion pathways have also been suggested. Both types of models suggest that a fundamental unit of behavior in cells is a feedback control circuit. In fact, negative and positive feedback of biosynthetic and other metabolic pathways seems so ubiquitous as to suggest that biological systems have evolved methods of feedback control for all levels of metabolism.

The basis for distinguishing between models for the control of macro-molecular synthesis and models for the control of metabolic fluxes is that a much longer time is required to synthetize a molecule of β-glactosidase than to convert a molecule of glucose into two molecules of acetyl-coenzyme A. This fact infers the existence of at least two separate time scales for cellular control and, hence, for biological oscillations. The fact that cells divide suggests the existence of at least one autonomous oscillator having a period on the same time scale as the cell cycle; this type of low-frequency oscillation could arise from the control circuits for macromolecular synthesis. On the other hand, the fact that intermediates in certain metabolic pathways can experience concentration oscillations suggests the existence of higher-frequency oscillators as well.

On this basis the cell can be viewed as a nonconservative system of chemical and diffusion reactions; some of these chemical reactions are undoubtedly governed by both deterministic and probabilistic components. A culture or group of cells can be viewed as a collection of these systems. The time course of each of the individual cells would reflect the particular set of initial states of the high-frequency oscillators at mitosis; the time course of the whole group of cells would reflect the initial states of the metabolic components and the deterministic and probabilistic features of the systems responsible for the synthesis of the macromolecular components. On top of this "noisy" behavior there exists the lower-frequency oscillator; it is common to assume that this type of oscillator predominates in the determination of the average cell size, age, and so on for the whole culture (1). The variance about these average values is due to the stochastic components mentioned above *and* to the variations of the initial states in the individual chemical processes and the integrated control circuits.

If, as suggested above, a low-frequency oscillator determines the time at which an individual cell divides, and higher-frequency oscillators contribute to the variance of the mean mitotic time in a cell population, it is clear that studies of the stability properties and oscillatory behavior of cellular control circuits are extremely important to the building of a model of cellular behavior. In this chapter we shall outline some methods for studying the stability properties of biological and biochemical control circuits; some of these methods have proved extremely useful in studies of nonbiological control systems. As we shall see, the complexity of biological control mechanisms provides a challange for the continued development of new methods for studying the stability properties of multicomponent, nonlinear, noncon-servative systems.

2. SOME PRELIMINARIES

A. Phase-Space Representation

To begin the discussion of the stability properties of biological control mechanisms, we consider the system of two first-order, autonomous differential equations:

$$\frac{dx}{dt} \equiv \dot{x} = X(x, y), \qquad \frac{dx}{dt} \equiv \dot{y} = Y(x, y), \tag{2.1}$$

where X and Y are Lipschitzian functions. Since the equations are autonomous, we can eliminate time by dividing the first by the second:

$$\frac{dx}{dy} = \frac{X(x, y)}{Y(x, y)}, \qquad Y(x, y) \neq 0. \tag{2.2}$$

Equation 2.2 is usually referred to as the *differential equation of integral curves*. It is clear from this equation that a geometrical representation of x and y in the plane of these variables will be independent of what happens in time. On the other hand, the *trajectories of the integral curves*

$$x(t) = \int_{t_0}^{t} X(x, y), \qquad y(t) = \int_{t_0}^{t} Y(x, y) \tag{2.3}$$

depend not only on the initial values of x and y but also on $t - t_0$. In other words $x(t)$ and $y(t)$ specify a certain motion; but since the system is autonomous, the replacement of t by $t + t_0$ changes the phase but not the differential equation of integral curves. Thus there are an infinite number of solutions corresponding to a given trajectory; for autonomous systems these solutions differ from each other only by the phase t_0.

This important property of autonomous differential equations makes it extremely convenient to investigate the behavior of equations 2.1 in the so-called *phase-space*, that is, in the plane of the variables x and y. The usefulness of investigating systems described by equations 2.1 in the phase-plane is best understood by considering a simple example.

The familiar differential equation for a harmonic oscillator is

$$\frac{d^2 x}{d\tau^2} = -x, \tag{2.4}$$

where the differentiation is with respect to the dimensionless time, $\tau = \omega t$.

We may write equation 2.4 as two first-order differential equations:

$$\frac{dx}{d\tau} = y, \qquad \frac{dy}{d\tau} = -x. \tag{2.5}$$

The differential equation of integral curves is

$$\frac{dx}{dy} = -\frac{y}{x}. \tag{2.6}$$

Equation 2.6 can be written in the form

$$x\,dx + y\,dy = 0. \tag{2.7}$$

Integration of equation 2.7 yields

$$x^2 + y^2 = r^2 = \text{constant}. \tag{2.8}$$

Equation 2.8 represents a circle in the phase-plane; this representation is illustrated in Fig. 1. We note here that there are an infinite number of solutions corresponding to a given radius, r, but different initial phase angles, φ_0. In addition there is a continuous family of circles that depend on r^2, the other constant of integration.

It should be obvious that, if the phase-plane representation is a circle, ellipse, or any other closed trajectory, the solution of the original differential equations is oscillatory. If the phase-plane representation is an inward winding spiral, asymptotic with the origin as time approaches infinity, the

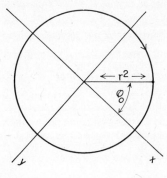

Fig. 1. Phase-space representation of equation 2.4.

solution is oscillatory, but in this case the oscillations are damped.* On the other hand, if the phase-plane representation is an outward winding spiral, the solution is oscillatory, but the amplitude of the oscillation increases with time. Finally, if the phase-plane representation is none of these possibilities, the solution is not oscillatory. Thus, from an inspection of the phase-plane representation of equations 2.1, it is possible to identify different types of motion inherent in these equations.

B. The Cauchy-Lipschitz Theorem

A theorem due to Cauchy and Lipschitz imposes a very important restriction on the phase-space representations of autonomous differential equations:

Theorem 1a. *Through every point of the phase-space there passes one and only one integral curve.*

Since the direction of motion of each point in the phase-space is completely defined by the original autonomous differential equations, it follows that two distinct integral curves cannot have a point in common, for they would then have to be tangent at this common point. This possibility is excluded by our original stipulation that the differential equations be Lipschitzian. With this argument in mind we restate Theorem 1a:

Theorem 1b. *Two distinct integral curves can have no point in common.*

Theorem 1b is a statement of the well-known fact that the trajectory of the integral curve cannot cross itself in the phase-space.

Any point (x_0, y_0) for which X and Y in equations 2.1 do not simultaneously vanish is called an *ordinary point* of the differential equations. A point (x^*, y^*) for which $X(x^*, y^*) = Y(x^*, y^*) = 0$ is called a *singular point*, an *equilibrium point*, or an *obvious solution* of the differential equations. The fact that a singular point is a stationary point to the flow is obvious from the fact that $\dot{x} = \dot{y} = 0$ when $x = x^*$ and $y = y^*$.

We now utilize these definitions to state another consequence of the Cauchy-Lipschitz theorem:

Theorem 1c. *The integral curve passing through a singular point consists just of that point itself.*

---

* Higgins (2) has suggested a more restrictive definition for damped oscillations. Since, however, this definition would make it impossible to employ phase-plane representations to distinguish between oscillatory systems with damping and nonoscillatory systems, it is not used here.

A further consequence of the Cauchy-Lipschitz theorem is

> **Theorem 1d.** *If the phase-space trajectory passes through an ordinary point, it cannot reach a singular point in finite time.*

It is clear from these definitions and theorems that singular points are extremely important to investigations of the stability properties of chemical control systems. It was recently proved (3) that at least one thermodynamic equilibrium position must exist for each closed chemical system. Since stationary points of flow correspond to thermodynamic equilibrium positions in closed chemical systems (4), a single singularity in a conservative chemical system can be identified as a stable point of thermodynamic equilibrium. However, if more than one singularity exists for a chemical system, at least one of the singularities must correspond to a stable point of thermodynamic equilibrium. Clearly, physically important singularities in chemical systems must be such that all the variables (i.e., the chemical concentrations) are nonnegative. These considerations lead us to

> **Theorem 2.** *The differential equations describing closed chemical systems must have at least one stable singularity for which all the variables are nonnegative.*

The singularities in open, nonconservative chemical systems may be points of thermodynamic equilibrium, or they may be special stationary states of minimum entropy production (5). From a thermodynamic point of view an open chemical system "degrades" the matter received from the exterior; it is this degradation that maintains the stationary flow through the system. Thus the properties of physically important singularities in nonconservative chemical systems can be quite different from the properties of the singularities corresponding to stable points of thermodynamic equilibrium.

C. Definition of Stability

We have seen in Theorems 1c and 1d that a trajectory passing through an ordinary point in the phase-space cannot reach a singular point in finite time. This statement does not preclude the possibility that a trajectory could approach a singular point as time approaches infinity. In fact we shall use this possibility for our definition of *asymptotic stability:*

1. If all trajectories in the phase-space approach a singular point as time approaches infinity, the singular point is said to possess *global asymptotic stability* or *asymptotic stability in the large.*

2. If all trajectories in a domain of the phase-space near a singular point approach the singularity as time approaches infinity, the singular point is said to be *asymptotically stable in the small.*

3. If all the trajectories in the phase-space approach a singular point as time approaches minus infinity, the singular point is said to be *totally unstable*.

This concept of stability is extremely important in the context of sustained oscillations. If, for example, it can be shown that a stationary state of a biological or biochemical control mechanism is asymptotically stable for a particular range of the biochemical variables, and it is known that these variables cannot leave this domain of the phase-space, then it can be guaranteed that the control system possesses a stable point of steady flow and, furthermore, that sustained oscillations cannot arise. Alternatively, if it can be shown that the stationary state is asymptotically stable in the large, then it is guaranteed that sustained oscillations will never occur.

For our definition of *stability* we shall use:

4. If all trajectories in the phase-space approach a neighborhood of a singular point as time approaches infinity, the singular point is said to be *globally stable* or *stable in the large*.

5. If all trajectories in a neighborhood of a singular point remain within another neighborhood of the singularity, the singular point is said to be *stable in the small*.

Clearly, sustained oscillations can occur around stable singular points but not around singularities possessing global asymptotic stability.

D. Stability Properties of Singular Points

A particular set of nonlinear differential equations may or may not possess a singular point at the origin. In the case of differential equations describing chemical reactions, the variables are necessarily nonnegative because chemical concentrations cannot be negative numbers. In general, the singular point will be the chemical concentrations at which there is a stationary state of flow for the chemical reactants. Regardless of whether the singular point describes a stationary state or thermodynamic equilibrium, systems with one singular point will not have the singularity at the origin; if they did, there would be no chemical reaction. However, it will always be possible to introduce a change of variables,

$$u = x - x^*, \qquad v = y - y^*, \tag{2.9}$$

such that the singular point is transformed to the new origin. Introduction of this transformation into equations 2.1 provides the transformed differential equations:

$$\dot{u} = X(u + x^*, v + y^*) \equiv U(u, v), \qquad \dot{v} = Y(u + x^*, v + y^*) \equiv V(u, v).$$
$$\tag{2.10}$$

The singular point of equations 2.10 corresponds to the singular point of equations 2.1, but now the singularity occurs at the origin $u = v = 0$. We note here that the transformed variables (unlike the original ones) can be negative; the minimum possible values are $u = -x^*$ and $v = -y^*$.

We now develop U and V in equations 2.10 into power series:

$$\dot{u} = a_1 u + a_2 v + U_2(u, v), \qquad \dot{v} = a_3 u + a_4 v + V_2(u, v), \qquad (2.11)$$

where U_2 and V_2 are power series in u and v beginning with terms of degree 2 or higher, and the a's are constants. We can now investigate equations 2.1 in the domain surrounding x^* and y^* by investigating equations 2.11 in the neighborhood surrounding the origin $u = v = 0$. In this domain the terms in U_2 and V_2 are at least second order in u and v, and these variables are small. Thus, assuming that the a's are not zero, there must be a neighborhood surrounding the singular point wherein the terms in U_2 and V_2 can be neglected. Restricting ourselves to this neighborhood, we abridge equations 2.11 and obtain:

$$\dot{u} = a_1 u + a_2 v, \qquad \dot{v} = a_3 u + a_4 v. \qquad (2.12)$$

In what follows we shall utilize these *abridged equations* to establish the nature of the singular point at the origin.

With the aid of a linear transformation of the variables

$$r = b_1 u + b_2 v \qquad \begin{vmatrix} b_1 & b_2 \\ b_3 & b_4 \end{vmatrix} \neq 0, \qquad (2.13)$$
$$s = b_3 u + b_4 v$$

we can reduce equations 2.12 to the canonical form:

$$\dot{r} = \mu_1 r, \qquad \dot{s} = \mu_2 s. \qquad (2.14)$$

In equation 2.13 the b's are the transformation constants. Since

$$\dot{r} = b_1 \dot{u} + b_2 \dot{v}, \qquad \dot{s} = b_3 \dot{u} + b_4 \dot{v}, \qquad (2.15)$$

we may substitute from equations 2.12–2.14 into equation 2.15 and obtain:

$$(b_1 u + b_2 v)\mu_1 = b_1(a_1 u + a_2 v) + b_2(a_3 u + a_4 v),$$
$$(b_3 u + b_4 v)\mu_2 = b_3(a_1 u + a_2 v) + b_4(a_3 u + a_4 v). \qquad (2.16)$$

Equations 2.16 may be written in the form

$$(b_1\mu_1 - a_1 b_1 - a_3 b_2)u + (b_2\mu_1 - a_2 b_1 - a_4 b_2)v = 0,$$
$$(b_3\mu_2 - a_1 b_3 - a_3 b_4)u + (b_4\mu_2 - a_2 b_3 - a_4 b_4)v = 0. \qquad (2.17)$$

Identifying the coefficients of u and v provides four relations:

$$b_1(a_1 - \mu_1) + a_3 b_2 = 0, \qquad b_2(a_4 - \mu_1) + a_2 b_1 = 0,$$
$$b_3(a_1 - \mu_2) + a_3 b_4 = 0 \qquad b_4(a_4 - \mu_2) + a_2 b_2 = 0. \tag{2.18}$$

Nontrivial solutions of equations 2.18 are possible only if μ_1 and μ_2 are the roots of

$$\begin{vmatrix} a_1 - \mu & a_2 \\ a_3 & a_4 - \mu \end{vmatrix} = 0. \tag{2.19}$$

Equation 2.19 is called the *characteristic equation* of equations 2.12; μ_1 and μ_2 are usually referred to as the *characteristic exponents*.

We now choose μ_1 and μ_2 as the two roots of the characteristic equation and illustrate that the nature of these roots determines the character of the singular point. A singular point is described as *nodal* if every integral curve has a limiting direction at the singularity.

Referring to the canonical differential equations (equations 2.14), we find that the differential equation of integral curves in the canonical variables is

$$\frac{ds}{dr} = \frac{\mu_2 s}{\mu_1 r} \equiv m \frac{s}{r}. \tag{2.20}$$

Since

$$\begin{vmatrix} a_1 & a_2 \\ a_3 & a_4 \end{vmatrix} \neq 0, \qquad \begin{vmatrix} b_1 & b_2 \\ b_3 & b_4 \end{vmatrix} \neq 0, \tag{2.21}$$

neither μ_1 nor μ_2 can be zero; hence m is neither zero nor infinite. If these roots are both real and of the same sign, $m > 0$, and the equation of integral curves is

$$s = C |r|^m. \tag{2.22}$$

Equation 2.22 describes a parabola in the phase-plane. By differentiating this equation with respect to r we obtain

$$\frac{ds}{dr} = mCr^{m-1} = \frac{mC}{r^{1-m}}. \tag{2.23}$$

Since $\dot{r} = \mu_1 r$, we may think of equation 2.23 as the differential equation of integral curves in the phase-plane of r and $\dot{r}$. We now consider the three situations: $m > 1$, $m < 1$, and $m = 1$.

For $m > 1$ equation 2.23 illustrates that every integral curve with the exception of the s axis approaches the singular point along the r axis. For $m < 1$ equation 2.23 illustrates that every integral curve with the exception of the r axis approaches the singular point along the s axis. Thus, for $m > 1$ or $m < 1$, the integral curves are parabolas converging to or radiating from the origin. For $m = 1$ the integral curves are half-lines converging to or radiating from the singularity. In all three situations these trajectories approach the singular point if μ_1 and μ_2 are both negative; in this case the singular point is called a *stable node*. On the other hand, if μ_1 and μ_2 are both positive, the trajectories radiate from the singular point; in this situation the singularity is called an *unstable node*.

Examples of stable and unstable nodes are illustrated in the phase-plane representations in Fig. 2. It is clear from this figure that stable nodes result when the roots of the characteristic equation are both negative real numbers; unstable nodes result when these roots are both positive real numbers. Solving the characteristic equation (equation 2.19) by quadratic formula yields

$$\mu_{1,2} = \frac{a_1 + a_4 \pm \sqrt{(a_1 + a_4) + 4(a_2 a_3 - a_1 a_4)}}{2} \tag{2.24}$$

The condition for nonzero real roots of the same sign is

$$0 < a_1 a_4 - a_2 a_3 \leq \frac{(a_1 + a_4)^2}{4}, \tag{2.25}$$

where the equal sign indicates the condition for identical real roots ($m = 1$).

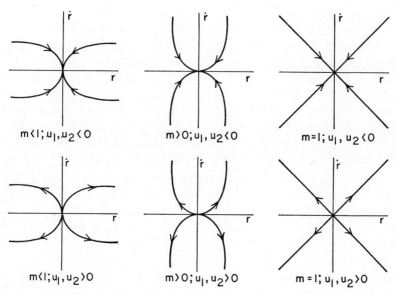

Fig. 2. Two-dimensional phase-space representation of a stable (upper three figures) and unstable (lower three figures) node.

The condition for a stable node (both roots negative) is

$$a_1 + a_4 < 0. \tag{2.26}$$

The condition for an unstable node (both roots positive) is

$$a_1 + a_4 > 0. \tag{2.27}$$

If both μ_1 and μ_2 are real and of opposite sign, $m < 0$, and the integral curve is

$$s \, |r|^{|m|} = C. \tag{2.28}$$

Equation 2.28 is hyperbolic. An illustration of the phase-plane representation of this type of singular point may be found in Fig. 3; in this figure $\mu_1 + \mu_2$

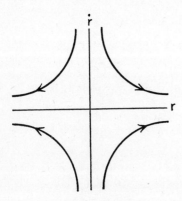

Fig. 3. Two-dimensional phase-space representation of a saddle point.

is zero. This type of singularity, characterized by one negative real characteristic exponent and one positive real characteristic exponent, is called a *saddle point*. Saddle points are always unstable singularities. From equation 2.24, the condition for real characteristic exponents of opposite sign is

$$a_1 a_4 - a_2 a_3 < 0. \tag{2.29}$$

If μ_1 and μ_2 are conjugate complex, the integral curves are spirals which tend either toward or away from the singular point. In this situation the singularity is called a *focus*. In order to illustrate the motion in the neighborhood of a focus, we follow the derivation suggested by Minorsky (6).

We set $u = \omega e^{i\varphi}$ and $v = \omega e^{-i\varphi}$ and obtain a new system of equations:

$$\dot{\omega} = \frac{\omega(\mu_1 + \mu_2)}{2}, \qquad \dot{\varphi} = \frac{\mu_1 - \mu_2}{2i}, \quad i = \sqrt{-1}. \tag{2.30}$$

The integral curves are

$$\omega = C_1 \exp\left(\frac{((\mu_1 + \mu_2)t}{2}\right), \qquad \varphi = \frac{(\mu_1 - \mu_2)(C_2 + t)}{2i}, \quad i = \sqrt{-1}, \quad (2.31)$$

where C_1 and C_2 are the constants of integration. Eliminating time from these equations, we obtain

$$\omega = C \exp\left(\frac{\mu_1 + \mu_2}{\mu_1 - \mu_2}\varphi i\right), \quad i = \sqrt{-1}, \qquad (2.32)$$

where C is a constant. Equation 2.32 illustrates that, if the real part of μ_1 and μ_2 is not zero, the phase-plane trajectory spirals around the singular point at the origin. As in the case of a nodal point, the stability of a focal

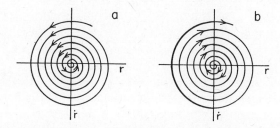

Fig. 4. Two-dimensional phase-space representation of (a) stable and (b) unstable focus.

point is determined by the sign of the real part of the characteristic exponents. If the real part of both characteristic exponents is negative, the phase-plane trajectory winds toward the origin, and the singularity is called a *stable focus*. Alternatively, if the real part of either characteristic exponent is positive, the trajectory winds away from the origin, and the singularity is called an *unstable focus*. The motion in the phase-space in the neighborhood of a stable focus is illustrated in Fig. 4a; the motion in the neighborhood of an unstable focus can be seen in Fig. 4b.

Returning to equation 2.19, we see that the condition for a focus is

$$0 < \frac{(a_1 + a_4)^2}{4} < a_1 a_4 - a_2 a_3. \qquad (2.33)$$

If μ_1 and μ_2 are purely imaginary,

$$\mu_1 + \mu_2 = 0 \qquad (2.34)$$

and equation 2.32 reduces to

$$\omega = C. \qquad (2.35)$$

Equation 2.35 illustrates that under these special conditions the integral curves in the neighborhood of a singular point are closed circles about the origin. This type of singularity is usually referred to as a *center;* the corresponding integral curve was illustrated in Fig. 1. A center is characterized by the fact that each trajectory arising from a particular set of initial conditions turns about the singularity with a characteristic radius. Since the trajectories in the neighborhood of a center turn about the singularity without either approaching it or receding indefinitely, sustained oscillations occur.

It is clear from equation 2.19 that the condition for purely imaginary characteristic exponents is

$$a_1 + a_4 = 0 < a_1 a_4 - a_2 a_3. \tag{2.36}$$

The conditions in equation 2.36 require that a_2 and a_3 be of opposite sign, and that $|a_2 a_3| > |a_1 a_4|$. A special case (usually associated with the harmonic oscillator) when these conditions are met is $a_1 = a_4 = 0$, $a_2 = -a_3 = 1$. The special case $a_1 = a_4 = 0$, $a_2 > 0$, $a_3 < 0$ characterizes a center.

We have seen that stable nodes arise when the characteristic exponents of equations 2.12 are negative real numbers; unstable nodes arise if the exponents are positive. Saddle points arise if one of the characteristic exponents is positive and the other negative. Unstable foci arise if the real part of one of the conjugate complex characteristic exponents is positive; if the real parts of μ_1 and μ_2 are negative, the singularity is a stable focus. These considerations lead us to the following generalization:

> **Theorem 3.** *If the real part of either of the characteristic exponents of equations 2.12 is positive, the singularity at the origin is unstable; if the real part of both characteristic exponents is negative, the singularity is asymptotically stable.*

Later we shall see that this theorem can be extended to systems involving more than two variables.

3.   LIMIT CYCLES

A.   The Properties of Limit Cycles

In 1881 Poincaré (7) showed that the type of differential equations we are considering,

$$\dot{x} = X(x, y), \qquad \dot{y} = Y(x, y), \tag{3.1}$$

sometimes result in a special type of solution represented by closed curves in the phase-space. Poincaré called these closed trajectories *limit cycles*. These closed curves are distinguished from the closed trajectories surrounding centers by the fact that limit cycles are *isolated* closed trajectories. We have seen that the radius of the closed trajectories surrounding a center depend on the initial conditions of the variables. In contrast, the radius of a limit cycle is the same for a whole range of initial conditions. In fact, every trajectory in the neighborhood of a stable limit cycle winds toward the limit cycle as time approaches infinity. Thus a stable limit cycle represents an asymptotically

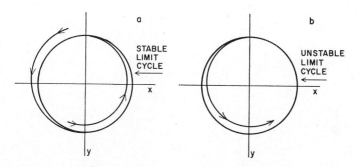

Fig. 5. Two-dimensional phase-space representation of (a) stable and (b) unstable limit cycle.

stable state of sustained oscillations in exactly the same way as stable singularities represent stable states of constant flow. The behavior of trajectories in the neighborhood of a stable limit cycle is illustrated in Fig. 5a.

It is also possible for a limit cycle to be unstable. Every trajectory in the neighborhood of an unstable limit cycle winds toward the limit cycle as time approaches minus infinity. The behavior of trajectories in the neighborhood of an unstable limit cycle can be seen in Fig. 5b.

Another important property of limit cycles is that they can arise only in systems described by nonconservative, nonlinear differential equations. This feature further distinguishes limit cycles from the closed trajectories surrounding a center.

A limit cycle may surround a singularity that is either stable or unstable. If the singularity is an unstable focus, the limit cycle closest to this singularity must be stable; in this situation sustained oscillations will arise spontaneously from the unstable state of rest. This situation is commonly referred to as *soft excitation* and is illustrated in Fig. 6a. Alternatively, if the singularity is a stable focus, the limit cycle closest to the singularity must be unstable; in this situation sustained oscillations will not arise spontaneously from the state of rest. However, if a stable limit cycle exists beyond the unstable cycle, sustained oscillations can arise if the state of rest is subject to sufficient perturbation; this perturbation must be large enough to cause the variables to move to the phase-space outside of the unstable limit cycle. This situation is commonly referred to as *hard excitation* and is illustrated in Fig. 6b.

It can be seen from Fig. 6 that a singularity can be surrounded by more than one limit cycle; the requirements of Theorem 1b are not violated as long as the limit cycles do not overlap. Thus a number of alternatively stable and unstable limit cycles can be enclosed in each other: if the innermost limit

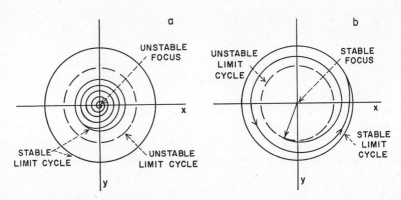

Fig. 6. Two-dimensional phase-space representation of alternating stable and unstable limit cycles surrounding (a) stable and (b) unstable focus.

cycle is stable, it encloses an unstable singular point; if the innermost limit cycle is unstable, it encloses a stable singularity. In Fig. 6a the singular point is an unstable focus, the innermost limit cycle is stable, the next is unstable, and so on. In Fig. 6b the singularity is a stable focus, the innermost limit cycle is unstable, the next is stable, and so on.

Figure 6 also illustrates why trajectories originating in one area of the phase-space can approach a particular limit cycle while trajectories originating in another area approach a different cycle and yet another range of initial conditions of the variables results in the approach to a singular point. If more than one limit cycle exists, the trajectories may originate in the phase-space between (1) a stable singularity and an unstable limit cycle, (2) an unstable singular point and a stable limit cycle, or (3) a stable and an unstable limit cycle. The trajectories originating in the phase-space between a stable singularity and an unstable limit cycle wind toward the singular point; all initial conditions in the phase-space between an unstable singularity and a stable limit cycle wind toward the limit cycle; all trajectories originating in the phase-space between an unstable limit cycle and a stable limit cycle wind toward the stable cycle. If multiple stable limit cycles exist, different ranges of initial conditions will lead to different but discrete oscillatory regimes. Thus widely different initial conditions of the variables will lead to identical oscillatory regimes provided the trajectories originate in a well-defined neighborhood of a stable limit cycle. On the other hand initial conditions that are only slightly different may mean that the trajectories originate on different sides of an unstable limit cycle; in this situation slightly different initial conditions can lead to completely different oscillatory regimes.

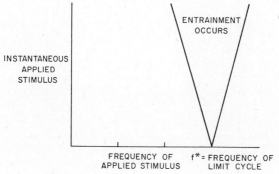

Fig. 7. Relationship among the amplitude of the instantaneous applied stimulus, the frequency of the instantaneous applied stimulus, and the occurrence of entrainment.

Another important feature of limit cycles is that they can synchronize to a periodic stimulus if the stimulus period is sufficiently close to the period of the limit cycle. This type of behavior, which is usually called *entrainment*, is illustrated in Fig. 7. It can be seen from this figure that entrainment occurs in weakly coupled systems only if the stimulus frequency is close enough to the limit cycle frequency, or the stimulus is strong enough to compensate for the difference.

In biological systems one expects to find a population of similar oscillators (e.g., cells or metabolic components). These oscillators are essentially autonomous and certainly self-sustaining; however, even if they are started simultaneously, the coherence will not persist in the absence of interactions because the individual periods cannot be expected to be exactly identical. On the other hand, if weak mutual interactions occur among the autonomous oscillators, coherence can arise and persist. Winfree (8) has shown that the mode of temporal organization of a population of self-sustaining, self-limiting oscillators depends on a quantity that he calls the *syntal index*. If the syntal index is negative, the physical situation is that each oscillator interferes with entrainment to any other oscillator. The effect is that coherence is self-repressing, and no spontaneous synchrony is possible. On the other hand, if the syntal index is between zero and unity, the oscillators run essentially independently, but there are abortive attempts to sustain coherence (apparently stimulated by the environmental noise). In this case the population maintains a statistical coherence of phases, but the oscillators are not strictly synchronized.

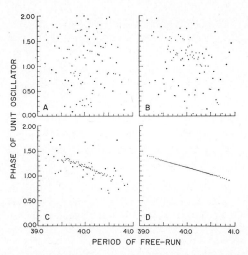

Fig. 8.   Relationship between time and entrainment. (A) The random initial states of 100 oscillators, (B) the communication between the oscillators 1000 arbitrary time units later, (C) the advanced entrainment of these systems at 2000 arbitrary time units, and (D) the nearly complete entrainment of the systems at 3000 time units. (From A. Winfree.)

If the syntal index is greater than unity, the pulsations of coherence become self-sustaining. Winfree (8) describes this situation as a temporal phenomenon analogous to crystallization or spontaneous combustion. Indeed, any small periodic influence is able to cause a few interacting oscillators to be coherent; this augments the influence rhythm to entrain a few more oscillators of nearby frequency. The process continues in this autocatalytic manner until all interacting periodic components within a certain frequency range are entrained. These events are represented graphically in Fig. 8.* In this figure 100 oscillators are located horizontally according to their *original* periods;

* I am indebted to Dr. Winfree for providing me with a copy of this figure from ref. 8.

vertically, each oscillator is located opposite its instantaneous phase. At zero time (Fig. 8a) each of the 100 oscillators is running at its own rate because the phases are initially random, and there is no net rhythm of influence in the population. At 1000 arbitrary time units (seconds, minutes, hours, days, etc.) later (Fig. 8b) we seen that the oscillators are in communication and the randomness in Fig. 8a is not stable. Now 100 individual rhythms are becoming synchronized, and the more coherent they become, the stronger is the influence rhythm that induces the entrainment. At 2000 arbitrary time units after time zero (Fig. 8c) some oscillators have completed as few as 49 cycles whereas faster ones have completed up to 51 cycles. Nevertheless, most of the oscillators are now at nearly the same stage of their cycles. They are falling into step. At 3000 arbitrary time units (Fig. 8d) almost perfect synchrony has been achieved. All the oscillators complete their cycles in the same period, and they remain in a fixed formation of relative phases with the originally faster oscillators anticipating a bit and the slower oscillators lagging behind.

Yet another important feature of limit cycle behavior arises when the differential equations contain a variable parameter. In biological systems this parameter could denote the steady-state concentration of a metabolic component or of any other constituent which itself can vary. Regardless of the mechanism, when this parameter varies the topological configuration of a group of concentric limit cycles varies too. At some critical value of the parameter it may happen that a stable and an unstable limit cycle coalesce and, in effect, disappear. If this occurs, the variables of the differential equations are suddenly freed of the restraints imposed by the existence of the original limit cycles; if another stable limit cycle exists, the variables come under its influence. This situation manifests itself as a sudden change in the amplitude of the oscillations of the variables.

The existence of multiple limit cycles, like the existence of multiple stationary states, suggests a basis for explaining the myriad of quasi-discontinuous events observed in biological systems. At present a good case can be made for the existence of biochemical systems for which multiple steady states are possible (see Section 7); it is also well known that such steady states can be subject to effective control through negative feedback mechanisms. On the other hand the search for biological and biochemical systems for which multiple limit cycles are possible has hardly begun. The insensitivity that biological limit cycles would have to a wide range of initial conditions and perturbations is particularly appealing in the context of biological clocks, evolutionary biology, macromolecular synthesis, and the like. Mechanisms for which multiple limit cycles and parametric variation are possible should remain high on the list of candidates for the explanation of quasi-discontinuous events in biological systems.

B. Conditions for the Nonexistence of Limit Cycles

It seems obvious that one cannot decide from the nature of a singular point that a surrounding limit cycle surely must exist. We have seen that stable singular points can be surrounded by concentric unstable and stable limit cycles. On the other hand, unless the system involves only two variables and is autonomous, unstable singular points are not necessarily surrounded by limit cycles even if they are focal points and the motion of the physical system is not totally unstable.* Thus we can never establish the sufficient conditions for the existence of a limit cycle from considerations of the linearized, abridged equations. However, we can establish some theorems that give necessary conditions for the existence of limit cycles; contrary to claims made elsewhere (9) these necessary conditions can be deduced solely from considerations of singular points and the abridged equations. To begin, we note an intuitively obvious theorem:

Theorem 4a.   *The Poincaré index of a closed trajectory is* $+1$.

The proof of this theorem can be found elsewhere (10). Since the Poincaré index of a regular point is zero (11), we have the important result:

Theorem 4b.   *A closed trajectory surrounds at least one singular point.*

Furthermore, since the Poincaré index of a node, focus, or center is $+1$, but the index of a saddle point is -1, we have

Theorem 4c.   *If a closed trajectory surrounds only one singular point, the singularity cannot be a saddle point.*

Theorem 4d.   *If a closed trajectory surrounds two singular points, neither singularity can be a saddle point.*

From Theorems 4c and 4d and inequality 2.29, it is clear that limit cycle behavior is not possible for systems involving one or two singular points provided

$$a_1 a_4 - a_2 a_3 < 0. \tag{3.2}$$

In terms of the original transformed differential equations (equations 2.10),

$$\dot{u} = U(u, v), \qquad \dot{v} = V(u, v). \tag{3.3}$$

---

* I wish to acknowledge the extensive work that Dr. P. Ponzo has done in this area and to thank him for his confirmation of this statement.

This means that the Jacobian

$$J = \begin{vmatrix} \dfrac{\partial U}{\partial u} & \dfrac{\partial U}{\partial v} \\[2ex] \dfrac{\partial V}{\partial u} & \dfrac{\partial V}{\partial v} \end{vmatrix}, \tag{3.4}$$

evaluated at $u = v = 0$, is positive for a focus, node, or center; in this case limit cycle behavior is possible. However, if J is negative, the singularity is a saddle point and limit cycle behavior is not possible. This argument leads us to

Theorem 5a. *A necessary condition for the existence of a limit cycle is*

$$\frac{\partial U}{\partial u} \cdot \frac{\partial V}{\partial v} - \frac{\partial U}{\partial v} \cdot \frac{\partial V}{\partial u} > 0.$$

Two obvious corollaries of Theorem 5a are

Theorem 5b. *If $\partial U/\partial u$ and $\partial V/\partial v$ are of opposite sign, two necessary conditions for the existence of a limit cycle are*

(1) $\partial U/\partial v$ and $\partial V/\partial u$ must be of opposite sign, and

(2) $|\partial U/\partial v \cdot \partial V/\partial u| > |\partial U/\partial u \cdot \partial V/\partial v|$.

Theorem 5c. *If $\partial U/\partial v$ and $\partial V/\partial u$ are of the same sign, two conditions necessary for the existence of a limit cycle are*

(1) $\partial U/\partial u$ and $\partial V/\partial v$ must have the same sign, and

(2) $\partial U/\partial v \cdot \partial V/\partial u < \partial U/\partial u \cdot \partial V/\partial v$.

The partial derivatives mentioned in Theorems 5a–5c are evaluated at the origin of equations 3.3. Hence the conditions necessary for the existence of a limit cycle can be established directly from the linearized differential equations:

$$\dot{u} = a_1 u + a_2 v, \qquad \dot{v} = a_3 u + a_4 v, \tag{3.5}$$

where $a_1 = \partial U/\partial u$, $a_2 = \partial U/\partial v$, $a_3 = \partial V/\partial u$, $a_4 = \partial V/\partial v$, and each partial derivative is evaluated at $u = v = 0$.

The negative criterion of Bendixson (12) is

Theorem 6a. *If the value of $\partial U/\partial u + \partial V/\partial v$ does not change sign within a region of the phase-space, no closed trajectory can exist within this region.*

An obvious corollary of Theorem 6a is

Theorem 6b. *A necessary condition for the existence of a limit cycle is that $\partial U/\partial u$ and $\partial V/\partial v$ must have opposite signs or be individually zero for some values of the variables within the region containing a closed trajectory.*

Theorem 6b refers to the signs of the partial derivatives evaluated at any portion of the phase-space available to the variables. If we wish, we can define a domain containing the singular point in such a manner that the partial derivatives are sign-definite throughout the phase-space in this domain. As we shall see in Section 6, there is a large class of commonly encountered chemical reactions for which this region is global. The advantage of restricting an analysis to this domain is that we can then make some rather strong statements about the necessary conditions for the existence of limit cycles. Higgins (13), for example, has devised some theorems that apply in the neighborhood of the singularities of certain chemical mechanisms; he refers to this neighborhood as the "region of definite character."

The disadvantage of restricting an analysis to a neighborhood of the singular point is that the results will not provide any information about whether or not limit cycles exist in the rest of the phase-space. Therefore, the conditions for limit cycles obtained from this type of analysis cannot always be used to exclude, and can never be used to guarantee, the existence of a limit cycle in the global phase-space.

If the roots of the characteristic equation are purely imaginary, equation 2.36 requires that

$$\frac{\partial U}{\partial u} = -\frac{\partial V}{\partial v}. \tag{3.6}$$

The partials in equation 3.6 are evaluated at $u = v = 0$. If we exclude the situation for a center, the partial derivatives in equation 3.6 must be of opposite sign. Consequently the necessary condition mentioned in Theorem 6b is always met if the characteristic exponents are purely imaginary.

If μ_1 and μ_2 are conjugate complex, the situation is considerably different. If the signs of $\partial V/\partial v$ and $\partial U/\partial u$ are different at the origin, the condition mentioned in Theorem 6b is again satisfied; in this case limit cycle behavior is clearly possible. However, if the signs of $\partial V/\partial v$ and $\partial U/\partial u$ are the same at the singular point, we can conclude only that limit cycles cannot occur in the neighborhood Higgins refers to as the region of definite character.

We can combine Theorems 5b and 6b and obtain:

Theorem 7. *Three necessary conditions for the existence of a limit cycle in a region of definite character are:*

(1) *$\partial U/\partial u$ and $\partial V/\partial v$ must be of opposite sign,*

(2) $\partial U/\partial v$ and $\partial V/\partial u$ must also be of opposite sign, and

(3) $|\partial U/\partial v \cdot \partial V/\partial u| > |\partial U/\partial u \cdot \partial V/\partial v|$.

Theorem 7 leads directly to the rules for oscillations suggested recently by Higgins (9).

Higgins (9) also makes the heurestic assumption that, if $\partial U/\partial u > 0$, then $\partial U/\partial u > |\partial V/\partial v|$; alternatively, if $\partial V/\partial v > 0$, then $\partial V/\partial v > |\partial U/\partial u|$. These circumstances would lead to the assumption that stable singular points cannot be surrounded by unstable limit cycles in chemical systems. It would be surprising if this assumption were correct; as we have seen, there are examples of other types of physical systems known to require hard excitation (14).

Another heuristic assumption that Higgins (15) makes is that, if trajectories cannot go off to infinity, and $\partial U/\partial u + \partial V/\partial v > 0$, then a limit cycle must exist. Clearly, if the variables represent chemical concentrations, finite bounds can be placed on the phase-space available to the trajectories. However, even in this situation the existence of an unstable singular point does not guarantee the existence of a surrounding limit cycle unless the physical system involves only two variables and is autonomous. It was pointed out earlier in this chapter that there are examples of other types of systems that wander aimlessly in a bounded phase-space. These systems do not violate the conditions in Theorem 1b, and they do not necessarily approach a limit cycle.

It would be extremely convenient if the existence and nonexistence of limit cycles in chemical systems could be established from studies of the properties of singular points. In two recent studies (16, 17) of the stability properties of biological and biochemical feedback systems, cases for which the singular point was asymptotically stable did not have limit cycles, and examples wherein the singular point was unstable did exhibit limit cycle behavior. If these results turn out to be general, the rules for the existence and the nonexistence of limit cycles in chemical systems will be very simple indeed.

4. CONTROL THEORY

A. Introduction

Classical control theory is concerned primarily with methods for studying the properties of linear control systems. One reason for this emphasis is that engineers are in a position to design their control circuits in such a manner that operation will occur in any domain of the phase-space that they wish to prescribe as a basis for their theory. The linear region is the obvious choice because the theory of linear differential equations is firmly established.

Biological scientists, on the other hand, have not designed the control systems they wish to investigate. It would be a far-reaching assumption to suppose that the physiological function of biological control systems could be studied by methods designed for use in the linear region. Biological control mechanisms are almost always nonlinear; the numerous examples of entrainment, limit cycle behavior, cooperative interactions, and the like observed in biological systems attest to the fact that these systems actually operate in the nonlinear region. In at least some situations it seems possible that the pressures of natural selection would be met by advantages inherent in nonlinear operation (18). In this sense biological control mechanisms have been designed strictly for integrated utility and with no regard for the convenience of a linear approximation. It is not surprising to find that in a large number of cases utility does not correspond to idealized behavior.

The following pages do not deal with open- and closed-loop gains, Laplace transformations, system transformations, and Nyquist and Bode diagrams. These concepts have been very useful to the engineer; unfortunately the tools of the linear method will not suffice for studies of biological control mechanisms. For the reader who wishes to learn the details of the linear method, the excellent book by Leonard Bayliss (19), *Living Control Systems*, is recommended.

Textbooks on control theory devote a great deal of space to defining terms and discussing synonyms. Usually, the first term defined is *system*. Grodins (20) defines a system as a collection of components arranged and interconnected in a definite way. If nothing is known about the details of a system, it is common to draw a rectangle and say that the system is represented by the rectangle. The rectangle is sometimes referred to as a *black box;* the blackness presumably emphasizes the fact that, since the components are not known, they cannot be seen. The addition of two arrows to the rectangle constitutes a *block diagram* of a system with *input* (what goes in) and *output* (what comes out):

The essential feature of a system is that a change of input causes a change in output. This sequence of events is usually called *cause and effect*.

In certain problems it may happen that the mathematical relationship between output and input is known. In linear systems it will be in the form

$$a_n \frac{d^n y}{dt^n} + a_{n-1} \frac{d^{n-1} y}{dt^{n-1}} + \cdots + a_1 \frac{dy}{dt} + y = z, \qquad (4.1)$$

where z is the input, y is the output, and the a_i are constants. Since equation

4.1 is a linear differential equation, we can take the Laplace transformation and obtain

$$(a_n s^n + a_{n-1} s^{n-1} + \cdots + a_1 s + 1)y(s)$$
$$= z(s) + \text{sum of all initial condition transform terms} \quad (4.2)$$

where s is the Laplace variable. Equation 4.2 can be solved algebraically for $y(s)$:

$$y(s) = \left[\frac{1}{a_n s^n + a_{n-1} s^{n-1} + \cdots + a_1 s + 1} \right] [z(s) + \text{sum I.C. terms}]. \quad (4.3)$$

According to equation 4.3, the output transform, $y(s)$, is equal to the product of two other transform terms. The first contains all of the essential information about the system itself; it is called the *system transform function*. The second contains the essential information about the disturbance applied to the system; it is called the *excitation function*. If all the initial conditions are zero, then the product on the right-hand side of equation 4.3 is called the *normal response transform;* usually this transform is represented by placing $G(s)$ in the rectangle. Thus a system can be represented in block diagram form as

$$\xrightarrow{z(s)} \boxed{\dfrac{1/K}{a_n s^n + \cdots + a_1 s + 1}} \xrightarrow{y(s)}$$

or, more simply, as

$$\xrightarrow{z(s)} \boxed{G(s)} \xrightarrow{y(s)}$$

The next term usually encountered in control theory is *cybernetics*. Wiener (21) describes cybernetics as the science of control and communication in the animal and the machine. The problem with this description is that what we mean by *control* is usually left to the imagination. Examination of usage suggests that "control" embodies two distinct concepts:

1. Control is a connection that makes two systems behave as though they were one.

2. Control is a restraint that makes a system behave as though it were not a system.

The first concept of control arises in *open-loop control systems:*

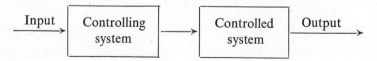

This block diagram illustrates that valves, throttles, and enzyme concentrations are all controlling systems. Since the two systems are connected in series, they behave as a single system.

The second concept of control arises in *closed-loop control systems:*

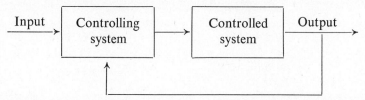

The closed-loop configuration in this diagram results from the fact that the output of the controlled system influences the controlling system. This situation is usually referred to as *feedback.*

Feedback can be either negative or positive. In *negative feedback* systems the output potentiates the input; in the classical theory the input and the output signals are compared (subtracted). Thus negative feedback causes the influence of a disturbance to a regulator to be minimized so that the output of the system is maintained nearly constant. *Positive feedback*, on the other hand, is added to the input. Clearly, systems possessing only positive feedback are likely to be unstable. However, if the main negative-feedback loops contain subloops of positive feedback, the entire control system can be stable and exhibit improved performance (22).

The usual representation of a negative feedback control system is as follows:

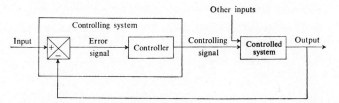

If the main purpose of this system is to follow a varying input, it is called a *servosystem.* If the main purpose is to keep the output close to a certain value, it is called a *regulator.*

Biological control systems include both the closed-loop and the open-loop configurations described above. Examples of closed-loop biological and biochemical control mechanisms with either negative feedback or negative and positive feedback are known (see Section 6).

One of the difficulties associated with experimental studies of biological control systems is the identification of the controlling and the controlled systems. In order to simplify this problem some individuals have adopted the concept of a "master reaction" or a "rate-limiting step" in metabolic sequences. In this view a single enzyme (usually one subject to feedback) is assumed to control the material flux through the entire metabolic sequence. However, as Hearon and his associates (23) pointed out some time ago, and Kacser and Burns (24) have emphasized recently, this concept cannot have a strict meaning in real biological systems. In even the simplest metabolic pathway, no single step can be completely rate limiting, for then its sensitivity coefficient (24) would be unity. Therefore metabolic flux control cannot be dependent on only one enzyme. Kacser and Burns have suggested that flux control is actually a systemic property: "the response characteristic of the whole system at one point of its phase-space."

It is certainly true that the investigation of closed-loop biological control systems promises to be far more difficult than current biochemical experiments seem to admit. When an "effect" is found to be connected by a feedback to a "cause," it is sometimes meaningless or hopeless, or both, to stipulate which came first. In order to obtain an idea of the behavior of the whole system, it is necessary to synthetize the separately established properties of its interconnecting parts into a functional unit. The establishment of the proper connections is no easy task, but quantitative formulation of the systemic behavior of the resultant biological control mechanisms is essential for our understanding of living systems.

A further complication for future experiments lies in the fact that well-controlled systems do not reflect large changes in the output when the system is perturbed. Kacser and Burns, for example, have illustrated a situation involving the *in vivo* alteration of three different enzymes that catalyze the sequential interconversion of closely related metabolites and are essential for the growth of *Neurospora crassa*. Large decreases in the *in vivo* activity of any one of these enzymes failed to produce a change in the growth rate of the organism. Since the enzymes and the associated metabolic sequence are known to be essential for growth, it would seem that this enzyme system is so well controlled that perturbations do not manifest themselves in the output (i.e., the growth rate). It is as though the system was unimportant to growth, although it is, of course, essential.

A series of consecutive chemical reactions (e.g., a sequence of metabolic reactions) is actually an open-loop control system. If a single chemical step is slow (either because a rate constant is small or the concentration of a component is low), this step influences the rate of all subsequent chemical reactions:

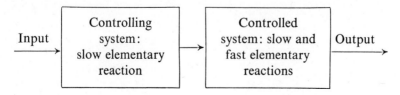

If we change the rate constant or the concentration of a component in the controlling system, the output of the controlled system also changes. In real chemical systems the controlling system may be a combination of relatively slow steps rather than a single slow elementary reaction. In general, however, the distinction between chemical components belonging in the controlling or the controlled systems will be completely arbitrary.

A series of consecutive chemical reactions with feedback can be represented as a closed-loop control system:

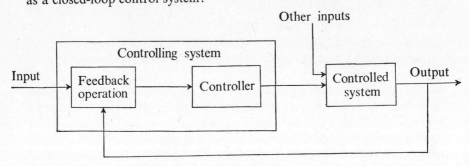

In this case the point at which the feedback interaction occurs might be well-defined, but here again the distinction between the chemical components in the controller and the components in the controlled system might be arbitrary. For example, suppose that each chemical step is described by

$$\dot{x}_i = k_{i-1}x_{i-1} - k_i x_i, \quad i = 2, \ldots, n+1, \tag{4.4}$$

and the feedback signal is described by

$$\dot{x}_1 = F(x_{n+1}) - k_1 x_1. \tag{4.5}$$

The k_i in equations 4.4 and 4.5 will be first-order rate constants, and $F(x_{n+1})$ will be the nonlinear feedback function. If $m < n+1$ is the subscript of the last component in the controller, the equation for the controlling system is

$$a_m \frac{d^m x_m}{dt^m} + a_{m-1} \frac{d^{m-1}x_m}{dt^{m-1}} + \cdots + a_1 \frac{dx_m}{dt} + x_m = F(x_{n+1}). \tag{4.6}$$

The equation for the controlled system is

$$b_{n-m+1} \frac{d^{n-m+1}x_{n+1}}{dt^{n-m+1}} + b_{n-m} \frac{d^{n-m}x_{n+1}}{dt^{n-m}} + \cdots + b_1 \frac{dx_{n+1}}{dt} + b_0 x_{n+1} = x_m. \tag{4.7}$$

The a_i and b_i in equations 4.6 and 4.7 are known combinations of the individual rate constants in equations 4.4 and 4.5. From equation 4.6 we identify $F(x_{n+1})$ as the input to the controlling system and x_m as the output; from equation 4.7 we identify x_m as the input to the controlled system and x_{n+1} as the output and the feedback signal. Obviously, we are free to select any integer between zero and $n + 1$ for m; this means that we can obtain any one of n possible combinations for the contents of the controller and the controlling systems. Since it is generally immaterial which value we select for m, we will usually use $m = 1$. Then equation 4.5 is the first-order differential equation for the controlling system, and equations 4.4 describe the controlled system. Equation 4.7 illustrates that these n first-order differential equations can be represented as a single nth-order equation.

B. The Lyapunov Direct Method

Biochemical control mechanisms involve complicated sets of chemical reactions (including diffusion processes) consisting of controlled concentrations and regulating components. The purpose of the regulator is to maintain the controlled variables continuously in a certain stationary state or in a state that varies in a prescribed way. Thus the control process consists of utilizing the regulator to prevent deviations from the desired state of the regulated concentrations. Since the modern theory of control is based on the foundation set by A. M. Lyapunov, this section contains a brief introduction to the so-called Lyapunov direct method.

To begin, consider the general system of n autonomous, nonlinear first-order differential equations:

$$\dot{x}_i = X_i(x_1, \ldots, x_n), \quad i = 1, \ldots, n. \tag{4.8}$$

As in the case involving two components, we can transform the origin of equations 4.8 and obtain

$$\dot{u}_i = U_i(u_1, \ldots, u_n), \quad i = 1, \ldots, n, \tag{4.9}$$

such that $U_i = 0$ when $u_i = 0$ for $i = 1, \ldots, n$. In the modern theory the singularity $u_i = 0$ $(i = 1, \ldots, n)$ is called an undisturbed motion of equations 4.9. The undisturbed motion is called *stable* if for all disturbances Δu_i satisfying the condition

$$\sum_{i=1}^{n} \Delta_i^2 \leq H \tag{4.10}$$

the motion of the differential equations is such that

$$\sum_{i=1}^{n} u_i^2 < A \tag{4.11}$$

for all possible values of the variables in the real (positive) time domain. This definition of stability is identical to the one made earlier; now, however, we can speak of the *H vicinity* and the *A vicinity* defined by inequalities 4.10 and 4.11. We can say that, if disturbances limited to the *H* vicinity do not result in variable excursions beyond the *A* vicinity, the motion of the system is stable. As before, the stability properties of stationary states and limit cycles are included in this definition of stability.

The nature of the stability of the motion in the *A* vicinity can be twofold. The motion will be stable in the sense either that inequalities 4.10 and 4.11 are satisfied, or that in addition

$$\lim_{t \to \infty} u_i(t) = 0. \tag{4.12}$$

In cases in which equation 4.12 is satisfied, the stability of the motion is called asymptotic. As in our previous definition, limit cycle behavior is not possible for *H*-vicinity disturbances of systems possessing asymptotic stability.

One of the important features of the Lyapunov theory is that it provides a method for the calculation of *H*. The magnitude of this number depends on the form of the so-called *Lyapunov V function* and the parameters of the differential equations that describe the system. After *H* has been calculated it is possible to establish the size of the region in which the stability of the system is assured. If *H* is not sufficiently large to include all of the phase-space accessible to the variables, we speak of *stability in the small;* if, on the other hand, *H* is sufficiently large, we speak of *stability in the large* or *global stability.*

As before, asymptotic stability in the small does not guarantee that a limit cycle cannot exist outside the *H*-vicinity. Global asymptotic stability, however, absolutely precludes the possibility of sustained oscillatory behavior of any kind.

The Lyapunov direct method involves the construction of a function of the variables whose Eulerian derivative has certain properties that assure stability or asymptotic stability. If the Lyapunov *V* function is unbounded and the properties of the Eulerian derivative ensure asymptotic stability, the stability is asymptotic and global.

A function is called *sign-invariant* if it is always of the same sign or is zero. An example of a positive-invariant function is

$$V = (u_1 + u_2)^2 + u_3^2. \tag{4.13}$$

This function is positive for all u_1, u_2, and u_3 except $u_1 = -u_2$, $u_3 = 0$, or $u_1 = u_2 = u_3 = 0$, in which case it is zero.

A sign-invariant function that is zero only at the origin is called *sign-definite*. An example of a positive-definite function is

$$V = u_1{}^2 + u_2{}^2 + u_3{}^2. \tag{4.14}$$

This function is positive for all u_1, u_2, and u_3 except at the origin, where it is zero.

If V is a sign-definite function, then $V = C = $ constant represents a one-parameter family of curves in the phase-space. If C is sufficiently small,

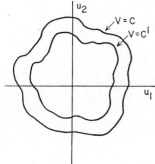

Fig. 9. Two-dimensional representation of a Lyapunov function.

each of these curves will be closed (25, 26), and they must contain the origin. The two-dimensional representation of this situation is illustrated in Fig. 9. Barashin and Krasovsky (27) have pointed out that closure of these surfaces is assured if the Lyapunov V function approaches infinity as

$$\sum_{i=1}^{n} u_i{}^2 \to \infty. \tag{4.15}$$

We see from Fig. 9 that, when C is decreased to C', the closed surface in the phase-space contracts toward the origin; thus decreasing C decreases the area of the domain defined by the Lyapunov V function. Obviously, if V is unbounded, the phase-space enclosed by the function will be global.

Regardless of the size of the region defined by the Lyapunov function, the surfaces surrounding the origin intersect all paths that lead from the origin to infinity. Thus, if all motion in the phase-space across the whole family of an unbounded Lyapunov function is from outside the surface into the region enclosed by the surface, the motion must possess global asymptotic stability. This point is of fundamental importance to what follows.

Theorem 8 (Lyapunov). *A motion is stable if it is possible to find a sign-definite function V having a Eulerian derivative that is sign-invariant and of sign opposite to that of V.*

Theorem 9 (Lyapunov). *A motion is asymptotically stable if it is possible to find a sign-definite function V having a Eulerian derivative that is sign-definite and of sign opposite to that of V.*

To prove Theorem 9 we assume that V is positive-definite. The Eulerian derivative is

$$\frac{dV}{dt} = \text{grad } V \cdot U = \sum_{i=1}^{n} \frac{\partial V}{\partial u_i} \cdot \dot{u}_i = \sum_{i=1}^{n} \frac{\partial V}{\partial u_i} \cdot U_i. \tag{4.16}$$

Since the partial derivatives in equation 4.16 are proportional to the directional cosines of the normal N to the surface defined by $V = C$, equation 4.16 can be written as (28)

$$\dot{V} = \sqrt{\sum_{i=1}^{n} (\partial V/\partial u_i)^2} \sum_{i=1}^{n} U_i \cos Nu_i, \tag{4.17}$$

where $\sum_{i=1}^{n} U_i \cos Nu_i$ represents the projection of the velocity of the representative point R on the normal to the $V = C$ surface. Since $\sum_{i=1}^{n} U_i \cos Nu_i < 0$ means that R moves in the phase-space along trajectories that intersect the family of surfaces inward, R cannot remain outside or on any surface $V = C$ ($C \neq 0$). Consequently, as time approaches infinity, the motion will approach the origin and will thus be asymptotically stable. A two-dimensional representation of this situation is given in Fig. 10.

The proof of Theorem 8 is identical to the proof of Theorem 9. However, the conditions of Theorem 8 allow $\sum_{i=1}^{n} U_i \cos Nu_i = 0$; this means that R may remain on some surface $V = C$. In this case the motion will be stable.

We have seen in a preceding section that the stability properties of singular points for systems involving two variables are closely related to the sign of the

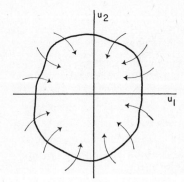

Fig. 10. Two-dimensional representation of trajectories intersecting a Lyapunov function toward the origin.

real part of the two roots of the characteristic equation. In each situation unstable singular points gave rise to a positive real part for one or both of the characteristic exponents, and stable singularities resulted in negative real parts for both μ_1 and μ_2. The situation in which the characteristic exponents are purely imaginary is more complex and cannot be evaluated by investigating the first approximation equations.

The procedures and some of the results that we obtained from the situation involving two variables can be used to investigate the singular points of the general system described by equations 4.9. If we develop the right-hand side of these equations into power series that converge in the vicinity of the origin, equations 4.9 can be written in the form

$$\dot{u}_i = \sum_{i=1}^{n} a_{ij} u_j + U_{2i}(u_1, \ldots, u_n), \quad i = 1, \ldots, n, \tag{4.18}$$

where $a_{ij} = \partial U_i / \partial u_j$ evaluated at $u_j = 0$, and the nonlinear functions U_{2i} do not contain terms of order less than 2. By restricting the analysis to the neighborhood of the origin, we obtain the system of abridged equations:

$$\dot{u}_i = \sum_{j=1}^{n} a_{ij} u_j, \quad i = 1, \ldots, n. \tag{4.19}$$

As in the case involving two variables, a linear one-to-one transformation provides the canonical equations:

$$\dot{z}_i = \mu_i z_i, \quad i = 1, \ldots, n. \tag{4.20}$$

The μ_i in equations 4.20 are the roots of the characteristic equation of the original abridged differential equations (equations 4.19):

$$|a_{ij} - \mu \delta_{ij}| = 0, \tag{4.21}$$

where $\delta_{ii} = 1$, $\delta_{ij} = 0$ for $i \neq j$. Note the similarity between equation 4.21 and the reciprocal of the normal response transform.

Equation 4.21 is an algebraic equation of degree n in the unknown μ; its solutions, $\mu_1, \ldots, \mu_n$, are called the characteristic exponents of the original linear differential equations (equations 4.19).

We now consider the positive-definite function:

$$V = \tfrac{1}{2} \sum_{i=1}^{n} z_i^2. \tag{4.22}$$

The Eulerian derivative,

$$\dot{V} = \sum_{i=1}^{n} \frac{\partial V}{\partial z_i} \dot{z}_i = \sum_{i=1}^{n} \mu_i z_i^2, \tag{4.23}$$

is negative-definite iff* all the characteristic exponents are negative. Consequently, for $\mu_i < 0$, $\dot{V}$ is a negative-definite function of the z_i in the neighborhood of the singular point $z_i = 0$, $i = 1, \ldots, n$. Extending these results to the case in which the characteristic exponents can be conjugate complex, we obtain three additional Lyapunov theorems:

> **Theorem 10** (Lyapunov). *If the real parts of all the characteristic exponents are negative, the undisturbed motion is asymptotically stable in the neighborhood of the origin regardless of the terms in U_{2i}.*

> **Theorem 11** (Lyapunov). *If among the roots of the characteristic equation there is at least one with a positive real part, the undisturbed motion is unstable.*

> **Theorem 12** (Lyapunov). *If one or more of the characteristic exponents is purely imaginary, the terms in U_{2i} can be chosen so that the system will be either stable or unstable.*

It is no coincidence that the characteristic equation is similar to the reciprocal of the normal response transform. If we write equations 4.19 as a single, nth-order linear differential equation, set all the initial conditions equal to zero, and take the Laplace transformation, we obtain the reciprocal of the characteristic equation. To obtain the inverse Laplace transform we would factor the characteristic equation and obtain the roots $\mu_1, \ldots, \mu_n$. This would lead to a sum of exponential terms for the solution of the original linear differential equations. The characteristic exponents $\mu_1, \ldots, \mu_n$ would appear as the coefficients of time in these exponentials. If the real parts of *all* these coefficients are negative, the exponentials will decay (with damped oscillations if the coefficients have imaginary parts) to zero as time approaches infinity; in this situation the motion is toward the singular point at the origin and is therefore stable. However, if one or more of the coefficients has a positive real part, the corresponding exponential term will grow as time grows larger; in this situation the motion will be away from the singularity at the origin and consequently will be unstable.

Since it is not always obvious whether or not all the roots of the characteristic equation possess negative real parts, we mention here the well-known Hurwitz criterion:

> **Theorem 13** (Hurwitz): *A necessary and sufficient condition that all roots of the equation*

$$\mu^n + A_1\mu^{n-1} + \cdots + A_{n-1}\mu + A_n = 0 \quad (A_i \text{ real})$$

---

* The abbreviation "iff" means "if and only if"; this usage was originally introduced by P. R. Halmos (*Measure Theory*, Princeton, N.J., 1964).

have negative real parts is that:

(1) $A_1 > 0$,

(2) $\begin{vmatrix} A_1 & A_3 \\ 1 & A_2 \end{vmatrix} > 0$,

(3) $\begin{vmatrix} A_1 & A_3 & A_5 \\ 1 & A_2 & A_4 \\ 0 & A_1 & A_3 \end{vmatrix} > 0$,

$(n-1)$ $\begin{vmatrix} A_1 & A_3 & A_5 & \cdots & 0 \\ 1 & A_2 & A_4 & \cdots & 0 \\ 0 & A_1 & A_3 & \cdots & 0 \\ 0 & 1 & A_2 & \cdots & 0 \\ \cdots & \cdots & \cdots & \cdots & 0 \\ \cdots & \cdots & \cdots & \cdots & A_n \end{vmatrix} > 0$.

If all $n - 1$ of these conditions are met, then the real parts of the n roots, $\mu_1, \ldots, \mu_n$, of the characteristic equation must be negative.

   The fundamental difficulty in the construction of Lyapunov V functions is that there are no general procedures to serve as guidelines for their selection. The selection of the "best" Lyapunov function will provide more stringent stability conditions than will the selection of an inferior V function. Although equation 4.22 could serve as a positive-definite Lyapunov function for any set of differential equations, it is clear that $\frac{1}{2} \sum z_i^2$ is not the optimal Lyapunov function for most nonlinear problems. Examples of Lyapunov functions appear from time to time in diverse sections of the scientific literature, especially the Russian literature, but the circle of investigators in the United States that has aquired the necessary practice in the construction of these functions remains small. Unfortunately, this situation seems especially prevalent in the biological sciences, where one of the most pressing needs for studies of nonlinear control systems exists. Subsequently in this chapter we will deal with a general method for the construction of Lyapunov functions for a class of differential equations that is of general importance to biochemical control mechanisms. In what follows in this section we will examine practical applications of some of the concepts we have considered thus far. The Lyapunov function described in what follows was kindly suggested to me by Gregory Dunkel.

C.   Stability Properties of "Prey-Predator" Equations

Many years ago Alfred Lotka (29, 30) illustrated that undamped oscillations could arise in the following equations:

$$\dot{x} = (a_1 - b_1 y)x, \qquad \dot{y} = -(a_2 - b_2 x)y, \qquad (4.24)$$

provided the constants, a_1, a_2, b_1, and b_2, are all of the same sign. Subsequently Volterra (31) used these equations in connection with his well-known prey-predator problem. The singularity $x = y = 0$ is easily identified for equations 4.24 as a saddle point; this singular point is therefore unstable. However, the singularity at $x = a_1/b_1$, $y = a_2/b_2$ is a center. Thus sustained oscillations occur around this singular point for all initial values of $x > 0$ and $y > 0$ except $x(0) = a_1/b_1$, $y(0) = a_2/b_2$; for these initial conditions the variables will not change with time.

Five years after Volterra published his theory, Kolmogorov (32) reported the results of his studies of the generalized form of equations 4.24:

$$\dot{x} = x \cdot X_1(x, y), \qquad \dot{y} = y \cdot Y_1(x, y). \qquad (4.25)$$

Kolmogorov showed that for certain X and Y the singularity at the transposed origin can be an unstable focus surrounded by a stable limit cycle. More recently, Higgins (33) has reported limit cycle behavior in analog computer simulations of a certain version of equations 4.25. Kerner (34) has reported studies of the multicomponent version of equations 4.24 in connection with his investigations of the dynamical behavior of ecosystems.

As illustrations of the methods we have discussed in this chapter, we will consider another version of equations 4.25:

$$\dot{x} = (a_1 - b_1 y - c_1 x)x \equiv X(x, y), \qquad \dot{y} = -(a_2 - b_2 x + c_2 y)y \equiv Y(x, y).$$
$$(4.26)$$

Equations 4.26 are identical to equations 4.24 except for the self-interaction terms ($c_1 \neq 0$, $c_2 \neq 0$). As in the case of equations 4.24 there are two singular points to consider:

$$x^{\#} = y^{\#} = 0 \qquad (4.27)$$

and

$$x^* = \frac{a_1 c_2 + a_2 b_1}{c_1 c_2 + b_1 b_2}, \qquad y^* = \frac{a_1 b_2 - a_2 c_1}{c_1 c_2 + b_1 b_2}. \qquad (4.28)$$

We assume in what follows that, if x and y represent chemical concentrations, x^* and y^* are nonnegative real numbers.

The partial derivatives of equations 4.26 are

$$\frac{\partial X}{\partial x} = a_1 - b_1 y - 2c_1 x, \qquad \frac{\partial X}{\partial y} = -b_1 x,$$

$$\frac{\partial Y}{\partial x} = b_2 y, \qquad \frac{\partial Y}{\partial y} = -a_2 + b_2 x - 2c_2 y. \tag{4.29}$$

If we evaluate these partials at the singularity $x = y = 0$, we obtain

$$\frac{\partial X}{\partial x}\bigg|_{0,0} = a_1, \qquad \frac{\partial X}{\partial y}\bigg|_{0,0} = 0,$$

$$\frac{\partial Y}{\partial x}\bigg|_{0,0} = 0 \qquad \frac{\partial Y}{\partial y}\bigg|_{0,0} = -a_2. \tag{4.30}$$

If $a_1 > 0$ and $a_2 > 0$ we are in the situation mentioned in Theorem 5b. Since the necessary conditions for the existence of a limit cycle are not met, sustained oscillations are not possible around the singular point $x = y = 0$.

In the linear region in the neighborhood of the singular point $x = y = 0$, the abridged differential equations are

$$\dot{x} = a_1 x, \qquad \dot{y} = -a_2 y, \tag{4.31}$$

The characteristic equation for equations 4.31 is

$$(a_1 - \mu)(a_2 + \mu) = 0, \tag{4.32}$$

and the characteristic exponents are $\mu_1 = a_1$ and $\mu_2 = -a_2$. If a_1 and a_2 are both real and have the same sign, the characteristic exponents are both real and have opposite signs. This identifies the singularity $x = y = 0$ as a saddle point. This singularity is therefore unstable.

Alternatively, we could have chosen a_1 and a_2 to be real but of opposite sign. Then the characteristic exponents would also be real but have the same sign. This identifies the singularity $x = y = 0$ as a node, stable if $a_1 < 0$ and $a_2 > 0$ but unstable if $a_1 > 0$ and $a_2 < 0$.

In order to investigate the stability properties of the singular point described in equations 4.28, we must transpose the origin of the original differential equations. First we make use of equations 4.28 to eliminate a_1 and a_2 from the differential equations:

$$\dot{x} = -c_1(x - x^*)x - b_1(y - y^*)x, \qquad \dot{y} = -c_2(y - y^*)y + b_2(x - x^*)y. \tag{4.33}$$

We now define the new variables,

$$u = x - x^*, \qquad v = y - y^*, \tag{4.34}$$

such that $u = v = 0$ at the singular point (x^*, y^*). Eliminating the old variables from equations 4.33 yields:

$$\dot{u} = -c_1(u + x^*)u - b_1(u + x^*)v \equiv U(u, v),$$
$$\dot{v} = -c_2(v + y^*)v + b_2(v + y^*)u \equiv V(u, v). \tag{4.35}$$

The partial derivatives of equations 4.35 are

$$\frac{\partial U}{\partial u} = -c_1 x^* - b_1 v - 2c_1 u, \qquad \frac{\partial U}{\partial v} = -b_1(u + x^*),$$

$$\frac{\partial V}{\partial u} = b_2(v + y^*), \qquad \frac{\partial V}{\partial v} = -c_2 y^* + b_2 u - 2c_2 v. \tag{4.36}$$

Evaluated at the singular point $u = v = 0$, these partial derivatives are

$$\left.\frac{\partial U}{\partial u}\right|_{0,0} = -c_1 x^*, \qquad \left.\frac{\partial U}{\partial v}\right|_{0,0} = -b_1 x^*,$$

$$\left.\frac{\partial V}{\partial u}\right|_{0,0} = b_2 y^*, \qquad \left.\frac{\partial V}{\partial v}\right|_{0,0} = -c_2 y^*. \tag{4.37}$$

Since we have assumed that x^* and y^* are positive numbers, we are in the situation mentioned in Theorem 5b whenever c_1 and c_2 are of opposite sign. In this case b_1 and b_2 must be of the same sign and $b_1 b_2 > |c_1 c_2|$ in order for sustained oscillations to be possible. From equations 4.28 we see that, if $c_1 < 0$ and all the other constants are positive, $b_1 b_2 > |c_1 c_2|$ is also the condition required for x^* and y^* to be nonnegative.

If all the constants in equation 4.26 are positive, we must examine the behavior of $\partial U/\partial u + \partial V/\partial v$ in order to decide whether or not we are in the situation mentioned in Theorem 7. Evaluation of this sum at $u = -x^*$, $v = -y^*$ provides

$$\left.\left(\frac{\partial U}{\partial u} + \frac{\partial V}{\partial v}\right)\right|_{-x^*, -y^*} = a_1 - a_2, \tag{4.38}$$

which is positive if $a_1 > a_2$. Evaluation of the same sum at $u = v = 0$ provides

$$\left.\left(\frac{\partial U}{\partial u} + \frac{\partial V}{\partial v}\right)\right|_{0,0} = -(c_1 x^* + c_2 y^*), \tag{4.39}$$

which is always negative. Thus, if $a_1 > a_2$, $\partial U/\partial u + \partial V/\partial v$ changes sign in the phase-space available to the variables. This change of sign occurs when

$$(2c_1 + b_2)u + (2c_2 + b_1)v + c_1 x^* + c_2 y^* = 0. \tag{4.40}$$

Equations 4.38–4.40 illustrate that the phase-space containing $x = y = 0$ and $u = v = 0$ is not a region of definite character. The negative Bendixson criterion (Theorem 6a) is not met throughout the phase-space containing $u = -x^*$, $v = -y^*$, and $u = v = 0$. In fact, the necessary condition mentioned in Theorem 6b is met; consequently the possibility of limit cycle behavior for equations 4.26 cannot be ruled out on this basis. We shall see later, however, that sustained oscillations cannot arise for any positive values of the constants in these equations.

In the linear region in the neighborhood of the singular point $u = v = 0$, the abridged differential equations are

$$\dot{u} = -c_1 x^* u - b_1 x^* v, \qquad \dot{v} = -c_2 y^* v + b_2 y^* u. \qquad (4.41)$$

The characteristic equation for these differential equations is

$$\begin{vmatrix} c_1 x^* + \mu & -b_1 x^* \\ b_2 y^* & c_2 y^* + \mu \end{vmatrix} = 0. \qquad (4.42)$$

The characteristic exponents are

$$\mu_{1,2} = -\frac{(c_1 x^* + c_2 y^*)}{2} \pm \sqrt{[(c_1 x^* + c_2 y^*)^2/4] - (c_1 c_2 + b_1 b_2) x^* y^*}. \qquad (4.43)$$

The conditions for nonzero real roots of the same sign is

$$0 < (c_1 c_2 + b_1 b_2) x^* y^* \leq \frac{(c_1 x^* + c_2 y^*)^2}{4}. \qquad (4.44)$$

The singular point is a node if these conditions are met. The node is stable if

$$-(c_1 x^* + c_2 y^*) < 0, \qquad (4.45)$$

and unstable if

$$-(c_1 x^* + c_2 y^*) > 0. \qquad (4.46)$$

The condition for nonzero real roots of opposite sign is

$$(c_1 c_2 + b_1 b_2) x^* y^* < 0. \qquad (4.47)$$

The singularity would be a saddle point if this condition could be met.

The conditions for conjugate complex roots are

$$0 < \frac{(c_1 x^* + c_2 y^*)^2}{4} < (c_1 c_2 + b_1 b_2) x^* y^*. \qquad (4.48)$$

The singular point is a focus if these conditions are met. The focus is stable if the condition in inequality 4.45 is met but unstable if the condition in inequality 4.46 is met.

Referring now to Theorem 10, we see that the condition for asymptotic stability of the linearized system (equations 4.41) is that the real parts of μ_1 and μ_2 be negative. This condition is always met if $c_1 > 0$ and $c_2 > 0$; under these conditions inequality 4.45 must be satisfied.

We return now to the nonlinear system (equations 4.35) and consider the positive-definite Lyapunov function:

$$V = \left(\frac{x^* e^{x/x^*}}{x}\right)^{\alpha} \left(\frac{y^* e^{y/y^*}}{y}\right)^{\beta} - e^{\alpha+\beta}. \tag{4.49}$$

This Lyapunov function has the desired property:

$$\lim_{x,y \to \infty} V = \infty. \tag{4.50}$$

The Eularian derivative of the Lyapunov function is

$$-\dot{V} = \left(\frac{x^* e^{x/x^*}}{x}\right)^{\alpha} \left(\frac{y^* e^{y/y^*}}{y}\right)^{\beta} \left[\frac{\alpha c_1}{x^*}(x - x^*)^2 + \frac{\beta c_2}{y^*}(y - y^*)^2\right]. \tag{4.51}$$

If we choose $\alpha = x^*/b_1$ and $\beta = y^*/b_2$, $\dot{V}$ is negative-definite provided $c_1 > 0$ and $c_2 > 0$. Referring now to Theorem 9, we see that this proves that the original differential equations (equations 4.26) possess global asymptotic stability for the singular point $x = x^*$, $y = y^*$.

On the basis of this result we can now make the strong statement: If c_1 and c_2 are positive and the product $b_1 b_2$ exceeds $c_1 c_2$, sustained oscillations cannot arise in systems described by equations 4.26.

5. STABILITY PROPERTIES OF CHEMICAL SYSTEMS

A. Thermodynamic Aspects in Closed Systems

The rate of entropy production due to any multistep chemical reaction can be written as the sum of the products of a generalized affinity of each step times the rate of each step. If the temperature, pressure, and volume are constant, the chemical system is closed, and the concentrations of the chemical constituents are dilute, this statement takes the form

$$\frac{dS}{dt} = \sum_{i=1}^{n} \dot{\xi}_i A_i \equiv V, \tag{5.1}$$

where S is the entropy,

$$\dot{\xi}_i = -\frac{\dot{C}_{i\rho}}{\zeta_{i\rho}} = k_i \prod_{\rho} C_{i\rho}^{\Delta_i \zeta_{i\rho}} - k_{-i} \prod_{\rho} C_{i\rho}^{(\Delta_i - 1)\zeta_{i\rho}}, \quad i = 1, \ldots, n, \tag{5.2}$$

and

$$A_i = K \ln \prod_\rho \frac{C_{i\rho}(\infty)^{\zeta_{i\rho}}}{C_{i\rho}^{\zeta_{i\rho}}}, \quad i = 1, \ldots, n. \tag{5.3}$$

Here $\dot{\xi}_i$ is the net rate of the ith step, $C_{i\rho}$ is the instantaneous concentration of the ρth component in the ith step, $C_{i\rho}(\infty)$ is the value of $C_{i\rho}$ at thermodynamic equilibrium, the $\zeta_{i\rho}$ are the stoichiometric numbers of the reactants participating in the forward ($\zeta_{i\rho} > 0$) and the reverse ($\zeta_{i\rho} < 0$) steps of the ith reaction, ξ_i is $[C_{i\rho}(\infty) - C_{i\rho}]/\zeta_{i\rho}$ for any ρ in the ith step, $\Delta_i = 1$ if $\zeta_{i\rho} > 0$ but $= 0$ if $\zeta_{i\rho} < 0$, k_i and k_{-i} are the rate constants for the forward and the reverse steps, respectively, of the ith reaction, A_i is proportional to the chemical affinity of the ith step, and K is a constant equal to the volume times the gas constant, R.

At thermodynamic equilibrium $C_{i\rho} = C_{i\rho}(\infty)$ and $\dot{C}_{i\rho} = 0$ for all i and ρ. From equations 5.2 and 5.3 we obtain

$$\dot{\xi}_i = A_i = 0, \quad i = 1, \ldots, n, \tag{5.4}$$

for every step when the system is at equilibrium. On this basis it is often assumed that, when a system is close to equilibrium, there will be an approximate linear relationship between the rate of each step and the chemical affinities:

$$\dot{\xi}_i = \sum_{j=1}^n L_{ij} A_j, \quad i = 1, \ldots, n. \tag{5.5}$$

Linear laws of this type are usually referred to as *phenomenological relations*, and the L_{ij} are the so-called *phenomenological constants*.

If the rate of the forward step of the ith reaction exceeds the rate of the reverse step, $C_{i\rho}$ is disappearing if $\zeta_{i\rho} > 0$ and appearing if $\zeta_{i\rho} < 0$. In this situation the net rate in equation 5.2 is positive, since the logarithmic term in equation 5.3 is greater than unity, A_i is also positive. In the opposite situation the ith reaction has overshot its equilibrium position, and the net rate is negative. Since in this case the logarithmic term is less than unity, A_i is also negative. Thus $\dot{\xi}_i$ and A_i are always of the same sign in each step. Substitution from equation 5.5 into equation 5.1 provides

$$V = \sum_{i=1}^n \sum_{j=1}^n L_{ij} A_i A_j > 0. \tag{5.6}$$

Equation 5.6 can be regarded as a positive-definite Lyapunov function of the A_i. Since there is a linear relationship between the logarithm of (dilute) concentrations of the reactants in a chemical step and the affinity of the step (equations 5.3), equation 5.6 can also be regarded as a positive-definite Lyapunov function of the $C_{i\rho}$.

The Eulerian derivative of equation 5.6 is

$$\dot{V} = \sum_{i=1}^{n} \sum_{j=1}^{n} (L_{ij} + L_{ji}) A_j \dot{A}_i. \tag{5.7}$$

If the principle of microscopic reversibility (35) is obeyed, equation 5.7 assumes the simple form:

$$\dot{V} = 2 \sum_{i=1}^{n} \dot{\xi}_i \dot{A}_i = 2 \sum_{j=1}^{n} \dot{\xi}_i \sum_{i=1}^{n} \frac{\partial A_j}{\partial \xi_i} \dot{\xi}_i. \tag{5.8}$$

Prigogine (36) has shown that the entropy change due to simultaneous fluctuations of the equilibrium concentrations of the chemical components of a multistep reaction is

$$\Delta S = \tfrac{1}{2} \sum_{j=1}^{n} \xi_j \sum_{i=1}^{n} \frac{\partial A_j}{\partial \xi_i} \xi_i. \tag{5.9}$$

If ΔS is sign-definite, the partial derivatives $\partial A_j / \partial \xi_i$ evaluated at equilibrium are the coefficients of a sign-definite form. Thus, if ΔS in equation 5.9 is sign-definite, $\dot{V}$ in equation 5.8 is also sign-definite and has the same sign. On the basis provided by Theorem 9 we arrive at the following conclusion:

> **Theorem 14.** *If the equilibrium position of a closed chemical system is asymptotically stable, the sign-definite rate of entropy change due to the reaction and the sign-definite entropy change due to fluctuations from the equilibrium position must be of opposite sign.*

In the usual situation entropy is produced during the course of the chemical reactions (equation 5.1), and entropy changes due to fluctuations from the equilibrium position must be negative. If we invert Theorem 14 we obtain

> **Theorem 15.** *The second law of thermodynamics requires (assumes) that the equilibrium position of any closed chemical system is asymptotically stable.*

The results summarized in Theorems 14 and 15 are independent of the order or the molecularity of the chemical reactions.

In a remarkable paper published in 1953 John Hearon (37) pointed out certain restrictions that thermodynamic principles impose on linear chemical systems. If such systems are described by

$$\dot{x}_i = \sum_{j=1}^{n} a_{ij} x_j, \quad i = 1, \ldots, n, \tag{5.10}$$

the a_{ij} are first-order rate constants and the x_i represents the concentrations of the n chemical components. The components in equation 5.10 are interrelated in an arbitrary manner, and the linearity suggested by this equation does not require a physical linearity of the system. The assumption of linearity means only that the reactions are first-order; they need not be unimolecular, and the sequence of reactions may be cyclic or branched.

If the volume of the system is constant, the conservation of mass in a closed chemical system requires that there exist positive numbers, b_i, such that

$$\sum_{i=1}^{n} b_i \dot{x}_i = \frac{d}{dt} \sum_{i=1}^{n} b_i x_i = 0. \tag{5.11}$$

Usually the x_i are expressed as moles per unit volume, and the b_i are the so-called stoichiometric coefficients. From equations 5.10 and 5.11 it follows that

$$\sum_{i=1}^{n} b_i \sum_{j=1}^{n} a_{ij} x_j = \sum_{j=1}^{n} \left(\sum_{i=1}^{n} b_i a_{ij} \right) x_j = 0. \tag{5.12}$$

Since equation 5.12 must be valid for all values of the x_j, the conservation of mass requires that the a_{ij} matrix be singular:

$$|a_{ij}| = 0. \tag{5.13}$$

The linear dependence of the rows of equation 5.13 can be expressed in terms of positive constants:

$$\sum_{i=1}^{n} b_i a_{ij} = 0, \quad j = 1, \ldots, n, \tag{5.14}$$

$$b_i > 0, \quad i = 1, \ldots, n. \tag{5.15}$$

We see from equations 5.14 and inequality 5.15 that all the a_{ij} cannot be of the same sign. The requirement that chemical concentrations be nonnegative demands that the a_{ij} $(i \neq j)$ be nonnegative also:

$$a_{ij} \geq 0, \quad i \neq j, \quad i, j = 1, \ldots, n. \tag{5.16}$$

If any of the off-diagonal a_{ij} $(i \neq j)$ were negative, it would be possible for one or more of the chemical concentrations to be negative.

The requirement resulting from the conservation of mass (equation 5.13) and the requirement implied by the fact that chemical concentrations cannot be negative (inequalities 5.16) lead us to the conclusion that the diagonal elements of $|a_{ij}|$ must be nonpositive:

$$a_{ii} = -\frac{1}{b_i} \sum_{j=1}^{n} b_j a_{ij} \leq 0, \quad i = 1, \ldots, n. \tag{5.17}$$

Equation 5.13 and inequalities 5.16 impose severe restrictions on the possible physical interpretations of equations 5.10. For example, since an off-diagonal element in $|a_{ij}|$ cannot be negative if the concentrations of the chemical components are to remain nonnegative, inhibition of chemical systems cannot be expressed by equations 5.10. Evidently the proper physical interpretation of equations 5.10 is that the rate of change of the ith chemical component is the sum of up to $n - 1$ positive terms and only one negative term. The negative term expresses the fact that the ith component disappears at a rate that is proportional to x_i; the positive term(s), if any, expresses the fact that the ith component is formed from one or more of the other x_j ($j \neq i$). This rate of appearance of x_i is ultimately balanced by its rate of disappearance at thermodynamic equilibrium.

According to the conservation of mass,

$$\sum_{i=1}^{n} b_i x_i(t) = \sum_{i=1}^{n} b_i x_i(0) = M = \text{constant.} \tag{5.18}$$

Since every $b_i > 0$, the maximum value possible for any x_i is less than M/b_i; since chemical concentrations are necessarily nonnegative, the minimum value possible for any x_i is zero. Thus, if the total mass of a system is finite, every x_i must also be physically bounded:

$$0 \leq x_i < \frac{M}{b_i}, \quad i = 1, \ldots, n. \tag{5.19}$$

The conditions in inequalities 5.19 express the fact that closed chemical systems described by equations 5.10 cannot be totally unstable. According to Theorem 11, this means that the real part of the characteristic exponents must be nonpositive. Furthermore, it can be shown (37) that equations 5.10 cannot have purely imaginary characteristic exponents if the equations describe closed chemical reactions.

The arguments outlined above are summarized in the following three theorems:

Theorem 16 (Hearon).   *The conservation of mass in closed, first-order systems requires that*
(1) *the $|a_{ij}|$ matrix is singular;*
(2) *for any given j, the a_{ij} cannot all have the same sign;*
(3) *the real parts of the characteristic exponents are negative.*

Theorem 17 (Hearon).   *The requirement that chemical concentrations be nonnegative ensures that the off-diagonal elements of $|a_{ij}|$ in closed, first-order chemical systems are nonnegative.*

Theorem 18 (Hearon).   *Theorems 16 and 17 require that the diagonal elements of $|a_{ij}|$ be nonpositive.*

Theorem 18 is sufficient to ensure that the characteristic exponents of equations 5.10 will be nonpositive real numbers when $n = 2$ or whenever there are only $2n - 2$ off-diagonal, nonzero a_{ij} $(i \neq j)$. For the general case the principle of microscopic reversibility (35) must be invoked in order to prove the reality of the characteristic exponents of equations 5.10. In particular, the requirement for finite standard free-energy changes implies that the zero elements in $|a_{ij}|$ must occur symmetrically about the diagonal. Thus:

> **Theorem 19** (Hearon). *For closed chemical systems obeying first-order rate laws and the principle of microscopic reversibility, the characteristic exponents of equations 5.10 are real nonpositive numbers; the concentrations of the chemical components in such systems cannot be periodic functions of time.*

Recently, Shear (38) has pointed out that this theorem can be extended to include systems obeying higher-order rate laws.

For open chemical systems equations 5.1 must be modified to include terms describing the exchange process for the components entering and leaving the system. This does not alter the form of equations 5.1 and 5.6, and an argument similar to the one given for closed systems could be presented for open systems. There is, however, the problem that the time-independent state for open chemical systems is usually not the thermodynamic equilibrium state; in fact, the stationary state of an open chemical system need not be in the neighborhood of the equilibrium state. Since the steady state of an open chemical system does not generally fall in the neighborhood for which the linear phenomenological relations are valid, we will not consider open systems here in this context. The reader who is interested in situations in which the phenomenological relations can be applied to open chemical systems should consult the excellent book by Prigogine (39).

B. Kinetic Aspects in Open Systems

In what follows we consider the open system of consecutive chemical reactions depicted in Scheme I:

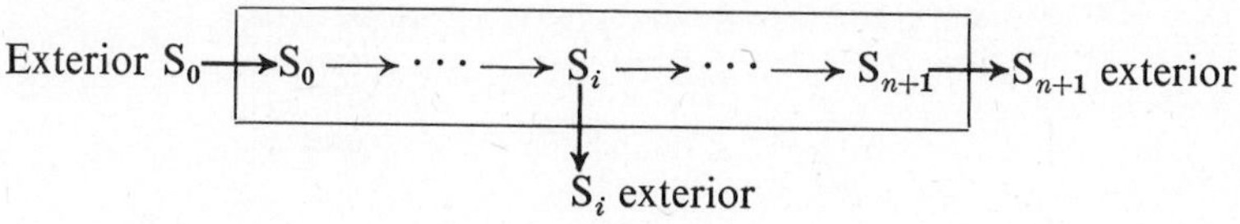

$$\text{Exterior } S_0 \longrightarrow S_0 \longrightarrow \cdots \longrightarrow S_i \longrightarrow \cdots \longrightarrow S_{n+1} \longrightarrow S_{n+1} \text{ exterior}$$

$$S_i \text{ exterior}$$

Scheme I

In these reactions the substance S_0 is maintained at a constant concentration by the exterior reservoir, and the formation of substance S_{i+1} from S_i $(i = 0, 1, \ldots, n)$ is again assumed to be first-order. Under these conditions the differential equations describing Scheme I are as follows:

$$\dot{s}_1 = C_0 - (K_1 + H_1)s_1,$$

$$\vdots$$

$$\dot{s}_i = K_{i-1}s_{i-1} - (K_i + H_i)s_i, \quad i = 2, \ldots, n, \qquad (5.20)$$

$$\vdots$$

$$\dot{s}_{n+1} = K_n s_n - H_{n+1}s_{n+1},$$

where s_i represents the molar concentration of the ith component, K_i is the first-order rate constant for the conversion of S_i to S_{i+1}, H_i is the first-order rate constant for the loss of the ith component to the exterior, and $C_0 = K_0 s_0$ is a constant.

In biological systems each of the biochemical reactions would be catalyzed by an enzyme:

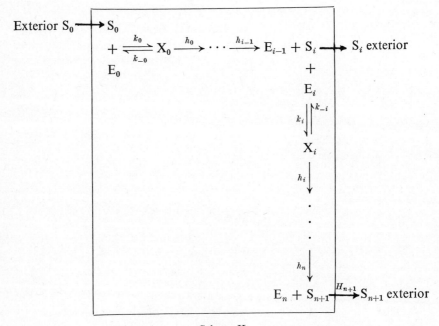

Scheme II

In Scheme II the formation of substance S_{i+1} from S_i is catalyzed by enzyme E_i via the enzyme-substrate compound, X_i. As in Scheme I the S_i may be lost to the exterior, but the E_i and X_i (the enzyme-containing species) are confined to the interior system. If S_0 is maintained at a constant concentration by a large external reservoir, the differential equations describing Scheme II are

$$\dot{x}_0 = k_0 s_0 e_0 - (k_{-0} + h_0)x_0,$$

$$\cdot$$
$$\cdot$$
$$\cdot$$

$$\left. \begin{aligned} \dot{s}_i &= h_{i-1}x_{i-1} - H_i s_i - k_i s_i e_i + k_{-i}x_i \\ \dot{x}_i &= k_i s_i e_i - (k_{-i} + h_i)x_i \end{aligned} \right\}, \quad i = 1, \ldots, n, \qquad (5.21)$$

$$\cdot$$
$$\cdot$$
$$\cdot$$

$$\dot{s}_{n+1} = h_n x_n - H_{n+1}s_{n+1},$$
$$0 = \dot{e}_i + \dot{x}_i, \quad i = 0, \ldots, n.$$

Equations 5.21 can be approximated by equations 5.20 if the following two conditions are met:

1. The $|\dot{x}_i| \ll |\dot{s}_i|$, $i = 1, \ldots, n$, during the period before the stationary state is reached. Under these conditions equations 5.21 become

$$\dot{s}_1 = \frac{{}^0 V_m}{{}^0 K_m + s_0} s_0 - \left(H_1 + \frac{{}^1 V_m}{{}^1 K_m + s_1} \right) s_1,$$

$$\cdot$$
$$\cdot$$
$$\cdot$$

$$\dot{s}_i = \frac{{}^{i-1} V_m}{{}^{i-1} K_m + s_{i-1}} s_{i-1} - \left(H_i + \frac{{}^i V_m}{{}^i K_m + s_i} \right) s_i, \qquad (5.22)$$

$$\cdot$$
$$\cdot$$
$$\cdot$$

$$\dot{s}_{n+1} = \frac{{}^n V_m}{{}^n K_m + s_n} s_n - H_{n+1}s_{n+1},$$

where ${}^i V_m = h_i e_i(0)$, ${}^i K_m = (k_{-i} + h_i)/k_i$, and $e_i(0) = x_i + e_i$ for $i = 0, 1, \ldots, n$.

2. The $s_i \ll {}^i K_m$, $i = 1, \ldots, n$. Under these conditions equations 5.22 reduce to equations 5.20; the physical meanings of C_0 and the K_i are

$$C_0 \equiv \frac{{}^0 V_m s_0}{{}^0 K_m + s_0}, \qquad K_i \equiv \frac{{}^i V_m}{{}^i K_m}, \quad i = 1, \ldots, n. \qquad (5.23)$$

It is clear from the above that Scheme I and equations 5.20 can be used to represent the consecutive series of enzyme-catalyzed, biochemical reactions in Scheme II if two conditions are satisfied. These conditions are not unreasonable for most *in vivo* metabolic sequences; nevertheless they should be viewed as idealized conditions in a first approximation. Whether or not these conditions are met in a particular case will not concern us here; clearly the validity of these approximations must be examined before equations 5.20 are used to represent Scheme II in actual situations. In what follows we shall assume that equations 5.20 are valid approximations for the phase-space available to the biochemical variables.

For the case $n = 1$ equations 5.20 are

$$\dot{s}_1 = C_0 - (K_1 + H_1)s_1, \qquad \dot{s}_2 = K_1 s_1 - H_2 s_2. \tag{5.24}$$

We introduce the change of variables:

$$u_1 = s_1 - s_1^*, \qquad u_2 = s_2 - s_2^*, \tag{5.25}$$

where

$$s_1^* = \frac{C_0}{K_1 + H_1}, \qquad s_2^* = \frac{K_1 C_0}{H_2(K_1 + H_1)}, \tag{5.26}$$

and s_1^* and s_2^* are the stationary-state concentrations of the two chemical components. In order to investigate the behavior in the neighborhood of the singular point $s_1 = s_1^*$, $s_2 = s_2^*$, we obtain the transformed version of equations 5.24:

$$\dot{u}_1 = -(K_1 + H_1)u_1, \qquad \dot{u}_2 = K_1 u_1 - H_2 u_2. \tag{5.27}$$

The characteristic equation for these differential equations is

$$\begin{vmatrix} -(K_1 + H_1) - \mu & 0 \\ K_1 & -H_2 - \mu \end{vmatrix} = 0, \tag{5.28}$$

and the characteristic exponents are

$$\mu_1 = -(K_1 + H_1), \qquad \mu_2 = -H_2. \tag{5.29}$$

Since K_1, H_1, and H_2 are first-order rate constants, they are nonnegative real numbers; thus the characteristic exponents are both nonpositive real numbers. If it is assumed that a chemical reaction occurs ($K_1 > 0$) and that the system is open ($H_2 > 0$), this identifies the singularity $s_1 = s_1^*$, $s_2 = s_2^*$ as a stable node.

For the general situation described by equations 5.24 it will always be possible to introduce the change of variables:

$$u_i = s_i - s_i^*, \qquad i = 1, \ldots, n + 1, \tag{5.30}$$

where

$$s_i^* = \frac{C_0 \prod_{j=1}^{i-1} K_j}{\prod_{j=1}^{i} (K_j + H_j)}, \quad i = 1, \ldots, n,$$

$$s_{n+1}^* = \frac{C_0 \prod_{j=1}^{n} K_j/(K_j + H_j)}{H_{n+1}},$$

(5.31)

and to investigate the behavior in the neighborhood of the singular point $s_i = s_i^*$ $(i = 1, \ldots, n + 1)$. The transformed version of equations 5.20 is

$$\dot{u}_1 = -(K_1 + H_1)u_1,$$

$$\cdot$$

$$\cdot$$

$$\dot{u}_i = K_{i-1}u_{i-1} - (K_i + H_i)u_i,$$

(5.32)

$$\cdot$$

$$\cdot$$

$$\dot{u}_{n+1} = K_n u_n - H_{n+1}u_{n+1}.$$

The characteristic equation for these differential equations is

$$(H_{n+1} + \mu) \prod_{i=1}^{n} (K_i + H_i + \mu) = 0,$$

(5.33)

and the characteristic exponents are

$$\mu_i = -(K_i + H_i), \quad i = 1, \ldots, n; \qquad \mu_{n+1} = -H_{n+1}.$$

(5.34)

Since the K_i and H_i are first-order rate constants, the characteristic exponents will always be negative real numbers. This result is sufficient to guarantee that the singularity $s_i = s_i^*$ $(i = 1, \ldots, n + 1)$ is asymptotically stable.

The preceding discussion indicates that chemical systems of the type depicted in Scheme I possess strong stability properties. However, these systems are clearly open-loop control systems because they are not regulated by any type of feedback. The stability properties ensure that, if the chemical systems are perturbed, they will return to a well-defined stationary or equilibrium state. If, for example, the stationary state is perturbed by the addition of one of the S_i, the system will return to the original stationary state. However, the lack of regulation in Scheme I is manifested by the fact that the return to the stable stationary state is not as efficient as it could be. If a system describable by equations 5.24 is in the stationary state $s_i = s_i^*$ $(i = 1, \ldots, n + 1)$ and one of the components, S_m $(0 < m < n + 1)$, is added, all the S_i $(i \leq m)$ will continue to be formed at the same rate. A return to the stable stationary state will be accomplished solely by the increased rate of disappearance of the S_i $(i \geq m)$. No control is exerted on the other S_i $(i \leq m)$ upstream from S_m to facilitate this return. It seems reasonable to suppose that a higher order of efficiency is an important ingredient of most biochemical control mechanisms.

We can introduce another type of control into Scheme I by including the reverse reaction for each of the chemical steps:

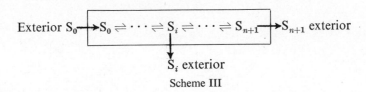

$$\text{Exterior } S_0 \longrightarrow S_0 \rightleftharpoons \cdots \rightleftharpoons S_i \rightleftharpoons \cdots \rightleftharpoons S_{n+1} \longrightarrow S_{n+1} \text{ exterior}$$

$$S_i \text{ exterior}$$

Scheme III

If we assume that the forward and the reverse steps in Scheme III are first-order, the differential equations describing the system are

$$\dot{s}_1 = C_0 - (K_1 + K_{-0} + H_1)s_1 + K_{-1}s_2,$$

$$\dot{s}_i = K_{i-1}s_{i-1} - (K_i + K_{-(i-1)} + H_i)s_i + K_{-i}s_{i+1}, \qquad (5.35)$$

$$\dot{s}_{n+1} = K_n s_n - (K_{-n} + H_{n+1})s_{n+1},$$

where K_{-i} is the first-order rate constant for the reverse of the ith step.

For the case in which $n = 1$ these equations are

$$\dot{s}_1 = C_0 - (K_1 + K_{-0} + H_1)s_1 + K_{-1}s_2,$$
$$\dot{s}_2 = K_1 s_1 - (K_{-1} + H_2)s_2. \tag{5.36}$$

We can again introduce the usual change of variables:

$$u_1 = s_1 - s_1^*, \qquad u_2 = s_2 - s_2^*, \tag{5.37}$$

but here

$$s_1^* = \frac{C_0(H_2 + K_{-1})}{H_2(H_1 + K_1 + K_{-0}) + K_{-1}(H_1 + K_{-0})},$$

$$s_2^* = \frac{C_0 K_1}{H_2(H_1 + K_1 + K_{-0}) + K_{-1}(H_1 + K_{-0})}. \tag{5.38}$$

If we eliminate s_1 and s_2 from equations 5.36, we obtain

$$\dot{u}_1 = -(K_1 + K_{-0} + H_1)u_1 + K_{-1}u_2,$$
$$\dot{u}_2 = K_1 u_1 - (K_{-1} + H_2)u_2. \tag{5.39}$$

The characteristic equation is

$$(K_1 + K_{-0} + H_1 + \mu)(K_{-1} + H_2 + \mu) - K_1 K_{-1} = 0. \tag{5.40}$$

Since all the K's and the H's in equations 5.39 are first-order rate constants, the roots of equation 5.40 are necessarily negative real numbers.* This identifies the singular point $s_1 = s_1^*$, $s_2 = s_2^*$ as a stable node.

For $n = 2$ the transformed linear differential equations are

$$\dot{u}_1 = -(K_1 + K_{-0} + H_1)u_1 + K_{-1}u_2,$$
$$\dot{u}_2 = K_1 u_1 - (K_2 + K_{-1} + H_2)u_2 + K_{-2}u_3, \tag{5.41}$$
$$\dot{u}_3 = K_2 u_2 - (K_{-2} + H_3)u_3.$$

Since S_1 and S_3 are not directly interconverted, the characteristic equation has exactly $2n$ off-diagonal elements:

$$\begin{vmatrix} -\alpha_1 - \mu & K_{-1} & 0 \\ K_1 & -\alpha_2 - \mu & K_{-2} \\ 0 & K_2 & -\alpha_3 - \mu \end{vmatrix} = 0, \tag{5.42}$$

where $\alpha_1 = K_1 + K_{-0} + H_1$, $\alpha_2 = K_2 + K_{-1} + H_2$, and $\alpha_3 = K_{-2} + H_3$.

---

* The fact that these roots are real may not be obvious. Equation 5.40 can be written in the form

$$\mu^2 + (\alpha + \beta)\mu + (\alpha\beta - K_1 K_{-1}) = 0,$$

where α, β, and $K_1 K_{-1}$ are real positive numbers. The roots of this equation are

$$\mu_{1,2} = \frac{-(\alpha + \beta) \pm \sqrt{(\alpha + \beta)^2 - 4\alpha\beta + 4K_1 K_{-1}}}{2}.$$

The roots of this equation are real if $(\alpha + \beta)^2 - 4\alpha\beta + 4K_1 K_{-1} > 0$. The minimum value of $F(\alpha, \beta) = (\alpha + \beta)^2 - 4\alpha\beta$ occurs when $\partial F/\partial\alpha = 0$ and $\partial^2 F/\partial\alpha^2 > 0$:

$$\frac{\partial F}{\partial\alpha} = 2(\alpha + \beta) - 4\beta = 0, \qquad \frac{\partial^2 F}{\partial\alpha^2} = +2 > 0.$$

Solving the first of these equations, we find that the minimum occurs when $\alpha = \beta$. Substitution of this result into $F(\alpha, \beta)$ reveals that the minimum value for F is zero. Therefore the minimum possible value for the quantity under the radical sign is $4K_1 K_{-1}$. Since this is a nonnegative number, μ_1 and μ_2 must be real. Furthermore, since $\alpha\beta > K_1 K_{-1}$, the characteristic exponents must be negative.

Expanding this determinant, we obtain the characteristic equation:

$$\mu^3 + (\alpha_1 + \alpha_2 + \alpha_3)\mu^2 + (\alpha_1\alpha_2 + \alpha_1\alpha_3 + \alpha_2\alpha_3)\mu +$$

$$(\alpha_1\alpha_2\alpha_3 - \alpha_1 K_2 K_{-2} - \alpha_3 K_1 K_{-1}) = 0. \quad (5.43)$$

Obviously, $\beta = \alpha_1\alpha_2\alpha_3 - \alpha_1 K_2 K_{-2} - \alpha_3 K_1 K_{-1}$ in equation 5.43 is less than $\alpha_1\alpha_2\alpha_3$:

$$0 < \beta < \alpha_1\alpha_2\alpha_3. \quad (5.44)$$

From Theorem 13, the necessary and sufficient conditions for the roots of equation 5.43 to have negative real parts are

$$\alpha_1 + \alpha_2 + \alpha_3 > 0,$$
$$(\alpha_1 + \alpha_2 + \alpha_3)(\alpha_1\alpha_2 + \alpha_1\alpha_3 + \alpha_2\alpha_3) > \alpha_1\alpha_2\alpha_3 - \alpha_1 K_2 K_{-2} - \alpha_3 K_1 K_{-1}. \quad (5.45)$$

In view of inequalities 5.44 and the fact that all the constants in equations 5.41 are nonnegative, both of the Hurwitz conditions (inequalities 5.45) are always met. This means that the singular point $s_1 = s_1^*$, $s_2 = s_2^*$, $s_3 = s_3^*$ is asymptotically stable for all chemical systems represented by Scheme III, $n = 2$.

The analysis described above could be carried out for equations 5.35 and any value of n. For reasons similar to those that we used to establish inequalities 5.45, the Hurwitz conditions for negative real parts of the characteristic exponents will be met for any value of n. This means that the stationary state of all chemical systems described by Scheme III and equations 5.35 is asymptotically stable.

The type of control introduced by the reversibility of each chemical step can be termed *mass action product control*. If a chemical reaction described by Scheme III is at the stationary state and the mth component, S_m ($0 < m < n + 1$), is added, S_m will not continue to be formed at the same net rate (as was the case for the irreversible systems described by Scheme I). The effect of the addition of S_m will be to reduce the net rate of the $(m - 1)$st reaction; this effect will be exerted through the reverse chemical reactions on each step upstream from S_m. Consequently, in this case the return to the stable stationary state will be accomplished by a decrease of the net rate of the reactions upstream from S_m, an increase of the rate of the reactions downstream from S_m, and an increased rate of loss of the components to the exterior. This is not the most efficient regulation possible, but there is a type of feedback control.

The prevalence of product inhibition in enzyme-catalyzed reactions suggests that some aspects of mass action product control may be an important component of biological control mechanisms (40). In what follows we will examine the effectiveness of this type of control mechanism by considering the simplified enzyme mechanism described in Scheme IV:

$$E + S \underset{k_{-1}}{\overset{k_1}{\rightleftharpoons}} ES \underset{k_{-2}}{\overset{k_2}{\rightleftharpoons}} EP \underset{k_{-3}}{\overset{k_3}{\rightleftharpoons}} E + P$$

Scheme IV

In Scheme IV E represents the free enzyme, S the substrate, ES the enzyme-substrate compound, EP the enzyme-product compound, and P the product; the k's are rate constants. The steady-state rate equation for such enzyme-catalyzed reactions is

$$\frac{d[P]}{dt} = \frac{V_f[S] - (K_s/K_p)V_r[P]}{K_s + [S] + (K_s/K_p)[P]}, \tag{5.46}$$

where the bracketed symbols indicate concentrations of the species,

$$V_f = \frac{k_2 k_3 e(0)}{k_2 + k_{-2} + k_3}, \qquad V_r = \frac{k_{-1}k_{-2}e(0)}{k_{-1} + k_2 + k_{-2}},$$

$$K_s = \frac{k_{-1}k_3 + k_{-1}k_{-2} + k_2 k_3}{k_1(k_2 + k_{-2} + k_3)}, \qquad K_p = \frac{k_{-1}k_3 + k_{-1}k_{-2} + k_2 k_3}{k_{-3}(k_{-1} + k_2 + k_{-2})},$$

and $e(0)$ is the total concentration of enzyme active centers. It is clear from equation 5.46 that inhibition of the steady-state rate by the product is more effective when

$$\frac{K_s}{K_p} = \frac{k_{-3}(k_{-1} + k_2 + k_{-2})}{k_1(k_2 + k_{-2} + k_3)} \tag{5.47}$$

is large (40).

Equation 5.46 reveals that there are actually two distinct aspects to mass action product inhibition. The negative term in the numerator contributes to the inhibition whenever the system is not far from thermodynamic equilibrium. This aspect of product inhibition can be neglected if

$$[S] \gg \frac{[P]}{K_{eq}}, \tag{5.48}$$

where $K_{eq} = k_1 k_2 k_3 / k_{-1} k_{-2} k_{-3}$ is the equilibrium constant for the overall reaction described in Scheme IV. We will refer to this aspect of product inhibition as the *reverse reaction* component (41). On the other hand, the last term in the denominator of equation 5.46 can contribute to the inhibition even if the system is remote from thermodynamic equilibrium. We will refer to this aspect of product inhibition as the *competitive* component (41).

The control properties of inhibitory products in enzyme-catalyzed reactions are illustrated in Fig. 11. In Fig. 11a, $V_f = V_r = K_s = [S] = 1$; this figure

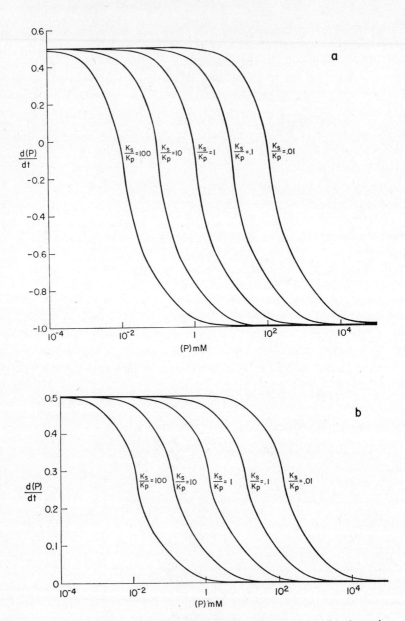

Fig. 11. The control properties of inhibitory products in enzyme-catalyzed reactions. (a) The reverse reaction component, and (b) the competitive aspect.

408

illustrates the reverse reaction component of mass action product inhibition. It can be seen from this illustration that the reverse reaction component is most effective when K_s/K_p (equation 5.47) is high. It is also apparent from Fig. 11a that this aspect of product inhibition is most effective when the reaction is close to thermodynamic equilibrium. In fact, at thermodynamic equilibrium the net rate of an isolated reaction (catalyzed or not) is zero, and the net formation of product is entirely stopped. Several authors have suggested (see, e.g., ref. 41) that the reverse reaction component of mass action product inhibition is not important in metabolic control systems, but Krebs (42) has recently pointed out several controlled metabolic systems wherein the reverse reaction component is important.

If the reactants (S and P in Scheme IV) of an enzyme in a metabolic sequence of catalyzed reactions are at their thermodynamic equilibrium concentrations, then the ratio [P]/[S] is invariant, but there is a net flux of material through the equilibrium step. Many enzymes of intermediary metabolism possess an activity sufficiently high to establish near-equilibrium concentrations of their reactants. However, the *individual* concentrations of the substrates and products of such enzymes can be either very low (because a preceding step is slow or a subsequent step is fast) or very high (because a preceding step is fast or a subsequent step is slow). In either situation the equilibrium determines the metabolic steady-state concentration of the product of the reaction. Since this product is the substrate of the next enzyme in the metabolic sequence, the equilibrium concentration of the product can play a role in the regulation of the overall rate of the metabolic sequence. For example, the equilibrium in the liver alcohol dehydrogenase reaction determines the steady-state concentration of acetaldehyde, which in turn is a factor in the regulation of the rate of the steps wherein acetaldehyde is converted to acetate (42).

The competitive aspect of mass action product inhibition is illustrated in Fig. 11b, in which $V_f = K_s = [S] = 1$ and k_{-2} is small compared to the other unimolecular rate constants. In this situation we can neglect the terms containing V_r, and equation 5.46 becomes

$$\frac{d[P]}{dt} = \frac{V_f[S]}{K_s + [S] + (K_s/K_p)[P]}, \tag{5.49}$$

where

$$V_f = \frac{k_2 k_3 e(0)}{k_2 + k_3}, \qquad K_s = \frac{k_3(k_{-1} + k_2)}{k_1(k_2 + k_3)}, \qquad \text{and} \qquad K_p = \frac{k_3}{k_{-3}}.$$

Figure 11b illustrates that the competitive aspect of product inhibition is also most effective when K_s/K_p is high (equation 5.47). Clearly, this aspect of product inhibition can be effective when the system is remote from thermodynamic equilibrium. The competitive component of mass action product inhibition is also undoubtedly of some biological significance (40–42).

Sel'kov (43) has recently suggested that the combined inhibition of an enzyme by its substrate and its product could constitute an important mechanism for the control of metabolic fluxes. The widespread existence of enzymes subject to both substrate and product inhibition suggests that control mechanisms involving these components may indeed by prevalent in metabolic pathways.

6. BIOLOGICAL AND BIOCHEMICAL FEEDBACK SYSTEMS

A. Introduction

The recent discoveries about the detailed biochemical operation of certain biological control mechanisms provide a sound basis for studying the stability properties of biological systems. Perhaps the most fundamental of these discoveries is that highly specific control mechanisms operate through feedback devices to regulate the concentration or activity of macromolecules. The establishment of specific closed loops in the biochemical organization of cells provides the detail necessary for an understanding of cellular control mechanisms and for studies of their stability properties.

There are at least three levels at which metabolites are known to act as feedback signals capable of altering the biological activities of macromolecules. Metabolites bearing no stereochemical relation to the substrate or the product of an enzyme can alter the enzyme activity. Often it is observed that enzymes inhibited in this manner synthetize substances that are distant precursors of the inhibitory metabolites. This situation, usually referred to as *end-product inhibition*, is illustrated in Scheme V:

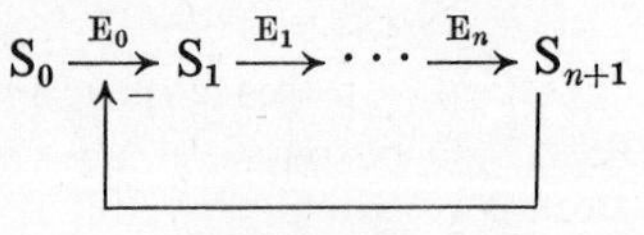

$$S_0 \xrightarrow{E_0} S_1 \xrightarrow{E_1} \cdots \xrightarrow{E_n} S_{n+1}$$

Scheme V

The negative sign beside the feedback arrow is intended to indicate that S_{n+1} *inhibits* the step catalyzed by E_0, and that the feedback signal is therefore negative. In certain situations it is known that end products can stimulate the *in vivo* activity of certain enzymes:

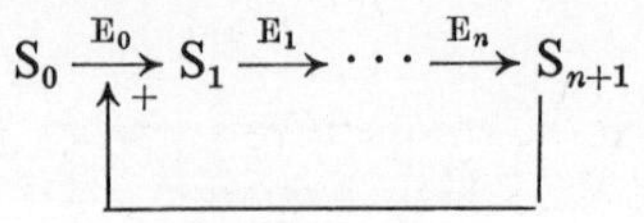

$$S_0 \xrightarrow{E_0} S_1 \xrightarrow{E_1} \cdots \xrightarrow{E_n} S_{n+1}$$

Scheme VI

The stimulation in Scheme VI can be effected by S_{n+1} if this metabolite relieves inhibition of E_0 by another compound, or if S_{n+1} acts directly to increase the catalytic efficiency of E_0. Note that in this diagram the positive sign beside the feedback arrow is used to indicate that positive feedback occurs.

Many examples of end-product inhibition of enzymes have been documented. In certain organisms or tissues each of the amino acids alanine, arginine, cysteine, cystine, glycine, histidine, isoleucine, leucine, lysine, proline, serine, threonine, tryptophan, and valine is known to inhibit at least one of the enzymes involved in its synthesis. Another example of an enzyme subject to end-product inhibition is glutamine synthetase, which is inhibited by at least eight different end products: histidine, tryptophan, alanine or glycine, brain adenylates, cytosine triphosphate, adenosine monophosphate, glucose amine or other amino sugars, and carbamyl phosphate. Each of these products is formed via a discrete metabolic pathway that in some way utilizes glutamine. Twelve moles of each end product react with 1 mole of glutamine synthetase, but any one of the products is capable of inhibiting no more than 50% of the total glutamine synthetase activity. Thus there are apparently up to ninety-six different sites for binding twelve molecules of each end product. The accumulative inhibition of glutamine synthetase by all eight products is complete (44).

Although end-product inhibition offers an efficient means of controlling metabolic fluxes when the concentrations of reactants are higher than normal, this control mechanism does not provide for economy of substances when the reactant concentrations are low. Clearly, positive rate laws provide some economy of metabolites during periods when the reactants are in short supply; it would appear, however, that end-product stimulation of an upstream enzyme (positive feedback) by *low* concentrations of a metabolic product would provide a basis for the stringent type of control necessary under these conditions. In this type of control mechanism the site on the enzyme to which the stimulatory product binds is saturated with stimulator when the metabolites are in their "normal" range. If the flux of material into the metabolic sequence decreases, the levels of the metabolites fall, and the activity of the controlling enzyme declines. The reason for this last event is twofold: the concentrations of the substrates of the controlling enzyme(s) decrease, and the concentration of the stimulatory product is also decreased. Alternatively, if the flux of material into the metabolic sequence increases, the activity of the controlling enzyme(s) again decreases. In this situation the decline in activity is due to the inhibition effected by the increased level of the inhibitory end product. Surprisingly, not many examples of end-product stimulation of enzymes have been reported.

The second level at which metabolites act as feedback signals involves the regulation of the biological activity of the messenger RNA of ribosomes.

This control mechanism is illustrated in Scheme VII:

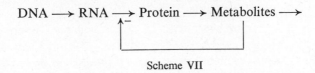

Scheme VII

Here DNA represents a genetic locus that synthesizes messenger RNA. The messenger encounters a ribosome, where its presence results in the synthesis of a particular protein which, in the example above, is an enzyme. The enzyme exerts its influence by catalyzing the conversion of one metabolite into another. Ultimately the metabolite capable of inhibiting the biological activity of the RNA is formed; this inhibitory step closes the loop diagrammed in Scheme VII.

The third level at which metabolites act as feedback signals involves the repression of the genetic activity of a particular locus of DNA. In some cases proteins act as repressors or aporepressors, and metabolites as derepressors or corepressors. The situation wherein metabolites act as repressors is illustrated in Scheme VIII:

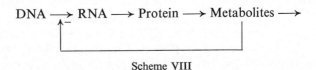

Scheme VIII

Here the protein is again an enzyme that exerts its influence by catalyzing the conversion of one metabolite into another. Ultimately the metabolite capable of repressing the biological activity of the DNA is formed. This feedback step acts as the control signal in Scheme VIII.

In higher-order living systems (e.g., mammals) control circuits also involve hormones that act as positive or negative feedback signals.

Studies concerning the question of how to generate sustained concentration oscillations in biochemical systems have been preoccupied with consideration of equations of the prey-predator type (equations 4.24 and 4.25). The difficulty with trying to arrange chemical mechanisms so that they are described by a prescribed set of differential equations is that the result is not always realistic from the chemical point of view. This has been true of the attempts to arrange biochemical systems to conform to the prey-predator equations; the type of bimolecular coupling between the autocatalytic components required to give the oscillatory behavior is rarely found in metabolic reactions. On the other hand, the existence of the closed-loop configurations in Schemes V–VIII suggests a large class of realistic biochemical mechanisms for which limit cycle behavior may be possible. It has been suggested (1) and verified (45) that sustained oscillations can arise "naturally" in controlled chemical reactions of this type.

B. The Fundamental Feedback Equations

The feedback control systems described in Schemes V–VIII are represented in the context of enzyme-catalyzed reactions in Scheme IX:

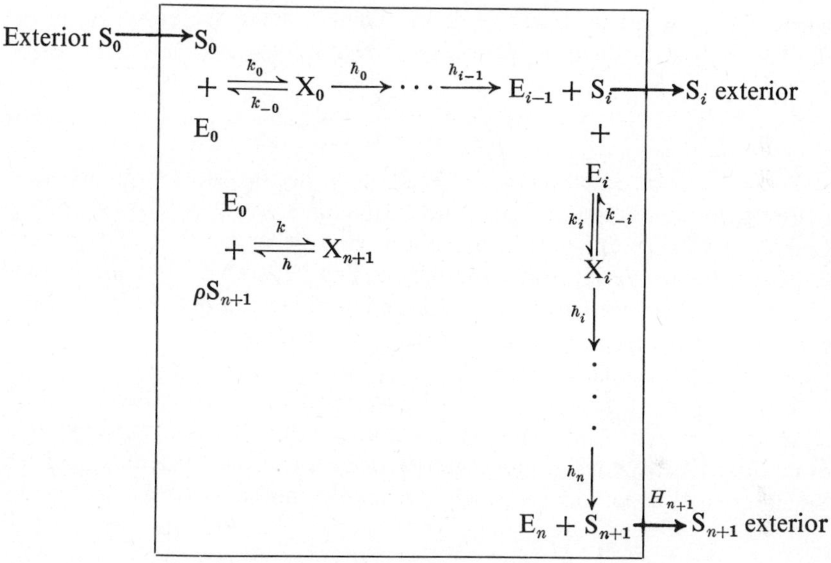

Scheme IX

Scheme IX is the same as Scheme II except for the reversible interaction between ρ molecules of S_{n+1} and the first enzyme in the sequence. This feedback step is inhibitory if X_{n+1} is less active than E_0; this is the situation actually depicted in Scheme IX, since X_{n+1} is totally inactive. In other situations X_{n+1} could be more active than E_0; in that case the rate of the step catalyzed by E_0 would be increased by the feedback signal, S_{n+1}. In what follows we shall refer to end-product inhibition as negative chemical feedback and to end-product stimulation as positive chemical feedback.

We now make the same assumptions that were used to represent Scheme II by equations 5.20, and in addition we assume that E_0, S_{n+1}, and X_{n+1} are in equilibrium with each other. Under these conditions the differential equations describing Scheme IX are

$$\dot{s}_1 = F(s_{n+1}) - (K_1 + H_1)s_1,$$

$$\vdots$$

$$\dot{s}_i = K_{i-1}s_{i-1} - (K_i + H_i)s_i, \tag{6.1}$$

$$\vdots$$

$$\dot{s}_{n+1} = K_n s_n - H_{n+1}s_{n+1},$$

where $F(s_{n+1})$ is the nonlinear feedback function describing the effect of the feedback signal on the rate of the step catalyzed by E_0:

$$F(s_{n+1}) = \frac{C_0}{1 + \alpha(s_{n+1})^\rho}, \tag{6.2}$$

and $\alpha = k/h$. In equation 6.2 $F(s_{n+1})$ is a negative-feedback function; however, in equations 6.1 $F(s_{n+1})$ could describe purely negative or purely positive feedback, or it could describe a mixture of negative and positive feedback. Figure 12 illustrates the relationship between $F(s_{n+1})$ and s_{n+1} for negative feedback systems (Fig. 12a) and positive feedback systems (Fig. 12b). Note that for purely negative feedback functions $\partial F/\partial s_{n+1} < 0$, but for purely positive feedback systems $\partial F/\partial s_{n+1} > 0$.

Without loss of generality (46) we now consider the situation in which $K_i \gg H_i$ $(i = 1, \ldots, n)$. This approximation amounts to supposing that practically all of the material loss to the exterior is in the form of S_{n+1}. Then the differential equations describing the control mechanism are

$$\dot{s}_1 = F(s_{n+1}) - K_1 s_1,$$
$$\dot{s}_i = K_{i-1}s_{i-1} - K_i s_i, \quad i = 2, \ldots, n + 1, \tag{6.3}$$

where $K_{n+1} = H_{n+1}$.

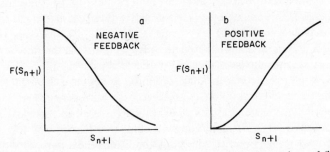

Fig. 12.   Typical relationship between a feedback variable and (a) negative and (b) positive feedback function.

Equations 6.3 are extremely important in the context of biological control mechanisms. These equations describe the idealized behavior of the general type of control system outlined in Schemes V–IX. The ubiquitous occurrence of these control circuits in biological systems suggests that studies of biological control should begin with considerations of equations 6.3. Goodwin (47), Morales and McKay (48), Sel'kov (49), Griffith (50, 51) and Walter (45, 46, 52–54) are among the investigators who have reported some of the properties of these equations.

C. Cooperativity in Enzyme-Catalyzed Reactions

Perhaps the single most popular control mechanism for which $F(s_{n+1})$ can reflect the type of feedback illustrated in Fig. 12 is the model named *allosteric* by Monod et al. (55). In this section we shall examine some of the more subtle aspects of this model.

In classical enzyme kinetics a hyperbolic relationship exists between the initial rate and the initial substrate concentration, regardless of whether or not the rate of turnover of the enzyme-substrate compound is slow compared to the reverse of the enzyme and substrate binding reaction (56, 57). For cases wherein the rate of turnover is slow, one says that *quasi-equilibrium* is maintained; for the simplest enzyme mechanism,

$$E + S \underset{k_{-1}}{\overset{k_1}{\rightleftharpoons}} X \underset{k_{-2}}{\overset{k_2}{\rightleftharpoons}} E + P,$$

Scheme X

the initial quasi-equilibrium rate is described by

$$v_{qe}(0) = \frac{V_m\,s(0)}{K_{qe} + s(0)}, \tag{6.4}$$

where V_m is $k_2 e(0)$ and K_{qe} is k_{-1}/k_1. In this case, since quasi-equilibrium among E, S, and X is maintained, the kinetic parameter, K, is also an equilibrium parameter for the dissociation of the enzyme-substrate compound (to free enzyme and substrate).

For cases wherein the rate of turnover is not slow, quasi-equilibrium is not maintained, and the initial, steady-state rate is described by

$$v_{ss}(0) = \frac{V_m\,s(0)}{K_m + s(0)}, \tag{6.5}$$

where K_m is $(k_{-1} + k_2)/k_1$. In both equations 6.4 and 6.5 the initial rate is described by a rectangular hyperbolic function of the initial substrate concentration, but in equation 6.5 the kinetic parameter, K, is not an equilibrium constant. Unlike K_{qe}, K_m depends on both the equilibrium constant for the dissociation of the enzyme-substrate compound and on the rate of turnover of the enzyme-substrate compound.

A sound theoretical basis for nonhyperbolic relationships between the initial rate and the initial substrate concentration for certain enzymes was developed by Jean Botts and her associates between 1957 and 1959 (58–60). In this formulation a sigmoidal relationship between the initial, steady-state rate and the initial substrate concentration was one of several possible relationships for enzymes involving ternary complexes between the enzyme, a substrate, and another molecule of the substrate or another substance. Botts (58) also pointed out that enzymes involving higher-order compounds could result in nonhyperbolic relationships between the initial steady-state rate and the initial substrate concentration. Subsequently, Monod et al. (55) suggested the now-well-known "allosteric" model as a theoretical basis for the sigmoidal relationship between oxygen binding to hemoglobin and oxygen concentration. In the general model the protein is composed of $n/2 = \rho$ subunits, each of which possesses one binding site for the ligand. Furthermore, each allosteric protein exists in two forms, R and T, such that the equilibrium constant for the reaction

$$T \rightleftharpoons R$$

is L. The equilibrium constant for the binding of the first molecule of ligand to the R form is ρK_r; the corresponding constant for the binding to the T form is ρK_t. We now define $\alpha = K_r s$ and $\beta = K_t/K_r$, where s is the concentration of the ligand. At equilibrium the fraction of the binding sites on the protein occupied by a molecule of ligand is

$$F = \frac{\alpha\beta(1 + \alpha\beta)^{\rho-1} + L\alpha(1 + \alpha)^{\rho-1}}{L(1 + \alpha)^{\rho} + (1 + \alpha\beta)^{\rho}}. \tag{6.6}$$

For hemoglobin the allosteric model is as shown in Scheme XI:

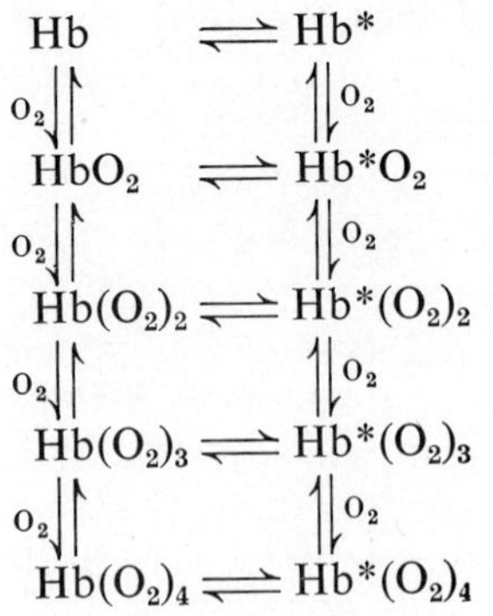

Scheme XI

and at equilibrium the fraction of binding sites occupied by oxygen is

$$F = \frac{\alpha\beta(1 + \alpha\beta)^3 + L\alpha(1 + \alpha)^3}{L(1 + \alpha)^4 + (1 + \alpha\beta)^4}.$$ (6.6a)

The model postulated for hemoglobin is a nonkinetic model wherein all the chemical components are assumed to be in thermodynamic equilibrium with one another. Specifically, hemoglobin is not an enzyme, and the hemoglobin-oxygen compounds (unlike enzyme-substrate compounds) do not "turn over" to form a product from the oxygen. There is only binding, no turnover. Therefore extrapolation of the results obtained from the theoretical hemoglobin model to enzyme systems should be limited to systems wherein quasi-equilibrium is strictly maintained. For the general enzyme system, we represent all forms of the enzyme by X's with odd subscripts to denote the T forms, and even subscripts to denote the R forms:

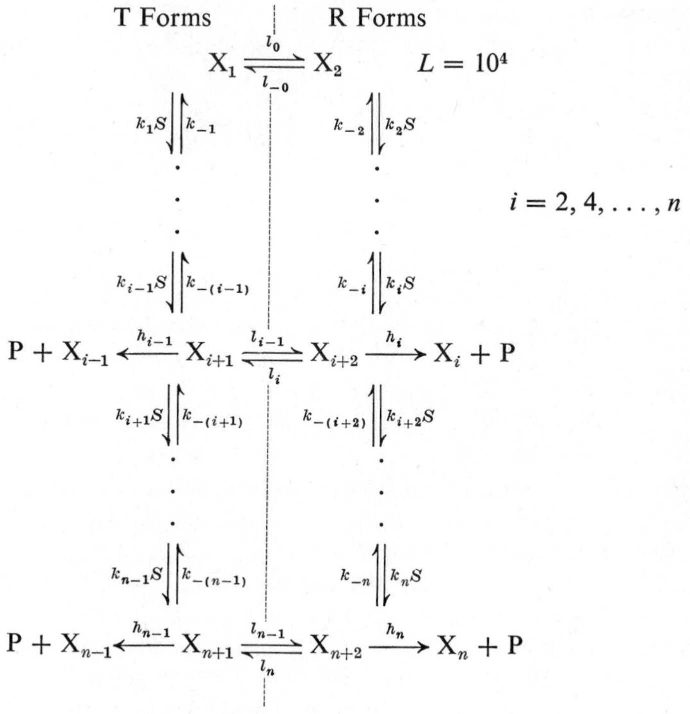

$$
\begin{array}{ccc}
\text{T Forms} & & \text{R Forms} \\[4pt]
X_1 \underset{l_{-0}}{\overset{l_0}{\rightleftharpoons}} X_2 & & L = 10^4 \\[6pt]
k_1 S \updownarrow k_{-1} & k_{-2} \updownarrow k_2 S & \\[4pt]
\vdots & \vdots & i = 2, 4, \ldots, n \\[4pt]
k_{i-1} S \updownarrow k_{-(i-1)} & k_{-i} \updownarrow k_i S & \\[4pt]
P + X_{i-1} \xleftarrow{h_{i-1}} X_{i+1} \underset{l_i}{\overset{l_{i-1}}{\rightleftharpoons}} X_{i+2} \xrightarrow{h_i} X_i + P & & \\[4pt]
k_{i+1} S \updownarrow k_{-(i+1)} & k_{-(i+2)} \updownarrow k_{i+2} S & \\[4pt]
\vdots & \vdots & \\[4pt]
k_{n-1} S \updownarrow k_{-(n-1)} & k_{-n} \updownarrow k_n S & \\[4pt]
P + X_{n-1} \xleftarrow{h_{n-1}} X_{n+1} \underset{l_n}{\overset{l_{n-1}}{\rightleftharpoons}} X_{n+2} \xrightarrow{h_n} X_n + P & &
\end{array}
$$

Scheme XII

In Scheme XII strict quasi-equilibrium means that

$$h_i \ll k_{-i}, l_i, \quad i = 1, \ldots, n.$$ (6.7)

Application of the allosteric model to enzyme-catalyzed reactions is implicitly limited to the conditions specified in inequalities 6.7.

A necessary condition for a sigmoidal relationship between the initial rate of an enzyme-catalyzed reaction and the initial substrate concentration is that there be an enzyme compound containing at least two molecules of the substrate, or at least one molecule of the substrate and one molecule of a modifier or cosubstrate. Weber (61) discusses an enzyme mechanism involving only binary compounds and suggests that cooperative effects can result in the absence of ternary (or higher-order) enzyme compounds, provided the binding by the protein forms is in "slow equilibrium." This conclusion is based on an argument that does not take into account the principle of microscopic reversibility: the cooperative effects disappear if detailed balance is maintained.

It is clear that more than one enzyme mechanism can result in cooperative effects. It is also well known that rectangular hyperbolic relationships between the initial rate and the initial substrate concentration result from certain enzyme mechanisms only if specific conditions are met; under other conditions sigmoidal or other nonhyperbolic relationships can result. For example, Botts and her associates (58, 60) have shown that random-sequence, multisubstrate enzyme mechanisms may result in a rectangular hyperbolic relationship between the initial rate and the initial concentration of one of the substrates if quasi-equilibrium is maintained, but the same mechanism may result in a nonhyperbolic relationship if quasi-equilibrium is not maintained. Similarly, it has been illustrated (60, 62) that enzyme mechanisms involving inhibitors that are not strictly competitive may result in a rectangular hyperbolic relationship between the initial rate and the initial substrate concentration if quasi-equilibrium is maintained, but the same mechanisms may result in a nonhyperbolic relationship if quasi-equilibrium is not maintained. In this situation it is possible for the nonhyperbolic relationship to be sigmoidal (58, 60).

It is also clear that the classical allosteric mechanism described in Scheme XII can result in a sigmoidal relationship between the initial rate and the initial substrate concentration if the values of the binding constants, $K_i = k_i/k_{-i}$ ($i = 1, \ldots, n$), and the transition constant, $L_0 = l_0/l_{-0}$, are in the correct range (55), and if the rate constants for the turnover of the enzyme-substrate compounds (the h_i, $i = 1, \ldots, n$) are all equal to one another and sufficiently small that quasi-equilibrium is maintained (inequalities 6.7 are maintained). This point is illustrated in the lowest curve in Fig. 13. In this figure $L = 10^4$, $\beta = 10^2$, $\rho = 4$, and the values of the h_i ($i = 1, \ldots, n$) are sufficiently small to guarantee that quasi-equilibrium is maintained. In curve 1 the h_i (i odd) equal the h_j (j even); in curve 2 the h_i (i odd) are twice the h_j (j even); in curve 5 the h_i (i odd) are five times the h_j (j even); in curve

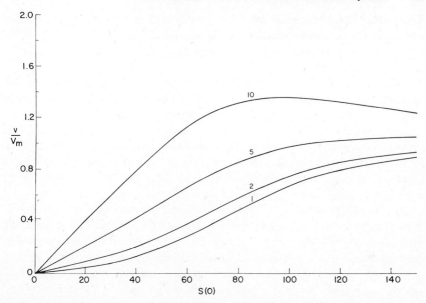

Fig. 13. Relationship between the initial quasi-equilibrium rate of an enzyme-catalyzed reaction and the initial substrate concentration. The numbers above each curve are the ratios of h_i (i odd) to h_j (j even) in Scheme XII.

10 the h_i (i odd) are ten times the h_j (j even); in each curve the h_j with the even subscripts are equal; and in all the curves the h_i with the odd subscripts are all equal. The plotted quantities are the relative initial rates* versus the initial substrate concentrations. It is clear from Fig. 13 that the sigmoidal relationship expected from the allosteric model when the rate constants for the turnover of the enzyme-substrate compound are equal or approximately equal (curves 1 and 2) can be lost when the turnover constants for the R forms of the enzyme are significantly lower than the constants for the T forms, $L = 10^4$ (curves 5 and 10), even though quasi-equilibrium is maintained for all the examples discussed.

It would be expected that the examples simulated in Fig. 13 should experience substrate inhibition when the h_i with odd subscripts are larger than the corresponding h_i with even subscripts. It is clear from Fig. 13 that an optimum in the plot of relative rate versus initial substrate concentration does accompany the loss of sigmoidal shape. However, the subsequent substrate inhibition is slight compared to the ratio of h_i/h_j (i odd, j even) and might easily be

---

* The relative initial, steady-state rate is the initial, steady-state rate at the indicated substrate concentration divided by the rate expected at infinite substrate concentration.

overlooked in experimental cases. For example, in curve 5, $h_i/h_j = 5$ (i odd, j even) but the ultimate substrate inhibition is only about 2% of the optimal rate. We can conclude from Fig. 13 that, even if quasi-equilibrium is maintained, an allosteric enzyme displaying a sigmoidal relationship between the initial rate and the initial substrate concentration in the absence of a modifier could be made to display a hyperbolic relationship between these same variables when a modifier that inhibits the h_j (j even) but not the h_i (i odd) is added to the system. Similarly, it is possible that an enzyme displaying a "normal" hyperbolic relationship between the initial rate and the initial substrate concentration in the absence of a modifier could be made to display a sigmoidal relationship when a modifier that stimulates the h_j (j even) but not the h_i (i odd) is added to the system.

In what follows we shall investigate what happens to the relationship between the initial rate and the initial substrate concentration for allosteric enzyme mechanisms described by Scheme XII when quasi-equilibrium is not maintained. The differential equations describing Scheme XII are:

$$\dot{x}_1 = l_{-0}x_2 + (k_{-1} + h_1)x_3 - (l_0 + k_1 s)x_1,$$

$$\dot{x}_2 = l_0 x_1 + (k_{-2} + h_2)x_4 - (l_{-0} + k_2 s)x_2,$$

$$\dot{x}_i = k_{i-2}s x_{i-2} + l_{i-j}x_{i+k} + (k_{-i} + h_i)x_{i+2}$$
$$- (k_{-(i-2)} + l_{i-2} + h_{i-2} + k_i s)x_i, \tag{6.8}$$

$$\dot{x}_{n+1} = k_{n-1}s x_{n-1} + l_n x_{n+2} - (l_{n-1} + k_{-(n+1)} + h_{n-1})x_{n+1},$$

$$\dot{x}_{n+2} = k_n s x_n + l_{n-1}x_{n+1} - (l_n + k_{-n} + h_n)x_{n+2},$$

$$\dot{p} = \sum_{i=1}^{n} h_i x_{i+2},$$

where $k = j = 1$ (i odd),

$\quad\quad j = 3, k = -1$ (i even),

$\quad\quad i = 3, 4, \ldots, n.$

The principle of detailed balance requires that

$$\left.\begin{array}{c} \dfrac{l_0}{l_{-0}} \cdot \dfrac{l_2}{l_1} \cdot \dfrac{k_2}{k_{-2}} \cdot \dfrac{k_{-1}}{k_1} \equiv 1 \\[2ex] \dfrac{l_{i-3}}{l_{i-2}} \cdot \dfrac{l_i}{l_{i-1}} \cdot \dfrac{k_i}{k_{-i}} \cdot \dfrac{k_{-(i-1)}}{k_{i-1}} \equiv 1 \end{array}\right\}, \quad i = 4, 6, \ldots, n. \tag{6.9}$$

From the law of conservation of mass it is necessary that

$$e(0) = \sum_{i=1}^{n} x_i,$$

$$s(0) = s + p + \sum_{i=3}^{n} x_i. \tag{6.10}$$

If, as usual, we assume (a) that the rates of change of the concentrations of the enzyme-containing species are small compared to the rate of disappearance of substrate or of appearance of product; (b) that $e(0) \ll s(0)$; and (c) that the initial rates are obtained during an early portion of the overall reaction so that $s(0) \approx s$, we can combine equations 6.8 and 6.10 to obtain the matrix:

$$\sum_{i=1}^{n+2} A_{i+j(n+3)} x_i = A_{(j+1)(n+3)}, \quad j = 0, 1, \ldots, n+2, \tag{6.11}$$

where

$$A_{(i)} = 1.0, \quad i = 1, 2, \ldots, n+2,$$

$$A_{(n+3)} = e(0),$$

$$A_{(n+4)} = l_0,$$

$$A_{(n+5)} = -[l_{-0} + k_2 s(0)],$$

$$A_{[2+(n+4)(1+2j)]} = 0, \quad i = 0, 1, \ldots, n+2,$$

$$A_{[3+i(n+4)]} = k_{-(i+1)} + h_{i+1}, \quad i = 1, 2, \ldots, n-1,$$

$$A_{[i(n+4)+j]} = 0, \quad i = 1, 2, \ldots, n; \quad j = 4, \ldots, n+2,$$

$$A_{[i(n+4)+n+3]} = k_i s(0), \quad i = 1, 2, \ldots, n,$$

$$A_{[2(n+4)(i+1)]} = 0, \quad i = 1, 2, \ldots, n/2,$$

$$A_{[1+i(n+4)]} = -(l_{i-1} + k_{-(i-1)} + h_{i-1} + k_{i+1} s(0),$$
$$i = 2, 3, \ldots, n-1,$$

$$A_{[2(1+(n+4)i)]} = l_{2i}, \quad i = 1, 2, \ldots, n/2,$$

$$A_{[(n+2)(n+2)-1]} = 0,$$

$$A_{[(n+2)(n+2)-3]} = -[k_{-(n-1)} + l_{n-1} + h_{n-1}],$$

$$A_{[(n+3)(n+2)-1]} = -(k_{-n} + l_n + h_n).$$

Equations 6.11 can be solved directly for the x_i ($i = 1, 2, \ldots, n$), and the results substituted into the equation for the initial, steady-state rate:

$$v_{ss}(0) = \sum_{i=1}^{n} h_i x_{i+2} \tag{6.12}$$

The relationship between the initial, steady-state rate for enzyme mechanisms described by Scheme XII and the initial substrate concentration can be calculated from equations 6.11 and 6.12. In order to obtain an idea of the

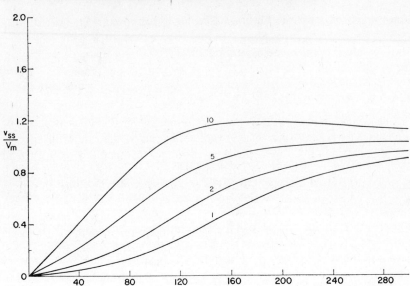

Fig. 14. Relationship between the initial steady-state rate of an enzyme-catalyzed reaction and the initial substrate concentration. The numbers above each curve are the ratios of h_i (i odd) to h_j (j even) in Scheme XII.

comparative values of these steady-state rates and the corresponding rates expected at quasi-equilibrium, we examine equations 6.4–6.6, 6.11, and 6.12. Since $K_m \geq K_{qe}$, it follows that the initial steady-state rate obtained at a particular initial substrate concentration has as a maximum value the corresponding initial, quasi-equilibrium rate obtained at the same initial substrate concentration.

It might appear from the above that, since quasi-equilibrium is a limiting case of the steady state, all sigmoidal relationships between quasi-equilibrium rates and initial substrate concentrations are themselves limiting cases of the corresponding sigmoidal relationships between initial steady-state rates and initial substrate concentrations. Indeed, this is clearly the situation for all cases wherein the microscopic rate constants except the $h_i = h_j$ ($i = 1, 2, \ldots,$ $n; j \neq i$) in Scheme XII are unchanged. For example, in Fig. 14 appears the relationship between the relative steady-state rate and the initial substrate concentration for the same values of the rate constants as were used to construct Fig. 13 except that in Fig. 14 the h_i are not sufficiently small to ensure that inequalities 6.7 will be maintained, and quasi-equilibrium is not maintained. Comparison of Figs. 13 and 14 illustrates that the quasi-equilibrium curves in Fig. 13 are limiting situations for each steady-state curve in Fig. 14. However, it is also clear from Fig. 14 that sigmoidal steady-state curves (curves 1 and 2 in Fig. 14), like the corresponding sigmoidal quasi-equilibrium curves (curves 1 and 2 in Fig. 13), can lose their sigmoidal nature if h_i/h_j (i odd, j even) is sufficiently different from unity (curve 10 in Fig. 14). For the case in curve 10, Fig. 14, $L = 10^4$ and h_i/h_j (i odd, j even) is 10. Note that the anticipated substrate inhibition at high substrate concentrations is even less apparent in the steady-state cases (Fig. 14) than in the quasi-equilibrium cases (Fig. 13). Taken together, Figs. 13 and 14 illustrate that, whether or not quasi-equilibrium is maintained, an allosteric enzyme (Scheme XII) displaying a sigmoidal relationship between the initial rate and the initial substrate concentration in the absence of a modifier could be made to display a hyperbolic relationship between the same variables when a modifier that inhibits the h_j (j even) but not the h_i (i odd) is added to the system. Similarly, it is possible that an enzyme displaying a rectangular hyperbolic relationship in the absence of a modifier could be made to display a sigmoidal relationship when a modifier that stimulates the h_j (j even) but not the h_i (i odd) is added. Particularly in situations in which quasi-equilibrium is not maintained, it will be difficult (perhaps impossible) to observe the concomitant inhibition expected at high substrate concentrations.

Figure 15 illustrates the effect of increasing remoteness from quasi-equilibrium on the sigmoidal relationship between initial rates and initial

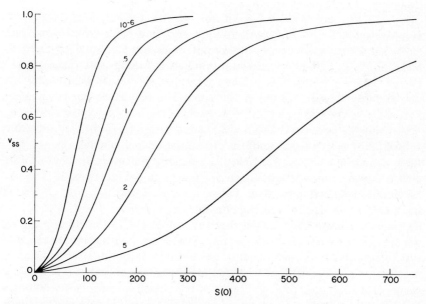

Fig. 15. Relationship between the initial rate of an enzyme-catalyzed reaction and the initial substrate concentration. The numbers above each curve are h_i/k_{-i} ($i = 1, 2, \ldots, n$) in scheme XII.

substrate concentrations. In the five curves all the rate constants except the h_i $(i = 1, 2, \ldots, n)$ remain unchanged. In the curve labeled 10^{-6} $h_i/k_{-i} = 10^{-6}$ $(i = 1, 2, \ldots, n)$, inequalities 6.7 are satisfied, and quasi-equilibrium is strictly maintained. In the remaining curves h_i/k_{-i} $(i = 1, 2, \ldots, n)$ is the value indicated for each curve (0.5, 1.0, 2.0, or 5.0), inequalities 6.7 are not satisfied, and quasi-equilibrium is not maintained. Thus the top curve in Fig. 15 represents the limiting case for the family of steady-state curves appearing under it. Note that in this situation the effect of becoming more remote from quasi-equilibrium is to increase the sigmoidal nature of the relationship between initial rates and initial substrate concentrations, while at the same time decreasing the binding or saturation of the substrate to the enzyme.

Figure 16 illustrates the effect of the ratios of h_i/h_j $(i$ odd, j even) on the relationship between the initial rate and the initial substrate concentration when quasi-equilibrium is not maintained, and it is the h_j with the even subscripts $(L = 10^4)$ that are not changed. In Fig. 16a the h_j $(j$ even) are twice the h_j $(j$ even) in Fig. 16b. In the curves labeled 1, the h_i $(i$ odd) equal the h_j $(j$ even); in the curves labeled 2, the h_i $(i$ odd) are twice the h_j $(j$ even); in curve 5 the h_i $(i$ odd) are five times the h_j $(j$ even); in curve 10 the h_i $(i$ odd) are ten times the h_j $(j$ even); in each curve the h_i with odd subscripts are equal; in all the curves in Fig. 16a the h_j with even subscripts are all equal to each other and are twice the values of the h_j with even subscripts in Fig. 16b. The curves in Fig. 16 illustrate that an allosteric enzyme (Scheme XII) displaying a sigmoidal relationship between the initial steady-state rate and the initial substrate concentration in the absence of a modifier could be made to display a hyperbolic relationship between these same variables and with no change of the asymptote when a modifier that stimulates the h_i $(i$ odd) but not the h_j $(j$ even) is added to the system. Similarly, an enzyme displaying a "normal" hyperbolic relationship could be made to exhibit a sigmoidal relationship between the initial steady-state rate and the initial substrate concentration with no change of the asymptote when a modifier that inhibits the h_i $(i$ odd) but not the h_j $(j$ even) is added to the enzyme system. These results, together with those illustrated in Fig. 14, are summarized in Table 1.

Figure 17 illustrates the effect of altering the individual k_j and k_{-j} $(j$ even) in such a manner that the binding constant, $K_j = k_j/k_{-j}$, is itself unchanged. Thus all the constants in the three curves in Fig. 17 are the same except k_j and k_{-j} $(j$ even), but their ratio, K_j, is also the same in each curve. In curve 1, h_j/k_{-j} $(j$ even) is 1; in curve 10, h_j/k_{-j} $(j$ even) is 10, but h_j is the same as in curve 1 and k_j is one tenth of the value used in curve 1; in curve 100 h_j/k_{-j} $(j$ even) is 100, and k_j is one hundredth of the value used in curve 1. Thus the effect illustrated in Fig. 17 is strictly on binding and therefore is due to the fact that quasi-equilibrium is not maintained. This effect of the steady state

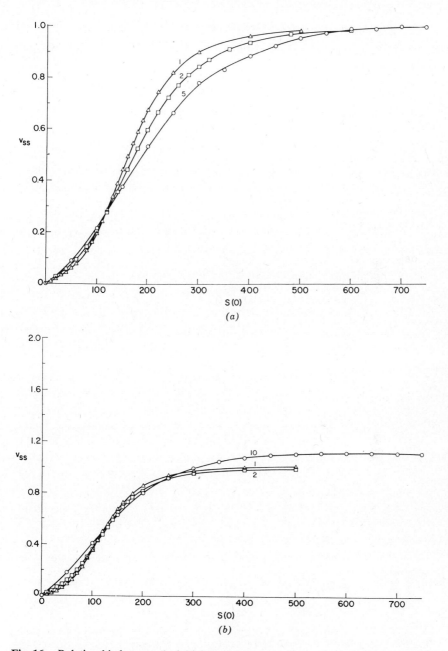

Fig. 16. Relationship between the initial steady-state rate of an enzyme-catalyzed reaction and the initial substrate concentration. The numbers above each curve are the ratios of h_i (i odd) to h_j (j even) in Scheme XII.

Tables 1.   Summary of Results Illustrated in Figs. 14 and 16

Modifier[a] Effect on Rate Constant for Turnover of Enzyme		Possible Effect on Initial Rate Versus Initial Substrate Concentration Curve	
h_i (i odd)	h_j (j even)		
None	Inhibit	Sigmoid	→ hyperbolic
Stimulate	None	Sigmoid	→ hyperbolic
None	Stimulate	Hyperbolic	→ sigmoid
Inhibit	None	Hyperbolic	→ sigmoid

[a] As Professor Manuel Morales pointed out at the 1970 Biophysical Society meetings, the effect on the rate constant need not be due to a chemical component. It is well known, for example, that a change in temperature can result in a decrease or an increase in the value of a rate constant.

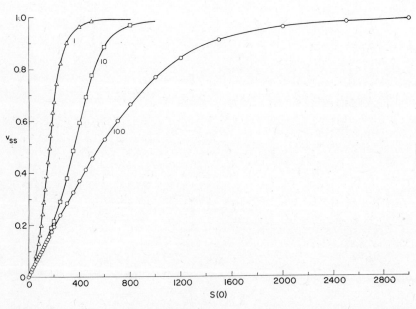

Fig. 17.   The effect of the individual k_j and k_{-j} (j even) on the sigmoidal or hyperbolic relationship between the initial steady-state rate of a reaction catalyzed by an allosteric enzyme and its initial substrate concentration. The numbers above each curve are the ratios of h_j to k_{-j} (j even) in Scheme XII, but in all the curves $K_j = k_j/k_{-j}$ is constant.

is to shift sigmoidal curve 1 (which is 50% saturated when the initial substrate concentration is about 160) to the rectangular hyperbola, curve 100 (which is 50% saturated when the initial substrate concentration is about 560). The effect in Fig. 17 is particularly remarkable in view of the fact that the mechanism, all the binding constants, and all the rate constants for the turnover of enzyme-substrate compounds are identical in the three curves. This result illustrates that a substance that effects neither the binding constants for an allosteric enzyme nor the rate constants for the turnover of the enzyme-substrate compounds can nevertheless cause a shift from a sigmoidal relationship between initial rates and initial substrate concentrations to a rectangular hyperbolic relationship or vice versa.

D. Systems with End-Product Inhibition and Substrate Inhibition

An important biological control mechanism for which $F(s_{n+1})$ can reflect negative and positive feedback has been described by Sel'kov (49). In this mechanism the enzyme, E_0, is subject to inhibition by its substrate, S_0, and to end-product inhibition by S_{n+1}:

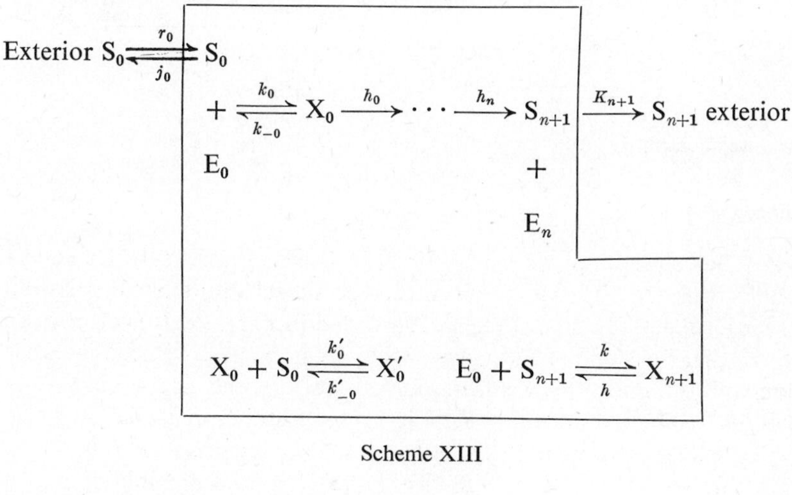

$$\text{Exterior } S_0 \underset{j_0}{\overset{r_0}{\rightleftharpoons}} S_0$$

$$+ \underset{k_{-0}}{\overset{k_0}{\rightleftharpoons}} X_0 \overset{h_0}{\longrightarrow} \cdots \overset{h_n}{\longrightarrow} S_{n+1} \overset{K_{n+1}}{\longrightarrow} S_{n+1} \text{ exterior}$$

$$E_0 \qquad\qquad + $$

$$E_n$$

$$X_0 + S_0 \underset{k'_{-0}}{\overset{k'_0}{\rightleftharpoons}} X'_0 \qquad E_0 + S_{n+1} \underset{h}{\overset{k}{\rightleftharpoons}} X_{n+1}$$

Scheme XIII

Sel'kov (49) describes the source of S_0 by

$$r = r_0 - j_0 s_0, \tag{6.13}$$

where r_0 is the maximum rate of introduction of S_0, and j_0 is a first-order decay constant. If we assume that the inhibitory steps in Scheme XIII are at equilibrium, we obtain the equation

$$\frac{r_0}{^0V_m} - \frac{j_0 s_0}{^0V_m} - \frac{s_0/^0K_m}{[1 + (s_0/^0K_m) + (K_0'/^0K_m)(s_0)^2](1 + \alpha s_{n+1})} = 0, \quad (6.14)$$

where $K_0' = k_0'/k_{-0}'$. Under these same conditions the nonlinear feedback function is

$$F(s_0, s_{n+1}) = \frac{^0V_m s_0}{[^0K_m + s_0 + K_0'(s_0)^2](1 + \alpha s_{n+1})}. \quad (6.15)$$

The function, $F(s_0, 0)$, is illustrated in Fig. 18a. Note that $\partial F/\partial s_0$ changes sign at $s_0 = s_0'$.

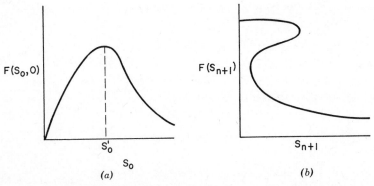

Fig. 18. Relationship between the feedback function in Scheme XIII and (a) s_0 and (b) s_{n+1}.

Sel'kov has solved equation 6.14 for s_0 at fixed s_{n+1} and substituted the result into equation 6.15. The result of this second substitution is the desired nonlinear feedback function, $F(s_{n+1})$, illustrated in Fig. 18b. It is clear from this figure that the slope of $F(s_{n+1})$ versus s_{n+1} can be positive even in situations when S_{n+1} inhibits E_0. Figure 19a illustrates that this effect occurs only when S_0 is inhibitory and $s_0 > s_0'$; Fig. 19b illustrates that the slope of $F(s_{n+1})$ versus s_{n+1} can be negative even when S_{n+1} increases the rate of the reaction catalyzed by E_0 iff $s_0 > s_0'$. In curves 1 S_0 does not inhibit E_0; in curves 2 S_0 inhibits E_0, but $s_0 < s_0'$; in curves 3 S_0 inhibits E_0 and $s_0 > s_0'$. It is clear from Fig. 19 that the slope of these plots depends on whether S_{n+1} is an inhibitor or an activator of E_0, on the concentration of S_0, and when $s_0 > s_0'$ on the concentration of S_{n+1} itself. If $s_0 > s_0'$, $\partial F/\partial s_{n+1}$ can be positive even if S_{n+1} inhibits E_0; correspondingly, this partial derivative can be negative even if S_{n+1} stimulates E_0.

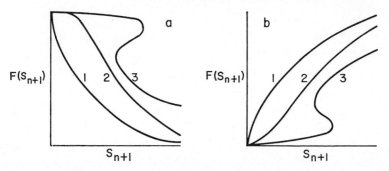

Fig. 19. Relationship between the feedback function in Scheme XIII and S_{n+1}; (a) when S_{n+1} inhibits E_0 and (b) when S_{n+1} stimulates E_0.

The widespread existence of substrate inhibition of enzymes and end-product feedback suggests that control mechanisms of the type described by Scheme XIII may be important in biological systems.

E. Linear Analysis of Biochemical Systems with Feedback

Returning now to equations 6.3, we consider the usual transformation:

$$u_i = s_i - s_i^*, \quad i = 1, \dots, n+1. \tag{6.16}$$

The s_i^* in equations 6.16 are the positive solutions of the algebraic equations:

$$
\begin{aligned}
0 &= F(s_{n+1}^*) - K_1 s_1^*, \\
0 &= K_{i-1} s_{i-1}^* - K_i s_i^*, \quad i = 2, \dots, n+1.
\end{aligned}
\tag{6.17}
$$

For the nonlinear feedback function defined in equation 6.2, we obtain

$$(s_1^*)^{\rho+1} + \frac{1}{\alpha}\left(\frac{K_{n+1}}{K_1}\right)^{\rho} s_1^* - \frac{1}{\alpha}\left(\frac{K_{n+1}}{K_1}\right)^{\rho} \frac{C_0}{K_1} = 0, \tag{6.18}$$

$$s_i^* = \frac{K_{i-1}}{K_i} s_{i-1}^*, \quad i = 2, \dots, n+1. \tag{6.19}$$

Since the s_i^* are chemical concentrations, we may limit the discussion to the nonnegative real roots of equation 6.18. We can obtain the s_i^* in radicals if ρ is 1, 2, or 3.

$\rho = 1$:

$$s_1^* = \frac{K_{n+1}}{2\alpha K_1}[\sqrt{1 + (4\alpha C_0/K_{n+1})} - 1],$$

$$s_i^* = \frac{K_{n+1}}{2\alpha K_i}[\sqrt{1 + (4\alpha C_0/K_{n+1})} - 1], \quad i = 2, \ldots, n+1.$$

(6.20)

$\rho = 2$:

$$s_1^* = \frac{K_{n+1}}{\alpha K_1}\sqrt[3]{\alpha^2 C_0/2K_{n+1}}(\sqrt[3]{1+A} + \sqrt[3]{1-A}),$$

$$s_i^* = \frac{K_{n+1}}{\alpha K_i}\sqrt[3]{\alpha^2 C_0/2K_{n+1}}(\sqrt[3]{1+A} + \sqrt[3]{1-A}), \quad i = 2, \ldots, n+1, \quad (6.21)$$

$$A = \sqrt{\tfrac{4}{27}(K_{n+1}^2/\alpha C_0^2) + 1}.$$

$\rho = 3$:

$$s_1^* = \frac{K_{n+1}}{\alpha K_1}[\sqrt{(M_{30} + M_{31})/2}$$

$$+ \sqrt{\sqrt{M_{30}^2 - M_{30}M_{31} + M_{31}^2} - (M_{30} + M_{31})/2}],$$

$$s_i^* = \frac{K_{n+1}}{\alpha K_i}[\sqrt{(M_{30} + M_{31})/2}$$

$$+ \sqrt{\sqrt{M_{30}^2 - M_{30}M_{31} + M_{31}^2} - (M_{30} + M_{31})/2}],$$

$$i = 2, \ldots, n+1, \quad (6.22)$$

$$M_{30} = \sqrt[3]{\alpha^4/16 + M_{32}\alpha^4/K_{n+1}}, \quad M_{31} = \sqrt[3]{\alpha^4/16 - M_{32}\alpha^4/K_{n+1}},$$

$$M_{32} = \sqrt{\alpha C_0^3/27K_{n+1} + K_{n+1}^2/256}$$

If $\rho > 3$, equation 6.18 cannot be solved in radicals (63), but we can still obtain the s_i^* by numerical methods provided all the coefficients in equations 6.18 and 6.19 are known.

The transformed differential equations are

$$\dot{u}_1 = G(u_{n+1}) - K_1 u_1,$$

$$\dot{u}_i = K_{i-1} u_{i-1} - K_i u_i, \quad i = 2, \ldots, n+1.$$

(6.23)

The transformed nonlinear function corresponding to the negative-feedback function described in equation 6.2 is

$$G(u_{n+1}) = \frac{C_0}{1 + \alpha(u_{n+1} + s_{n+1}^*)^2} - K_1 s_1^*. \tag{6.24}$$

The linearization of equations 6.23 in the neighborhood of the singularity $u_i = 0$ $(i = 1, \ldots, n + 1)$ gives the abridged equations:

$$\dot{u}_1 = Au_{n+1} - K_1 u_1,$$
$$\dot{u}_i = K_{i-1} u_{i-1} - K_i u_i, \quad i = 2, \ldots, n + 1. \tag{6.25}$$

The linearized feedback term is

$$A = \frac{\partial G}{\partial u_{n+1}}\bigg|_{u_{n+1}=0} = -\frac{\alpha \rho C_0 (s_{n+1}^*)^{\rho-1}}{[1 + \alpha(s_{n+1}^*)^\rho]^2}. \tag{6.26}$$

Since s_{n+1}^* is completely defined by α, ρ, C_0, and the K_i, A is also completely defined by these constants.

The characteristic equation for equations 6.25 is

$$\prod_{i=1}^{n+1} (K_i + \mu) - A \prod_{i=1}^{n} K_i = 0. \tag{6.27}$$

If we set $\mu = 0$, we obtain

$$K_{n+1} = A. \tag{6.28}$$

In what follows we shall refer to equation 6.28 as the *equation of zero roots*.

Alternatively, if we set $\mu = i\omega$ $(i = \sqrt{-1})$, we obtain what we shall refer to as the *neutrality equation*. The neutrality equations for n equal to 1, 2, 3, and 4 are as follows.

$n = 1$:

$$K_1 = -K_2. \tag{6.29}$$

$n = 2$:

$$\frac{(K_1 + K_2)(K_1 + K_3)(K_2 + K_3)}{K_1 K_2} + A = 0. \tag{6.30}$$

$n = 3$:

$$\frac{(K_3 + K_4)(K_2^2 + K_2 K_3 + K_2 K_4 + K_3 K_4)[K_1^2(K_1 + K_2 + K_3 + K_4) + K_1 K_2 K_3 + K_1 K_2 K_4 + K_1 K_3 K_4 + K_2 K_3 K_4]}{K_1 K_2 K_3 (K_1 + K_2 + K_3 + K_4)^2} + A = 0. \tag{6.31}$$

$n = 4$:

$$\sqrt{B^2 + 4C} - B + 2A = 0, \tag{6.32}$$

$$B = \frac{K_5(2a_0 + a_3{}^2a_4 - a_2a_3 - 2a_1a_4)}{a_0},$$

$$C = \frac{K_5{}^2(a_1a_2a_3a_4 - a_0a_3{}^2a_4 - a_1{}^2a_4{}^2 - a_1a_2{}^2 + a_0a_2a_3 + 2a_0a_1a_4 - a_0{}^2)}{a_0{}^2},$$

$$a_0 = K_1K_2K_3K_4K_5, \tag{6.33}$$

$$a_1 = K_1K_2K_3K_4 + K_1K_2K_3K_5 + K_1K_2K_4K_5 + K_1K_3K_4K_5 + K_2K_3K_4K_5,$$

$$a_2 = K_1K_2K_3 + K_1K_2K_4 + K_1K_2K_5 + K_1K_3K_4 + K_1K_3K_5 + K_1K_4K_5$$
$$+ K_2K_3K_4 + K_2K_3K_5 + K_2K_4K_5 + K_3K_4K_5,$$

$$a_3 = K_1K_2 + K_1K_3 + K_1K_4 + K_1K_5 + K_2K_3 + K_2K_4$$
$$+ K_2K_5 + K_3K_4 + K_3K_5 + K_4K_5,$$

$$a_4 = K_1 + K_2 + K_3 + K_4 + K_5.$$

The neutrality equations for $n > 4$ become progressively complicated.

With the aid of equations 6.28–6.32 we can partition the plane of the parameters, A and any one K_i, into areas for which at least one characteristic exponent changes sign in its real part. In Fig. 20 appears the line of zero roots (solid line) and the neutrality line (dashed line) for $n = 1, 2, 3,$ and 4. In Fig. 20a all the $K_i = 1$ ($i = 1, \ldots, n$) and K_{n+1} is varied; in Fig. 20b all the $K_i = 1$ ($i = 1, \ldots, e - 1, e + 1, \ldots, n + 1$) except K_e ($e \leq n$), which is varied. In the shaded areas of Fig. 20 the real part of at least one of the characteristic exponents is positive; the shaded areas therefore denote regions wherein the singularity $u_i = 0$ must be unstable. For values of A and K_{n+1} or A and K_e corresponding to the unshaded areas the singularity is asymptotically stable if the real parts of all the other characteristic exponents are also negative.

Since in the systems we are considering the K_i are positive, first-order rate constants, we will limit our discussion to the first and fourth quadrants of Fig. 20. For the case involving two variables ($n = 1$) there is no shaded area for any positive K_i and negative A. Furthermore, it is easy to show*

* For $n = 1$ the characteristic equation is
$$\mu^2 + (K_1 + K_2)\mu + K_1K_2 - A = 0.$$
The characteristic exponents are
$$\mu_{1,2} = \frac{-(K_1 + K_2)}{2} + \frac{\sqrt{(K_1 + K_2)^2 - 4(K_2 - A)}}{2}$$
If $4(K_2 - A) > (K_1 + K_2)^2$, the characteristic exponents are complex conjugate, but the real part is the negative sum of K_1 and K_2. If $4(K_2 - A) < (K_1 + K_2)^2$ and $A < 0$, $\sqrt{(K_1 + K_2)^2 - 4(K_2 - A)} < K_1 + K_2$, and both characteristic exponents are negative real numbers.

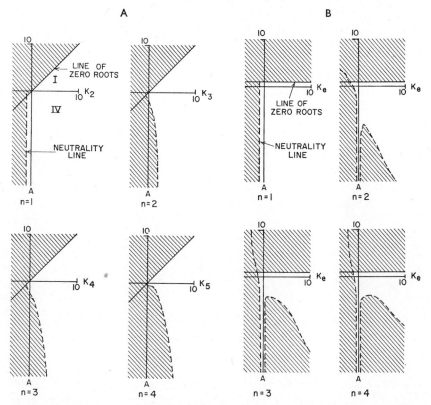

Fig. 20. Partition of the parameter space for an enzyme-catalyzed system with negative feedback (Scheme IX) into areas of stability and instability (slashed area).

that, when $n = 1$, the real parts of *both* roots of equation 6.27 are necessarily negative when K_1 and K_2 are positive and $A < 0$. Thus linear negative-feedback systems are always asymptotically stable in the neighborhood of the singularity $u_i = 0$ $(i = 1, \ldots, n + 1)$ when $n = 1$. However, there is a large area of instability available to systems involving two components if they possess positive feedback.

For systems involving three variables $(n = 2)$ there is a small area of instability in the fourth quadrant of Fig. 20. This means that linear negative-feedback systems can be unstable when $n = 2$. The area of instability available to positive-feedback systems when $n > 1$ is the same as the area available to the two-component systems. However, for negative-feedback systems the area of instability in the fourth quadrant becomes progressively larger as the number of variables increases from two to three to four. In fact, Fig. 20 suggests (45) that the potential for oscillatory behavior of the limit cycle type in open chemical systems involving negative feedback becomes greater as the number of kinetically important chemical steps becomes larger.

It is possible to examine the stability properties of equations 6.25 by examining the sign of the real parts of the characteristic exponents for specific n. We use Theorem 13 (the Hurwitz criterion) to ascertain that for $n = 1$ the roots of the characteristic equation* both have negative real parts provided

$$K_1 + K_2 > 0 \quad \text{and} \quad K_1(K_2 - A) > 0. \tag{6.33}$$

Since K_1 and K_2 are positive rate constants, the first condition is always satisfied. If $A < 0$ (negative feedback), the second condition is also invariably met. Thus, when $n = 1$, the negative-feedback systems are always asymptotically stable in the neighborhood of the singular point $u_1 = 0$, $u_2 = 0$. However, if $A > 0$ (positive feedback), inequality 6.33 infers that the area of stability is limited to $K_2 < A$. This area of stability corresponds to the area in the first quadrant of Fig. 20a under the line $A = K_2$. As predicted from this figure, instability results when $A > K_2$.

For systems involving three components ($n = 2$) the characteristic equation is

$$\mu^3 + (K_1 + K_2 + K_3)\mu^2 + (K_1K_2 + K_1K_3 + K_2K_3)\mu + K_1K_2(K_3 - A) = 0. \tag{6.34}$$

The roots of equation 6.34 have negative real parts iff

$$K_1 + K_2 + K_3 > 0,$$
$$(K_1 + K_2)(K_1 + K_3)(K_2 + K_3) + K_1K_2A > 0, \tag{6.35}$$
$$K_1K_2(K_3 - A) > 0.$$

Since the K_i are all positive rate constants, the first inequality is always satisfied. If $A < 0$ (negative feedback), the third condition is also met. The second condition can be written in the form:

$$(K_1 + K_2 + K_3)(K_1K_2 + K_1K_3 + K_2K_3) - K_1K_2K_3 + K_1K_2A > 0. \tag{6.36}$$

There is a theorem (64) that states:

Theorem 20. $(K_1 + K_2 + K_3)(K_1K_2 + K_1K_3 + K_2K_3) \geq 9K_1K_2K_3$ *if K_1, K_2, and K_3 are positive real numbers.*

Substitution from this theorem into inequality 6.36 yields

$$8K_1K_2K_3 + K_1K_2A > 0. \tag{6.37}$$

---

* The characteristic equation may be found in the preceding footnote.

If we take advantage of the fact that in the negative-feedback systems under discussion here

$$-A < \rho K_3, \qquad (6.38)$$

we obtain the condition first derived by Griffith (50):

$$(8 - \rho)K_1 K_2 K_3 > 0. \qquad (6.39)$$

Inequality 6.39 illustrates that the second Hurwitz inequality (inequality 6.36) is always satisfied provided the stoichiometry of the feedback inhibition is less than 9 ($\rho < 9$). However, if $\rho > 8$, the chemical systems described by equations 6.2 and 6.3 (Scheme IX) involving three components ($n = 2$) can have unstable singularities. The shaded area in the fourth quadrant of Fig. 20, $n = 2$, corresponds to versions of the nonlinear feedback function in equation 6.2, for which $\rho > 8$.

If $A > 0$ (positive feedback), the second condition in inequalities 6.35 is always satisfied, and the third condition is met if

$$K_3 > A. \qquad (6.40)$$

Inequality 6.40 describes the area in the first quadrant of Fig. 20, $n = 2$, for which equations 6.25 can be unstable.

For systems involving four components ($n = 3$) the characteristic equation is

$$\mu^4 + (K_1 + K_2 + K_3 + K_4)\mu^3 + (K_1 K_2 + K_1 K_3 + K_1 K_4 + K_2 K_3$$
$$+ K_2 K_4 + K_3 K_4)\mu^2 + (K_1 K_2 K_3 + K_1 K_2 K_4 + K_1 K_3 K_4$$
$$+ K_2 K_3 K_4)\mu + K_1 K_2 K_3 (K_4 - A) = 0. \quad (6.41)$$

The roots of this equation have negative real parts iff

$$K_1 + K_2 + K_3 + K_4 > 0,$$
$$(K_1 + K_2 + K_3 + K_4)(K_1 K_2 + K_1 K_3 + K_1 K_4 + K_2 K_3 + K_2 K_4 + K_3 K_4)$$
$$+ (K_1 + K_2)(K_1 K_2 + K_1 K_3 + K_1 K_4 + K_2 K_3 + K_2 K_4)$$
$$+ (K_3 + K_4)(K_1 K_3 + K_1 K_4 + K_2 K_3 + K_2 K_4 + K_3 K_4) > 0, \quad (6.42)$$
$$(K_3 + K_4)(K_2^2 + K_2 K_3 + K_2 K_4 + K_3 K_4)[K_1^2(K_1 + K_2 + K_3 + K_4)$$
$$+ K_1 K_2 K_3 + K_1 K_2 K_4 + K_1 K_3 K_4 + K_2 K_3 K_4]$$
$$+ (K_1 + K_2 + K_3 + K_4)^2 K_1 K_2 K_3 A > 0,$$
$$K_1 K_2 K_3 (K_4 - A) > 0.$$

Since K_1, K_2, K_3, and K_4 are positive rate constants, the first two conditions in inequalities 6.42 are always fulfilled. If $A < 0$ (negative feedback), the fourth condition is also always satisfied, and the third condition is the criterion for asymptotic stability of the linearized chemical control system.

Alternatively, if $A > 0$ (positive feedback), the third condition is always met, and the fourth condition is the criterion for asymptotic stability. Note that, as in systems involving three components, the line of zero roots determines the border of the stability region for systems involving positive feedback, and the neutrality line determines the border for systems involving negative feedback.

The derivation of the conditions for which the systems involving more than three components $(n > 2)$ can be unstable is obviously more difficult than the derivation of inequality 6.39. Viniegra and Martinez (65) have suggested a way to obtain an idea of the minimum value of ρ for which the linearized multicomponent system can be unstable. In this method we consider the special case in which

$$K_i = K, \quad i = 1, \ldots, n + 1,$$
$$C_0 = K s_{n+1}^*. \tag{6.43}$$

For this special situation the characteristic equation is

$$\prod_{i=1}^{n+1} (K + \mu)^{n+1} + C K^{n+1} = 0, \tag{6.44}$$

where $C = -A/K$ is a positive real number that must be less than ρ:

$$0 < C = \frac{\rho \alpha (s_{n+1}^*)^\rho}{[1 + \alpha (s_{n+1}^*)^\rho]^2} < \rho. \tag{6.45}$$

Viniegra and Martinez (65) report that the principal root of equation 6.45 has the form

$$\mu = K[(-1)^{1/m} C^{1/m} - 1] = K\left(C^{1/m} \cos \frac{\pi}{m} - 1 + \sin \frac{\pi}{m}\right), \tag{6.46}$$

where $m = n + 1$. This root has the smallest projection on the real axis; thus the necessary and sufficient condition for instability of the system is met when the real part of μ becomes positive, as occurs when

$$\rho > C > \sec^m \frac{\pi}{m}. \tag{6.47}$$

Viniegra and Martinez also report that a necessary condition for instability when the K_i $(i = 1, \ldots, n + 1)$ are not necessarily equal is

$$\rho > C > \frac{K_{\min}^m}{\prod\limits_{i=1}^{m} K_i} \sec^m \frac{\pi}{m}. \tag{6.48}$$

According to inequality 6.47, the conditions for unstable singularities in the negative-feedback systems described by equations 6.2 and 6.3 depend on whether or not ρ and n are sufficiently large. Since the secant is never less than 1, there will not be unstable singularities when $\rho = 1$ no matter how many components exist in the control system. Similarly, since the secant of $\pi/2$ is infinitely large, there will not be unstable singularities when $n = 1$ no matter how large the stoichiometry of the feedback inhibition is. For all other values of $n > 1$ and $\rho > 1$ it should be possible to have unstable singularities in the negative-feedback systems described by equations 6.2 and 6.3. The necessary and sufficient conditions for the case when all the $K_i = K$ and $C_0 = Ks_{n+1}^*$ (equations 6.43) are listed in Table 2.

The results in this section suggest that nonlinear biochemical control systems involving negative feedback can exhibit limit cycle behavior if there are a sufficient number of kinetically important chemical steps between the step wherein the feedback component is formed and the inhibited step (45). In order to test this idea we will consider the effect of increasing the number of components in an example system described by equations 6.2 and 6.3:

$$\rho = 4, \qquad \alpha = 0.08, \qquad C_0 = 5.1, \qquad K_i = 1$$
$$(i = 1, \ldots, n) \qquad \text{and} \qquad K_{n+1} = 0.6.$$

Substitution of these values into equations 6.18, 6.19, and 6.26 provides the numerical value of the linearized negative feedback term:

$$A = \frac{\partial G}{\partial u_{n+1}}\bigg|_{u_{n+1}=0} = -1.7277. \tag{6.49}$$

For the case involving four components ($n = 3$) the characteristic equation is

$$\mu^4 + 3.6\mu^3 + 4.8\mu^2 + 2.8\mu + 2.3277 = 0. \tag{6.50}$$

Table 2. Conditions for Unstable Singularities in the Negative-Feedback Systems Described by Equations 6.2 and 6.3

When n is:	ρ must be larger than:
2	8
3	4
4	3
5	2
6	2
7	2
8 or larger	1

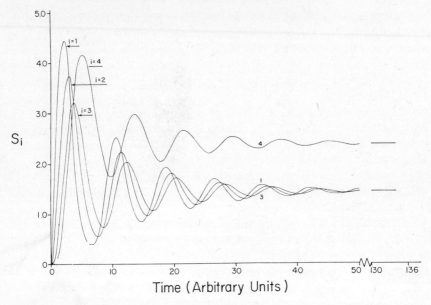

Fig. 21. $s_i(t)$ for the components of an enzyme-catalyzed system involving four variables and negative feedback (Scheme IX).

The roots of equation 6.50 have negative real parts if

$$\begin{vmatrix} 3.6 & 2.8 & 0 \\ 1 & 4.8 & 2.3277 \\ 0 & 3.6 & 2.8 \end{vmatrix} > 0. \tag{6.51}$$

By expanding the determinant, we find that inequality 6.51 is satisfied. Thus we are in the situation in which equations 6.25 are asymptotically stable. We cannot decide on this basis whether or not the nonlinear system (equations 6.2 and 6.3) is asymptotically stable in all the phase-space available to the variables. We do know, however, that, if we start the chemical system in the neighborhood of the singularity $s_i = s_i^*$ ($i = 1, 2, 3, 4$), sustained oscillations will not arise. In Fig. 21 appears the actual time course of the S_i ($i = 1, 2, 3, 4$) in the system under discussion here. Since the variables approach the singularity from their minimum possible values, this figure illustrates that the nonlinear system involving four components possesses global asymptotic stability.*

* The stability is global because of the requirements of Theorem 1b.

The characteristic equation for the case involving five components ($n = 4$) is

$$\mu^5 + 4.6\mu^4 + 8.4\mu^3 + 7.6\mu^2 + 3.4\mu + 2.3277 = 0. \tag{6.52}$$

It can be verified from Theorem 13 that the roots of equation 6.52 all have negative real parts. Thus we are again in the situation in which equations 6.25 are asymptotically stable. In Fig. 22 appears the actual time course of the S_i ($i = 1, 2, 3, 4, 5$) in the system under discussion here; this figure illustrates that the nonlinear system involving five components possesses global asymptotic stability.

The characteristic equation for the case involving six components ($n = 5$) is

$$\mu^6 + 5.6\mu^5 + 13\mu^4 + 16\mu^3 + 11\mu^2 + 4\mu + 2.3277 = 0. \tag{6.53}$$

All of the conditions in Theorem 13 for the roots of equation 6.53 to have negative real parts are not met. Thus in this case we are in a situation in which equations 6.25 are not stable. In Fig. 23 appears the time course for the S_i ($i = 1, 2, 3, 4, 5, 6$) in the system under discussion here. In this case the oscillations are self-sustaining and self-limiting. This limit cycle behavior is further illustrated in the phase-plane representations in Fig. 24. Note that, if the system is initiated in the neighborhood of the singularity $s_i = s_i^*$ ($i = 1, 2, 3, 4, 5, 6$), the variables wind outward to the surrounding limit cycle. If the system is initiated in the phase-space outside the limit cycle [$s_i(0) = 0$], the variables wind inward toward the same closed trajectory. clearly, the introduction of the sixth component is sufficient to cause a stable system (Fig. 22) to become unstable. Figures 23 and 24 represent the first report of *bona fide* limit cycle behavior in biochemical control mechanisms possessing purely negative feedback (45).

F. Oscillations in Chemical Control Systems Involving Two Components

The negative Bendixson criterion (Theorem 6a) provides a convenient basis for establishing conditions for the nonexistence of limit cycles in certain two-component chemical systems. Consider any system described by

$$\dot{x} = f(y) - K_1 x \equiv X(x, y), \qquad \dot{y} = g(x) - K_2 y \equiv Y(x, y). \tag{6.54}$$

No matter what the function $f(y)$ is (as long as it does not include x as an argument), and no matter what the function $g(x)$ is (as long as it does not contain y as an argument),

$$\frac{\partial X}{\partial x} + \frac{\partial Y}{\partial y} = -(K_1 + K_2) \tag{6.55}$$

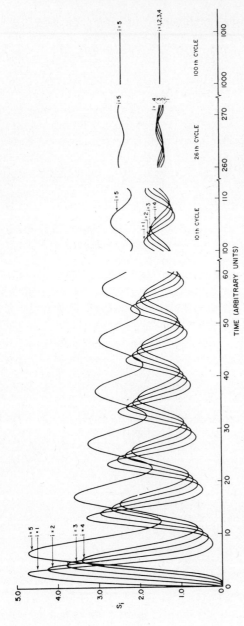

Fig. 22. $s_i(t)$ for the components of the system in Fig. 21 but involving five variables (Scheme IX).

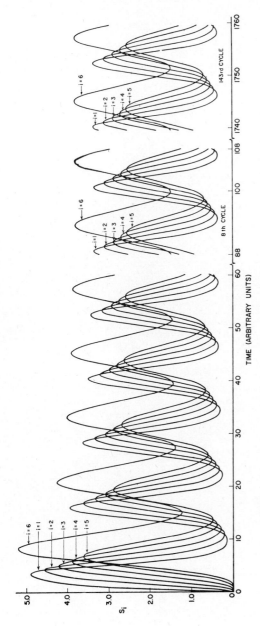

Fig. 23. $s_i(t)$ for the components of the system in Figs. 21 and 22 but involving six variables (Scheme IX).

441

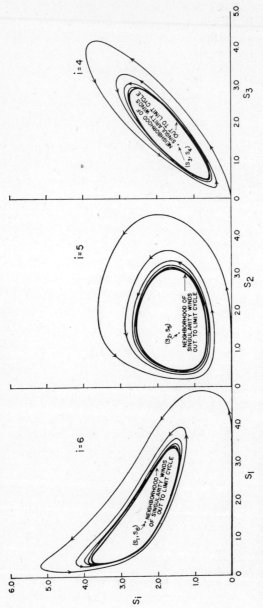

Fig. 24. Phase-space representation of the limit cycle in Fig. 23.

for all values of x and y. Since the right-hand side of equation 6.55 is sign-definite, no limit cycles can exist for equations 6.54.

In equations 6.54 the decay terms for the two components are first-order. In the general situation when the decay terms are of (integer) order m_1 and m_2 ($m > 0$), the differential equations describing the system are

$$\dot{x} = f(y) - K_1 x^{m_1} \equiv X(x, y), \qquad \dot{y} = g(x) - K_2 y^{m_2} \equiv Y(x, y). \quad (6.56)$$

In this case

$$\frac{\partial X}{\partial x} + \frac{\partial Y}{\partial y} = -(m_1 K_1 x^{m_1-1} + m_2 K_2 y^{m_2-1}). \quad (6.57)$$

Since chemical concentrations are always nonnegative, the right-hand side of equation 6.57 is sign-definite and no limit cycles can exist for equations 6.56.

If the decay terms for x and y are hyperbolic functions of the type usually associated with enzyme-catalyzed reactions,

$$\dot{x} = f(y) - \frac{{}^1V_m x}{{}^1K_m + x} \equiv X(x, y), \qquad \dot{y} = g(x) - \frac{{}^2V_m y}{{}^2K_m + y} \equiv Y(x, y). \quad (6.58)$$

In this case we have

$$\frac{\partial X}{\partial x} + \frac{\partial Y}{\partial y} = -\left[\frac{{}^1V_m {}^1K_m}{({}^1K_m + x)^2} + \frac{{}^2V_m {}^2K_m}{({}^2K_m + y)^2}\right]. \quad (6.59)$$

Since the right-hand side of equation 6.59 is sign-definite, no limit cycles can exist for chemical control systems described by equations 6.58.

As an alternative to the above, it is possible that the decay term for x is described by the ratio of two quadratic functions (58). In this case the differential equations describing the dynamical behavior of the control mechanism are (54)

$$\dot{x} = f(y) - \frac{a_1 x + a_2 x^2}{\alpha_0 + \alpha_1 x + \alpha_2 x^2} \equiv X(x, y),$$

$$\dot{y} = g(x) - \frac{{}^2V_m y}{{}^2K_m + y} \equiv Y(x, y),$$

$$(6.60)$$

where a_1, a_2, α_0, α_1, and α_2 are nonnegative coefficients. For equations 6.60 we obtain

$$\frac{\partial X}{\partial x} + \frac{\partial Y}{\partial y} = -\left[\frac{\alpha_0(a_1 + 2a_2 x) + (a_2 x - a_1 \alpha_2)x^2}{(\alpha_0 + \alpha_1 x + \alpha_2 x^2)^2} + \frac{{}^2K_m {}^2V_m}{({}^2K_m + y)^2}\right]. \quad (6.61)$$

If

$$\frac{a_2}{\alpha_2} > \frac{a_1}{\alpha_1}, \tag{6.62}$$

equation 6.61 always describes a negative function, and according to the negative Bendixson criterion no closed trajectories can exist for equations 6.60.

Yet another possibility is that the decay term for x is described by the ratio of two general polynomials of the same order and with nonnegative coefficients.* In this case the differential equations describing the dynamical behavior of the control system are

$$\dot{x} = f(y) - \frac{\sum\limits_{i=1}^{n} a_i x^i}{\alpha_0 + \sum\limits_{i=1}^{n} \alpha_i x^i} \equiv X(x, y),$$

$$\dot{y} = g(x) - \frac{{}^{2}V_m y}{{}^{2}K_m + y} \equiv Y(x, y). \tag{6.63}$$

In this case the negative Bendixson criterion is always satisfied if

$$\frac{a_i}{\alpha_i} > \frac{a_{i-1}}{\alpha_{i-1}}, \quad i = 2, 3, \ldots, n. \tag{6.64}$$

Inequalities 6.64 are always satisfied if

$$\frac{\sum\limits_{i=1}^{n} a_i x^i}{\alpha_0 + \sum\limits_{i=1}^{n} \alpha_i x^i} \quad \text{versus } x$$

does not experience an extremum (as, e.g., in the allosteric model in which the interactions are positive). If the conditions in inequalities 6.64 are satisfied for this or any other reason, no limit cycles can exist for equations 6.63.

From the preceding discussion it is possible to formulate the generalization:

Theorem 21. *No limit cycles are possible for two-component chemical control systems described by*

$$\dot{x} = f_1(y) - f_2(x), \qquad \dot{y} = g_1(x) - g_2(y),$$

---

* Many mechanisms lead to this type of rate equation; perhaps the most commonly discussed one is the allosteric model (55) in which the "cooperativity" is the same as the order of the polynomials.

regardless of the nature of $f_1(y)$ and $g_1(x)$ if $f_2(x)$ and $g_2(y)$ represent rate laws that are strictly positive order for all nonnegative values of x and y.

Confirmation of this result has been reported independently by Ličko (66).

G. Aiserman's Method

Equations 6.2 and 6.3 permit the use of a method first suggested by M. A. Aiserman (67) to establish a range of the nonlinear feedback function, $G(u_{n+1})$, for which the asymptotic stability of the singularity $s_i = s_i^*$ ($i = 1, \ldots, n + 1$) can be guaranteed. The method requires that the linearized system be asymptotically stable. When this condition is met, it is possible to find an interval for the linear feedback term such that the linear system remains asymptotically stable. It seems reasonable to assume that, if the linear system remains asymptotically stable when the linear feedback term is between $(A - A_1)u_{n+1}$ and $(A + A_2)u_{n+1}$, the nonlinear system will also be stable if $G(u_{n+1})$ is between $G_1 = (A - A_1)u_{n+1}$ and $G_2 = (A + A_2)u_{n+1}$. This assumption has been proved for systems involving two components ($n = 1$); for larger n it seems plausible to suppose that its application is more general (68).

Recently, the Aiserman method was applied to the nonlinear biochemical feedback systems described by equations 6.2 and 6.3 (52).

We have seen that linearized chemical feedback systems involving negative feedback and only two components are always asymptotically stable; according to Theorem 21 the corresponding nonlinear negative-feedback systems cannot possess a limit cycle. Since chemical control systems should not be totally unstable, we have a complete picture of the stability properties and dynamic behavior of two-component chemical feedback systems described by equations 6.2 and 6.3. Therefore, in what follows, we use the Aiserman method to examine the stability properties of chemical feedback systems involving positive feedback ($A > 0$) and two components. To begin, we consider the negative-definite Lyapunov function:

$$W = -M(u_1^2 + u_2^2), \tag{6.65}$$

where M is a positive constant. Since the linear system is asymptotically stable, there must exist a positive-definite function,

$$2V = au_1^2 + 2bu_1u_2 + cu_2^2, \tag{6.66}$$

such that the Eulerian derivative of V is

$$\dot{V} = \frac{\partial V}{\partial u_1} \dot{u}_1 + \frac{\partial V}{\partial u_2} \dot{u}_2 = W. \tag{6.67}$$

Substitution of $\partial V/\partial u_1$ and $\partial V/\partial u_2$ from equation 6.66, $\dot{u}_1$ and $\dot{u}_2$ from equations 6.25, and W from equation 6.65 into equation 6.67 yields

$$(Ab - cK_2)u_2{}^2 + (cK_1 - bK_1 - aK_1 - bK_2)u_1 u_2$$
$$+ (bK_1 - aK_1)u_1{}^2 = -Mu_2{}^2 - Mu_1{}^2. \tag{6.68}$$

By equating the coefficients of like terms in equation 6.68 we obtain three relationships from which we can calculate a, b, and c from A, M, K_1, and K_2:

$$\begin{aligned}
\Delta a &= M[K_2(K_1 + K_2) + K_1(K_1 - A)], \\
\Delta b &= M(K_1{}^2 + AK_2), \\
\Delta c &= M[K_1(K_1 + K_2) - A(K_1 - A)], \\
\Delta &= K_1(K_1 + K_2)(K_2 - A).
\end{aligned} \tag{6.69}$$

Equations 6.66 and 6.69 completely determine the Lyapunov function, V. Since we have chosen $W = \dot{V}$ as a negative-definite function, V must be positive-definite when the system is asymptotically stable.

We now examine the range of the linear feedback term for which $\dot{V}$ remains negative-definite. If δ is added to A, $\delta u_2(\partial V/\partial u_1)$ must be added to W to maintain the equalities in equation 6.67:

$$\begin{aligned}
\dot{V} &= \frac{\partial V}{\partial u_1}(A + \delta)u_2 - K_1 u_1 + \frac{\partial V}{\partial u_1} \dot{u}_2 \\
&= W + \delta u_2 \frac{\partial V}{\partial u_1} \\
&= (\delta b - M)u_2{}^2 + \delta a u_1 u_2 - Mu_1{}^2.
\end{aligned} \tag{6.70}$$

In order to obtain an idea of the sign of equation 6.70, consider the general polynomial:

$$\dot{V} = -\sum_{k=1}^{n} \alpha_i x_i{}^2 - qy^2 - \sum_{k=1}^{n} \beta_i x_i y, \tag{6.71}$$

where the α_i, β_i, and q are the coefficients of the polynomial. From the theory of quadratic forms, the necessary and sufficient conditions for $\dot{V}$ to be negative are

$$\begin{vmatrix} \alpha_1 & 0 & \cdots & 0 & \beta_1/2 \\ 0 & \alpha_2 & \cdots & 0 & \beta_2/2 \\ \cdot & \cdot & & \cdot & \cdot \\ \cdot & \cdot & \cdot & \cdot & \cdot \\ \cdot & \cdot & \cdot & \cdot & \cdot \\ 0 & \cdots & \cdots & \alpha_n & \beta_n/2 \\ \beta_1/2 & \beta_2/2 & \cdots & \beta_n/2 & q \end{vmatrix} > 0, \tag{6.72}$$

$$\alpha_1 > 0, \qquad \alpha_1\alpha_2 > 0, \qquad \cdots, \prod_{i=1}^{n} \alpha_i > 0.$$

For equations 6.70 these conditions are

$$\begin{vmatrix} M & -\delta a/2 \\ -\delta a/2 & M - \delta b \end{vmatrix} > 0, \tag{6.73}$$

$$M > 0.$$

The last condition is always met because we defined M as a positive number for the specification of a negative-definite Lyapunov function. If we expand the determinant in inequalities 6.73, we obtain

$$a^2\delta^2 + 4bM\delta - 4M^2 < 0. \tag{6.74}$$

Inequality 6.74 describes the range of δ for which $\dot{V}$ remains negative-definite. The condition in inequality 6.74 is satisfied provided δ is in the interval $-\delta_1 < \delta < \delta_2$, where

$$-\delta_1 = \frac{2\Delta}{a^2}(-b - \sqrt{a^2 + b^2}),$$

$$\delta_2 = \frac{2\Delta}{a^2}(-b + \sqrt{a^2 + b^2}). \tag{6.75}$$

If, for all values of u_2, $G = G(u_2)$ is between the limits $G_1 = (A - \delta_1)u_2$ and $G_2 = (A + \delta_2)u_2$, the singularity $s_1 = s_1^*$, $s_2 = s_2^*$ of the nonlinear control system is asymptotically stable for any set of initial conditions. This means that, if $G(u_2)$ remains within these limits, no sustained oscillations are possible for the chemical systems described by equations 6.3, $n = 1$.

For three-component chemical control systems involving negative feedback of the type described by equation 6.2, equations 6.25 are asymptotically stable if $\rho < 9$. For chemical systems involving three components and positive feedback, equations 6.25 are asymptotically stable if $K_3 < A$. In what follows we use the Aiserman method to examine the stability properties of three-component chemical feedback systems for which the conditions for asymptotic stability of the linearized system are satisfied. To begin, we again consider a negative-definite function:

$$W = -M(u_1^2 + u_2^2 + u_3^2). \tag{6.76}$$

Again, M is an arbitrary positive constant. Since the linear system is asymptotically stable, there must exist a positive-definite function,

$$2V = au_1^2 + 2bu_1u_2 + 2cu_1u_3 + 2du_2u_3 + eu_2^2 + fu_3^2, \tag{6.77}$$

such that the Eulerian derivative of V is

$$\dot{V} = \frac{\partial V}{\partial u_1}\,\dot{u}_1 + \frac{\partial V}{\partial u_2}\,\dot{u}_2 + \frac{\partial V}{\partial u_3}\,\dot{u}_3 = W. \tag{6.78}$$

Using the same method as in the case involving two components, we obtain three of the constants in equation 6.77:

$$\Delta a = M\{K_2K_3(K_1 + K_2)(K_1 + K_3)(K_2 + K_3) + K_1^2K_3(K_1 + K_3)(K_2 + K_3)$$
$$+ K_1^2K_2^2(K_1 + K_3) + K_1^2K_2^3 + K_1K_2[K_3(K_1 + K_3) + K_2(K_1 + K_2)$$
$$+ K_2K_3 + K_1^2]A\},$$

$$\Delta b = M\{K_1^2[K_3(K_2 + K_3)(K_1 + K_3) + K_2^2(K_1 + K_3) + K_2^3]$$
$$+ [K_1^3K_2 - K_2^2K_3(K_2 + K_3) + K_1K_2^2A]A\},$$

$$\Delta c = M\{K_1^2K_2^2(K_1 + K_2) - [K_1^2K_2^2 + K_1^2K_3(K_2 + K_3) + K_1^3K_3 \tag{6.79}$$
$$+ K_2K_3(K_1 + K_2)(K_2 + K_3) + K_1K_2K_3A]A\},$$

$$\Delta = \{K_1K_2K_3(K_1 + K_2)(K_1 + K_3)(K_2 + K_3) + K_1K_2[K_1K_2(K_1 + K_2)$$
$$+ K_1K_3(K_1 + K_3) + K_2K_3(K_2 + K_3) + K_1K_2K_3 - K_1K_2A]A\}.$$

As before, we examine the range of the linearized feedback term for which $\dot{V}$ remains negative-definite. In this case, if δ is added to A, $\delta u_3(\partial V/\partial u_1)$ must be added to W to maintain the equalities in equation 6.78:

$$\dot{V} = (\delta c - M)u_3^2 + \delta au_1u_3 + \delta bu_2u_3 - Mu_1^2 - Mu_2^2. \tag{6.80}$$

From inequalities 6.72, the necessary and sufficient conditions for $\dot{V}$ to be negative are

$$
\begin{vmatrix}
M & 0 & -\delta a/2 \\
0 & M & -\delta b/2 \\
-\delta a/2 & -\delta b/2 & M - \delta c
\end{vmatrix} > 0,
\tag{6.81}
$$

$$M > 0, \qquad M^2 > 0.$$

The last two conditions in inequalities 6.81 are always met. If we expand the determinant in inequality 6.81, we obtain

$$
(a^2 + b^2)\delta^2 + 4Mc\delta - 4M^2 < 0.
\tag{6.82}
$$

Inequality 6.82 describes the range of δ for which $\dot{V}$ remains negative-definite for the systems involving three components. This condition is satisfied provided δ is in the interval $-\delta_1 < \delta < \delta_2$, where

$$
-\delta_1 = \frac{2\Delta}{a^2 + b^2}(-c - \sqrt{a^2 + b^2 + c^2}),
$$

$$
\delta_2 = \frac{2\Delta}{a^2 + b^2}(-c + \sqrt{a^2 + b^2 + c^2}).
\tag{6.83}
$$

The terms a, b, c, and $\Delta = M$ in equations 6.83 are defined in equations 6.79.

Here again, if for all values of u_3, $G = G(u_3)$ is between the limits $G_1 = (A - \delta_1)u_3$ and $G_2 = (A + \delta_2)u_3$, the singularity $s_1 = s_1^*$, $s_2 = s_2^*$, $s_3 = s_3^*$ is asymptotically stable for any initial conditions of the nonlinear biochemical control systems described by equations 6.3, $n = 2$.

An example will serve to illustrate the use of this method for investigations of the stability properties of nonlinear biochemical control systems. Consider the situation in which the constants in equations 6.2 and 6.3 are

$$n = 2, \qquad \alpha = 0.1, \qquad C_0 = 1.1, \qquad K_1 = K_2 = K_3 = 1,$$

and

$$\rho = 1, 2, 3, \text{ or } 4.$$

Since $\rho < 9$, the linearized systems are asymptotically stable. The values of the linear feedback terms, the Δ, a, b, and c (calculated from equations 6.79), and δ_1 and δ_2 (calculated from equations 6.83) are listed in Table 3.

In Fig. 25 is shown a comparison of G, G_1, and G_2. In Fig. 25a $\rho = 1$, $G_1 = -0.9894u_3$, $G_2 = 0.6701u_3$, and $G = G(u_3)$ defined in equation 6.24; in Fig. 25b $\rho = 2$, $G_1 = -1.1177u_3$, $G_2 = 0.7136u_3$, and $G = G(u_3)$; in Fig. 25c $\rho = 3$, $G_1 = -1.1839u_3$, $G_2 = 0.7156u_3$, and $G = G(u_3)$; in Fig. 25d

Table 3. Calculations for the Control System Involving Three Components

ρ	$-A$	$\Delta = M$	a	b	c	δ_1	δ_2
1	1/11	6.287	13.55	6.901	1.264	0.8985	0.7610
2	20/121	7.130	13.99	6.807	0.6503	0.9524	0.8789
3	300/1331	7.527	14.35	6.724	0.1460	0.9585	0.9410
4	4000/14,641	7.838	14.64	6.652	−0.2602	0.9594	0.9910

$\rho = 4$, $G_1 = -1.2326u_3$, $G_2 = 0.7178u_3$, and $G = G(u_3)$. It can be seen from Fig. 25 that G remains within the prescribed limits, G_1 and G_2 for each ρ and all possible $u_3 = s_3 - s_3^*$. Therefore asymptotic stability of the singularity $s_1^* = s_2^* = s_3^* = 1$ is guaranteed for all initial conditions of these nonlinear control systems.

For biochemical control systems involving more than three components and negative feedback of the type described by equation 6.2, equations 6.25 are asymptotically stable if the conditions in Table 2 are met. For systems involving more than three components and positive feedback, equations 6.25 are asymptotically stable if $K_{n+1} < A$. It is possible to utilize the Aiserman method to examine the stability properties of these multicomponent

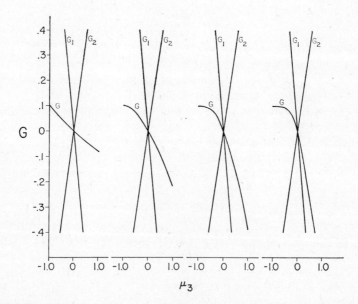

Fig. 25. Comparison of nonlinear feedback function G with G_1 and G_2, calculated utilizing Aiserman's method.

biochemical feedback systems for which the conditions for the asymptotic stability of the linearized systems are satisfied. For example, the systems involving four components ($n = 3$) have been examined elsewhere (52). In this case the Eulerian derivative of the Lyapunov function remains negative-definite if δ is in the interval $-\delta_1 < \delta < \delta_2$, where

$$-\delta_1 = \frac{2\Delta}{a^2 + b^2 + c^2}(-d - \sqrt{a^2 + b^2 + c^2 + d^2}),$$

$$\delta_2 = \frac{2\Delta}{a^2 + b^2 + c^2}(-d + \sqrt{a^2 + b^2 + c^2 + d^2}).$$

(6.84)

The meanings of a, b, c, d, and Δ can be found in Table 4. Thus, if $G(u_4)$ defined in equation 6.24 is between the limits $G_1 = (A - \delta_1)u_4$ and $G_2 = (A + \delta_2)u_4$ for all values of u_4, the singularity $s_1 = s_1^*$, $s_2 = s^*$, $s_3 = s_3^*$, $s_4 = s_4^*$ is asymptotically stable for all initial conditions of the nonlinear system involving four components (equations 6.3, $n = 3$).

The Aiserman method provides a convenient means of establishing a range for which the asymptotic stability of nonlinear control systems can be guaranteed. However, this method does not always provide the strongest possible stability conditions; hence it will not always be possible to guarantee the asymptotic stability of systems possessing such stability by these means. As an example, consider the case discussed by Morales and McKay (48) and illustrated in Fig. 21. The constants in equations 6.2 and 6.3 are

$$n = 3, \qquad \rho = 4, \qquad \alpha = 0.08, \qquad C_0 = 5.1,$$

$$K_1 = K_2 = K_3 = 1, \qquad \text{and} \qquad K_4 = 0.6.$$

We have already established that the corresponding linear system ($A = -1.7277$) is asymptotically stable. From Table 4 we obtain $M = \Delta = 24.155$, $a = 101.806$, $b = 77.651$, $c = 11.667$, and $d = 64.809$. From these values of a, b, c, d, and Δ and equations 6.84, we obtain that $\delta_1 = 0.2314$ and $\delta_2 = 0.6101$. If, for all values of u_4, $G = G(u_4)$ defined in equation 6.24 is between the limits $G_1 = -1.9591u_4$ and $G_2 = -1.1176u_4$, the nonlinear control system is asymptotically stable for any initial conditions of the variables.

In Fig. 26, $G_1 = -1.9591u_4$, $G_2 = -1.1176u_4$, and $G(u_4)$ defined in equation 6.24 are plotted versus u_4. It can be seen from this figure that $G(u_4)$ does not remain within the prescribed limits, G_1 and G_2, for the whole range of u_4. Therefore it is not possible to decide by means of the Aiserman method whether or not the nonlinear system remains stable for the domain accessible to the biochemical variables, or whether or not sustained oscillations can arise in this system.

Table 4. Definitions of the Constants in Equations 6.84

$\Delta b = M$

$-K_1$	$K_1 K_2$	K_2^2	0	0	0
$-K_2$	$-K_3(K_2 + K_3)$	0	0	K_3^2	$K_2 K_3$
0	K_1	$-(K_1 + K_3)$	K_3	0	0
A	0	0	$-K_1(K_1 + K_4)$	K_1^2	0
0	0	0	0	$K_2 + K_4$	$-K_2$
K_3	0	$K_4 A$	$K_3 A$	0	$K_4(K_3 + K_4)$

$\Delta c = M$

$-K_2(K_1 + K_2)$	$K_1 K_2$	$-K_1$	0	0	0
0	$-K_3(K_2 + K_3)$	$-K_2$	0	K_3^2	$K_2 K_3$
0	K_1	0	K_3	0	0
$-K_1 A$	0	A	$-K_1(K_1 + K_4)$	K_1^2	0
A	0	0	0	$K_2 + K_4$	$-K_2$
0	0	K_3	$K_3 A$	0	$K_4(K_3 + K_4)$

$$\Delta d = M \begin{vmatrix} -K_2(K_1+K_2) & K_1K_2 & K_2^2 & -K_1 & 0 & 0 \\ 0 & -K_3(K_2+K_3) & 0 & -K_2 & K_3^2 & K_2K_3 \\ 0 & K_1 & -(K_1+K_3) & 0 & 0 & 0 \\ -K_1A & 0 & 0 & A & K_1^2 & 0 \\ A & 0 & 0 & 0 & K_2+K_4 & -K_2 \\ 0 & 0 & K_4A & K_3 & 0 & K_4(K_3+K_4) \end{vmatrix}$$

$$\Delta = \begin{vmatrix} -K_2(K_1+K_2) & K_1K_2 & K_2^2 & 0 & 0 & 0 \\ 0 & -K_3(K_2+K_3) & 0 & 0 & K_3^2 & K_2K_3 \\ 0 & K_1 & -(K_1+K_3) & K_3 & 0 & 0 \\ -K_1A & 0 & 0 & -K_1(K_1+K_4) & K_1^2 & 0 \\ A & 0 & 0 & 0 & K_2+K_4 & -K_2 \\ 0 & 0 & K_4A & K_3A & 0 & K_4(K_3+K_4) \end{vmatrix}$$

$$\Delta a = \frac{\Delta}{K_1} M + \Delta b$$

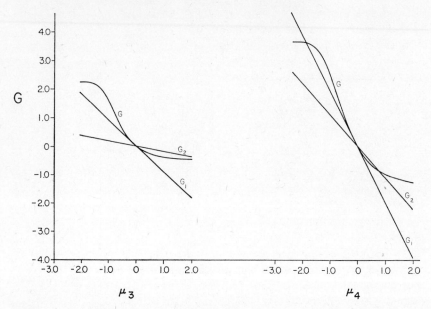

Fig. 26. Comparison of the nonlinear feedback function from the enzyme-catalyzed system in Fig. 21 (Scheme IX) with G_1 and G_2, calculated utilizing Aiserman's method.

H.   The Lur'e Algorithm

In this discussion of biochemical control mechanisms we have been concerned with regulated concentrations whose uncontrolled motion is described by a linear set of differential equations.

$$\dot{u}_2 = -K_2 u_2, \qquad \dot{u}_i = K_{i-1} u_{i-1} - K_i u_i, \quad i = 3, \ldots, n+1. \quad (6.85)$$

We have seen that the reason chemical systems of this type are inherently stable is that the rate constants are positive real numbers.

Equations 6.85 are a specification of the general set of linear differential equations:

$$\dot{u}_i = \sum_{j=2}^{n+1} B_{ij} u_j, \quad i = 2, \ldots, n+1, \quad (6.86)$$

where $B_{ii} = -K_i$, $B_{i-1,i} = K_i$ ($B_{12} = 0$), and all other $B_{ij} = 0$.

In this chapter we have assumed that the regulated concentrations are subject to the action of a regulating agent; in most cases we have described the motion of our biochemical actuator by means of the nonlinear first-order differential equation:

$$\dot{u}_1 + K_1 u_1 = G(u_{n+1}). \quad (6.87)$$

We have used the nonlinear function, $G(u_{n+1})$, to describe positive or negative feedback by S_{n+1}, the last chemical component in our systems. The complete chemical control system is therefore represented in the familiar form, equations 6.23. These equations are a specification of the general set of control equations:

$$\dot{u}_1 = G(u_{n+1}) - K_1 u_1,$$

$$\dot{u}_i = \sum_{j=2}^{n+1} B_{ij} u_j + b_i u_1, \quad i = 2, \dots, n+1. \tag{6.88}$$

In equations 6.88 the connection to the regulating agent is described by the constants, b_i $(i = 2, \dots, n+1)$. In the biochemical control systems described by equations 6.23, $b_2 = K_1$ and all the other $b_i = 0$ $(i = 3, \dots, n+1)$.

In most engineering problems it is impossible to rigorously define the function G; as a consequence it is also impossible to carry out the correct linearization of equations 6.88. Problems similar to those encountered in nonbiological systems should make it at least as difficult to establish the exact nature of the feedback function in biological control mechanisms. However, it should be possible to establish experimentally whether $G(u_{n+1})$ describes positive or negative feedback. In the method outlined below we will obtain sufficiency conditions for the asymptotic stability of the nonlinear control systems described by equations 6.23; the only knowledge about the feedback function that is required is whether it describes a positive- or a negative-feedback circuit.

The first and fundamental problem in control theory is to determine all of the values of the constants in the differential equations which are required to guarantee the stability of the stationary state. We have seen how this can be done in the case of linear systems, and how the linearization of systems limits the domain of the stability to the phase-space in the neighborhood of the stationary state. With the aid of the Lyapunov direct method it is possible to obtain stability criteria that are considerably stronger and, in certain cases, unbounded.

It is not essential to transform equations 6.23 into a canonical form to solve this problem in control theory. However, some of the difficulties of constructing unbounded positive-definite Lyapunov functions are avoided if a canonical form is used. In what follows our canonical variables will be determined with the aid of the equation

$$z_i = \sum_{j=2}^{n+1} C_{ij} u_j + u_1, \quad i = 2, \dots, n+1, \tag{6.89}$$

where the z_i are the canonical variables, and the C_{ij} are the transformation constants.

The transformation described in equations 6.89 is due to Lur'e (69); a more complete description of this transformation and its use in establishing sufficiency criteria in engineering control problems can be found in the excellent book by Letov (70). A discussion of the application of some of these methods to biological control mechanisms has been published elsewhere (52).

If we differentiate equations 6.89 and substitute the value for the $\dot{u}_j$ $(j = 1, \ldots, n + 1)$ from equations 6.88, we obtain

$$\dot{z}_i = \sum_{j=2}^{n+1} C_{ij} \left(\sum_{k=2}^{n+1} B_{ik} u_k + b_i u_1 \right) + G(u_{n+1}) - K_1 u_1, \quad i = 2, \ldots, n + 1. \quad (6.90)$$

If we choose the transformation constants according to

$$-\lambda_i C_{ij} = \sum_{k=2}^{n+1} B_{kj} C_{ik}, \quad i, j = 2, \ldots, n + 1, \quad (6.91)$$

$$K_1 - \lambda_i = \sum_{k=2}^{n+1} b_k C_{ik}, \quad i = 2, \ldots, n + 1, \quad (6.92)$$

the canonical equations will be in the form

$$\dot{z}_i = \lambda_i z_i + G(u_{n+1}), \quad i = 1, \ldots, n + 1. \quad (6.93)$$

The λ_i in equations 6.91–6.93 are the transformation parameters; these constants must be chosen to be the roots of

$$\begin{vmatrix} B_{22} + \lambda & B_{32} & \cdots & B_{n+1,2} \\ \cdots & \cdots & \cdots & \cdots \\ B_{2,n+1} & B_{3,n+1} & \cdots & B_{n+1,n+1} + \lambda \end{vmatrix} = 0. \quad (6.94)$$

For the chemical systems described by equations 6.23 the λ are chosen to be the roots of

$$\begin{vmatrix} -K_2 + \lambda & 0 & \cdots & 0 & 0 \\ K_2 & -K_3 + \lambda & \cdots & 0 & 0 \\ \cdots & \cdots & \cdots & \cdots & \cdots \\ 0 & 0 & \cdots & K_n & -K_{n+1} + \lambda \end{vmatrix} = 0. \quad (6.95)$$

If we expand this determinant, we obtain

$$\prod_{i=2}^{n+1} (\lambda - K_i) = 0. \quad (6.96)$$

Since the K_i are positive real numbers, all the roots of equation 6.96 are also positive and real. In fact,

$$\lambda_i = K_i, \quad i = 2, \ldots, n + 1 \tag{6.97}$$

for the chemical systems described by equations 6.23.

From equations 6.97 and the definition of the B_{ij} and the b_i, it is clear that we must choose the transformation constants such that

$$-K_i C_{ij} = -K_j C_{ij} + K_j C_{i,j+1}, \quad C_{i,n+2} = 0, \quad i, j = 2, \ldots, n + 1, \tag{6.98}$$

$$K_1 - K_i = K_1 C_{i,1}, \quad i = 2, \ldots, n + 1, \tag{6.99}$$

in order to reduce the original differential equations to the desired canonical form:

$$\dot{z}_i = -K_i z_i + G(u_{n+1}), \quad i = 1, \ldots, n + 1. \tag{6.100}$$

Since the roots of equation 6.96 are simple, equations 6.91 can always be solved if the K_i are different; this assures that the transformation described in equations 6.89 exists, and that it is not singular. This latter circumstance is important because it ensures that we can return to equations 6.89 and solve them for the original variables, u_i ($i = 2, \ldots, n + 1$). In particular we need to express u_{n+1}, the argument of the nonlinear function, in terms of the canonical variables:

$$u_{n+1} = \begin{vmatrix} C_{22} & C_{23} & \cdots & z_2 - z_1 \\ C_{32} & C_{33} & \cdots & z_3 - z_1 \\ \cdots & \cdots & \cdots & \cdots \\ C_{n+1,2} & C_{n+1,3} & \cdots & z_{n+1} - z_1 \end{vmatrix} = \sum_{i=1}^{n+1} D_i z_i, \tag{6.101}$$

where $z_1 = u_1$ and the D_i are known constants. Differentiating equation 6.101 with respect to time, we obtain

$$\dot{u}_{n+1} = \sum_{i=1}^{n+1} D_i \dot{z}_i. \tag{6.102}$$

Substitution from equations 6.100 yields

$$\dot{u}_{n+1} = \sum_{i=1}^{n+1} - D_i[K_i z_i + G(u_{n+1})] = -\sum_{i=1}^{n+1} D_i K_i z_i - G(u_{n+1}) \sum_{i=1}^{n+1} D_i. \tag{6.103}$$

If we define $L_i = D_i K_i$ ($i = 1, \ldots, n + 1$) and $r = \sum_{i=1}^{n+1} D_i$, the collective canonical equations assume the form

$$\dot{z}_i = -K_i z_i + G(u_{n+1}), \quad i = 1, \ldots, n + 1,$$

$$u_{n+1} = \sum_{i=1}^{n+1} D_i z_i, \tag{6.104}$$

$$\dot{u}_{n+1} = \sum_{i=1}^{n+1} L_i z_i - r G(u_{n+1}).$$

The last equation is not essential, but it will be useful in some of our forth-coming applications.

For chemical control systems described by equations 6.23,

$$D_i = \frac{-\prod_{j=1}^{n} K_j}{\prod_{j=1}^{n+1} (K_i - K_j), \quad i \neq j}, \quad i = 1, \ldots, n + 1, \tag{6.105}$$

$$L_i = \frac{K_i \prod_{j=1}^{n} K_j}{\prod_{j=1}^{n+1} (K_i - K_j), \quad i \neq j}, \quad i = 1, \ldots, n + 1, \tag{6.106}$$

$$r = 0. \tag{6.107}$$

If we substitute these quantities into equations 6.104, the collective canonical equations assume the final form:

$$\dot{z}_i = -K_i z_i + G(u_{n+1}), \quad i = 1, \ldots, n + 1,$$

$$u_{n+1} = \sum_{i=1}^{n+1} \frac{-\prod_{j=1}^{n} K_j}{\prod_{j=1}^{n+1} (K_i - K_j), \quad i \neq j} z_i, \tag{6.108}$$

$$\dot{u}_{n+1} = \sum_{i=1}^{n+1} \frac{K_i \prod_{j=1}^{n} K_j}{\prod_{j=1}^{n+1} (K_i - K_j), \quad i \neq j} z_i.$$

Equations 6.23 have the singular point, $u_i = 0$ $(i = 1, \ldots, n + 1)$; this solution corresponds to the positive real solution, $s_i = s_i^*$ $(i = 1, \ldots, n + 1)$, of equations 6.3. The corresponding singular point for the canonical equations is $z_i = 0$ $(i = 1, \ldots, n + 1)$. All three of these singularities have a one-to-one correspondence; the asymptotic stability of the singularity

$z_i = 0$ $(i = 1, \ldots, n + 1)$ assures that the other singularities are also asymptotically stable. The one-to-one correspondence of these singular points can be established through the transformation equations themselves. For this purpose we next consider methods for obtaining the inverse Lur'e transformation.

We may write equations 6.89 in the form

$$\sum_{j=2}^{n+1} E_{kj}(\lambda_i)u_j = \frac{H_k(\lambda_i)}{K_1 - \lambda_i}(z_i - u_1), \quad i = 2, \ldots, n + 1, \qquad (6.109)$$

where $E_{kj}(\lambda_i)$ are cofactors of the elements of a certain kth row of the determinant in equations 6.95, and $H_k(\lambda_i)$ represents the determinant in equations 6.95 except that the elements of the kth row are replaced by K_1 in the first column and by zeros in the other columns.

Following the procedure outlined by Lur'e (69), we consider the determinant:

$$\Delta(\lambda_2, \ldots, \lambda_{n+1}) = \begin{vmatrix} E_{k,2}(\lambda_2) & \cdots & E_{k,k}(\lambda_2) & \cdots & E_{k,n+1}(\lambda_2) \\ \cdots & \cdots & \cdots & \cdots & \cdots \\ E_{k,2}(\lambda_{n+1}) & \cdots & E_{k,k}(\lambda_{n+1}) & \cdots & E_{k,n+1}(\lambda_{n+1}) \end{vmatrix}$$

$$(6.110)$$

From Cramer's rule,

$$u_j = \frac{\Delta_j}{\Delta}, \quad j = 2, \ldots, n + 1. \qquad (6.111)$$

We now let Δ_{ij} represent the cofactors of the elements of the ith row and the jth column of the determinant in equation 6.110. Substitution into equation 6.111 yields

$$\Delta_j = \sum_{i=2}^{n+1} \frac{H_k(\lambda_i)}{K_1 - \lambda_i}(z_i - u_1)\Delta_{ij}, \quad j = 2, \ldots, n + 1. \qquad (6.112)$$

Lur'e proved that

$$\Delta(\lambda_2, \ldots, \lambda_{n+1}) = E'(\lambda_i)\Delta_{ij} \equiv \frac{\partial E}{\partial \lambda}\Delta_{ij}. \qquad (6.113)$$

Substitution from equations 6.112 and 6.113 into equations 6.111 yields

$$u_j = \sum_{i=2}^{n+1} \frac{H_k(\lambda_i)(z_i - u_1)}{(K_1 - \lambda_i)E'(\lambda_i)}, \quad j = 2, \ldots, n + 1,$$

$$(6.114)$$

$$u_1 = z_1.$$

Equations 6.114 provide the basis for the inverse Lur'e transformation and can be used to express the results of stability studies of the canonical equations in terms of the variables in equations 6.23. We may then express our results in terms of chemical concentrations by utilizing the transformation $s_i = s_i^* + u_i$ ($i = 1, \ldots, n+1$).

Methods for the construction of Lyapunov functions for the Lur'e-transformed differential equations have been described elsewhere (70). In what follows, this general method is applied to the chemical and biological control systems described by equations 6.3. To begin, we consider the Lyapunov function:

$$V = \varphi_1 + \varphi_2 \pm \int_0^{u_{n+1}} G(u_{n+1})\,du_{n+1}, \tag{6.115}$$

where

$$\varphi_1 = \sum_{i=1}^{n+1} \sum_{j=1}^{n+1} \frac{a_i a_j}{K_i + K_j}\, z_i z_j, \tag{6.116}$$

$$\varphi_2 = \tfrac{1}{2} \sum_{i=2}^{n+1} A_i z_i^2. \tag{6.117}$$

From the obvious equality

$$\frac{1}{K_i + K_j} = \int_0^\infty e^{-(K_i + K_j)\tau}\, d\tau \tag{6.118}$$

and equation 6.116, we obtain

$$\varphi_1 = \int_0^\infty \left(\sum_{j=1}^{n+1} a_j z_j e^{-K_j \tau} \right)^2 d\tau. \tag{6.119}$$

The integrand in equation 6.119 is the square of a real number provided the a_j are real numbers. Consequently, for any real a_j ($j = 1, \ldots, n+1$) φ_1 is positive-invariant for all values of the canonical variables. Furthermore, if the rate constants are all different, φ_1 cannot be zero except when $z_i = 0$ ($i = 1, \ldots, n+1$). Thus, if all the rate constants are different and all the a_j are real numbers, φ_1 is a positive-definite function of the canonical variables.

For the control systems we are considering, φ_2 is a positive-definite function if the A_i are positive real numbers. However, the A_i may be as small as desired.

For certain nonlinear functions (including functions describing positive feedback),

$$G(u_{n+1}) = 0 \quad \text{for } |u_{n+1}| \leq u_{n+1}^*; \qquad G(u_{n+1})u_{n+1} > 0 \quad \text{for } |u_{n+1}| > u_{n+1}^*. \tag{6.120}$$

However, for other nonlinear functions (including functions describing negative feedback),

$$G(u_{n+1}) = 0 \quad \text{for } |u_{n+1}| \leq u_{n+1}^*; \qquad G(u_{n+1})u_{n+1} < 0 \quad \text{for } |u_{n+1}| > u_{n+1}^*. \tag{6.121}$$

Thus, in what follows, if G describes positive feedback or otherwise fulfills the conditions in 6.120, we use the plus sign in equation 6.115 to arrive at the conditions described in inequalities 6.120; if G describes negative feedback, we use the minus sign and arrive at the same conditions. Satisfaction of the conditions in inequalities 6.120 ensures that the integral

$$\pm \int_0^{u_{n+1}} G(u_{n+1})\, du_{n+1} > 0 \tag{6.122}$$

and that

$$\lim_{u_{n+1} \to \infty} \int_0^{u_{n+1}} \pm G(u_{n+1})\, du_{n+1} = \infty. \tag{6.123}$$

The plus signs are used in inequality 6.122 and equation 6.123 if G describes positive feedback or otherwise fulfills the conditions in inequalities 6.120; the minus signs are used if G satisfies the conditions in inequalities 6.121.

It is clear from equations 6.117, 6.119, 6.122, and 6.123 that the Lyapunov function described in equation 6.115 is positive-definite if we choose the plus sign for positive feedback functions and the minus sign for systems involving negative feedback. In either case equation 6.123 guarantees that the Lyapunov function approaches infinity as u_{n+1} approaches infinity. The validity of equation 6.123 enables us to investigate the global asymptotic stability of control systems described by equations 6.3.

In view of these properties of V and Theorem 9 we may conclude that the singularity $s_i = s_i^*$ $(i = 1, \ldots, n+1)$ possesses global asymptotic stability if the Eulerian derivative of V is negative-definite:

$$\dot{V} = \dot{\varphi}_1 + \dot{\varphi}_2 \pm G(u_{n+1})\dot{u}_{n+1}. \tag{6.124}$$

Differentiating equations 6.119 and 6.120 with respect to time, we obtain

$$\dot{V} = \sum_{i=1}^{n+1} A_i z_i \dot{z}_i + \sum_{i=1}^{n+1} \sum_{j=1}^{n+1} \frac{a_i a_j}{K_i + K_j} (z_i \dot{z}_j + z_j \dot{z}_i) \pm G(u_{n+1})\dot{u}_{n+1}. \tag{6.125}$$

Substitution of the derivatives from equation 6.108 into equation 6.125 yields

$$\dot{V} = -\sum_{i=1}^{n+1} A_i K_i z_i^2 - \sum_{i=1}^{n+1} \sum_{j=1}^{n+1} \frac{a_i a_j}{K_i + K_j} (K_j z_i z_j + K_i z_i z_j)$$

$$+ G(u_{n+1}) \left[\sum_{i=1}^{n+1} A_i z_i + \sum_{i=1}^{n+1} \sum_{j=1}^{n+1} \frac{a_i a_j}{K_i + K_j} (z_i + z_j) \right.$$

$$\left. \pm \sum_{i=1}^{n+1} \frac{K_i \prod_{j=1}^{n} K_j}{\prod_{j=1}^{n+1} (K_i - K_j)} z_i, \quad i \neq j \right]. \tag{6.126}$$

By taking into account that

$$\sum_{i=1}^{n+1} \sum_{j=1}^{n+1} \frac{a_i a_j}{K_i + K_j} (K_j z_i z_j + K_i z_i z_j) = \sum_{i=1}^{n+1} \sum_{j=1}^{n+1} a_i a_j z_i z_j = \left(\sum_{i=1}^{n+1} a_i z_i \right)^2 \quad (6.127)$$

and

$$\sum_{i=1}^{n+1} \sum_{j=1}^{n+1} \frac{a_i a_j}{K_i + K_j} (z_i + z_j) = 2 \sum_{i=1}^{n+1} \sum_{j=1}^{n+1} \frac{a_i a_j}{K_i + K_j} z_j, \quad (6.128)$$

we can reduce equations 6.126 to

$$\dot{V} = - \sum_{i=1}^{n+1} A_i K_i z_i^2 - \left(\sum_{i=1}^{n+1} a_i z_i \right)^2$$

$$+ G(u_{n+1}) \sum_{j=1}^{n+1} \left[A_j \pm \frac{K_j \prod_{i=1}^{n} K_i}{\prod_{i=1}^{n+1} (K_j - K_i), \quad i \neq j} + 2 \sum_{i=1}^{n+1} \frac{a_i}{K_i + K_j} \right] z_j.$$

$$(6.129)$$

The first two terms in equation 6.129,

$$- \sum_{i=1}^{n+1} A_i K_i z_i^2 \quad \text{and} \quad - \left(\sum_{i=1}^{n+1} a_i z_i \right)^2,$$

are obviously negative-definite. Thus $\dot{V}$ will be negative-definite if the last terms in equation 6.129 are nonpositive:

$$A_j \pm \frac{K_j \prod_{i=1}^{n} K_i}{\prod_{i=1}^{n+1} (K_j - K_i), \quad i \neq j} + 2 \sum_{i=1}^{n+1} \frac{a_i}{K_i + K_j} = 0, \quad j = 1, \dots, n + 1.$$

$$(6.130)$$

Since

$$\left| \frac{\prod_{i=1}^{n} K_i}{\prod_{i=1}^{n+1} (K_j - K_i), \quad i \neq j} \right| > 1, \quad j = 1, \dots, n + 1, \quad (6.131)$$

we may select positive A_j $(j = 1, \dots, n + 1)$ such that

$$A_j \ll K_j, \quad j = 1, \dots, n + 1. \quad (6.132)$$

By selecting the A_j in this manner, we simplify equations 6.130 to

$$\pm \frac{K_j \prod_{i=1}^{n} K_i}{\prod_{i=1}^{n+1} (K_j - K_i), \quad i \neq j} + 2 \sum_{i=1}^{n+1} \frac{a_i}{K_i + K_j} = 0, \quad j = 1, \dots, n + 1. \quad (6.133)$$

When equations 6.133 are satisfied, the negative-definite Eulerian derivative is

$$\dot{V} = -\sum_{i=1}^{n+1} A_i K_i z_i^2 - \left(\sum_{i=1}^{n+1} a_i z_i\right)^2. \tag{6.134}$$

If the $n+1$ relations described by equations 6.133 are satisfied by real a_i $(i = 1, \ldots, n+1)$, it is possible to construct a sign-definite Lyapunov function, V, having a sign-definite Eulerian derivative of opposite sign to V. In this case Theorem 9 can be utilized to guarantee that the singularity $s_i = s_i^*$ $(i = 1, \ldots, n+1)$ possesses global asymptotic stability. Thus satisfaction of equations 6.133 by real numbers for the a_i $(i = 1, \ldots, n+1)$ guarantees that sustained oscillations will not arise in the nonlinear biochemical control systems described by equations 6.3.

The canonical forms of the control equations for systems involving two components $(n = 1)$ are

$$\dot{z}_1 = G(u_2) - K_1 z_1,$$

$$\dot{z}_2 = G(u_2) - K_2 z_2,$$

$$u_2 = \frac{K_1}{K_1 - K_2}(z_2 - z_1), \tag{6.135}$$

$$\dot{u}_2 = \frac{K_1}{K_1 - K_2}(K_1 z_1 - K_2 z_2).$$

In this case there are two sufficiency criteria for the global asymptotic stability of systems involving negative feedback:

$$-\frac{K_1^2}{K_1 - K_2} + 2a_1\left(\frac{a_1}{2K_1} + \frac{a_2}{K_1 + K_2}\right) = 0, \tag{6.136}$$

$$\frac{K_1 K_2}{K_1 - K_2} + 2a_2\left(\frac{a_2}{2K_2} + \frac{a_1}{K_1 + K_2}\right) = 0. \tag{6.137}$$

If we divide equation 6.136 by K_1 and equation 6.137 by K_2 and add the results, we obtain

$$\frac{a_1^2}{K_1^2} + \frac{2a_1 a_2}{K_1 K_2} + \frac{a_2^2}{K_2^2} = 0 \tag{6.138}$$

or

$$\left(\frac{a_1}{K_1} + \frac{a_2}{K_2}\right)^2 = 0. \tag{6.139}$$

From equation 6.139 we obtain the condition

$$a_1 = -\frac{K_1 a_2}{K_2}, \tag{6.140}$$

which can always be satisfied by real a_1 and a_2.

Addition of equations 6.136 and 6.137 yields

$$\frac{a_1^2}{K_1} + \frac{4a_1 a_2}{K_1 + K_2} + \frac{a_2^2}{K_2} = 2\varphi_1(a_1, a_2) = K_1. \tag{6.141}$$

Since φ_1 is a positive-definite function (equation 6.119), K_1 must also be positive. Thus we have deduced the condition

$$K_1 > 0 \tag{6.142}$$

as a requirement for global asymptotic stability of the negative-feedback system involving two components.

If we multiply equation 6.136 by K_1 and equation 6.137 by K_2 and add the result, we obtain

$$(a_1 + a_2)^2 = K_1(K_1 + K_2). \tag{6.143}$$

Since $(a_1 + a_2)^2$ is a positive-invariant function, $K_1(K_1 + K_2)$ must be positive. Thus we have deduced the condition

$$K_1 + K_2 > 0 \tag{6.144}$$

as an additional requirement for the global asymptotic stability of the negative-feedback systems involving two components.

Equations 6.140 and 6.143 can be solved directly for a_1 and a_2:

$$a_{1,2} = \pm \frac{K_1\sqrt{K_1(K_1 + K_2)}}{K_1 - K_2}. \tag{6.145}$$

Since K_1 and K_2 are always positive real numbers, the sufficiency criteria, $K_1 > 0$ and $K_1 + K_2 > 0$, will always be satisfied because $a_{1,2}$ in equation 6.145 is always real in chemical control systems. Thus the negative chemical feedback circuits involving two components and described by equations 6.3 will always possess global asymptotic stability. This result is consistent with Theorem 21.

The two sufficiency criteria for systems involving positive feedback are

$$\frac{K_1^2}{K_1 - K_2} + 2a_1\left(\frac{a_1}{2K_1} + \frac{a_2}{K_1 + K_2}\right) = 0, \tag{6.146}$$

$$-\frac{K_1 K_2}{K_1 - K_2} + 2a_2\left(\frac{a_2}{2K_2} + \frac{a_1}{K_1 + K_2}\right) = 0. \tag{6.147}$$

If we divide equation 6.146 by K_1 and equation 6.147 by K_2 and add the results, we obtain the same relationship (equation 6.140) as we found for the case involving negative-feedback functions. However, addition of equations 6.146 and 6.147 yields

$$2\varphi_1(a_1, a_2) = -K_1. \tag{6.148}$$

Since φ_1 is a positive-definite function (equation 6.119), K_1 must be negative. Thus we have deduced the condition

$$K_1 < 0 \tag{6.149}$$

as a requirement for the global asymptotic stability of the positive-feedback systems involving two components. Obviously, this condition is never met in chemical systems since K_1 is a nonnegative first-order rate constant. This means that we can never guarantee that the chemical control systems involving two components and described by equations 6.3 will possess global asymptotic stability when the nonlinear function describes positive feedback (or otherwise satisfies the conditions of inequalities 6.120). This result is also consistent with Theorem 21.

These results of investigations (52) using the direct Lyapunov method illustrate that there is a fundamental difference between two-component control systems for which the feedback function satisfies the conditions in inequalities 6.120 and similar systems for which inequalities 6.121 are relevant. Clearly there is an important difference between the stability properties and the potential for oscillatory behavior for the two possible classes of feedback functions. If the feedback function describes negative feedback (or otherwise satisfies inequalities 6.121), the biochemical control systems involving two components and described by equations 6.3 cannot be unstable, and they cannot exhibit limit cycle behavior. However, if the feedback function describes positive feedback (or otherwise satisfies the conditions in inequalities 6.120), a physically important singularity of the biochemical control mechanism involving two components and described by equations 6.3 can be unstable, and the systems can exhibit limit cycle behavior. Whether or not instabilities and limit cycles occur will depend on the nature of the positive-feedback function, but such behavior is possible for any set of values of the first-order rate constants (K_1 and K_2) in equations 6.3.

The conditions for global asymptotic stability of the two-component biochemical control systems can be compared to the conditions for which the corresponding linear system is asymptotically stable. For $n = 1$ the first Hurwitz condition (inequalities 6.33) is identical to inequality 6.144. For systems involving negative feedback $A < 0$, and the second Hurwitz condition can be written in the form:

$$K_1 > 0, \qquad K_1 K_2 > 0. \tag{6.150}$$

However, if K_1 and K_2 must be of the same sign, we know from inequality 6.144 that they must both be positive. The first condition in inequalities 6.150 expresses the fact that K_1 must be positive and is identical to the condition derived from the direct Lyapunov method for the nonlinear system and described in inequality 6.142.

For systems involving positive feedback $A > 0$, and the second Hurwitz condition (inequalities 6.33) can be written in the form:

$$K_1 < 0 \quad \text{if} \quad K_2 < A. \tag{6.151}$$

The condition in inequality 6.151 is identical to the condition described in inequality 6.149.

The canonical forms of the control equations for chemical systems involving three components ($n = 2$) are

$$\dot{z}_1 = G(u_3) - K_1 z_1,$$

$$\dot{z}_2 = G(u_3) - K_2 z_2,$$

$$\dot{z}_3 = G(u_3) - K_3 z_3,$$

$$u_3 = \frac{K_1 K_2}{(K_1 - K_2)(K_1 - K_3)(K_2 - K_3)} \tag{6.152}$$
$$\times [(K_3 - K_2)z_1 + (K_1 - K_3)z_2 + (K_2 - K_1)z_3],$$

$$\dot{u}_3 = \frac{K_1 K_2}{(K_1 - K_2)(K_1 - K_3)(K_2 - K_3)}$$
$$\times [K_1(K_2 - K_3)z_1 + K_2(K_3 - K_1)z_2 + K_3(K_1 - K_2)z_3].$$

The three sufficiency criteria for systems involving negative feedback are

$$-\frac{K_1^2 K_2}{(K_1 - K_2)(K_1 - K_3)} + 2a_1\left(\frac{a_1}{2K_1} + \frac{a_2}{K_1 + K_2} + \frac{a_3}{K_1 + K_3}\right) = 0, \tag{6.153}$$

$$-\frac{K_1 K_2^2}{(K_2 - K_1)(K_2 - K_3)} + 2a_2\left(\frac{a_2}{2K_2} + \frac{a_1}{K_1 + K_3} + \frac{a_3}{K_2 + K_3}\right) = 0, \tag{6.154}$$

$$-\frac{K_1 K_2 K_3}{(K_3 - K_1)(K_3 - K_2)} + 2a_3\left(\frac{a_3}{2K_3} + \frac{a_1}{K_1 + K_3} + \frac{a_2}{K_2 + K_3}\right) = 0. \tag{6.155}$$

In this case we multiply equation 6.153 by K_1, equation 6.154 by K_2, and equation 6.155 by K_3. Addition of the resulting equations yields the condition

$$K_1 K_2 > 0, \tag{6.156}$$

which must be met if the a_i ($i = 1, 2, 3$) in equations 6.153–6.155 are real numbers.

If, on the other hand, we divide equation 6.153 by K_1, equation 6.154 by K_2, and equation 6.155 by K_3 and add the resulting equations, we obtain

$$\frac{a_1^2}{K_1^2} + \frac{2a_1 a_2}{K_1 K_2} + \frac{2a_1 a_3}{K_1 K_3} + \frac{a_2^2}{K_2^2} + \frac{2a_2 a_3}{K_2 K_3} + \frac{a_3^2}{K_3^2} = 0. \tag{6.157}$$

Equation 6.157 is the perfect square:

$$\left(\frac{a_1}{K_1} + \frac{a_2}{K_2} + \frac{a_3}{K_3}\right)^2 = 0. \tag{6.158}$$

Thus we have the relation

$$-a_1 = \frac{K_1}{K_2} a_2 + \frac{K_1}{K_3} a_3, \tag{6.159}$$

which can always be satisfied by real a_1, a_2, and a_3.

Addition of equations 6.153, 6.154, and 6.155 yields

$$-K_1 K_2 \left[\frac{K_1}{(K_1 - K_2)(K_1 - K_3)} + \frac{K_2}{(K_2 - K_1)(K_2 - K_3)} \right.$$

$$\left. + \frac{K_3}{(K_3 - K_1)(K_3 - K_2)} \right] + 2\varphi_1(a_1, a_2, a_3) = 0. \tag{6.160}$$

Equations 6.159 and 6.160 provide another relationship among a_1, a_2, and a_3:

$$\left(\frac{a_2}{a_3}\right)^2 + 2 \frac{K_2(K_1 - K_3)[K_2(K_1 + K_3) + K_3(K_1 + K_2)]}{K_3(K_1 - K_2)(K_1 + K_3)(K_2 + K_3)} \frac{a_2}{a_3}$$

$$+ \frac{(K_1 - K_3)^2 K_2^2 (K_1 + K_2)}{(K_1 - K_2)^2 K_3^2 (K_1 + K_3)} = 0. \tag{6.161}$$

From equations 6.161 we obtain for a_2/a_3:

$$\frac{a_2}{a_3} = -\frac{K_2(K_1 - K_3)}{K_3(K_1 - K_2)} \left\{ \frac{K_2(K_1 + K_3) + K_3(K_1 + K_2)}{(K_1 + K_3)(K_2 + K_3)} \right.$$

$$\left. \pm \sqrt{\left[\frac{K_2(K_1 + K_3) + K_3(K_1 + K_2)}{(K_1 + K_3)(K_2 + K_3)}\right]^2 - \frac{K_1 + K_2}{K_1 + K_3}} \right\}. \tag{6.162}$$

According to equation 6.162 we are still free to choose real a_1, a_2, and a_3 provided

$$[K_2(K_1 + K_3) + K_3(K_1 + K_2)]^2 > (K_1 + K_3)(K_1 + K_2)(K_2 + K_3)^2. \tag{6.163}$$

Inequality 6.163 can be written in the form

$$(K_3 - K_2)[K_2{}^2(K_1 + K_3) - K_3{}^2(K_1 + K_2)] > 0. \qquad (6.164)$$

Since the conditions in inequality 6.164 can never be satisfied by real positive K_1, K_2, and K_3,* the sufficiency criteria for global asymptotic stability can never be satisfied by real a_i $(i = 1, 2, 3)$ for chemical systems involving negative feedback. This means that limit cycle behavior cannot be excluded on this basis for any set of K_1, K_2, and K_3 for certain negative-feedback functions (e.g., the function in equation 6.2 when $\rho > 8$).

For three-component systems described by equations 6.3 and involving positive feedback, the condition in inequality 6.164 remains the same, and the condition in inequality 6.156 is inverted. Thus for positive-feedback systems we have the requirements

$$(K_3 - K_2)[K_2{}^2(K_1 + K_3) - K_3{}^2(K_1 + K_2)] > 0, \qquad (6.165)$$

$$K_1 K_2 < 0. \qquad (6.166)$$

Neither of these conditions is ever satisfied if K_1, K_2, and K_3 are rate constants. This means that the sufficiency criteria for global asymptotic stability of the control systems involving three components and positive feedback cannot be satisfied by real a_i $(i = 1, 2, 3)$.

Conclusions similar to those obtained for systems involving three components and positive or negative feedback have been formulated (52) for systems involving more than three components.

These results from the direct Lyapunov method illustrate that there is a fundamental difference between two-component and three-component

---

* If, for example, $K_3 > K_2$ we have that

$$K_2{}^2(K_1 + K_3) > K_3{}^2(K_1 + K_2)$$

or

$$K_1 K_2{}^2 + K_2{}^2 K_3 > K_1 K_3{}^2 + K_2 K_3{}^2.$$

According to our original proposition the first term on the left-hand side of the inequality is less than the first term on the right, and the last term on the left is also less than the last term on the right. Thus the left-hand side of the inequality cannot be larger than the right side (the inequality cannot be satisfied). On the other hand, if $K_2 > K_3$, we have that

$$K_1 K_3{}^2 + K_2 K_3{}^2 > K_1 K_2{}^2 + K_2{}^2 K_3.$$

Again, according to the original proposition the individual terms on the left are less than the corresponding terms on the right. Thus the left side of the inequality cannot be larger than the right side (the inequality cannot be satisfied). Finally, if $K_2 = K_3$, the left side of inequality 6.164 is zero and the inequality cannot be satisfied. Consequently, regardless of whether K_3 is greater than, less than, or equal to K_2, inequality 6.164 can never be satisfied.

systems involving negative feedback. Clearly, there is an important difference between the stability properties and the potential for oscillatory behavior in negative-feedback systems involving more than two, rather than just two, components. A biochemical control system that involves negative feedback and two components cannot be unstable and cannot exhibit limit cycle behavior. However, if the system involves three or more components, the only physically important singularity of the control mechanism can be unstable, and the system can exhibit limit cycle behavior. Whether or not instabilities and limit cycles occur in a system involving three or more components will depend on the nature of the feedback function, G, but such behavior is possible for any set of values of the first-order rate constants, K_i ($i = 1, \ldots, n + 1$), in equations 6.3.

The results of these investigations are summarized in Theorem 22:

> **Theorem 22.** *For control systems described by equation 6.3, there is a fundamental difference between mechanisms involving two components and feedback functions described by inequalities 6.121 and mechanisms involving more than two components or feedback functions described by inequalities 6.120. Two-component systems involving feedback of the type described by inequalities 6.121 cannot be unstable or exhibit limit cycle behavior, but systems involving more components or feedback functions of the type described by inequalities 6.120 can exhibit limit cycle behavior regardless of the values of the K_i ($i = 1, \ldots, n + 1$) in equations 6.3.*

I. The Occurrence of Limit Cycle Behavior in Controlled Biochemical Systems

In order to obtain an idea of when limit cycles actually arise in control systems described by equations 6.2 and 6.3, I (54) have obtained the numerical solution for these equations for a wide range of n, ρ, C_0, α, and K_{n+1}. All of these solutions were obtained for the case in which $K_i = 1$ ($i = 1, 2, \ldots, n$). The results of over 2500 simulations representing most combinations of all possible (integer) values of $n = 2$ to $n = 7$, all possible (integer) values of $\rho = 1$ to $\rho = 10$, three values of α in the range 0.01–1.0, three values of C_0 in the range 0.5–50, and six values of K_{n+1} in the range 0.05–10 are summarized in Fig. 27. In this figure the squares at the intersection of specific values of $C_0 = 0.5$, 5.0, or 50 (vertical axis) and specific values of $K_{n+1} = 0.05$, 0.1, 0.5, 1.0, 5.0, or 10 (horizontal axis) are filled in if a limit cycle arises in the control systems described by equations 6.2 and 6.3. If the square at the intersection is filled in with slashes leaning to the left, the limit cycle was found only when $\alpha = 1.0$; if the intersection square is filled with

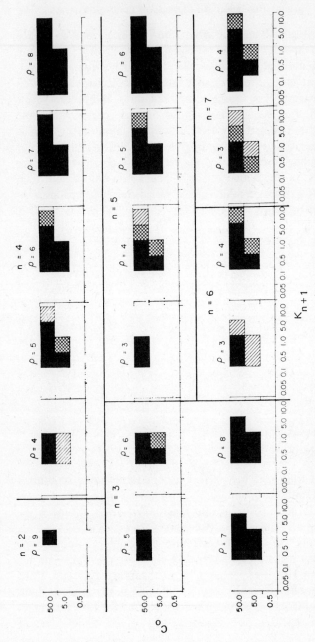

Fig. 27. Relationship between the parameter values in an enzyme-catalyzed system with negative feedback (Scheme IX) and the occurrence of limit cycle behavior.

slashes leaning to the left and the right (diamonds), limit cycles were found when $\alpha = 1.0$ and $\alpha = 0.1$; if the intersection square is filled in with solid darkening, limit cycles were found when $\alpha = 1.0$, $\alpha = 0.1$, and $\alpha = 0.01$. If the intersection is not filled, the systems involving the indicated values of C_0 and K_{n+1} were simulated for $\alpha = 1.0$, $\alpha = 0.1$, and $\alpha = 0.01$ and the indicated values of n and ρ, but a limit cycle did not arise. In each of these cases equations 6.2 and 6.3 possess global asymptotic stability. The values of n and ρ used in each case are identified as inserts in the figure; for each value of n all 54 possible combinations of the indicated values of C_0, K_{n+1}, and α were simulated for all possible (integer) values of ρ less than those indicated on the figure. The results obtained from these lower values of ρ were that no limit cycles arose, and equations 6.2 and 6.3 possessed global asymptotic stability.

Several important generalizations can be drawn from Fig. 27. First, if a limit cycle occurs for a set of values of n, ρ, α, and C_0, then a limit cycle will also occur for larger values of any one of these constants provided none of the remaining constants is decreased. Second, at higher C_0 limit cycles are most likely to occur when $K_{n+1} = K_i$ $(i = 1, 2, \ldots, n)$, but at lower C_0 limit cycles are more likely to occur when $K_{n+1} < K_i$ $(i = 1, 2, \ldots, n)$. Third, limit cycles do not arise unless both C_0 and K_{n+1} are sufficiently large. Finally, Fig. 27 indicates that the result reported by Viniegra and Martinez (65) might be correct for a wider set of conditions than they actually considered. No limit cycles were found for any set of α, C_0, and K_{n+1} unless the conditions indicated in Table 2 were satisfied.

In Fig. 28 appears the relationship between the amplitude (A_1) of the oscillation of the biochemical variable (S_1), the period of the oscillation, and the number of variables in equations 6.2 and 6.3. In Fig. 28b (both curves) and Fig. 28a $\rho = 9$, $\alpha = 1.0$, and $C_0 = 50$; in Fig. 28b (curve I) and Fig. 28a $K_{n+1} = 1.0$, but in curve II in Fig. 28b $K_{n+1} = 0.5$. The corresponding concentrations of S_1 at the singularity are $s_1^* = 1.475$ when $K_{n+1} = 1.0$ and $s_1^* = 0.7915$ when $K_{n+1} = 0.5$. As can be seen from Fig. 28, both the amplitude and the period increase as the number of components in the control system increases. The near linearity of the relationships at lower amplitudes and periods can be used to extrapolate to zero amplitude or period to obtain an idea of the minimum number of variables for which oscillations occur. In Figs. 28a and 28b $(K_{n+1} = 1.0)$ this extrapolation intersects the horizontal axis between two and three components; in Fig. 28b, curve II $(K_{n+1} = 0.5)$, the extrapolation intersects the horizontal axis between three and four components. This result corresponds to the fact that oscillations were not observed for these systems when $K_{n+1} = 0.5$ unless there were at least four components, but exactly the same systems could exhibit limit cycle behavior when there were only three components and $K_{n+1} = 1.0$.

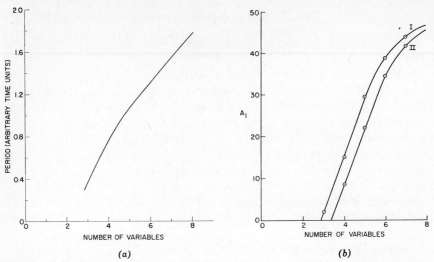

Fig. 28. Relationship between the amplitude and period of oscillations of S_1 in Scheme IX and the number of variables in equations 6.2 and 6.3.

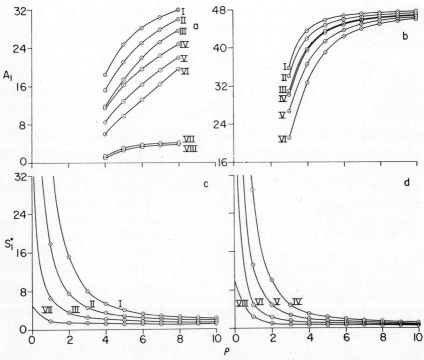

Fig. 29. Relationship between the amplitude of oscillations of S_1 in Scheme IX, steady-state level of S_1, and the stoichiometry of the feedback inhibition described in Equations 6.2 and 6.3.

472

In Fig. 29 appears the relationship between A_1, s_1^*, and ρ. From Figs. 29a and 29b it is clear that A_1 is a continuously increasing function of ρ, but Figs. 29c and 29d illustrate that s_1^* is a continuously decreasing function of ρ. In all four figures $\alpha = 0.01$ in curves I and IV, $\alpha = 0.1$ in curves II and V, $\alpha = 1.0$ in curves III, VI, VII, and VIII; $C_0 = 50$ in curves I–VI, $C_0 = 5.0$ in curves VII and VIII; $K_{n+1} = 0.5$ in curves IV, V, VI, and VIII, $K_{n+1} = 1.0$ in curves I, II, III, and VII. In Fig. 29a $n = 4$, and in Fig. 29b $n = 7$.

An interesting feature indicated by Figs. 28 and 29 is that A_1 becomes considerably larger than s_1^* if there is a sufficient number of components in the control system, or if the feedback stoichiometry of the inhibition is sufficiently large. This result implies that the biochemical variables spend considerably more time slightly below the singularity (because they cannot assume negative values) than in the immediate vicinity slightly above s_1^*. The biological significance of this result could include a possible chemical basis for the explanation of sustained rhythmic binary logic and precisely timed "spike" functions in biological systems (see Section 7.D).

7. BINARY VARIABLES IN BIOCHEMICAL CONTROL SYSTEMS

A. Introduction

Any chapter dealing with biological control theory must eventually mention the possible origins of step functions in biochemical reactions. Experimental evidence from biological systems clearly suggests that "threshold concentrations" exist for certain substances. For example, it is sometimes observed that within a specific concentration range a particular compound has little or no effect in a biological system, whereas a slightly different concentration of the same substance has a nearly maximal effect. This type of situation is illustrated in Fig. 30.

An entire theory of gene action based on postulated threshold effects (71) has not been made obsolete by recent discoveries in molecular biology. The

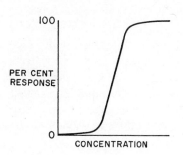

Fig. 30. Illustration of a biological or biochemical threshold effect.

observation that during the cycle of certain cells an event seems to either occur or not occur suggests the use of a binary digital model to serve as an "off/on" switch. Thus Sugita and his associates (72–74) have suggested the use of a binary variable as the function controlling RNA synthesis during the cell cycle; similarly Goodwin (1) has suggested the use of a binary variable as the function controlling cell division itself. A similar type of function could serve to describe other aspects of the activity of a specific gene or metabolite during cell cycles. The imposition of a cycle (such as that observed during cell division) on the action of a step function suggests the need for a

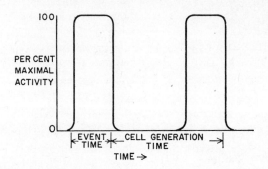

Fig. 31.   Illustration of a biological, quasi-discontinuous oscillating variable.

model describing sustained rhythmic behavior in biological systems. This well-known type of biological behavior is illustrated in Fig. 31. The type of activity represented in this figure could refer to the systems responsible for cell division, differentiation, DNA replication, RNA synthesis, or the synthesis or activity of an enzyme.

B.   Systems Possessing Multiple Stable Stationary States

The possibility that certain types of chemical reactions possess the potential for two or more stable stationary states or multiple limit cycles suggests a possible basis for binary events in biological systems. Spangler and Snell (75) have pointed out that multiple stationary states are possible in certain systems involving cross interactions between enzyme-catalyzed sequences; they suggest that such systems provide the essential ingredients for a binary switch. Sel'kov (49) suggests that the systems represented in Figs. 18 and 19 are also capable of exhibiting more than one singularity for which all the variables are nonnegative real numbers.

It is not difficult to construct a reasonable biochemical (or genetic) system for which multiple stationary states of the chemical constituents are possible (76). Consider, for example, a sequence of two biochemical catalyzed reactions wherein the first step is not only inhibited by the product of the second step (S_2) but also is stimulated by the end product:

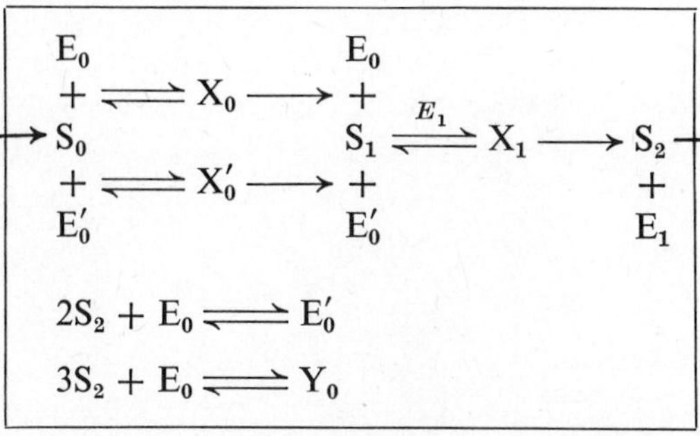

Scheme XIV

If we make the usual assumptions about these constituents, the differential equations describing the control mechanism in Scheme XIV are

$$\dot{s}_1 = \frac{C_0 + \beta s_2^2}{1 + \alpha s_2^3} - K_1 s_1, \tag{7.1}$$

$$\dot{s}_2 = K_1 s_1 - K_2 s_2.$$

The stationary-state solution of equations 7.1 are the roots of the polynomial:

$$\alpha K_2 s_2^4 - \beta s_2^2 + K_2 s_2 - C_0 = 0. \tag{7.2}$$

According to the Descartes rule of signs there can be as many as three real, nonnegative roots for equations 7.2. The stability properties of each of these singularities can be studied by the methods outlined in the previous sections of this chapter. If we select $K_1 = 1$, $K_2 = 10$, $C_0 = 5$, $\beta = 5$, and $\alpha = 0.01$, then $s_1^* = 10 s_2^* = 8.881$ and $s_1^* = 10 s_2^* = 58.655$ are asymptotically stable singular points and $s_1^* = 10 s_2^* = 12.057$ is an unstable singularity (76).

In Fig. 32 appears $s_1(t)$ when $s_1(0) = 0$ for two nearly identical initial conditions of the other variable. For the lower curve $s_2(0) = 31.3$. $s_1(t)$ rises rapidly, overshoots its nearest stationary state, $s_1^* = 10 s_2^* = 8.881$, and subsequently appears to approach its *unstable* stationary state, $s_1^* = 10 s_2^* = 12.057$. Instead, however, a maximum value slightly less than 12.057 is reached and ultimately $s_1(t)$ falls back toward its lower stable stationary state, $s_1^* = 10 s_2^* = 8.881$. In the upper curve $s_2(0) = 31.4$. Again $s_1(t)$ rises

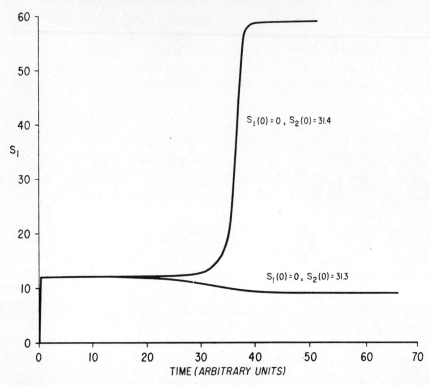

Fig. 32. Illustration of a biochemical system (Scheme XIV) possessing alternating stable, unstable, and stable stationary states.

rapidly and appears to approach the unstable stationary state, $s_1^* = 10s_2^* = 12.057$. In this case however, $s_1(t)$ eventually leaves the neighborhood of the unstable singular point and rises rapidly toward the higher stable stationary state, $s_1^* = 10s_2^* = 58.655$.

In Fig. 33 appears the phase-plane relationship between s_1 and s_2. This figure illustrates the stationary-state destinations for any initial condition of the variables. Clearly, any set of $s_1(0)$, $s_2(0)$ in the phase-space above the *separatrix line* connecting $s_1(0) = 13.1$, $s_2(0) = 0$ and $s_1(0) = 0$, $s_2(0) = 31.3$ through $s_1^* = 10s_2^* = 12.057$ will approach the stationary state $s_1^* = 10s_2^* = 58.655$; any set of $s_1(0)$, $s_2(0)$ in the space below this separatrix will approach the stationary state $s_1^* = 10s_2^* = 8.881$.

In cases like the one illustrated in Figs. 32 and 33 there are two or more stable stationary states. It is clearly possible for the variables to be at any of these stationary states. If one of these states denotes "on" and the other "off,"

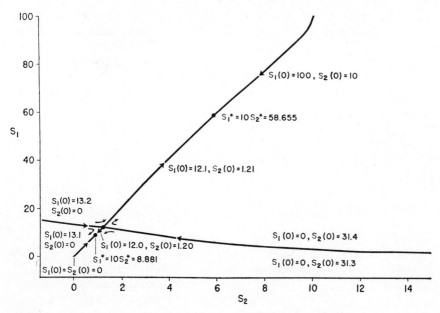

Fig. 33. Phase-space representation of the system illustrated in Fig. 32.

the binary state of the system will depend only on the factors that determine which stationary state the variables approach. If the variables are initially below the separatrix, they will approach the lower stable stationary state; however, if the variables are initially above the separatrix, they will approach the higher stable singularity. Furthermore, it is conceivable that a sufficient perturbation of the system could cause the components to move in the phase-space from the attracting influence of one of the stable singularities into the attracting influence of the other stable point. Behavior of this type is suggestive of the kind of events that must occur when cells differentiate.

In systems capable of sustained oscillations, it is conceivable that a chemical component could switch from one limit cycle to another. We have already examined how the coalescence of stable and unstable limit cycles can result when a parameter reaches a critical value; the general result is that both limit cycles disappear, and the variable "jumps" to the nearest stable limit cycle or stationary state.

The transient time during which a binary switch is neither "on" nor "off" should be as short as possible. This transient time can be thought of as a "noisy" period during which probabilistic events occur. If we are speaking, for example, of a binary variable that determines when a cell divides, the standard deviation from the average division time for a synchronized culture of cells might reflect the length of time that the binary variable is neither "on" nor "off." We would expect small standard deviations from the average if the transient switching time is completed in a relatively short time (is stringently controlled), but larger deviations if the switching is slow. Since the coefficient of variance of the average division time for a culture of *Escherichia coli* is in the range of 10% when the control is stringent (77), we would expect the transient time for the binary variable that determines when an *E. coli* cell divides to be less than 10–15 min. This is relatively fast on the time scale associated with cell division frequencies, which are in the range of $0.01/\text{min}$.

C. Nonlinear Systems Coupled to Oscillators

In order to achieve the sustained cyclic nature of an "off/on" switch of the type illustrated in Fig. 31 we could envision a system composed of a rapidly responding, quasi-discontinuous function coupled to an autonomous sustained oscillator. An obvious candidate for the oscillator is a biochemical system of the type described in Schemes V–IX. A possible biochemical model for the step function is a series of enzymes possessing rate laws that are sigmoidal functions of the substrate concentrations (78). If we connect the two pathways by assuming that one of the metabolites is a branch-point substance (i.e., a substance capable of entering both pathways), the integrated control mechanism is represented by the chemical reactions depicted in Scheme XV (79). The enzymes in Scheme XV with subscripts less than or

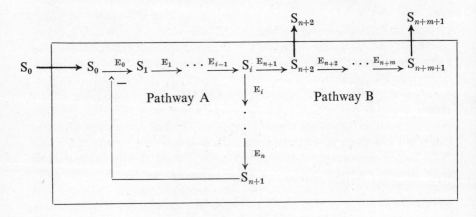

$$S_{n+1}$$

Scheme XV

equal to n are the catalysts for the series of steps with feedback; these reactions result in the sustained oscillations of the system. In what follows we shall refer to this component of Scheme XV as "Pathway A." The dynamical behavior of Pathway A is described by equations 6.2 and 6.3. Each step catalyzed by each enzyme with a subscript greater than n is assumed to be a sigmoidal function of the concentration of the substrate for that enzyme. In what follows we shall refer to this component of Scheme XV as "Pathway B."

Although the results that follow are independent of the mechanistic basis for the sigmoidal rate laws in Pathway B, the allosteric model (55) fits the needs of our system. In this model the ith enzyme is composed of p_i subunits, each of which possesses one binding site for the appropriate substrate. Furthermore, each allosteric enzyme exists in two forms, R_i and T_i, such that the equilibrium constants for the reactions

$$T_i \rightleftharpoons R_i, \quad i = n + 1, \ldots, m + n, \tag{7.3}$$

are L_i. The equilibrium constants for the binding of the first molecule of S_i to the R form of E_i are $p_i K_{ri}$; the corresponding constants for the binding to the T form of E_i are $p_i K_{ti}$. We now define

$$\alpha_i = K_{ri} s_i \quad \text{and} \quad \beta_i = \frac{K_{ti}}{K_{ri}} \quad \text{for } i = n + 1, \ldots, m + n \tag{7.4}$$

and assume approximate equilibrium between each S_i and R_i, S_i and T_i, and R_i and T_i. The fraction of active sites of E_i bound by S_i is

$$F_i = \frac{\alpha_i \beta_i (1 + \alpha_i \beta_i)^{p_i-1} + L_i \alpha_i (1 + \alpha_i)^{p_i-1}}{L_i (1 + \alpha_i)^{p_i} + (1 + \alpha_i \beta_i)^{p_i}}, \quad i = n + 1, \ldots, m + n \tag{7.5}$$

The quasi-equilibrium rate of each step catalyzed by a sigmoidal enzyme is

$$v_i = {}^i V m F_i, \tag{7.6}$$

where ${}^i V m$ is the maximum rate obtained at high s_i for the ith enzyme.

Under conditions wherein Pathway A is capable of sustained oscillations ($n = 5$, $\rho = 4$, $C_0 = 5.1$, $\alpha = 0.08$, $K_1 = K_2 = K_3 = K_4 = K_5 = 1$, and $K_6 = 0.6$) and S_{n+m+1} in Pathway B is a quasi-discontinuous variable ($m = 5$, $L_i = 0.0001$, $\beta_i = 0.01$, and $p_i = 4$ for $i = 6, 7, 8, 9$, and 10), the integrated system described in Scheme XV can serve as the source of sustained rhythmic binary logic (79). This example is illustrated in Fig 34, where S_{n+m+1} ($n = 5$, $m = 5$) exhibits a sustained rhythm of being very nearly zero ("off") or very nearly maximal ("on"). The transition from "off" to "on" or vice versa

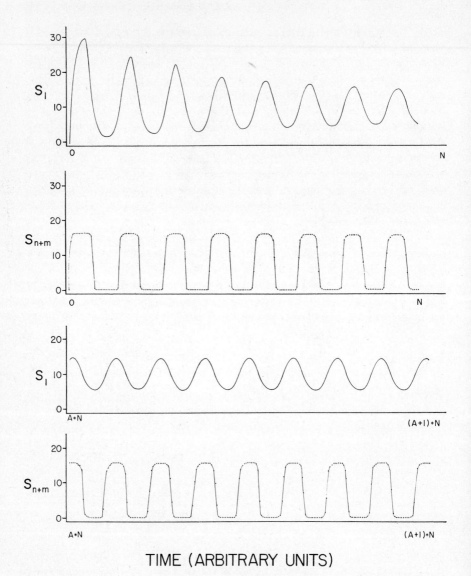

TIME (ARBITRARY UNITS)

Fig. 34. Example of a quasi-discontinuous, oscillating biochemical variable arising from Scheme XV.

occurs in a time interval that is quite short compared to the period of the rhythm.

D. High-Order, Nonlinear Biochemical Oscillators

An even simpler chemical basis for well-timed quasi-discontinuous functions in biological systems was inferred in Figs. 28 and 29. These figures suggest that, if there are a sufficient number of components in a negative-feedback control system consisting of a linear sequence of enzyme-catalyzed reactions, or if the stoichiometry of the feedback is sufficiently high, it is possible to obtain a quasi-binary function from the control system alone. An example is illustrated in Figs. 35–37, which show numerical solutions of equations 6.2 and 6.3 for the cases wherein $K_i = 1$ ($i = 1, \ldots, n + 1$), $\alpha = 1$, $C_0 = 5$, $\rho = 6$, and $n = 3$ (Fig. 35), 11 (Fig. 36), and 19 (Fig. 37). As expected, the limit cycle for the system involving four components surrounds $s_1 = s_1^*$ in a relatively symmetrical fashion. However, the same system involving 12 first-order enzyme-catalyzed steps has a very unsymmetrical limit cycle. Finally, the same system involving 20 components possesses an extremely square-shaped wave form. This example (Fig. 37) approximates a binary function of the type illustrated in Fig. 31.

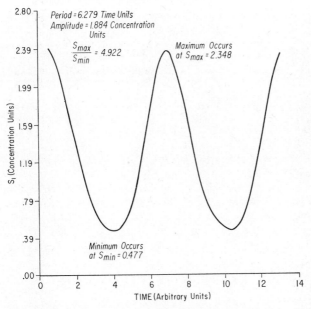

Fig. 35. Example of a symmetrical oscillation arising from the enzyme-catalyzed system with negative feedback in Scheme IX.

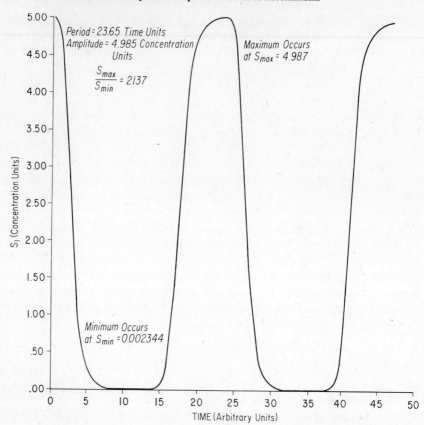

Fig. 36. Example of a quasi-discontinuous oscillation arising from the enzyme-catalyzed system with negative feedback in Scheme IX.

E. Systems Capable of Different Types of Dynamical Behavior

Another situation wherein a chemical system is capable of discrete binary behavior arises, if because of a reasonably small perturbation on the system, the dynamical behavior can be changed in a qualitative manner. For example, consider the feedback system consisting of a linear sequence of enzyme-catalyzed reactions depicted in Scheme V. Under certain circumstances each rate law for each step in such a set of reactions is first order with respect to the substrate for that step. It is also possible, however, for each rate law to be a hyperbolic (56, 57), sigmoidal (55, 58), or some other function of the substrate concentration. Furthermore, it is possible that one or more of these

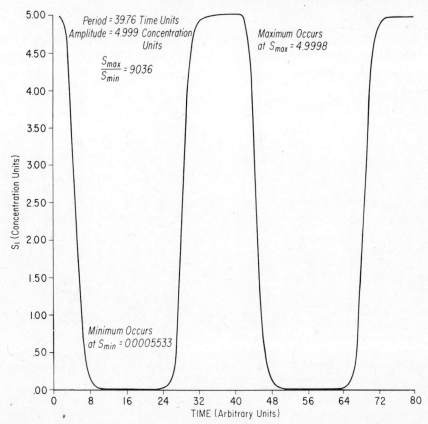

Fig. 37.   Example of a quasi-discontinuous oscillation arising from the enzyme-catalyzed system with negative feedback in Scheme IX.

rate laws can be altered even while the chemical reaction is in progress. A change in rate law from first order to hyperbolic and vice versa can be induced by a perturbation that causes a sufficiently large change in the ratio of the classical Michaelis constant (57) to the concentration of the substrate for the enzyme. Similarly, a change in rate law from hyperbolic to sigmoidal and vice versa can be induced by a conformational change on the surface of the enzyme (55), by the addition of a substance that binds the enzyme simultaneously with the substrate (58), or by any perturbation that alters one or more of the rate constants associated with the system (80). The question that we wish to consider here is whether or not a change in the rate law(s) in a system of the type depicted in Scheme V can induce a qualitative change in the dynamical behavior of the whole control system.

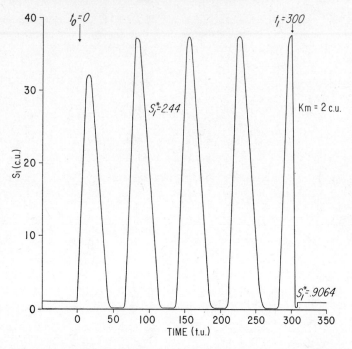

Fig. 38.   Illustration of a shift of dynamical behavior of an enzyme-catalyzed system with negative feedback (Scheme IX). The shift is due to a change of the rate law. In this figure $K_m = 2$, and the transient response to the change is fast.

The answer to this question is "yes" (81). Figures 38–40 contain simulations of a series of six sigmoidal enzymes. The rate law for each enzyme is described by equation 7.5; the maximum velocities are each equal to the individual Michaelis constants, $L = 0.0001$, $\beta = 0.01$, and the cooperativity for each step is $p = 4$. In these figures the feedback function is the usual one (equation 6.3), and $C_0 = 5$, $\alpha = 0.1$, and $\rho = 4$. In Fig. 38 the Michaelis constant is 2 concentration units, so the ratio of s_1^*/K_m is slightly over 2; in Fig. 39 the Michaelis constant is 16 concentration units, so s_1^*/K_m is a little more than 0.1; in Fig. 40 the Michaelis constant is 64 concentration units, so s_1^*/K_m is about 0.03. In all of the figures the stationary state for the sigmoidal system possesses global asymptotic stability. At time t_0 a perturbation is applied to the system. This perturbation is such that the rate laws for all six enzymes are rapidly changed from sigmoidal to hyperbolic functions. Although the Michaelis constant, maximum velocity, C_0, α, ρ, and n are not altered, the change in the form of the rate laws causes the stationary state to become unstable. It is an important feature of this system that the complete change in

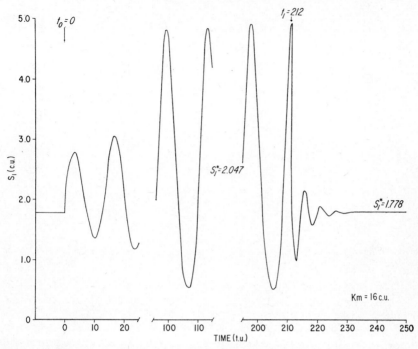

Fig. 39. Same as Fig. 38 except $K_m = 16$. The transient response to the change of the rate law is slower.

the dynamical behavior occurs for a wide range of enzyme saturation (although it certainly will not occur for the entire range of saturation). The limit cycle that arises for the hyperbolic system is reached very rapidly when the Michaelis constant is 2, not so rapidly when the Michaelis constant is 16, and rather slowly when the Michaelis constant is 64.

At time t_1 another perturbation is applied to the oscillating hyperbolic systems depicted in Figs. 38–40. This time the perturbation is such that it causes all of the rate laws to be rapidly changed from rectangular hyperbolas back to the original sigmoidal functions. This perturbation returns the global asymptotic stability to the systems in all three figures. As before, the transients are fastest for the smallest Michaelis constant and slowest for the largest one. Other simulations not included here illustrate that similar responses can be obtained in multienzyme systems with negative-feedback control when only one or two of the rate laws are changed by the perturbation.

The type of behavior depicted in Figs. 38–40 (a switching involving a qualitative change in the type of dynamical behavior exhibited by a

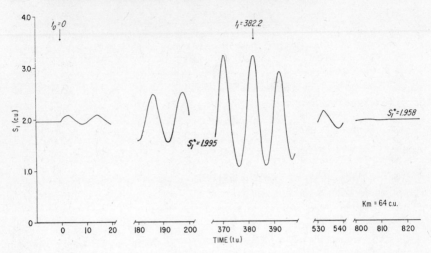

Fig. 40. Same as Figs. 38 and 39 except $K_m = 64$. The transient response in this case is the slowest.

biochemical control system) could also be a mechanism whereby cells change characteristics (differentiate) (81). At present it is a matter of speculation as to what extent the various types of mechanisms we have discussed play a role as control mechanisms or quasi-binary functions in biological processes. A possible role merits consideration, however, since the main components of the models are well known to most biologists and biochemists. It is not unreasonable to expect that, if any of the models we have discussed could be useful, they would be readily provided by natural selection.

ACKNOWLEDGMENTS

This work was supported in part by Grants GB 7110 and GB 20612 from the National Science Foundation. Part of the work was completed while the author was a Public Health Service Career Development Awardee, National Institutes of Health Award K3-GM-11,237, at the University of Tennessee Medical Units, Biochemistry Department, Memphis, Tennessee.

I am grateful to my wife Vivienne for the tangible and intangible assistance she provided. Appreciation is due also to the participants of the 1968 Theoretical Biology Colloquium, which was held in Traverse City, Michigan. Many of my colleagues at this colloquium contributed to some of the ideas developed in this chapter.

REFERENCES

1. B. C. Goodwin, *Temporal Organization in Cells*, Academic Press, London, 1963, p. 16.
2. J. Higgins, *J. Ind. Eng. Chem.*, **59**, 20 (1967).
3. J. Wei and C. D. Pratter, *Advan. Catalysis*, **13**, Section VIIC (1962).
4. C. F. Walter, *Enzyme Kinetics: Open and Closed Systems*, Ronald Press, New York, 1966, p. 14.
5. I. Prigogine, *Thermodynamics of Irreversible Processes*, Interscience-John Wiley, New York, 1967.
6. N. Minorsky, *Nonlinear Oscillations*, Van Nostrand, Princeton, N.Y., 1962, p. 21.
7. H. Poincaré, *J. Math.*, **7**, 3 (1881).
8. A. Winfree, *J. Theoret. Biol.*, **16**, 15 (1967).
9. J. Higgins, *J. Ind. Eng. Chem.*, **59**, 32 (1967).
10. E. A. Coddington and N. Levinson, *Theory of Ordinary Differential Equations*, New York, 1955, p. 399.
11. N. Minorsky, *Nonlinear Oscillations*, Van Nostrand, Princeton, N.J., 1962, p. 78.
12. I. Bendixson, *Acta Math.*, **24** (1901).
13. J. Higgins, *J. Ind. Eng. Chem.*, **59**, 54 (1967).
14. N. Minorsky, *Nonlinear Oscillations*, Van Nostrand, Princeton, N.J., 1962, p. 75.
15. J. Higgins, *J. Ind. Eng. Chem.*, **59**, 58 (1967).
16. C. F. Walter, *Biophys. J.*, **9**, 863 (1969).
17. C. F. Walter, *Proc. Biophys. Soc.*, **13**, a-211 (1969).
18. R. Bernhard, *J. Theoret. Biol.*, **7**, 532 (1964).
19. L. Bayliss, *Living Control Systems*, W. H. Freeman, San Francisco, 1966.
20. F. Grodins, *Control Theory and Biological Systems*, Columbia University Press, New York, 1963, p. 1.
21. N. Wiener, *Cybernetics*, John Wiley, New York, 1948.
22. K. E. Machlin, *Soc. Exptl. Biol. Symp.*, **18**, 421 (1964).
23. J. Hearon, S. Bernhard, S. Friess, D. Botts, and M. Morales, in *The Enzymes* (Eds.: P. Boyer, H. Lardy, and K. Myrbäck), Academic Press, New York, 1959, Chapter 2.
24. H. Kacser and J. Burns, *Proc. Quant. Biol. Metal. Symp.*, **3**, 11 (1968).
25. N. P. Erugin, *Appl. Math. Mech. USSR*, **15**, 2 (1951).
26. N. P. Erugin, *Appl. Math. Mech. USSR*, **16**, 5 (1952).
27. E. A. Barashin and N. N. Krasovsky, *Dokl. AN USSR*, **36**, 3 (1953).
28. A. M. Letov, *Stability in Nonlinear Control Systems*, Princeton University Press, Princeton, N.J., 1961, p. 8.
29. A. Lotka, *J. Phys. Chem.*, **14**, 271 (1910).
30. A. Lotka, *J. Phys. Chem.*, **42**, 1595 (1920).
31. V. Volterra, *Theorie Mathematique de la Lutte pour la Vie*, Gauthiers-Villars, Paris, 1931.
32. A. N. Kolmogorov, *Glorn. Inst. Ital. degli Attuari*, **14** (1936).
33. J. Higgins, *Proc. Natl. Acad. Sci. U.S.*, **51**, 989 (1964).

34. E. H. Kerner, *Bull. Math. Biophys.*, **19**, 121 (1957).

35. L. Onsager, *Phys. Rev.*, **37**, 405 (1931).

36. I. Prigogine and R. Defay, *Chemical Thermodynamics*, Desoer, Liege, 1950, Chapter 15.

37. J. Z. Hearon, *Bull. Math. Biophys.*, **15**, 121 (1953).

38. D. Shear, *J. Theoret. Biol.*, **16**, 212 (1967).

39. I. Prigogine, *Thermodynamics of Irreversible Processes*, Interscience-John Wiley, New York, 1967.

40. C. F. Walter and E. Frieden, *Advan. Enzymol.*, **25**, 167 (1963).

41. A. Koch, *J. Theoret. Biol.*, **15**, 75 (1966).

42. Sir Hans Krebs, Lecture delivered at First Symposium on Control of Metabolism, Miami, 1969, p. 1 of abstracts.

43. E. E. Sel'kov, in *Control Processes of Biological and Chemical Systems* (Ed.: V. A. Vavilian), Moscow, 1967, p. 93.

44. E. Stadtman, Lecture delivered at First Symposium on Control of Metabolism, Miami, 1969.

45. C. F. Walter, *Biophys. J.*, **9**, 863 (1969).

46. C. F. Walter, *J. Theoret. Biol.*, **23**, 23 (1969).

47. B. Goodwin, *Advan. Enzyme Regulation*, **3**, 425 (1965).

48. M. F. Morales and D. McKay, *Biophys. J.*, **7**, 621 (1967).

49. E. E. Sel'kov, in *Control Processes of Biological and Chemical Systems* (Ed.: V. A. Vavilian), Moscow, 1967, p. 81.

50. J. S. Griffith, *J. Theoret. Biol.*, **20**, 202 (1968).

51. J. S. Griffith, *J. Theoret. Biol.*, **20**, 209 (1968).

52. C. F. Walter, *J. Theoret. Biol.*, **23**, 39 (1969).

53. C. F. Walter, *Proc. Biophys. Soc.*, **13**, A-211 (1969).

54. C. F. Walter, *J. Theoret. Biol.*, **27**, 259 (1970).

55. J. Monod, J. Wyman, and J. Changeux, *J. Mol. Biol.*, **12**, 88 (1965).

56. V. Henri, *Lois Générales de l'Action des Diastases*, Hermann, Paris, 1903.

57. G. E. Briggs and J. B. S. Haldane, *Biochem. J.*, **19**, 383 (1925).

58. J. Botts, *Trans. Faraday Soc.*, **54**, 593 (1958).

59. J. Botts and C. Drain, Conference on the Chemistry of Muscular Contraction, Igaku Shoin, Tokyo, 1957.

60. J. Hearon, S. Bernhard, S. Friess, D. Botts, and M. Morales, in *The Enzymes* (Eds.: P. Boyer, H. Lardy, and K. Myrbäck), Academic Press, New York, 1959, p. 89.

61. G. Weber, in *Molecular Biophysics* (Eds.: B. Pullman and M. Weissbluth), Academic Press, New York, 1966, p. 369.

62. C. F. Walter, *Biochemistry*, **1**, 652 (1962).

63. R. Ruffini, *Riflessioni Intorno Soluzione delle Equazioni Algebraiche Generali*, Modena, 1813.

64. G. H. Hardy, J. E. Littlewood, and G. Polya, *Inequalities*, Cambridge University Press, 1952.

65. G. Viniegra-Gonzalez and H. Martinez, *Proc. Biophys. Soc.*, **13**, A-210 (1969).

66. V. Ličko, *Proc. Biophys. Soc.*, **9**, A-102 (1969).

67. M. A. Aiserman, *Lectures on Theory of Automatic Regulation*, Moscow, 1958.

68. N. Minorsky, *Nonlinear Oscillations*, Van Nostrand, Princeton, N.J., 1962, p. 148.

69. A. L. Lur'e, *Certain Nonlinear Problems in the Theory of Automatic Control*, Gostek-hizdat, Moscow, 1951.

70. A. M. Letov, *Stability in Nonlinear Control Systems*, Princeton University Press, Princeton, N.J., 1961.

71. R. Goldschmidt, *Physiological Genetics*, McGraw-Hill, New York, 1938.

72. M. Sugita, *J. Theoret. Biol.*, **1**, 415 (1961).

73. M. Sugita, *Proc. Quant. Biol. Metal. Symp.*, **3**, 33 (1967).

74. M. Yčas, M. Sugita, and A. Bensam, *J. Theoret. Biol.*, **9**, 444 (1965).

75. R. A. Spangler and F. M. Snell, *J. Theoret. Biol.*, **16**, 381 (1967).

76. C. F. Walter, *Comput. Biol. Med.*, **1**, 215 (1970).

77. H. Kubitscheck, *Proc. Berkeley Symp. Math. Statistics Probability*, **5**, 549 (1965).

78. C. F. Walter, R. Parker, and M. Yčas, *J. Theoret. Biol.*, **15**, 208 (1967).

79. C. F. Walter, *Proc. Quant. Biol. Metal. Symp.*, **3**, 38 (1967).

80. C. F. Walter, *Biogenesis—Evolution—Homeostasis* Springer-Verlag, Heidelberg-New York, in press.

81. C. F. Walter, *First European Biophys. Cong.*, **YIII/7**, 459 (1971).

CHAPTER **12**

Control Mechanisms in Lipid Mobilization and Transport

RODOLFO PAOLETTI AND
CESARE R. SIRTORI

Institute of Pharmacology and Pharmacognosy,
University of Milan, Milan, Italy

491

1. INTRODUCTION

The main source of metabolic energy, adipose tissue triglycerides (TG), are mobilized in the form of free fatty acids (FFA) and free glycerol. The FFA are transported to the liver and to other peripheral tissues as a plasma albumin complex (1). The metabolic fate of FFA is oxidation (complete to CO_2 or partial to ketone bodies) or esterification into glycerides, phospholipids, and sterol esters, a process which occurs mainly in the liver. Newly formed liver glycerides stimulate lipoprotein (LP) release from the liver to plasma, and transport to the peripheral tissues and back to adipose tissue. The presence of FFA in mammalian blood was observed by Szent-Györgyi and Tominaga (2) in 1924, but the biological significance of these physiological plasma components has been established only by Goodman and Gordon (3). The significance of FFA as a transport form of lipids in energy homeostasis was supported by the demonstration of their rapid turnover in plasma (4, 5). Plasma FFA may well account for the lipid substrates used by peripheral tissues in oxidating processes. Large amounts of FFA are rapidly released in emergency conditions, such as physical exercise (6, 7), or psychological stress (8). The mechanism of regulation of FFA release is under nervous and hormonal control. The underlying mechanisms have been largely clarified by the discovery of the cyclic 3′,5′-adenosine monophosphate (cAMP) system, a compound accumulated in adipose tissue cells after hormonal stimulation.

2. CONTROL MECHANISMS IN ADIPOSE TISSUE

Adipose tissue metabolism is altered by many different hormones: isolated fat cells respond to peptide hormones, catecholamines, thyroid hormones, prostaglandins, and steroids (9).

A. First Messengers in Lipolysis

A hormone-sensitive enzyme system has been isolated from adipose tissue in a particulate fraction with the sedimentation characteristics of microsomes starting from tissue homogenized in 0.25 M sucrose. This lipolytic activity has a pH optimum between 6 and 7 and is inhibited by sodium fluoride, whereas it is not affected by EDTA, phosphate, protamine, and high salt concentrations (10). The properties of this enzyme differ in several respects from those of the LP lipase, also present in adipose tissue, which has an alkaline pH optimum and is not inhibited by fluoride. The fatty acid acceptor of both enzymes is albumin.

With regard to the behavior of hormone-sensitive lipase during centrifugation, a recent classification has been proposed by Biale et al. (11). In addition to the microsomal lipases, another lipase is localized in adipose tissue with a pH optimum between 4 and 5 and is bound to particles sedimenting with the mitochondrial fractions. Vaughan et al. (12) have also shown a monoglyceride lipase in adipose tissue with an optimal pH of 8; this lipase is stable in acetone powder prepared from adipose tissue.

In *in vitro* experiments with adipose tissue, epinephrine and norepinephrine (13, 14), ACTH (15, 16), and glucagon (17, 18) induced a prompt increase in the rate of lipolysis. Glycerol accumulates in the medium (19), and FFA both in the medium (where albumin is present) and in the tissue (15). Similar effects have been observed also by using perfused adipose tissue (20, 21) or suspensions of free fat cells (22). In all these experimental conditions, the increase of FFA levels in adipose tissue is due to an acceleration of TG breakdown with little effect on FFA re-esterification.

There are species differences in the response to single hormones in adipose tissue cells: in rabbits, epinephrine and norepinephrine, hormones very active in rats, are without any effect on FFA mobilization (23) *in vitro* and *in vivo*, whereas fat cell preparations of rabbits show a strong lipolytic response to ACTH (24). In adrenalectomized and hypophysectomized animals, ACTH maintains its extraadrenal lipolytic effects, which are even exacerbated: rapid increase of lipid mobilization and severe hypoglycemia occur (25).

B. The Second Messenger in Lipolysis

Both lipolysis and phosphorylase are activated in adipose tissue by ACTH, glucagon, TSH, and epinephrine (26), suggesting that these hormones ("first messengers") may act on a cyclase and accumulate cAMP in the tissue. Caffeine, an inhibitor of phosphodiesterase, has been found by Vaughan and Steinberg (27) to act synergistically with epinephrine in stimulating lipolysis. Klainer (28) has found in adipose tissue a cyclase sensitive to epinephrine. These earlier observations opened the way for Butcher et al. (29, 96), who compared directly the levels of cAMP to the rate of lipolysis in the epididymal fat pad. Both the intracellular cAMP levels and FFA release are increased after addition to the incubation medium of low concentrations of epinephrine. At higher concentrations the rate of FFA release reaches a maximum, whereas cAMP continues to accumulate, indicating that, for a certain range of concentrations of epinephrine, cAMP is a limiting factor of the lipolytic response. Inhibitors of phosphodiesterase, the enzyme hydrolyzing cAMP, such as theophylline and caffeine, act synergistically in regard to both cAMP levels and lipolytic effect. On the other hand, a beta adrenergic blocking agent, dichloroisopropyl-arterenol, antagonizes the effects of epinephrine on cAMP accumulation and lipolysis.

The effects of other hormones on adenylcyclase in adipose tissue have been more recently investigated (30–32), and in all cases the results support the cAMP hypothesis regarding the control of lipolysis. The addition to adipose tissue cells of a lipid-soluble derivative of cAMP, N^6-2′-O-dibutyryl cAMP, stimulates both FFA and glycerol release from adipose cells or fat pads (29). This derivative is more active than cAMP because it is also resistant to phosphodiesterase cleavage. It is believed that cAMP increases lipolysis through the activation of specific protein kinases. Recently, Steinberg and Huttunen (33) have shown that a cAMP-dependent protein kinase from rabbit muscle increases lipolytic activity in rat adipose tissue. The cAMP-dependent protein kinases have been separated into catalytic and regulatory subunits; cAMP acts by binding the regulatory subunit, causing it to dissociate from the catalytic subunit (34–36).

Insulin, which decreases lipolysis in adipose tissue, has been found also by Butcher (37) to lower cAMP levels in fat cells. Insulin could act by inhibiting the combination of cAMP with the regulatory subunit, therefore reducing the dissociation of the subunit, or by interacting directly with the catalytic site (38). Senft et al. (39) and Loten and Sneyd (40) have observed activation of phosphodiesterase in adipose tissue by insulin, a finding not confirmed by others. Insulin, however, inhibits the effects of exogenous cAMP, but not of dibutyryl cAMP, in liver and adipose tissue, supporting an effect on phosphodiesterase (39).

In a recent investigation, Steinberg and Huttunen (33) were able to partially purify the hormone-sensitive lipase in rat adipose tissue and to characterize its activation, demonstrating a close correlation between lipase activation and transfer of phosphate from the gamma position of ATP. This may represent a physiological mechanism of the hormone-stimulated activation. Lipase preparations from adipose tissue previously exposed to epinephrine were not activated, whereas preparations made from non-stimulated fat pads were readily activated. In homogenates, activation could be obtained by addition of ATP and cAMP; in purified preparations, by addition of ATP, cAMP, and exogenous protein kinase.

C. Prostaglandins

Prostaglandins are present in adipose tissue and, as shown originally by Steinberg et al. (41), are able to antagonize the lipolytic effects of epinephrine, norepinephrine, ACTH, TSH, and glucagon. The lipolytic effect of theophylline is also inhibited by prostaglandin E_1 (PGE_1) (42), while the effects of cAMP and dibutyryl cAMP are not influenced by this hormone (43). The antilipolytic effect of PGE_1 is also evident in the dog, but is much less obvious in man, where probably the doses tested, which are free of side effects, are insufficient (44, 45). The negative results in man are probably related (46), to a catecholamine release from the adrenal medulla, as pointed out by Berti et al. (47). The effect of PGE_1 differs from that of adrenolytic agents: PGE_1 antagonizes spontaneous lipolysis *in vitro* (41) and reduces increased lipolysis in alloxan diabetic rats (48). The inhibition by prostaglandins of FFA mobilization induced by norepinephrine is extremely specific: the antilipolytic effects are observed with doses unable to change the hyperglycemia elicited by norepinephrine (49).

The mechanism of action of prostaglandins in adipose tissue is still unknown. According to Muhlbachovà et al. (42), PGE_1 inhibits lipolysis elicited by norepinephrine in a noncompetitive way, whereas Stock et al. (50) believe there is a competitive inhibition. This apparent discrepancy may be explained by the observation that, according to Stock et al. (50), the slopes of regression of norepinephrine-induced lipolysis in the presence of different concentrations of PGE_1 did not increase further if the concentration of PGE_1 was raised above the values able to inhibit lipolysis by approximately 50% (51), in sharp contrast with other lipolytic agents. In animals deficient in essential fatty acids and therefore unable to synthesize prostaglandins (52), the lipolytic agents are more active than in control animals, indicating that endogenous prostaglandins may play a physiological role in the regulation of FFA release. Similarly, norepinephrine, theophylline, and cAMP are more active on isolated fat cells obtained from animals deficient in essential fatty acids than on cells from normal rats (Table 1).

D. Drug Interactions with the Regulation of Lipolysis

The adrenolytic agents are inhibitors of the lipolytic effect of catecholamines in isolated adipose tissue. α-Adrenolytics and methoxamine derivatives are much less active than β-adrenolytics in blocking the lipolytic effects of norepinephrine and epinephrine, while β-adrenolytics are not only more active but also show a clear-cut competitive inhibition (53). Although Butoxamine and isopropylmethoxamine are less active than the β-adrenolytics, they are also competitive inhibitors (54). The β-adrenolytics are also

Table 1.   FFA Release from Isolated Fat Cells of Control and Essential-Fatty-Acid (EFA)-Deficient Rats

Agent	Concentration of Agent, M	Microequivalents/mM TG/hr ($\pm$S.E.)	
		Controls	EFA-Deficient
Norepinephrine	1×10^{-7}	6.6 ± 3.1	18.8 ± 4.8
	3.16×10^{-7}	17.3 ± 4.2	36.8 ± 1.2
	5×10^{-7}	19.8 ± 3.0	40.1 ± 2.6
	1×10^{-6}	21.4 ± 1.1	39.1 ± 2.1
	3.16×10^{-6}	22.3 ± 1.4	38.8 ± 0.3
Theophylline	1×10^{-4}	0.87 ± 0.14	12.9 ± 0.57
	3.16×10^{-4}	4.63 ± 0.24	40.2 ± 1.5
	5×10^{-4}	25.5 ± 0.07	79.5 ± 1.0
	1×10^{-3}	44.1 ± 0.6	86.6 ± 1.5
3'5'-Cyclic AMP[a]	...	3 ± 0.6	3 ± 0.1
(dibutryl ester)	4.5×10^{-4}	48 ± 2	63 ± 2
	9×10^{-4}	123 ± 6	229 ± 6
	1.8×10^{-3}	147 ± 8	307 ± 16

[a] The gas phase was air in this case, whereas for norepinephrine and theophylline it was 95% O_2–5% CO_2.

able to inhibit the lipolytic effects of ACTH, dibutyryl cAMP, and theophylline, but in a noncompetitive way. The inhibition of dibutyryl cAMP suggests a second point of attack of β-adrenolytics, different from the cyclase and interfering with the hormone-sensitive lipase (53). α-Adrenolytics are equally as active as noncompetitive inhibitors for catecholamines, ACTH, theophylline, and dibutyryl cAMP (55). These results have been explained by Birnbaumer and Rodbell (30), who have shown that α-adrenolytics do not inhibit the activation of adenylcyclase, but rather prevent the activation of the hormone-sensitive lipase.

Other inhibitors of lipolysis are α-methyl-DOPA and α-methylmetatyrosine, which deplete norepinephrine in adipose tissue (56). α-Methyl-DOPA is converted to α-methylnorepinephrine (cobefrine) (57), while α-methylmetatyrosine is converted to α-methyl-β-hydroxy-m-tyramine (metaraminol) (58). Cobefrine and metaraminol act as false transmitter substances, but the former is still effective in promoting lipolysis, whereas metaraminol is almost devoid of lipolytic activity. Cobefrine, although practically as active as norepinephrine in promoting lipolysis, cannot be released by reserpine analogs as easily as the natural transmitter (56).

The blood glucose-lowering agents, particularly tolbutamide, reduce plasma FFA levels in man (59); *in vivo* these drugs have a direct action on adipose tissue, decreasing lipolysis and FFA release (60). *In vitro* the same effect has been shown in the absence of glucose and lipolytic hormones (61). The effect is similar to the antilipolytic action of insulin in adipose tissue *in vitro*, even in the absence of glucose and lipolytic hormones; however, it is not related to the liberation of insulin bound to albumin or present in adipose tissue, because the antilipolytic effect of the sulphonylureas is still present in the absence of albumin or in the presence of anti-insulin serum.

Nicotinic acid inhibits the release of FFA induced by catecholamines (62), by fasting (63), and in alloxan diabetic rats (64). The inhibition is evident against catecholamines, ACTH, growth hormone, glucocorticoids, and theophylline (65, 66). Solyom and Puglisi (67) have shown that the glucose uptake in adipose tissue is increased without a concomitant rise in triglyceride synthesis. Peterson et al. (68) have suggested that nicotinic acid may act at a site identical or closely associated to the beta receptor, and therefore its antilipolytic action may be exerted through a depression of cAMP synthesis. A nicotinic acid analog, β-pyridylcarbinol (BPC), has pronounced vasodilatatory properties and a marked depressor effect on plasma FFA levels (69). The norepinephrine-induced lipolysis is also readily antagonized by this agent.

3. CONTROL MECHANISMS OF PLASMA FREE FATTY ACID RELEASE AND TRANSPORT

A. Sympathetic Regulations *in vivo*

The hormone-sensitive lipase in adipose tissue is under the control of the central nervous system through the sympathetic innervation of the adipose tissue. Demonstration of such control arises from the clinical observation that fat accumulates in patients with localized paralysis (70), and that the lipid content increases in denervated intrascapular fat bodies (71). The sympathetic nerve endings have a rich supply of norepinephrine (72), which decreases after denervation and is depleted by reserpine (73). Electrical stimulation of the postganglionic sympathetic fibers innervating dog subcutaneous adipose tissue is followed by rapidly increased mobilization of glycerol and FFA (74).

Several conditions stimulate FFA mobilization through activation of the sympathetic fibers. After birth, in man and sheep (75, 76), there is a rapid increase of plasma FFA levels, which is blocked by hexamethonium, indicating a neurogenic mechanism. Similarly, adaptation to upright posture in human subjects induces an increased plasma FFA concentration. This rate is not altered by adrenalectomy, although it is prevented by adrenergic blocking agents (77). Evidence is also present in man and other species regarding the effect of physical exercise, psychological stress, and exposure to cold. In the last of these experimental conditions, the body temperature can be maintained only if plasma FFA levels are increased. Chemically sympathectomized rats cannot raise their plasma FFA levels, and body temperature drops very rapidly (78).

B.   Hormonal Regulation

Thyroid hormones modulate the glycerophosphate cycle and lipolysis in adipose tissue and increase oxygen consumption. A deficiency of thyroid hormones reduces the lipolytic effect of epinephrine and other hormones (79–81), but not glucose oxidation (82). These effects are mediated by the intracellular concentration of cAMP. There is evidence that, although cAMP stimulates lipolysis, it is not involved in glucose oxidation mediated by epinephrine; dibutyryl cAMP, while stimulating lipolysis, has only negligible effects on glucose oxidation, and ACTH, which induces cAMP accumulation in adipose tissue to the same extent as epinephrine, has small effect on the conversion of labeled glucose into CO_2. It may be concluded, therefore, that only the effect of thyroid hormones on lipolysis is mediated by cAMP accumulation. This is also in agreement with investigations by Berti and Usardi (83), showing that epididymal fat pads, removed from hypothyroid animals and stimulated *in vitro* through the postganglionic sympathetic fibers, display a normal release of norepinephrine, but no increase in lipolysis. A short pretreatment with triiodothyronine restores the normal lipolytic response.

Growth hormone secretion is involved in the excess FFA mobilization during diabetes and the increased lipolysis during exercise and fasting. An increase in growth hormone secretion is a specific response to a shortage of carbohydrate supply (84). The action of growth hormone requires a lag period (85, 86), and this may explain the frequently negative results obtained after addition of growth hormone to adipose tissue *in vitro*. Fain et al. (87) were able to accelerate lipolysis *in vitro* by adding growth hormone and a glucocorticoid. The lipolytic action of growth hormone does not involve a cycloheximide-sensitive process (88), but the potentiation of growth hormone by glucocorticoids is inhibited by cycloheximide (89). The addition of theophylline potentiates the lipolytic action of growth hormone and glucocorticoids, suggesting an implication of the cAMP system. On the other hand, growth hormone and glucocorticoids increase the sensitivity of adipose tissue cells to catecholamines or theophylline without affecting the maximal response (90), but do not influence the response of fat cells to dibutyryl cAMP. More recently Moskowitz and Fain (91) reported an increased level of cAMP in adipose tissue cells treated with growth hormone and gluco-corticoids, an effect inhibited by puromycin. The proteins made under the influence of growth hormone facilitate the accumulation of cAMP elicited in adipose tissue by epinephrine and theophylline.

Insulin has been shown to influence the metabolism of adipose tissue cells not only by altering the permeability of the cell membrane to glucose, but also with independent actions. Perry and Bowen (92) observed an inhibitory action of insulin on the lipolytic effect of epinephrine, ACTH and growth hormone, added to adipose tissue in absence of glucose. Jungas and Ball (93) have shown a depression of glycerol and fatty acid production induced by epinephrine, glucagon or TSH, when added in absence of glucose. The antagonism of insulin for epinephrine is dependent upon its concentrations for the low doses of epinephrine (93). The effect of insulin may be ascribed to a rapid and complete re-esterification of the FFA, as determined by Jungas and Ball (94) in starved-refed rats (95). Insulin may antagonize epinephrine through an opposite effect on the adenylcyclase system, as shown independently by Butcher et al. (96) and Jungas (97). Recent studies by Vaughan and Murad (31) and Cryer et al. (98) suggest that insulin does not act directly on the adenylcyclase: when membranes containing the adenylcyclase system are isolated, the accumulation of cAMP is stimulated by epinephrine, but is not inhibited by insulin. Also the effect on phospho-diesterase suggested by Senft et al. (39) has been challenged by Ball (99). It should be noted that the antilipolytic action of insulin disappears when glucose is present (99). Moreover, this action has not been clearly related to any effect on the cAMP levels.

C. Physicochemical Aspects of Free Fatty Acid Transport

Free fatty acids mobilized from adipose tissue are bound to serum albumin. Three classes of binding sites for FFA anions have been described with different association constants. There are two sites for each albumin molecule in the first class, which bind FFA quite strongly. The other classes have weaker affinities for the anions but contain a larger number of binding sites (100). Low-density LP and erythrocytes can bind fatty acids, but they do not compete with albumin until the mole ratio of FFA to albumin is greater than 2 (101, 102). Neither LP nor erythrocytes compete with the first class of binding sites of albumin. At physiological FFA levels the molar ratio of FFA to albumin is below 1, and less than 2% of fatty acids would be bound to low-density LP. α-Lipoproteins are involved in the binding of plasma FFA when the binding capacity of albumin is saturated; electro-phoretically a change in α-LP mobility may be seen in dogs under cate-cholamine or FFA infusion (103). This may be important in subjects with hyperlipidemia with reduced levels of α-LP (104); the decreased binding of FFA to albumin may be significant for the possible role of FFA in thrombogenesis (105).

The FFA-albumin complex supplies FFA to peripheral tissue, particularly to muscle and heart, where they are largely oxidized to CO_2 or are taken up by the peripheral organs and deposited as triglycerides. The amount of FFA taken up peripherally depends on the molar ratio of FFA to albumin and not on the total concentration of FFA, as shown in an *in vitro* system by Spector and Steinberg (106). The FFA uptake in the presence of a fixed amount of FFA increases progressively as the amount of albumin decreases. Steinberg (107) suggests that FFA are taken up as the anions in free solution, rather than in the FFA-albumin complex. This is supported by the *in vivo* observation that the FFA leave the blood flow more rapidly than albumin, ruling out the possibility of a direct carrier mechanism. The total amount of FFA taken up *in vitro* by the cells during the first minute of incubation far exceeds the amount present at zero time, implying an extremely rapid dissociation of the albumin-FFA complex (108).

The FFA binding to albumin may be affected by a variety of other com-pounds, including hormones and drugs. Clofibrate (*p*-chlorophenoxy-isobutyrate) greatly reduces this binding because it competes with albumin sites normally involved in FFA uptake and transport (109). This explains the constant lowering of plasma FFA levels in animals and man after treat-ment with clofibrate (110). Nicotinic acid is less effective at this site because it has a very short half-life in human blood, whereas that of clofibrate exceeds 12 hr.

D.   Effects of Free Fatty Acids on Liver Lipoprotein Synthesis

Hyperlipemia is a well-known consequence of prolonged FFA hyperse-cretion, as induced by lack of insulin or excess of epinephrine. The excess FFA are rapidly taken up by the liver, stored as triglycerides, and then reappear as very-low-density LP (VLDL)-bound triglycerides. Epinephrine, at doses able to increase plasma FFA levels for an extended period of time, also elevates serum LP levels (111). In perfused livers from diabetic rats release of triglycerides is proportional to FFA levels but is less than in controls. The rate of *de novo* fatty acid synthesis by the perfused liver is directly related to the rate of triglyceride release as VLDL into the perfusion fluid (112). This explains the poor release of triglycerides from diabetic livers, where fatty acid synthesis is greatly depressed.

4. CONTROL OF LIVER AND PLASMA LIPIDS

A. Lipid Deposition in the Liver

As previously mentioned, excessive mobilization of FFA from adipose tissue may induce fatty liver. In dogs under constant intravenous infusion of norepinephrine, increased FFA plasma levels are followed by a considerable rise in liver triglycerides (113, 114). Fatty liver occurs when there is an imbalance between the uptake of fatty acids by the liver and the ability of

Table 2.   Protective Effect of Adrenalectomy[a] on ACTH[b]-Induced Fatty Liver

Group	Plasma FFA, μeq/ml[c]	Liver TG, mg/g[c]
Controls	0.23 ± 0.01	3.7 ± 0.45
Controls + ACTH	0.53 ± 0.07	16.0 ± 1.90
Adrenalectomized	0.25 ± 0.03	2.2 ± 0.50
Adrenalectomized + ACTH	0.23 ± 0.005	4.3 ± 0.97

[a] Adrenalectomy was performed 11 days before the experiment; controls were sham operated.
[b] ACTH: 30 iu/kg injected 8 times every 30 min.
[c] Mean ± S.E.

the liver to dispose of them. The imbalance can be produced by a very prolonged hypersecretion of FFA (114), exceeding the ability of the liver to dispose of them, or by a decreased capacity of the liver to dispose of FFA when the organ is functionally damaged. When the isolated liver is perfused with a mixture of red blood cells, albumin, and saline containing different concentrations of FFA, there is rapid uptake of FFA from the perfusate and a conversion into liver triglycerides (115). The uptake of FFA is greater when the liver is obtained from a fasting animal (116).

The protection observed in adrenalectomized animals against fatty liver after cold exposure, starvation, or ACTH treatment is clearly explained by the essential role of the corticoids in facilitating FFA mobilization by catecholamines. In adrenalectomized animals the decreased rate of FFA release from adipose tissue allows the liver to release completely the FFA taken up from plasma (Table 2) (117).

B. Liver Triglyceride and Lipoprotein Synthesis

Plasma contains four major forms of LP: chylomicrons, very low density (VLDL), low density (LDL), and high density (HDL) (Tables 3 and 4).

Table 3.   Physical Characteristics of Lipoproteins

Lipoprotein	$S_f{}^a$	Density, g/ml	Mobility[b]	Molecular Weight	Size, Å
Chylomicrons	>400	<0.95	Origin	$10^3-10^4 \times 10^6$	750–10,000
VLDL	20–400	0.95 –1.006	Pre-β	$5-10 \times 10^6$	250–800
LDL	0–12	1.019–1.062	β	$2-2.5 \times 10^6$	205–220
HDL	. . .	1.063–1.21	α	200,000	75–100

[a] S_f = Svedberg units (10^{-13} cm/sec/dyne/g) in a NaCl solution of density 1.063 g/ml at 26°C.
[b] Paper electrophoresis.

Two species of LP are particularly related to triglyceride transport: chylomicrons and VLDL (118). The VLDL constitute a family of macromolecules rich in endogenous triglycerides with an S_f range of 20–400, a diameter of 250–800 Å, and a molecular weight of $5-10 \times 10^6$. They migrate in electrophoresis with a mobility of pre-beta or alpha₂. In composition (Table 4) they consist mainly of triglycerides (50–70% of the molecule by weight), with 15–20% of phospholipids, 15–20% of free and esterified cholesterol, and only 7–12% of proteins. They have been recently fractionated into four constituents (119, 120). Of the VLDL protein, 40% is identical with the major protein component of LDL; 50% of human VLDL is made up of three smaller proteins. The function of VLDL is to transport endogenous triglycerides; the liver is the major site of synthesis (121) but the intestine has also been shown to produce VLDL (122, 123). Particles 300–1000 Å in diameter have been isolated from the Golgi apparatus of the liver (124). These particles contain a protein identical to the plasma VLDL apoprotein (125), suggesting that the Golgi apparatus of the liver stores preformed VLDL particles for release into the plasma. The major precursors for hepatic

Table 4.   Composition of Plasma Lipoproteins (Per Cent of Dry Weight)

Constituent	Chylomicrons	VLDL	LDL	HDL
Protein	1–2	10	25	45–55
Triglyceride	80–90	50–70	10	3
Free cholesterol	1–3	10	8	15
Esterified cholesterol	2–4	5	35	22
Phospholipids	3–6	15–20	22	30
Carbohydrate	?	<1	~1	<1

glyceride synthesis are glucose and FFA (126). The increased outputs of FFA from adipose tissue and carbohydrate feeds are, not surprisingly, stimulants of VLDL synthesis.

The LDL, which are in an S_f subclass of 0–12, are the major cholesterol-bearing LP. They contain, by weight, 8 % free and 35 % esterified cholesterol, 22 % phospholipid, 10 % triglyceride, and 20–25 % protein (127). They are spherical molecules of mean diameter 205–220 Å and with the mobility of β-LP on electrophoresis. Apparently, six peptide fractions have been obtained from LDL (128). The molecules of apoproteins vary in shape according to the amount of lipid present: with increasing lipid content the shape changes from the alpha to the beta conformation (129). Increase in LDL concentrations are common in hypothyroidism (139) and in early atherosclerosis, where apparently a tissue component (131) may be responsible for a reduced rate or removal of LDL from plasma (132) and for the accumulation of this form of LP on the arterial walls (133).

The HDL have a density of 1.063–1.210 and are composed of 4–5 protein subunits, bound to variable amounts of phospholipids and neutral lipids (134, 135). The protein is believed to extend on the surface of the complex along irregularities formed by the polar heads of the lipids, exposed to the external aqueous environment (136). The HDL contain an apoprotein with carboxy-terminal alanine, which is common with that of the VLDL. This fact, as well as the acute increase in plasma HDL levels produced by LP lipase, suggests that the HDL may derive from the metabolism of the VLDL (137). One of the apoproteins of the HDL, apoLP-glu, functions as the cofactor for LP lipase. The mechanism of activation, however, remains unknown. The observation that phospholipids enhance this activation suggests that apoproteins, triglycerides, and phospholipids form an artificial LP in which the triglycerides are more readily available to the LP lipolytic enzymes than in a simple emulsion. It seems, therefore, that apoLP-glu is an obligatory factor for the normal clearing of plasma triglycerides (138).

C. Lipoprotein Removal from the Plasma

The VLDL are removed with a half-life in plasma of 6–12 hr (139). They are broken down by a process involving triglyceride hydrolysis with LP lipase, with resultant conversion into smaller LP; labeling VLDL with ^{125}I has revealed that this form of LP breaks down to LDL (140, 141), suggesting that the latter is the remnant of VLDL catabolism. The plasma concentration of VLDL is, therefore, under the influence of synthesis, stimulated by FFA and carbohydrates, and of the rate of catabolism, which is activated by insulin, LP lipase, and triglyceride concentration (142, 143). Several of the VLDL apoproteins exchange between chylomicrons and LDL, and are also found in HDL (141). Similar rapid exchange takes place for triglycerides, free cholesterol, and phospholipids between the LP species. It has been suggested that all the LDL may derive from VLDL, but the former are not utilized for the synthesis of either chylomicrons, VLDL, or HDL (144).

These interrelationships between the LP families may provide an explanation for some of the LP patterns observed in human hyperlipoproteinemia (145). In Type I (deficiency of LP lipase) and Type V (deficiency of LP lipase combined with an increase of VLDL) there are low concentrations of LDL and HDL, suggesting a decreased clearance of chylomicrons and VLDL, with a concomitant diminished formation of small LP forms. In Type II hyperlipoproteinemia (increased LDL), a defect in the clearance of these LP remnants is present, in addition to the possible tissue defects mentioned above. In Type III, in which an LDL rich in glycerides is accumulated, the best explanation is that a deficiency occurs in the conversion of an intermediate-density LP fraction to LDL (146). In Type IV (increased VLDL) the mechanism of induction is less clear because of the technical difficulties in performing kinetic studies.

D.   Hormonal and Drug Control of Plasma Lipoproteins

The levels of LP-bound cholesterol and phospholipids change according to sex, age, and other physiological conditions. These levels, similar in the sexes or higher in women than in men of comparable age (147), rise during pregnancy (148), vary with the menstrual cycle (149), and constantly increase with age (150). Women have more α- and less β-LP than men of comparable age, and high density α-LP occurs in greater concentration in the serum of eunuchs (151). Estrogens have been used to lower β-LP; and coronary patients given estrogens show an increase in α-cholesterol and a decrease in β-cholesterol, whereas the opposite occurs after androgen treatment (152). Estrogens constantly increase the serum lipids carried by α-LP, as well as by VLDL: plasma phospholipids (bound mainly to α-LP) and triglycerides (bound to VLDL) are both increased in plasma. Cholesterol, which is bound mostly to the LDL fractions, may diminish significantly (151).

Correspondingly, androgens tend to increase human serum cholesterol, while decreasing serum triglycerides (153). After estrogen treatment an increased number of α-LP particles are present, as indicated by the lower cholesterol/protein ratio of the purified LP. Androgens increase the cholesterol/protein ratio of α-LP, indicating that the number of particles is decreased, and the amount of cholesterol of the individual particles is increased (153). Testosterone administration in dogs diminishes the incorporation of ^{14}C lysine into the α-apoLP (154). Oral contraceptives increase plasma triglycerides in premenopausal women. An increase of pre-β-LP has been also shown, particularly with a mixture of ethynodiol diacetate and mestranol (Ovulen); phospholipids are also raised by 19 % (155). In this case the primary mechanism seems to be an increased hepatic synthesis of triglycerides (156). An increase of FFA is also present in women taking oral contraceptives (157).

Insulin, whose effects on adipose tissue were described in Section 3.B, is increased in the plasma of women taking oral contraceptives (158). Elevated triglycerides and pre-β-LP levels are related to this drug-induced hyperinsulinism (159).

Thyroid hormones and D-thyroxine lower plasma glycerides (160). Only D-thyroxine significantly lowers β-LP, serum cholesterol, and total lipids; L-thyroxine, however, reduces liver cholesterol (161).

Heparin has been extensively investigated for its effects on plasma LP, after the observation of its activation on LP lipase (162). Heparin itself, or a closely related substance, forms an integral part of the enzyme: when LP lipase is inactivated by pyrophosphate, this inactivation can be reversed by adding heparin. The enzyme is probably located in the walls of the capillaries (163), thus explaining its rapid release by substances, such as dextran sulfate, not leaving the circulation (164). The LP lipase is inhibited by many factors or experimental conditions. Inhibitors are contained in platelets (165) and in white blood cells (166). Inhibition is also induced by protamine (167), toluidine blue (168), and nonionic detergents (169, 170).

In addition to D-thyroxine, several drugs have been used to treat human hyperlipoproteinemias. Clofibrate is the compound of choice for lowering triglycerides, which are elevated in Types III, IV and V of these diseases, according to Fredrickson et al. (171). There are indications that this compound may act on LP synthesis (172, 173), but the mechanism of action is still nor completely understood. Nicotinic acid, which reduces triglycerides, probably via an inhibition of lipolysis and reduction of plasma levels of FFA (174), displays also a hypocholesterolemic action (175). There is evidence that this drug can cause a marked reduction in LDL protein synthesis (144). This can be explained by the fact, pointed out earlier, that LDL are a breakdown product of VLDL proteins, and nicotinic acid can decrease adipose tissue FFA flux, hepatic endogenous triglyceride synthesis, and, hence, LDL production.

SUMMARY

The release of free fatty acids (FFA) from adipose tissue is controlled by first messengers (hormones such as glucagon, GH, and catecholamines) activating the enzyme adenylcyclase, which, in turn, transforms ATP into 3'5'-cyclic AMP (cAMP), the second messenger. The cAMP activates the

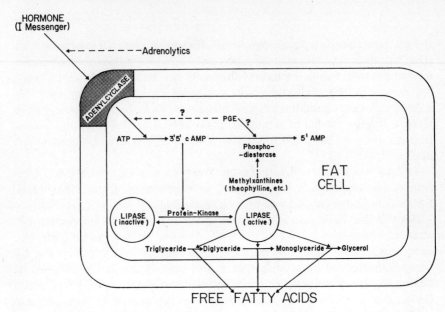

Fig. 1. Regulation of lipolysis in adipocytes: (– – – –) inhibition; (——) stimulation.

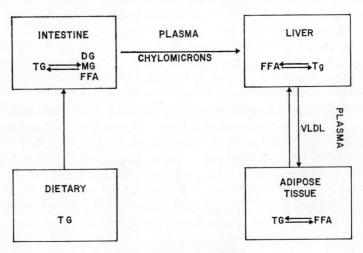

Fig. 2. Absorption and transport of triglycerides.

lipase, converting triglycerides into di- and monoglycerides and releasing FFA. A pharmacologic control is possible at various stages of this process. The adrenolytic agents interfere with the activation of adenylcyclase by the first messenger. Prostaglandin E_1 either blocks the action of adenylcyclase or stimulates the inactivation of cAMP by phosphodiesterase. Methylxanthines (theophylline, caffeine) inhibit phosphodiesterase, and thus potentiate the second messenger. The regulation of lipolysis in fat cells is schematized in Fig. 1.

After being released, FFA are bound to albumin and possibly to low density lipoproteins. They may be taken up by peripheral tissues and oxidized to CO_2, or taken up by the liver and transformed into triglycerides. These reenter the circulation, bound to the very low density lipoproteins (VLDL). Exogenous triglycerides, absorbed with the diet, constitute, instead, chylomicrons. An excessive uptake of FFA by the liver may imbalance the capacity of the organ to dispose of them, and may induce fatty liver. The same can occur after toxic agents have impaired the ability of the liver to synthesize lipoproteins. The transport of FFA and TG is schematized in Fig. 2.

Lipoproteins are synthesized by the liver under the major influence of the uptake of glucose and FFA. The largest lipoproteins, the VLDL, appear, in fact, to be the precursors totally or in large part, of the other lipoprotein classes. VLDL transport mainly triglycerides, the smaller low density

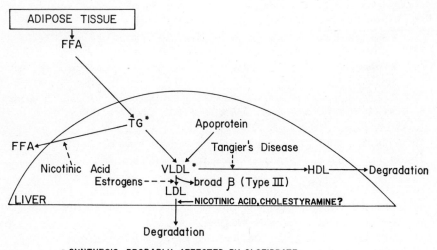

Fig. 3. Metabolism of lipoproteins and its genetic and pharmacological control. Tangier's disease is the genetic deficiency of HDL: (– – – –) inhibition; (———) stimulation.

lipoproteins (LDL) are the major vehicle for cholesterol and the high density lipoproteins (HDL) have a large protein constituent and carry mostly phospholipids. A peculiar lipoprotein is present in the human Type III hyperlipoproteinemia: it appears to be an LDL with abnormal electro-phoretic mobility (broad beta) and rich also in glycerides. Other hyperlipo-proteinemias involve an increase in chylomicrons (Type I), LDL (Type II), VLDL (Type IV), and of both chylomicrons and VLDL (Type V). A control of lipoprotein synthesis and/or transformation and degradation can be accomplished with a variety of drugs and hormones. Estrogens increase the VLDL while markedly decreasing the LDL, therefore exerting a hypo-cholesterolemic effect. An opposite action is that of androgens. Nicotinic acid lowers plasma lipids by interfering with FFA mobilization, decreasing LDL synthesis and stimulating its breakdown. Clofibrate mainly lowers VLDL by inhibiting triglyceride and lipoprotein synthesis. The metabolism and the pharmacologic control of lipoproteins is schematized in Fig. 3.

REFERENCES

1. A. E. Renold and G. F. Cahill, Jr., *Handbook of Physiology*, Section 5; "Adipose Tissue" (Eds.: A. R. Renold and S. F. Cahill, Jr.), American Physiological Society, Washington, D.C., 1965.

2. A. Szent-Györgyi and K. Tominaga, *Biochem. Z.*, **146**, 226 (1924).

3. D. S. Goodman and R. S. Gordon, *Am. J. Clin. Nutr.*, **6**, 669 (1958).

4. R. J. Havel and D. S. Fredrickson, *J. Clin. Invest.*, **35**, 1025 (1956).

5. S. Laurell, *Acta Physiol. Scand.*, **41**, 158 (1957).

6. A. Basu, R. Passmore, and J. A. Strong, *Quart. J. Exptl. Physiol.*, **45**, 312 (1960).

7. L. A. Cobb and W. P. Johnson, *J. Clin. Invest.*, **42**, 800 (1963).

8. M. Bogdonoff, E. H. Estes, and D. Trout, *Proc. Soc. Exptl. Biol. Med.*, **100**, 503 (1959).

9. M. Rodbell, in "Adipose Tissue: Regulation and Metabolic Function," *Hormone and Metabolic Research*, Suppl. 2, Academic Press, New York, 1970, p. 1.

10. M. A. Rizack, *J. Biol. Chem.*, **236**, 657 (1961).

11. Y. Biale, E. Gorin, and E. Shafrir, *Biochim. Biophys. Acta*, **152**, 28 (1968).

12. M. Vaughan, I. E. Berger, and D. Steinberg, *J. Biol. Chem.*, **239**, 401 (1964).

13. R. S. Gordon and A. Cherkes, *Proc. Soc. Exptl. Biol. Med.*, **97**, 150 (1958).

14. J. E. White and F. L. Engel, *J. Clin. Invest.*, **37**, 942 (1958).

15. J. E. White and F. L. Engel, *Proc. Soc. Exptl. Biol.*, No. 1, **99**, 375 (1958).

16. N. Freinkel, *J. Clin. Invest.*, **40**, 476 (1961).

17. J. H. Hagen, *J. Biol. Chem.*, **236**, 1023 (1961).

18. M. Vaughan, *J. Lipid Res.*, **2**, 293 (1961).

19. W. S. Lynn, R. M. Macleod, and K. H. Brown, *J. Biol. Chem.*, **235**, 1904 (1960).

20. R. J. Ho and H. C. Meng, *J. Lipid Res.*, **5**, 203 (1964).

21. R. O. Scow, in *Handbook of Physiology*, Section 5: "Adipose Tissue" (Eds.: A. R. Renold and S. F. Cahill, Jr.), American Physiological Society, Washington, D.C., 1965, p. 437.

22. M. Rodbell, *J. Biol. Chem.*, **239**, 375 (1964).

23. B. Desbals, R. Agid, and P. Desbals, *J. Physiol. (Paris)*, **61**, 264 (1969).

24. B. Desbals, P. Desbals, and R. Agid, in "Adipose Tissue: Regulation and Metabolic Functions," *Hormone and Metabolic Research*, Suppl. 2, Academic Press, New York, 1970, p. 28.

25. B. Desbals, Doctoral thesis, Toulouse, 1967.

26. M. Vaughan, *J. Biol. Chem.*, **235**, 3049 (1960).

27. M. Vaughan and D. Steinberg, *J. Lipid Res.*, **4**, 193 (1963).

28. L. M. Klainer, Y. M. Chi, S. L. Freidberg, T. W. Rall, and E. W. Sutherland, *J. Biol. Chem.*, **237**, 1239 (1962).

29. R. W. Butcher, J. E. Pike, and E. W. Sutherland, in *Proceedings of the Second Nobel Symposium, Stockholm* (Eds.: S. Bergström and B. Samuelsson), Interscience-John Wiley, New York, 1966, p. 321.

30. L. Birnbaumer and M. Rodbell, *J. Biol. Chem.*, **244**, 3477 (1969).

31. M. Vaughan and F. Murad, *Biochemistry*, **8**, 3092 (1969).

32. H. P. Bär and O. Hechter, *Proc. Natl. Acad. Sci. U.S.*, **63**, 350 (1969).

33. D. Steinberg and J. K. Huttunen, in *Book of Abstracts*, International Conference on Cyclic AMP, 1971, p. 51.

34. M. A. Brostrom, E. M. Reimann, D. A. Walsh, and E. G. Krebs, *Advan. Enzyme Regulation*, **8**, 191 (1970).

35. A. Kumon, H. Yamamura, and Y. Nishizura, *Biochem. Biophys. Res. Commun.*, **41**, 1290 (1970).

36. E. M. Reimann, C. O. Brostrom, J. D. Corbin, C. A. King, and E. G. Krebs, *Biochim. Biophys. Res. Commun.*, **42**, 187 (1971).

37. A. T. Butcher, in "Adipose Tissue: Regulation and Metabolic Functions," *Hormone and Metabolic Research*, Suppl. 2, Academic Press, New York, 1970, p. 5.

38. J. H. Exton, S. B. Lewis, R. J. Ho, and C. R. Park, in *Physiology and Pharmacology of Cyclic AMP* (P. Greengard, R. Paoletti, and G. A. Robison), Raven Press, New York, in press.

39. G. Senft, G. Schultz, K. Munske, and M. Hoffman, *Diabetologia*, **11**, 322 (1968).

40. E. G. Loten and J. G. T. Sneyd, *Biochem. J.*, **120**, 187 (1970).

41. D. Steinberg, M. Vaughan, P. J. Nestel, and S. Bergström, *Biochem. Pharmacol.*, **12**, 764 (1963).

42. E. Muhlbachova, A. Solyom, and L. Puglisi, *European J. Pharmacol.*, **1**, 321 (1967).

43. D. Steinberg, *Ann. N.Y. Acad. Sci.*, **139**, 897 (1967).

44. S. Bergström, L. A. Carlson, and L. Orö, *Acta Physiol. Scand.*, **60**, 170 (1964).

45. S. Bergström, L. A. Carlson, L. G. Ekelund, and L. Orö, *Proc. Soc. Exptl. Biol. Med.*, **118**, 110 (1965).

46. D. Steinberg, *Ann. N.Y. Acad. Sci.*, **139**, 897 (1967).

47. F. Berti, N. Kabir, R. Lentati, M. M. Usardi, P. Mantegazza, and R. Paoletti, *Progr. Biochem. Pharmacol.*, **3**, 110 (1967).

48. K. Stock and E. Westermann, *Naunyn-Schmiedebergs Arch. Exptl. Path. Pharmakol.*, **254**, 354 (1966).

49. F. Berti, R. Lentati, and M. M. Usardi, *Boll. Soc. Ital. Biol. Sper.*, **41**, 1327 (1965).

50. K. Stock, A. Aulich, and E. Westermann, *Life Sci.*, **7**, 113 (1968).

51. M. Blecher, N. S. Merlino, J. T. Roane, and P. D. Flynn, *J. Biol. Chem.*, **244**, 3423 (1969).

52. A. Bizzi, E. Veneroni, S. Garattini, L. Puglisi, and R. Paoletti, *European J. Pharmacol* **2**, 48 (1967).

53. K. Stock and E. Westermann, in "Adipose Tissue: Regulation and Metabolic Functions," *Hormone and Metabolic Research*, Suppl. 2, Academic Press, New York, 1970, p. 47.

54. K. Stock and E. Westermann, *Life Sci.*, **5**, 1667 (1966).

55. M. Wenke, E. Muhlbachova, D. Schusterova, K. Elisova, and S. Hynie, *Intern. J. Neuropharmacol.*, **3**, 283 (1964).

56. E. Westermann, K. Stock, and P. Bieck, *Progr. Biochem. Pharmacol.*, **3**, 233 (1967).

57. E. Muscholl and K. H. Rahn, *Pharmacol. Clin.*, **1**, 19 (1968).

58. A. Carlsson and M. Lundquist, *Acta Physiol. Scand.*, **54**, 87 (1962).

59. E. L. Bierman, T. N. Roberts, and V. P. Dole, *Proc. Soc. Exptl. Biol. Med.*, **95**, 437 (1957).

60. D. B. Stone and J. D. Brown, *Diabetes*, **15**, 314 (1966).

61. D. B. Stone, J. D. Brown, and C. P. Cox, *Am. J. Physiol.*, **210**, 26 (1966).

62. L. A. Carlson and L. Orö, *Acta Med. Scand.*, **172**, 641 (1962).

63. T. Farkas, R. Vertua, M. M. Usardi, and R. Paoletti, *Life Sci.*, **3**, 821 (1964).

64. L. A. Carlson and J. Östman, *Acta Med. Scand.*, **177**, 631 (1965).

65. L. A. Carlson, *Acta Med. Scand.*, **173**, 719 (1963).

66. J. N. Fain, V. P. Kovacev, and R. O. Scow, *J. Biol. Chem.*, **240**, 3522 (1966).

67. A. Solyom and L. Puglisi, *Biochem. Pharmacol.*, **15**, 41 (1966).

68. M. J. Peterson, C. C. Hillman, and J. Ashmore, *Mol. Pharmacol.*, **4**, 1 (1968).

69. A. Crastes de Paulet, P. Barjon, and P. Descombs, *Progr. Biochem. Pharmacol.*, **3**, 256 (1967).

70. D. Z. Goering, *Ges. Anat.* (*Z. Konstit*)(*Lehre*), **8**, 312 (1922).

71. E. W. Hausberger, *Z. Mikrosk. Anat. Forsch.*, **36**, 231 (1934).

72. R. Paoletti, R. L. Smith, R. P. Maickel, and B. B. Brodie, *Biochem. Biophys. Res. Commun.*, **5**, 424 (1961).

73. R. L. Sidman, M. Perkins, and N. Veiner, *Nature*, **193**, 36 (1962).

74. B. Fredholm and S. Rosell, *J. Pharmacol. Exptl. Therap.*, **159**, 1 (1968).

75. C. M. Van Duyne, H. R. Parker, R. J. Havel, and L. W. Holm, *Am. J. Physiol.*, **199**, 987 (1960).

76. M. Novak, V. Melicar, P. Hahn, and O. Koldovsky, *Physiol. Bohemoslov.*, **10**, 488 (1961).

77. J. T. Hamlin, III, R. B. Hickler, and R. G. Haskins, *J. Clin. Invest.*, **39**, 606 (1960).

78. A. Gilgen, R. P. Maickel, O. Nikodyevich, and B. B. Brodie, *Life Sci.*, **12**, 709 (1962).

79. A. F. Debons and I. L. Schwartz, *J. Lipid Res.*, **2**, 86 (1961).

80. H. M. Goodman and G. A. Bray, *Am. J. Physiol.*, **210**, 1053 (1966).

81. G. A. Bray, *J. Lipid Res.*, **8**, 300 (1967).

82. G. A. Bray and H. M. Goodman, *J. Lipid Res.*, **9**, 714 (1968).

83. F. Berti and M. M. Usardi, *Biochem. Pharmacol.*, **14**, 357 (1965).

84. S. M. Glick in *Frontiers of Endocrinology*, Oxford University Press, 1969, p. 141.

85. M. S. Raben and C. H. Hollenberger, *J. Clin. Invest.*, **38**, 484 (1959).

86. B. Winkler, R. Steele, N. Altszuler, and R. C. De Bodo, *Am. J. Physiol.*, **206**, 174 (1964).

87. J. N. Fain, D. J. Galton, and V. P. Kovalev, *Mol. Pharmacol.*, **2**, 237 (1966).

88. H. M. Goodman, *Endocrinology*, **82**, 1027 (1968).

89. A. B. Caldwell and J. N. Fain, "Adipose Tissue: Regulation and Metabolic Functions," in *Hormone and Metabolic Research*, Suppl. 2, Academic Press, New York, 1970, p. 3.

90. J. N. Fain, *Endocrinolgy*, **82**, 825 (1968).

91. J. Moskowitz and J. N. Fain, *J. Biol. Chem.*, **245**, 1101 (1970).

92. W. F. Perry and H. F. Bowen, *J. Biochem. Physiol.*, **40**, 749 (1962).

93. R. L. Jungas and E. G. Ball, *Biochemistry*, **2**, 586 (1963).

94. R. L. Jungas and E. G. Ball, *Biochemistry*, **3**, 1696 (1964).

95. R. L. Jungas, in "Diabetes," *Excerpta Med.*, p. 334, 1969.

96. R. W. Butcher, J. G. I. Sneyd, C. R. Park, and E. W. Sutherland, *J. Biol. Chem.*, **241**, 1651 (1966).

97. R. L. Jungas, *Proc. Natl. Acad. Sci. U.S.*, **56**, 757 (1966).

98. P. E. Cryer, L. Jaret, and D. M. Kipnis, *Biochim. Biophys. Acta*, **177**, 586 (1969).

99. E. G. Ball, in "Adipose Tissue: Regulation and Metabolic Functions," *Hormone and Metabolic Research*, Suppl. 2, Academic Press, New York, 1970, p. 102.

100. D. S. Goodman, *Science*, **125**, 1296 (1957).

101. D. S. Goodman, *J. Am. Chem. Soc.*, **80**, 3892 (1958).

102. D. S. Goodman, *J. Clin. Invest.*, **37**, 1729 (1958).

103. M. J. Lipson and S. Naimi, *J. Lipid Res.*, **12**, 994 (1971).

104. A. V. Nichols, *Advan. Biol. Med. Phys.*, **11**, 109 (1967).

105. J. C. Hoak, W. E. Connor, and E. D. Warner, *Arch. Path.*, **63**, 791 (1966).

106. A. A. Spector and D. Steinberg, *J. Biol. Chem.*, **240**, 3747 (1965).

107. D. Steinberg, *Progr. Biochem. Pharmacol.*, **3**, 139 (1967).

108. H. H. Ott, J. B. Hayes, G. Ott, and R. J. Gorelon, Jr., in *Protides of Biological Fluids*, Elsevier, Amsterdam, 1966, p. 301.

109. J. M. Thorp and A. M. Barrett, *Progr. Biochem. Pharmacol.*, **2**, 337 (1967).

110. B. Rifkind, T. Bagg, and B. Bronte-Stewart, *Progr. Biochem. Pharmacol.*, **2**, 358 (1967).

111. E. Shafrir and D. Steinberg, *J. Clin. Invest.*, **39**, 310 (1960).

112. H. G. Windmueller and A. E. Spaeth, *Arch. Biochem. Biophys.*, **122**, 362 (1967).

113. E. B. Feigelson, W. W. Pfaff, and D. Steinberg, *Clin. Res.*, **8**, 239 (1960).

114. E. B. Feigelson, W. W. Pfaff, A. Karmen, and D. Steinberg, *J. Clin. Invest.*, **40**, 2171 (1961).

115. P. J. Nestel and D. Steinberg, *J. Lipid Res.*, **4**, 461 (1963).

116. H. Heimberg, I. Weinstein, M. Klausner, and M. L. Watkins, *Am. J. Physiol.*, **202**, 353 (1962).

117. R. Paoletti, L. Puglisi, and M. M. Usardi, in *Pharmacology of Hormonal Polypeptides and Proteins*, Plenum Press, New York, 1968, p. 425.

118. R. I. Levy, R. S. Lees, and D. S. Fredrickson, *J. Clin. Invest.*, **45**, 63 (1966).

119. W. V. Brown, R. I. Levy, and D. S. Fredrickson, *Biochim. Biophys. Acta*, **200**, 573 (1970).

120. W. V. Brown, R. I. Levy, and D. S. Fredrickson, *J. Biol. Chem.*, **245**, 6588 (1970).

121. H. G. Windmueller and R. I. Levy, *J. Biol. Chem.*, **242**, 2246 (1967).

122. H. G. Windmueller and R. I. Levy, *J. Biol. Chem.*, **243**, 4878 (1968).

123. R. K. Ockner and A. L. Jones, *J. Lipid Res.*, **11**, 284 (1970).

124. R. W. Mahley, R. L. Hamilton, and V. S. Lequire, *J. Lipid Res.*, **10**, 433 (1969).

125. R. W. Mahley, T. P. Bersot, V. S. Lequire, R. I. Levy, H. G. Windmueller, and W. M. Brown, *Science*, **168**, 380 (1970).

126. E. Nikkilä, *Advan. Lipid Res.*, **7**, 63 (1969).

127. J. H. Bragdon, R. J. Havel, and F. Boyle, *J. Lab. Clin. Med.*, **48**, 36 (1956).

128. B. Shore and V. Shore, in *Atherosclerosis*, Springer Verlag, New York, 1970, p. 144.

129. A. M. Gotto, R. I. Levy, and D. S. Fredrickson, *Proc. Natl. Acad. Sci. U.S.*, **60**, 1436 (1968).

130. R. H. Furman, R. P. Howard, K. Kakshani, and L. N. Norcia, *Am. J. Clin. Nutr.*, **9**, 73 (1961).

131. T. Watanabe, K. Tanaka, and N. Yanai, *Acta Pathol. Japon.*, **18**, 319 (1968).

132. T. Langer, W. Strober, and R. I. Levy, *J. Clin. Invest.*, **48**, 49a (1969).

133. P. J. Scott and P. J. Hurley, *Israel J. Med. Sci.*, **5**, 631 (1969).

134. A. Scanu, *J. Biol. Chem.*, **242**, 711 (1967).

135. V. Shore and B. Shore, *Biochemistry*, **6**, 1962 (1967).

136. F. A. Vandenheuvel, *Can. J. Biochem. Physiol.*, **40**, 1299 (1962).

137. J. C. La Rosa, R. I. Levy, W. V. Brown, and D. S. Fredrickson, *Am. J. Physiol.*, **220**, 785 (1971).

138. J. C. La Rosa, R. I. Levy, P. Herbert, S. E. Lux, and D. S. Fredrickson, *Biochem. Biophys. Res. Commun.*, **41**, 57 (1970).

139. D. Gitlin, D. G. Cornwell, D. Nakasato, J. L. Oncley, W. L. Hughes, and C. A. Janeway, *J. Clin. Invest.*, **37**, 172 (1958).

140. T. Langer, D. Bilheimer, and R. I. Levy, *Circulation*, **42**, III, 7 (1970).

141. D. W. Bilheimer, S. Eisenberg, R. T. Levy, *Biochem. J.*, 1972, in press.

142. R. J. Havel et al., *J. Clin. Invest.*, **49**, 2017 (1970).

143. S. H. Quarfordt, A. Frank, D. M. Shames, M. Berman, and D. Steinberg, *J. Clin. Invest.*, **49**, 2281 (1970).

144. T. Langer and R. I. Levy, in *Proceedings of the Workshop on Metabolic Effects of Nicotinic Acid*, Huber, Bern, 1971.

145. D. S. Fredrickson, *Brit. Med. J.*, **2**, 187 (1971).

146. S. Eisenberg, personal communication, 1971.

147. H. A. Eder, in *Hormones and Atherosclerosis*, Academic Press, New York, 1959, p. 335.

148. J. P. Peters, H. Heinemann, and E. B. Man, *J. Clin. Invest.*, **30**, 388 (1951).

149. M. F. Oliver and G. S. Boyd, *Clin. Sci.*, **12**, 217 (1953).

150. D. Adlersberg, H. E. Schaefer, A. G. Steinberg, and C. Wong, *J. Am. Med. Assoc.*, **162**, 619 (1956).

151. R. H. Furman, B. Alaupovic, and R. P. Howard, *Progr. Biochem. Pharmacol.*, **2**, 215 (1967).

152. M. F. Oliver and G. S. Boyd, *Lancet*, **ii**, 1273, 1956.

153. R. H. Furman, S. S. Sanbar, P. Alaupovic, R. H. Bradford, and R. P. Howard, *J. Lab. Clin. Med.*, **63**, 193 (1964).

154. Q. E. Crider, R. H. Bradford, and R. H. Furman, *Circulation*, **32**, 8 (1965).

155. B. A. Sachs, L. Wolfman and N. Herzig, *Obstet. Gynec.*, **34**, 530 (1969).

156. E. Zorilla, M. Hulse, A. Hernandez, and H. Gershberg, *J. Clin. Endocrinol.*, **28**, 1793 (1968).

157. W. R. Hazzard, M. J. Sproger, J. D. Bagdade, and E. L. Bierman, *Clin. Res.*, **16**, 267 (1968).

158. S. S. C. Yen and P. Vela, *J. Clin. Endocrinol.*, **28**, 1564 (1968).

159. B. A. Sachs and L. Wolfmann, *Circulation*, **38**, 169 (1968).

160. M. M. Best and C. H. Duncan, *J. Am. Med. Assoc.*, **137**, 37 (1964).

161. M. Herold, J. Cahn, N. Barré, and E. Gauthier, *Progr. Biochem. Pharmacol.*, **2**, 105 (1967).

162. D. S. Robinson, *Advan. Lipid Res.*, **1**, 133 (1963).

163. D. S. Robinson and J. E. French, *Quart. J. Exptl. Physiol.*, **42**, 151 (1957).

164. D. S. Robinson, P. M. Harris, and C. R. Ricketts, *Biochem. J.*, **71**, 286 (1959).

165. C. Hollett and P. J. Nestel, *Am. J. Physiol.*, **199**, 803 (1960).

166. L. L. Fekete, W. F. Lever, and E. Klein, *J. Lab. Clin. Med.*, **52**, 680 (1958).

167. W. D. Brown, *Quart. J. Exptl. Physiol.*, **37**, 75 (1952).

168. J. H. Bragdon and R. J. Havel, *Am. J. Physiol.*, **177**, 128 (1954).

169. A. Kellner, J. W. Correl, and A. T. Ladd, *J. Exptl. Med.*, **93**, 373 (1951).

170. A. M. Scanu, P. Oriente, J. M. Szajewski, J. L. McCormack, and I. H. Page, *J. Exptl. Med.*, **114**, 279 (1961).

171. D. S. Fredrickson, R. I. Levy, and R. S. Lees, *New Engl. J. Med.*, **276**, 32, 94, 148, 215, 173 (1967).

172. D. L. Azarnoff, D. R. Tucker, and C. A. Barr, *Metabolism*, **14**, 959 (1965).

173. P. J. Nestel, E. Z. Hirsch, and F. A. Couzens, *J. Clin. Invest.*, **44**, 891 (1965).

174. L. A. Carlson and L. Orö, *Acta Med. Scand.*, **172**, 641 (1962).

175. T. Leigh, D. S. Plast, and J. M. Thorp, in *Third International Symposium on Drugs Affecting Lipid Metabolism*, Plenum Press, New York, 1969, p. 138.

Index

515